A Course in Time Series Analysis

WILEY SERIES IN PROBABILITY AND STATISTICS
PROBABILITY AND STATISTICS SECTION

Established by WALTER A. SHEWHART and SAMUEL S. WILKS

Editors: *Noel A. C. Cressie, Nicholas I. Fisher, Iain M. Johnstone, J. B. Kadane, David W. Scott, Bernard W. Silverman, Adrian F. M. Smith, Jozef L. Teugels; Vic Barnett, Emeritus, Ralph A. Bradley, Emeritus, J. Stuart Hunter, Emeritus, David G. Kendall, Emeritus*

A complete list of the titles in this series appears at the end of this volume.

A Course in Time Series Analysis

Edited by

DANIEL PEÑA
Universidad Carlos III de Madrid

GEORGE C. TIAO
University of Chicago

RUEY S. TSAY
University of Chicago

A Wiley-Interscience Publication
JOHN WILEY & SONS, INC.
New York · Chichester · Weinheim · Brisbane · Singapore · Toronto

This book is printed on acid-free paper. ∞

Copyright © 2001 by John Wiley & Sons, Inc. All rights reserved.

Published simultaneously in Canada.

No part of this publication may be reproduced, stored in a retrieval system or transmitted in any form or by any means, electronic, mechanical, photocopying, recording, scanning or otherwise, except as permitted under Sections 107 or 108 of the 1976 United States Copyright Act, without either the prior written permission of the Publisher, or authorization through payment of the appropriate per-copy fee to the Copyright Clearance Center, 222 Rosewood Drive, Danvers, MA 01923, (978) 750-8400, fax (978) 750-4744. Requests to the Publisher for permission should be addressed to the Permissions Department, John Wiley & Sons, Inc., 605 Third Avenue, New York, NY 10158-0012, (212) 850-6011, fax (212) 850-6008, E-Mail: PERMREQ @ WILEY.COM.

For ordering and customer service, call 1-800-CALL-WILEY.

Library of Congress Cataloging in Publication Data:

A course in time series analysis / editors, Daniel Peña, George C. Tiao, Ruey S. Tsay.
    p. cm.—(Wiley series in probability and statistics)
  Includes bibliographical references and index.
  ISBN 0-471-36164-X (cloth : alk. paper)
  1. Time-series analysis. I. Peña, Daniel, 1948– II. Tiao, George C., 1933– III. Tsay, Ruey S., 1951– IV. Series.

QA280 .C66 2001
519.5′5—dc21   00-038131

Printed in the United States of America.

10 9 8 7 6 5 4 3 2

Contents

Preface	xv
About ECAS	xvi
Contributors	xvii

1. **Introduction**   1
   D. Peña and G. C. Tiao

   1.1. Examples of time series problems, 1
       1.1.1. Stationary series, 2
       1.1.2. Nonstationary series, 3
       1.1.3. Seasonal series, 5
       1.1.4. Level shifts and outliers in time series, 7
       1.1.5. Variance changes, 7
       1.1.6. Asymmetric time series, 7
       1.1.7. Unidirectional–feedback relation between series, 9
       1.1.8. Comovement and cointegration, 10
   1.2. Overview of the book, 10
   1.3. Further reading, 19

PART I   BASIC CONCEPTS IN UNIVARIATE TIME SERIES

2. **Univariate Time Series: Autocorrelation, Linear Prediction, Spectrum, and State-Space Model**   25
   G. T. Wilson

   2.1. Linear time series models, 25
   2.2. The autocorrelation function, 28
   2.3. Lagged prediction and the partial autocorrelation function, 33

- 2.4. Transformations to stationarity, 35
- 2.5. Cycles and the periodogram, 37
- 2.6. The spectrum, 42
- 2.7. Further interpretation of time series acf, pacf, and spectrum, 46
- 2.8. State-space models and the Kalman Filter, 48

3. Univariate Autoregressive Moving-Average Models   53
G. C. Tiao

- 3.1. Introduction, 53
  - 3.1.1. Univariate ARMA models, 54
  - 3.1.2. Outline of the chapter, 55
- 3.2. Some basic properties of univariate ARMA models, 55
  - 3.2.1. The ψ and π weights, 56
  - 3.2.2. Stationarity condition and autocovariance structure of z_t, 58
  - 3.2.3. The autocorrelation function, 59
  - 3.2.4. The partial autocorrelation function, 60
  - 3.2.5. The extended autocorrelaton function, 61
- 3.3. Model specification strategy, 63
  - 3.3.1. Tentative specification, 63
  - 3.3.2. Tentative model specification via SEACF, 67
- 3.4. Examples, 68

4. Model Fitting and Checking, and the Kalman Filter   86
G. T. Wilson

- 4.1. Prediction error and the estimation criterion, 86
- 4.2. The likelihood of ARMA models, 90
- 4.3. Likelihoods calculated using orthogonal errors, 94
- 4.4. Properties of estimates and problems in estimation, 98
- 4.5. Checking the fitted model, 101
- 4.6. Estimation by fitting to the sample spectrum, 104
- 4.7. Estimation of structural models by the Kalman filter, 105

5. Prediction and Model Selection   111
D. Peña

- 5.1. Introduction, 111
- 5.2. Properties of minimum mean-square error prediction, 112
  - 5.2.1. Prediction by the conditional expectation, 112
  - 5.2.2. Linear predictions, 113

CONTENTS

5.3. The computation of ARIMA forecasts, 114
5.4. Interpreting the forecasts from ARIMA models, 116
    5.4.1. Nonseasonal models, 116
    5.4.2. Seasonal models, 120
5.5. Prediction confidence intervals, 123
    5.5.1. Known parameter values, 123
    5.5.2. Unknown parameter values, 124
5.6. Forecast updating, 125
    5.6.1. Computing updated forecasts, 125
    5.6.2. Testing model stability, 125
5.7. The combination of forecasts, 129
5.8. Model selection criteria, 131
    5.8.1. The FPE and AIC criteria, 131
    5.8.2. The Schwarz criterion, 133
5.9. Conclusions, 133

6. Outliers, Influential Observations, and Missing Data   136
D. Peña

6.1. Introduction, 136
6.2. Types of outliers in time series, 138
    6.2.1. Additive outliers, 138
    6.2.2. Innovative outliers, 141
    6.2.3. Level shifts, 143
    6.2.4. Outliers and intervention analysis, 146
6.3. Procedures for outlier identification and estimation, 147
    6.3.1. Estimation of outlier effects, 148
    6.3.2. Testing for outliers, 149
6.4. Influential observations, 152
    6.4.1. Influence on time series, 152
    6.4.2. Influential observations and outliers, 153
6.5. Multiple outliers, 154
    6.5.1. Masking effects, 154
    6.5.2. Procedures for multiple outlier identification, 156
6.6. Missing-value estimation, 160
    6.6.1. Optimal interpolation and inverse autocorrelation function, 160
    6.6.2. Estimation of missing values, 162
6.7. Forecasting with outliers, 164
6.8. Other approaches, 166
6.9. Appendix, 166

7. **Automatic Modeling Methods for Univariate Series**   **171**
   V. Gómez and A. Maravall
   - 7.1. Classical model identification methods, 171
     - 7.1.1. Subjectivity of the classical methods, 172
     - 7.1.2. The difficulties with mixed ARMA models, 173
   - 7.2. Automatic model identification methods, 173
     - 7.2.1. Unit root testing, 174
     - 7.2.2. Penalty function methods, 174
     - 7.2.3. Pattern identification methods, 175
     - 7.2.4. Uniqueness of the solution and the purpose of modeling, 176
   - 7.3. Tools for automatic model identification, 177
     - 7.3.1. Test for the log-level specification, 177
     - 7.3.2. Regression techniques for estimating unit roots, 178
     - 7.3.3. The Hannan–Rissanen method, 181
     - 7.3.4. Liu's filtering method, 185
   - 7.4. Automatic modeling methods in the presence of outliers, 186
     - 7.4.1. Algorithms for automatic outlier detection and correction, 186
     - 7.4.2. Estimation and filtering techniques to speed up the algorithms, 190
     - 7.4.3. The need to robustify automatic modeling methods, 191
     - 7.4.4. An algorithm for automatic model identification in the presence of outliers, 191
   - 7.5. An automatic procedure for the general regression–ARIMA model in the presence of outlierw, special effects, and, possibly, missing observations, 192
     - 7.5.1. Missing observations, 192
     - 7.5.2. Trading day and Easter effects, 193
     - 7.5.3. Intervention and regression effects, 194
   - 7.6. Examples, 194
   - 7.7. Tabular summary, 196

8. **Seasonal Adjustment and Signal Extraction Time Series**   **202**
   V. Gómez and A. Maravall
   - 8.1. Introduction, 202
   - 8.2. Some remarks on the evolution of seasonal adjustment methods, 204

8.2.1. Evolution of the methodologic approach, 204
8.2.2. The situation at present, 207
8.3. The need for preadjustment, 209
8.4. Model specification, 210
8.5. Estimation of the components, 213
    8.5.1. Stationary case, 215
    8.5.2. Nonstationary series, 217
8.6 Historical or final estimator, 218
    8.6.1. Properties of final estimator, 218
    8.6.2. Component versus estimator, 219
    8.6.3. Covariance between estimators, 221
8.7. Estimators for recent periods, 221
8.8. Revisions in the estimator, 223
    8.8.1. Structure of the revision, 223
    8.8.2. Optimality of the revisions, 224
8.9. Inference, 225
    8.9.1. Optical Forecasts of the Components, 225
    8.9.2. Estimation error, 225
    8.9.3. Growth rate precision, 226
    8.9.4. The gain from concurrent adjustment, 227
    8.9.5. Innovations in the components (pseudoinnovations), 228
8.10. An example, 228
8.11. Relation with fixed filters, 235
8.12. Short-versus long-term trends; measuring economic cycles, 236

PART II ADVANCED TOPICS IN UNIVARIATE TIME SERIES

9. Heteroscedastic Models   249
R. S. Tsay

9.1. The ARCH model, 250
    9.1.1. Some simple properties of ARCH models, 252
    9.1.2. Weaknesses of ARCH models, 254
    9.1.3. Building ARCH models, 254
    9.1.4. An illustrative example, 255
9.2. The GARCH Model, 256
    9.2.1. An illustrative example, 257
    9.2.2. Remarks, 259

9.3. The exponential GARCH model, 260
  9.3.1. An illustrative example, 261
9.4. The CHARMA model, 262
9.5. Random coefficient autoregressive (RCA) model, 263
9.6. Stochastic volatility model, 264
9.7. Long-memory stochastic volatility model, 265

10. Nonlinear Time Series Models: Testing and Applications 267
R. S. Tsay

10.1. Introduction, 267
10.2. Nonlinearity tests, 268
  10.2.1. The test, 268
  10.2.2. Comparison and application, 270
10.3. The Tar model, 274
  10.3.1. U.S. real GNP, 275
  10.3.2. Postsample forecasts and discussion, 279
10.4. Concluding remarks, 282

11. Bayesian Time Series Analysis 286
R. S. Tsay

11.1. Introduction, 286
11.2. A general univariate time series model, 288
11.3. Estimation, 289
  11.3.1. Gibbs sampling, 291
  11.3.2. Griddy Gibbs, 292
  11.3.3. An illustrative example, 292
11.4. Model discrimination, 294
  11.4.1. A mixed model with switching, 295
  11.4.2. Implementation, 296
11.5. Examples, 297

12 Nonparametric Time Series Analysis: Nonparametric Regression, Locally Weighted Regression, Autoregression, and Quantile Regression 308
S. Heiler

12.1 Introduction, 308
12.2 Nonparametric regression, 309
12.3 Kernel estimation in time series, 314
12.4 Problems of simple kernel estimation and restricted approaches, 319

CONTENTS   xi

    12.5   Locally weighted regression, 321

    12.6   Applications of locally weighted regression to time series, 329

    12.7   Parameter selection, 330

    12.8   Time series decomposition with locally weighted regression, 336

13. Neural Network Models   348
K. Hornik and F. Leisch

    13.1.   Introduction, 348

    13.2.   The multilayer perceptron, 349

    13.3.   Autoregressive neural network models, 354

            13.3.1.   Example: Sunspot series, 355

    13.4.   The recurrent perceptron, 356

            13.4.1.   Examples of recurrent neural network models, 357

            13.4.2.   A unifying view, 359

PART III   MULTIVARIATE TIME SERIES

14. Vector ARMA Models   365
G. C. Tiao

    14.1.   Introduction, 365

    14.2.   Transfer function or unidirectional models, 366

    14.3.   The vector ARMA model, 368

            14.3.1.   Some simple examples, 368

            14.3.2.   Relationship to transfer function model, 371

            14.3.3.   Cross-covariance and correlation matrices, 371

            14.3.4.   The partial autoregression matrices, 372

    14.4.   Model building strategy for multiple time series, 373

            14.4.1.   Tentative specification, 373

            14.4.2.   Estimation, 378

            14.4.3.   Diagnostic checking, 379

    14.5.   Analyses of three examples, 380

            14.5.1.   The SCC data, 380

            14.5.2.   The gas furnace data, 383

            14.5.3.   The census housing data, 387

    14.6.   Structural analysis of multivariate time series, 392

            14.6.1.   A canonical analysis of multiple time series, 395

14.7. Scalar component models in multiple time series, 396
- 14.7.1. Scalar component models, 398
- 14.7.2. Exchangeable models and overparameterization, 400
- 14.7.3. Model specification via canonical correlation analysis, 402
- 14.7.4. An illustrative example, 403
- 14.7.5. Some further remarks, 404

15. Cointegration in the VAR Model   408
S. Johansen

15.1. Introduction, 408
- 15.1.1. Basic definitions, 409

15.2. Solving autoregressive equations, 412
- 15.2.1. Some examples, 412
- 15.2.2. An inversion theorem for matrix polynomials, 414
- 15.2.3. Granger's representation, 417
- 15.2.4. Prediction, 419

15.3. The statistical model for $I(1)$ variables, 420
- 15.3.1. Hypotheses on cointegrating relations, 421
- 15.3.2. Estimation of cointegrating vectors and calculation of test statistics, 422
- 15.3.3. Estimation of β under restrictions, 426

15.4. Asymptotic theory, 426
- 15.4.1. Asymptotic results, 427
- 15.4.2. Test for cointegrating rank, 427
- 15.4.3. Asymptotic distribution of $\hat{\beta}$ and test for restrictions on β, 429

15.5. Various applications of the cointegration model, 432
- 15.5.1. Rational expectations, 432
- 15.5.2. Arbitrage pricing theory, 433
- 15.5.3. Seasonal cointegration, 433

16. Identification of Linear Dynamic Multiinput/Multioutput Systems   436
M. Deistler

16.1. Introduction and problem statement, 436
16.2. Representations of linear systems, 438
- 16.2.1. Input/output representations, 438

- 16.2.2. Solutions of linear vector difference equations (VDEs), 440
- 16.2.3. ARMA and state-space representations, 441
- 16.3. The structure of state-space systems, 443
- 16.4. The structure of ARMA systems, 444
- 16.5. The realization of state-space systems, 445
  - 16.5.1. General structure, 445
  - 16.5.2. Echelon forms, 447
- 16.6. The realization of ARMA systems, 448
- 16.7. Parametrization, 449
- 16.8. Estimation of real-valued parameters, 452
- 16.9. Dynamic specification, 454

INDEX 457

Preface

This book is based on the lectures of the ECAS' 97 Course in Time Series Analysis held at El Escorial, Madrid, Spain, from September 15 to September 19, 1997. The course was sponsored by the European Courses in Advanced Statistics (ECAS). In accordance with the objectives of ECAS, the lectures are directed to both researchers and teachers of statistics in academic institutions and statistical professionals in industry and govermment, with the goal of presenting an overview of the current status of the area. In particular, different approaches to time series analysis are discussed and compared. In editing the book, we have worked hard to uphold ECAS' objectives. In addition, special efforts have been made to unify the notation and to include as many topics as possible, so that readers of the book can have an overview of the current status of time series research and applications.

The book consists of three main components. The first component concern basic materials of univariate time series analysis presented in the first eight chapters. It includes recent developments in outlier detection, automatic model selection, and seasonal adjustment. The second component addresses advanced topics in univariate time series analysis such as conditional heteroscedastic models, nonlinear models, Bayesian analysis, nonparametric methods, and neural networks. This component represents current research activities in univariate time series analysis. The third and final component of the book concerns with multivariate time series, including vector ARMA models, cointegration, and linear systems.

The book can be used as a principal text or a complementary text for courses in time series. A basic time series course can be taught from the first part of the book that presents the basic material that can be found in the standard texts in time series. This part also includes topics not normally covered in these texts, such as the extended and inverse autocorrelation function, the decomposition of the forecast function of ARIMA models, a detailed analysis of outliers and influential observations and automatic methods for model building and model based seasonal adjustment. For a basic course this book should be complemented with some of the excellent texts available. The book would be very well suited for an advanced course in which some of the basic material can be quickly reviewed using the first part, that skips many details and concentrates in the main concepts of general applicability. Then the course can concentrate in the topics in Parts 2 and 3. If the scope of the course is more in methodological extensions of univariate linear models the material in Part 2 can be useful, whereas if the objective is to introduce multivariate modeling Part 3 will be appropriate. To facilitate the use of the book as a text, all the time series data used in this book can be down loaded from the web address: *http://gsbwww. uchicago.edu/fac/ruey.tsay/teaching/ecas/*

We are grateful to all people who have made this book possible: (1) to the 11 authors of the chapters of the book who have been extremely helpful in the timely revisions of the drafts of the chapters and have made a big effort to unify the presentation and (2) to the organizers of the course and all the students from many different countries in four continents that made this one week of lectures a very enjoyable experience for all the participants. We are very grateful to our host in the Monastery of El Escorial, father Agustin Alonso, who did his best to make our staying in the monastery an unforgettable experience. The success of the course was in large part due to the enthusiastic work in all the organization details of Ana Justel, Regina Kaiser, Juan Romo, Esther Ruiz, and María Jesús Sánchez. In the preparation of the book we are also grateful to Monica Benito for her help in organizing the index and the references in the book.

The Editors

ABOUT ECAS

ECAS is a foundation of Statistical Societies within Europe that, according to its constitution, was founded in order to foster links and to promote cooperation between statisticians in Europe. In order to achieve these aims, courses on an advanced level covering varying aspects of statistics are organized every 2 years in different countries of Europe. In 1999 Statistical Societies members of ECAS belongs to the following countries: Austria, Belgium, Denmark, France, Finland, Germany, Italy, Portugal, Spain, Sweden, Switzerland, The Netherlands, and the United Kingdom.

The first ECAS course was held in Capri, Italy, on Multidimensional Data Analysis in 1987. Subsequent courses were held on robustness in statistics in 1989 in the castle Reisenburg, Germany; on experimental design in 1991 in Sète, France; on the analysis of categorical data in 1995 in Leiden, The Netherlands; on longitudinal data analysis and repeated measures in 1995 in Milton Keynes, United Kingdom; on time series analysis in 1997 in San Lorenzo del Escorial, Spain; and on environmental statistics in 1999 in Garpenberg, Sweden.

A Council has the overall responsibility for ECAS. Its members are nominated by the statistical societies of participating countries. The Presidents of ECAS have been Jean Jacques Debrosque (Belgium, 1987–1993) and Siegfried Heiler (Germany, 1994–1997). The current President is Daniel Peña (Spain, 1998–2001).

Contributors

Manfred Deistler
Institut für Ökonometrie, Operations Research und Systemtheorie
Technische Universität Wien
Wien, Austria

Víctor Gómez
Dirección General de Presupuestos
Ministerio de Hacienda
Madrid, Spain

Siegfried Heiler
Fakultaet fuer Mathematik und Informatik
Universität Konstanz
Konstanz, Germany

Kurt Hornik
Institut für Statistik
Technische Universität Wien
Wien, Austria

Søren Johansen
Economic Department
European University Institute
Florence, Italy

Friedrich Leisch
Institut für Statistik
Technische Universität Wien
Wien, Austria

Agustín Maravall
Servicio de Estudios
Banco de España
Madrid, Spain

Daniel Peña
Departamento de Estadística y Econometría
Universidad Carlos III de Madrid
Madrid, Spain

George C. Tiao
Graduate School of Business
University of Chicago
Chicago, IL, USA

Ruey S. Tsay
Graduate School of Business
University of Chicago
Chicago, IL, USA

G. Tunnicliffe Wilson
Department of Mathematics and Statistics
Lancaster University
Lancaster, UK

A Course in Time Series Analysis

CHAPTER 1

Introduction

Daniel Peña
Universidad Carlos III de Madrid

George C. Tiao
University of Chicago

1.1. EXAMPLES OF TIME SERIES PROBLEMS

Data in business, economics, engineering, environment, medicine, and other areas of scientific investigations are often collected in the form of time series, that is, a sequence of observations taken at regular intervals of time such as hourly temperature readings, daily stock prices, weekly traffic volume, monthly beer consumption, and annual growth rates. The main objectives of time series modeling and analysis are (1) understanding the dynamic or time-dependent structure of the observations of a single series—*univariate* time series analysis and (2) ascertaining the leading, lagging, and feedback relationships among several series—*multivariate* time series analysis.

Knowledge of the dynamic structure will help produce accurate forecasts of future observations and design optimal control schemes. This chapter presents first a number of univariate and multivariate time series data sets arisen from various scientific disciplines. These data examples are used to introduce and illustrate the following:

- Stationary versus nonstationary series
- Linear versus nonlinear dynamic relationship
- Homogeneity versus heterogeneity in variance
- Unidirectional versus feedback relation between series

A Course in Time Series Analysis, Edited by Daniel Peña, George C. Tiao, and Ruey S. Tsay.
ISBN 0-471-36164-X. © 2001 John Wiley & Sons, Inc.

- Outlier, level shift, structural change, and intervention
- Comovement and cointegration

These concepts motivate many of the topics discussed in this book.

1.1.1. Stationary series

Figure 1.1a shows a series of the yield of 70 consecutive batches of a chemical process given in Box and Jenkins (1976). The observations fluctuate about a fixed mean level with constant variance over the observational period. In other words, the overall behavior of the series remains the same over time. Such a series is called a *stationary series*. A formal definition of stationarity will be given later.

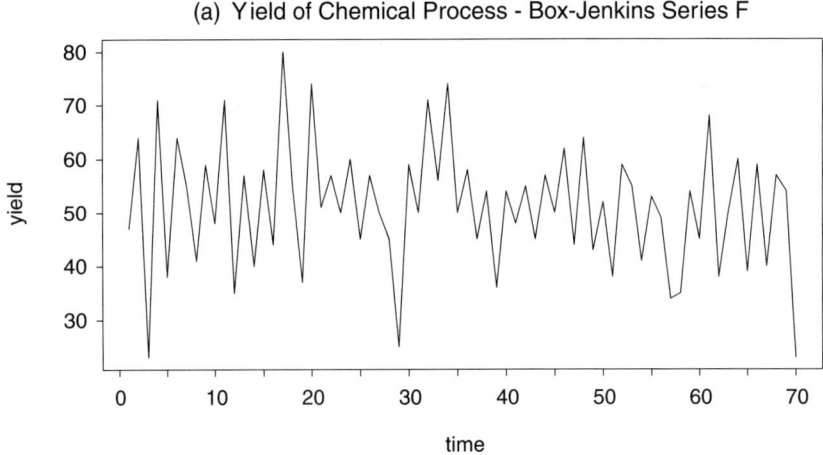

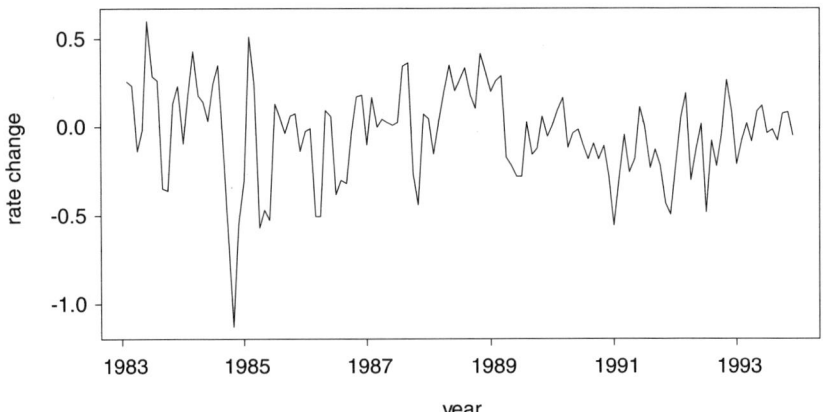

FIGURE 1.1 Two examples of stationary series.

1.1. EXAMPLES OF TIME SERIES PROBLEMS

As another example of stationary series, Figure 1.1b gives a series of the month to month changes in the interest rates of 90-day U.S. Treasury bills (T-bills) from 1983 to 1993. Except for the sharp dip near the end of 1984, this series appears to be quite stationary with a mean level close to zero over time.

In practice, temporal changes (week to week, month to month, or quarter to quarter) of many economic time series often exhibit this kind of stationary behavior. Good examples are stock returns and changes in exchange rates.

1.1.2. Nonstationary series

Instead of month to month changes, if we look at the series of monthly rates of the 90-day T-bills themselves, we see a vastly different behavior. This is shown in Figure 1.2a.

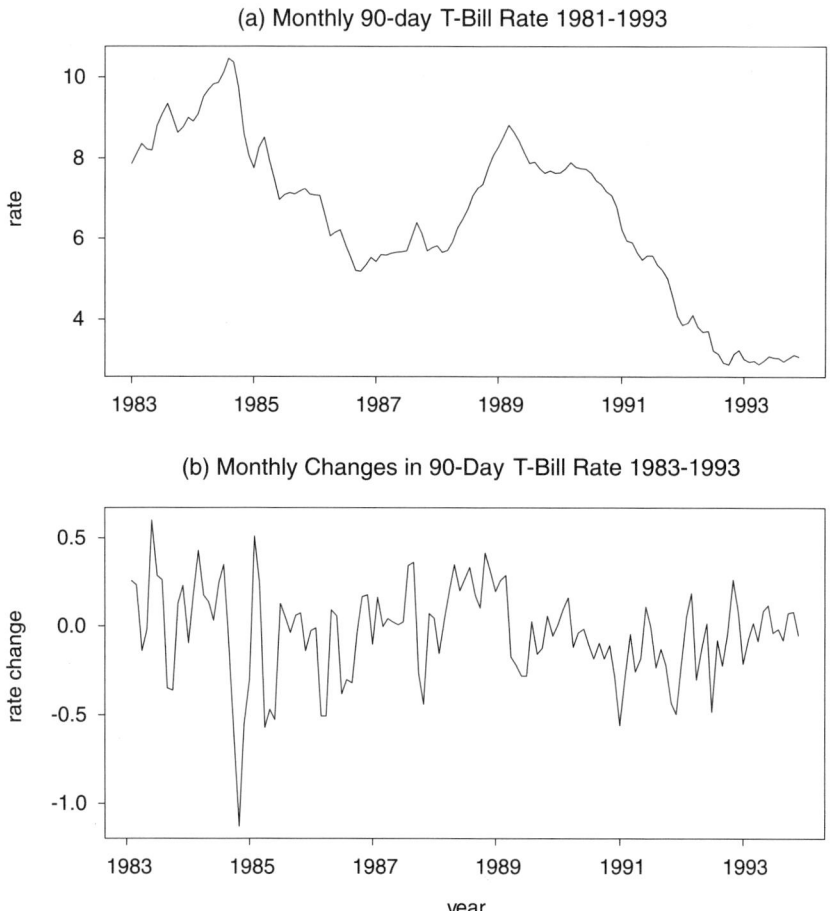

FIGURE 1.2 A nonstationary series and its first difference.

This series does not seem to have a mean level and exhibits a drifting or wandering behavior. It is clearly not a stationary series. Financial time series such as stock prices, prices of derivatives, and exchange rates often behave in this manner. However, by taking successive differences of the observations, we obtain the series of monthly changes in Figure 1.1b, which is reproduced in Figure 1.2b for easy comparison. This example shows that a drifting nonstationary series can be transformed into a stationary one by the differencing operation. The series in Figure 1.2b is called the first difference of the series in Figure 1.2a. In practice, sometimes the first difference series may not be stationary and it may be necessary to difference the series again to make it stationary.

Figure 1.3a shows quarterly data of U.S. real GNP (gross national product) over the period 1946–1991. The series shows an exponential growth. By taking a logarithmic

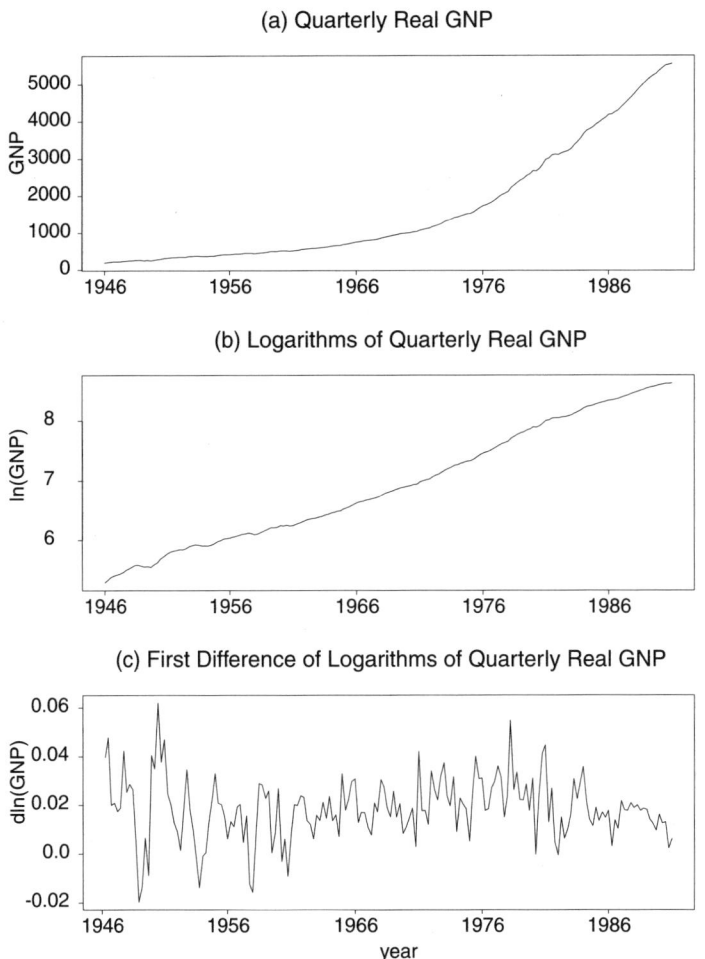

FIGURE 1.3 Quarterly U.S. real GNP 1946–1991.

1.1. EXAMPLES OF TIME SERIES PROBLEMS

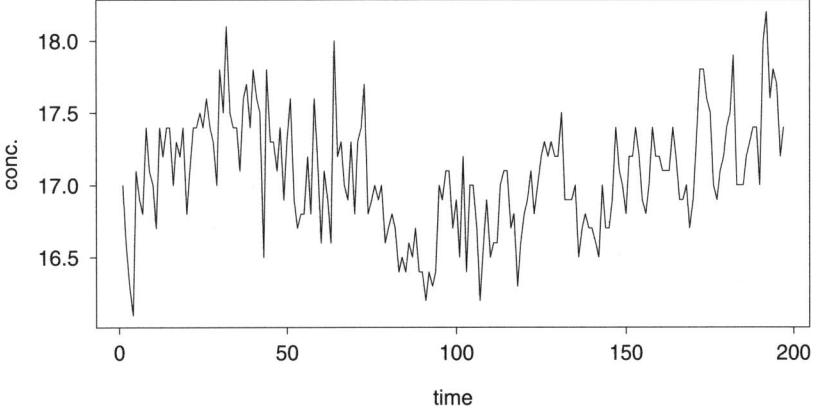

FIGURE 1.4 Concentration readings of a chemical process: Box–Jenkins series A.

transformation of the observations, we see a persistent linear growth in Figure 1.3b. The first difference series of the logged data is shown in Figure 1.3c, which is fairly stationary, although there appears to have some changes in the variability of the series over the data period.

Figure 1.4 presents series A and Figure 1.5 shows series C of Box and Jenkins (1976). The first appears to lie in the gray area of a stationary or a nonstationary series. For the second, differencing may be called for. Both have been used in the literature by other authors to illustrate novel methods for modeling time series.

1.1.3. Seasonal series

Time series data in business, economics, environment, and other disciplines often exhibit a strong cyclical or seasonal behavior. Modeling and analyzing such series

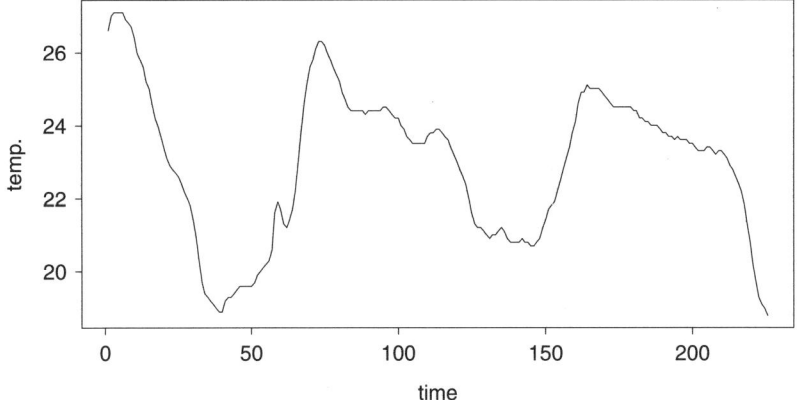

FIGURE 1.5 Temperature readings: Box–Jenkins series C.

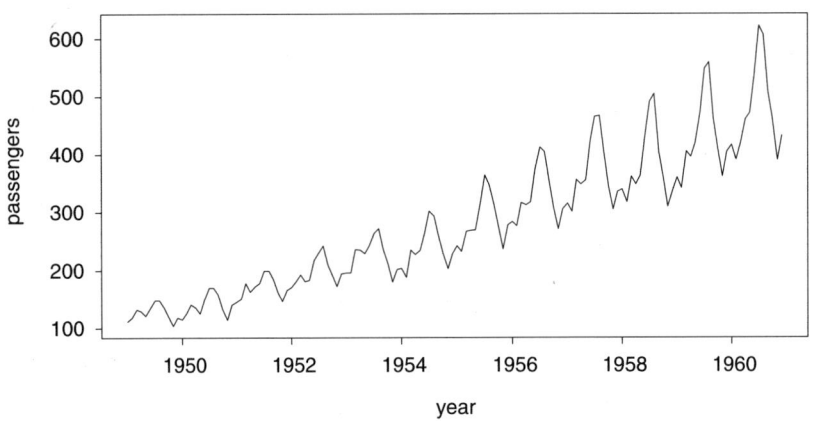

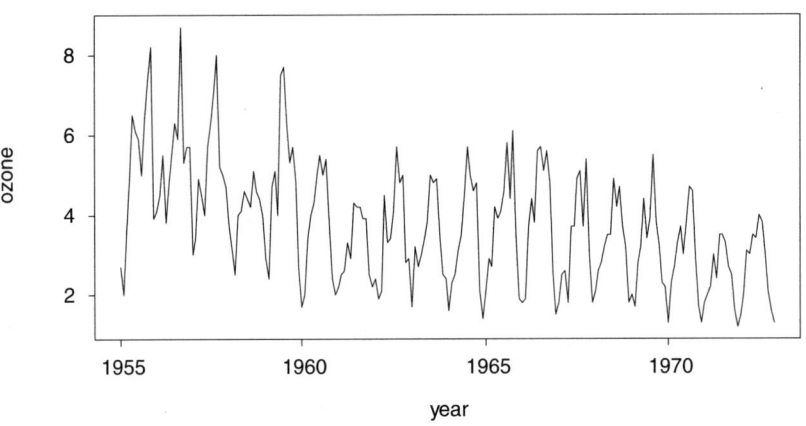

FIGURE 1.6 Two examples of seasonal series.

is an important topic in time series study. Figure 1.6a shows monthly international airline passenger totals in 1949–1959, which were used by Box and Jenkins (1976) to illustrate their innovative seasonal models. In practice, the user of the data may wish to remove the seasonality from the series in order to discern the "underlying trend," and this has led to the vast literature on seasonal decomposition and seasonal adjustment, which will be discussed later.

Figure 1.6b shows monthly averages of ozone in downtown Los Angeles during the period 1955–1972. Ambient ozone is an indicator of air pollution and is strongly seasonal: high in the summer months and low in the winter. In addition to seasonal cycles, there appears to be a level shift in the beginning of the sixth year and a down trend in the last 7 years of the data. The level shift may be associated with changes in

1.1. EXAMPLES OF TIME SERIES PROBLEMS

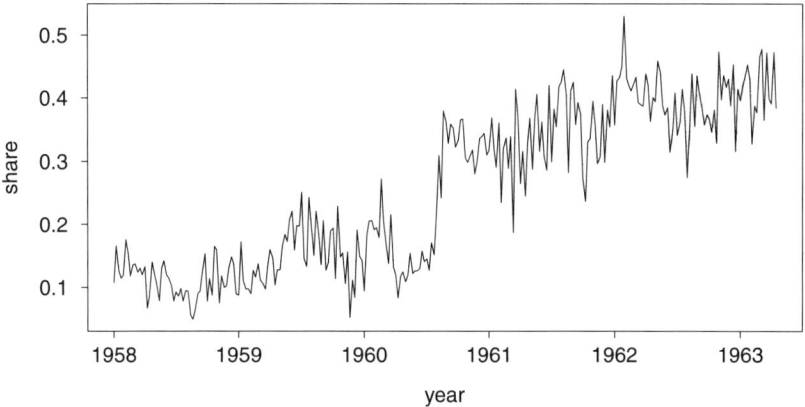

FIGURE 1.7 The Crest market share weekly data 1958–1963.

the traffic pattern and/or changes in the composition of gasoline sold in Los Angeles, and the down trend may be the result of progressively more stringent air quality standards at that time. This series was used by Box and Tiao (1975) to motivate intervention analysis in time series.

1.1.4. Level shifts and outliers in time series

Another example of a level shift is shown in Figure 1.7, the weekly market share data of Crest toothpaste from January 1958 to April 1963. In August 1960, the American Dental Association publicly endorsed Crest, and this led to a substantial jump in its market share as it is clearly seen in the figure. If the timing of this event is known, as in this case, the intervention analysis techniques can be applied to estimate its effect. In practice, such interventions are often unknown to the investigator, and detection of level shifts, outliers, and other types of structural changes becomes an important problem in time series analysis.

1.1.5. Variance changes

Figure 1.8 shows monthly returns of value-weighted S&P (Standard and Poor) 500 stocks from 1926 to 1991. While the mean level stayed close to zero over the entire period, it is clear that changes in the variance, called *volatility* in the finance literature, occurred. There has been an intense interest in modeling data of this kind in recent (at the time of writing) years, and some of the methods will be discussed later in the book.

1.1.6. Asymmetric time series

Two time series are shown in Figure 1.9. The first, in panel (a) is a series of annual sunspot numbers from 1700 to 1979; the other in panel (b), shows seasonally adjusted quarterly U.S. unemployment rates from 1948 to 1993. Both series share a common

8   INTRODUCTION

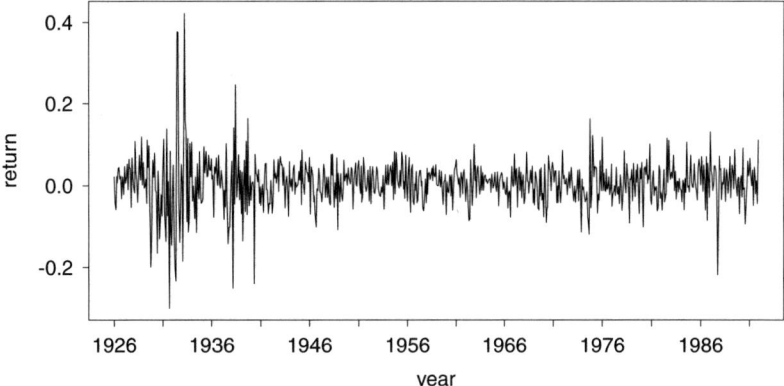

FIGURE 1.8   Value-weighted S&P 500 returns 1926–1991.

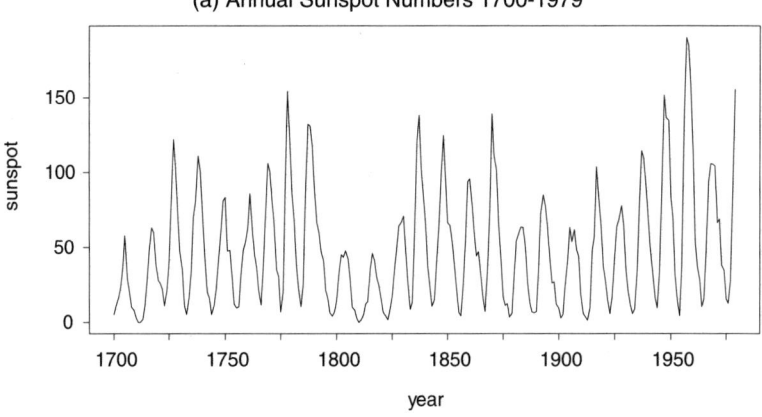

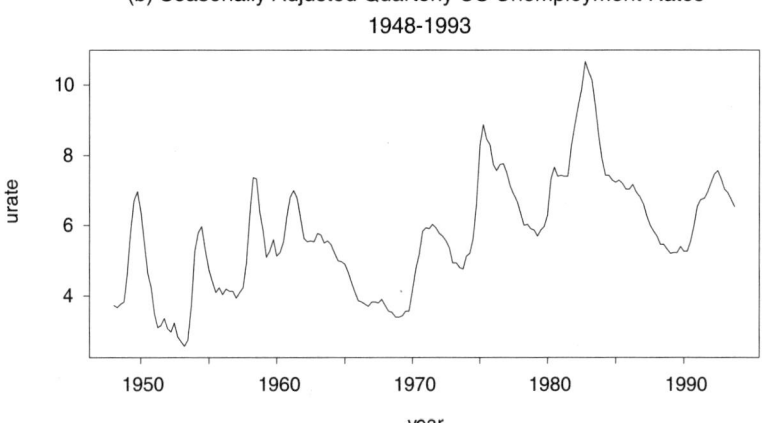

FIGURE 1.9   Two nonlinear time series.

1.1. EXAMPLES OF TIME SERIES PROBLEMS

feature; asymmetry in the rise and fall of the observations. Another way to express this asymmetric behavior is to say that these series are not time-reversible. The first will be used to illustrate nonlinear time series models, which is one of the most important research topics in time series analysis.

All the examples given above are univariate time series, and models will be introduced to relate the observations to their own past history. In practice, the principal purpose of modeling univariate series is on forecasting future observations of the series using its own past values. In the following examples, we now turn to consider several time series jointly.

1.1.7. Unidirectional–feedback relation between series

Figure 1.10 shows two series given in Box and Jenkins (1976), the input gas rate and the output CO_2 of a chemical reactor. The data were taken in 9-s intervals. These two

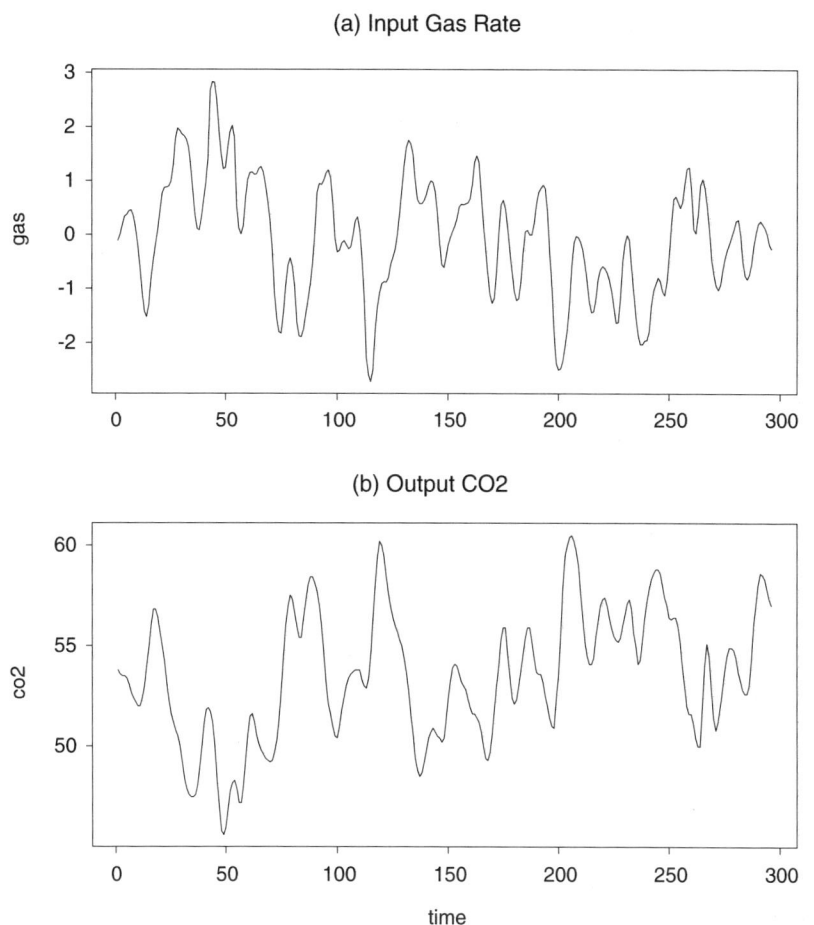

FIGURE 1.10 The gas furnace data: Box–Jenkins series J.

series are clearly related; when one goes up the other comes down. The figure also shows that, as expected, the input series leads the output series by several periods. Intuitively, the input series should therefore help to produce more efficient forecasts of the output series than just using the past values of the output. This example was used by Box and Jenkins to introduce their *transfer function* modeling techniques, but has also been used by others to illustrate modeling several series together.

The three series in Figure 1.11 are quarterly *Financial Times* stock, car production, and commodity indices over the period 1952–1967. They were first used by Coen et al. (1969) to establish an apparent regression relationship between stock index, car production index, which lagged six periods, as well as commodity index, which lagged seven periods. This result has led to many criticisms in the literature and has become a notorious example showing the importance of careful checking of the independence assumption of the residuals in regression of time series data.

1.1.8. Comovement and cointegration

Figure 1.12 gives five series, consisting of annual hog supply, hog prices, corn supply, corn prices and farm wages for the period 1867–1948. The data were given in Quenouille (1957), and all five series appear to be nonstationary. Box and Tiao (1977) used the data to illustrate their canonical analysis of multiple time series showing that linear combinations of nonstationary series can be stationary. Figure 1.13 shows five linearly transformed series, the first two of which are apparently stationary. This phenomenon has become known as "cointegration" (Engle and Granger 1986) and has been one of the most intensely studied topics in the econometrics literature in the last 10 years.

The three series in Figure 1.14 represent monthly logged flour price indices over the 9-year period 1972–1980 at the commodity exchanges of Buffalo, Minneapolis, and Kansas City, respectively. The example was used by Tiao and Tsay (1989) to illustrate their scalar component model technique in multiple time series model specification. The three series move in tandem as they should be, but they are not found to be cointegrated. This raises the interesting question as to how to characterize comovement in multiple time series.

As an further example of comovement of economic time series, Figure 1.15 shows three monthly series of 3-month, 6-month, and 9-month interest rates on bank deposits in Taiwan from 1961 to 1989. Again the three series move largely in tandem, but there is no cointegration. Tiao et al. (1993) employed this data set to illustrate the usefulness and limitations of various dimension reduction techniques including principal component, canonical correlation, and scalar component model methods.

1.2. OVERVIEW OF THE BOOK

The book is organized in three parts. Parts 1 and 2 concentrate on univariate time series models and Part 3, on multivariate models. A model for a univariate time series, z_t,

1.2. OVERVIEW OF THE BOOK

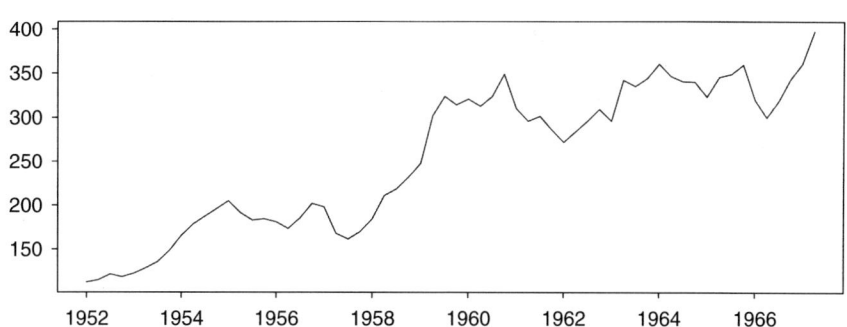

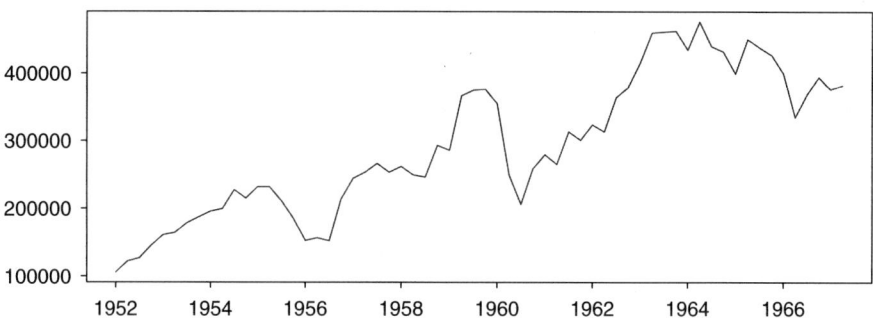

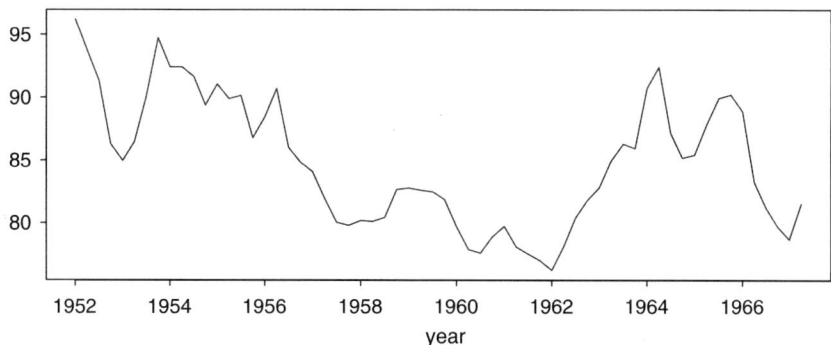

FIGURE 1.11 Quarterly *Financial Times* data 1952–1967.

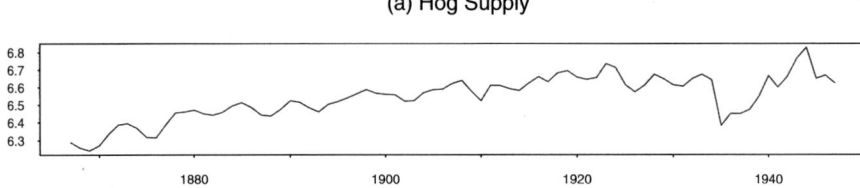

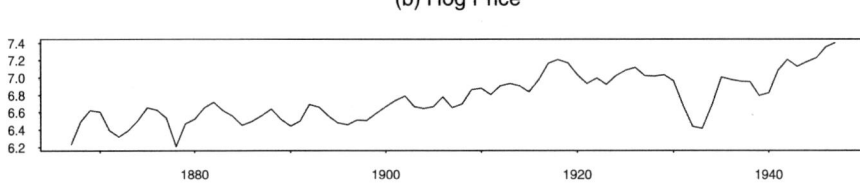

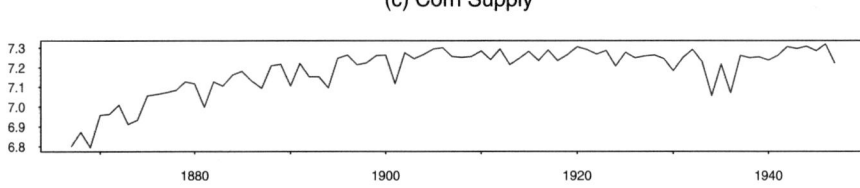

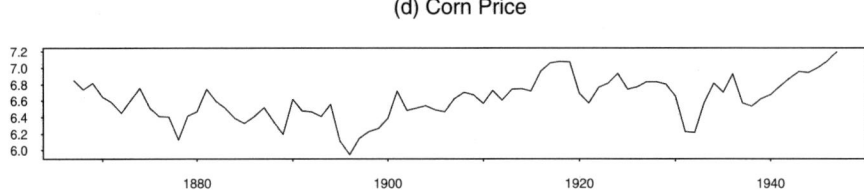

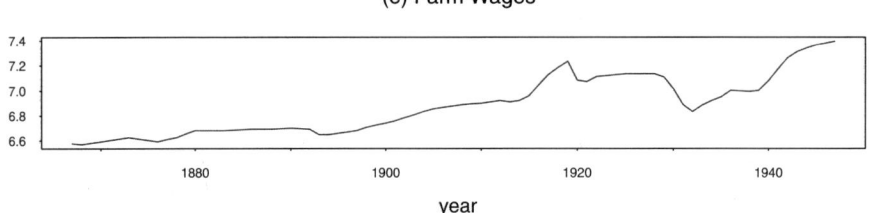

FIGURE 1.12 Quenouille's hog data 1867–1948.

1.2. OVERVIEW OF THE BOOK

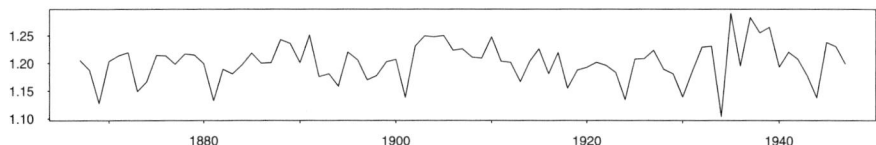

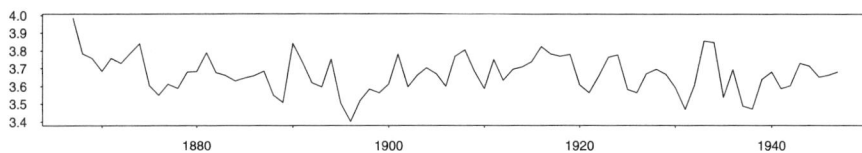

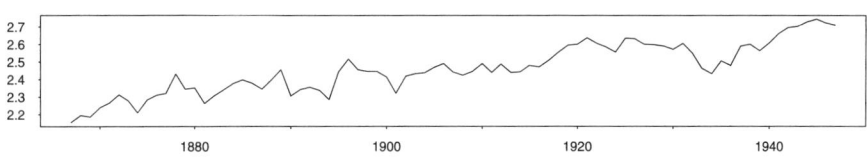

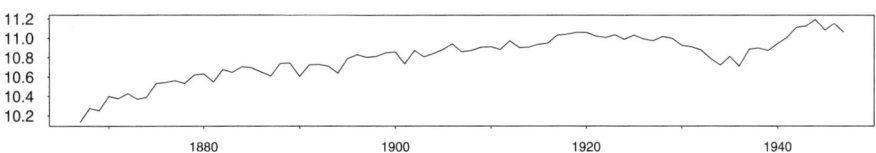

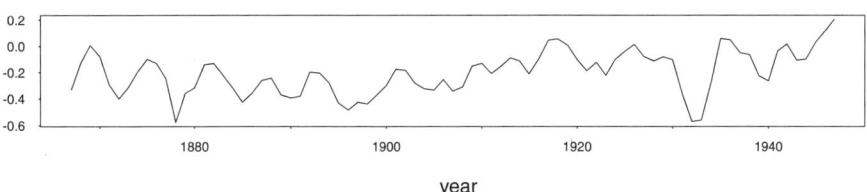

FIGURE 1.13 Linearly transformed hog data.

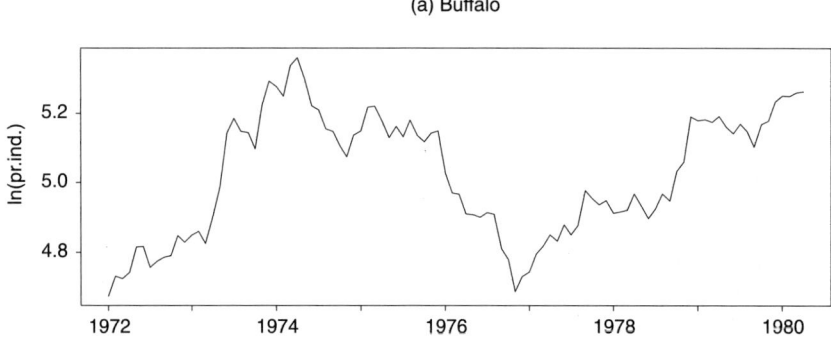

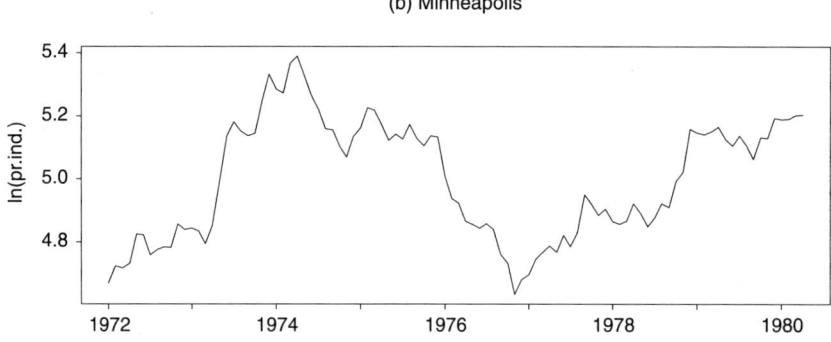

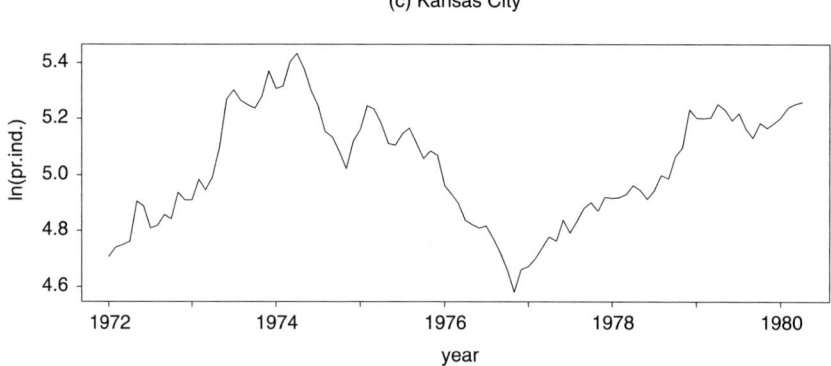

FIGURE 1.14 Logarithms of monthly flour price indices 1972–1980.

1.2. OVERVIEW OF THE BOOK

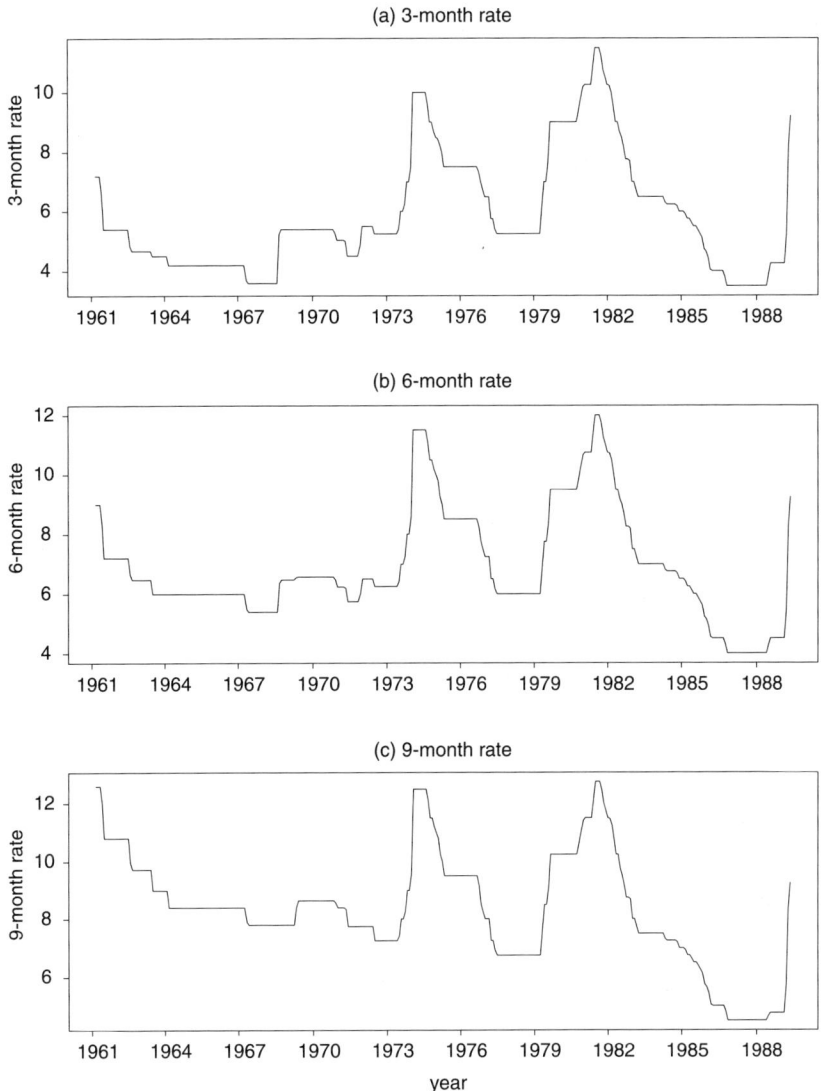

FIGURE 1.15 Taiwan's interest rate data 1961–1989.

takes usually the additive decomposition form

$$z_t = f(z_{t-1}, \ldots, z_1) + a_t \tag{1.1}$$

where $f(z_{t-1}, \ldots, z_1)$ is a function of the past values of the series, to be determined from the data, and a_t is a sequence of independent and identically distributed (iid) variables. The time series model can be seen as a way to decompose the data into a systematic part (the signal part), which depends on past values and therefore can be forecasted, $f(z_{t-1}, \ldots, z_1)$, and a noise part, a_t, which is independent from previous values and therefore it is unpredictable from its past. In some cases obtaining the structure of the function f is the main objective of the analysis whereas in other cases our interest is mostly in obtaining forecasts.

Part 1, on basic concepts for univariate time series modeling, presents the main ideas and tools for building univariate time series models. The presentation emphasizes linear autoregressive integrated moving-average (ARIMA) models, in which it is assumed that the function f has a linear form, that is, it can be written as

$$f(z_{t-1}, \ldots, z_1) = \pi_1 z_{t-1} + \cdots + \pi_{t-1} z_1 \qquad (1.2)$$

and a key problem is how to approximate the sequences of weights $(\pi_1, \pi_2, \ldots .)$ by using a small number of parameters. This is the idea of ARIMA models studied by Box and Jenkins (1976) in a landmark book on time series analysis. In addition to ARIMA models, this part also presents a brief analysis of two alternative time series approaches. The first one is the spectral approach in which it is assumed that the function f can be represented as a sum of sine and cosine waves. The second is the state space approach in which the evolution of the series is assumed to be a linear function of some unobserved factors or states as

$$z_t = \mu_t + a_t,$$

where μ_t is the mean, and a_t has the same interpretation as before. Then we have to assume some equation for the evolution of the mean as, for instance

$$\mu_t = \mu_{t-1} + u_t,$$

where u_t is another sequence of iid variables. State-space models and ARIMA models are closely related, but the latter provide more flexibility as we do not need to determine the evolution of the state variables; it is identified from the data.

Part 2, on advanced topics for univariate time series, covers Chapters 9–13, and includes more sophisticated time series models. The first generalization is to assume that the variability of the noise process is not constant but depends on past values of the process. This allows for particular forms of heteroscedasticity in a_t that have many applications in financial data. A second generalization is to assume some parametric nonlinear structures for the function f, and then we have nonlinear time series models. A third generalization is to try to estimate the function f without assuming a priori any parametric structure. This can be carried out by the nonparametric approach, if we have a large sample size, or we can try to approximate f by a general method as neural network models.

The third part of the book deals with multivariate models. In the simplest case of a bivariate system the two components series can be split into a dependent response time series, y_t, and an independent input series x_t. The first model equation will describe the dynamic evolution of the response as a function of the input variable in

1.2. OVERVIEW OF THE BOOK

what is called a *dynamic regression model*

$$y_t = g(x_t, \ldots, x_1) + f(y_{t-1}, \ldots, y_1) + a_t \qquad (1.3)$$

where, as before, a_t is a sequence of independent and identically distributed (iid) variables and the functions f and g are to be determined from the data. In addition to this equation, the time series model has to specify the model for the evolution of the univariate independent series x_t, that will be a univariate time series model. In the general case, we cannot say that the variable x_t causes y_t or vice versa because there is feedback between the two variables. The time series model will describe this situation by two dynamic regression equations. In general, letting $\mathbf{z}_t$ be a vector of k related time series, a multivariate time series model takes the form

$$\mathbf{z}_t = \mathbf{f}(\mathbf{z}_{t-1}, \ldots, \mathbf{z}_1) + \mathbf{a}_t \qquad (1.4)$$

where now $\mathbf{f}$ is a vector of functions of the past values of all the components of the vector time series to be determined from the data and $\mathbf{a}_t$ is a sequence of vector variables without lag dependency. This model will be equivalent to k dynamic regression equations, in which each series is explained as a function of the past of all the other series and its own past. Note, however, that the noises from different equations are usually correlated. If we assume that f has a linear form, we obtain the *multivariate ARIMA* models and the *linear system models*. A key problem in multivariate modeling is finding simplifying structures in order to reduce the number of parameters in the model and facilitate its interpretation. An interesting structure with clear economic meaning is that linear combinations of the components of the vector time series are more stable over time than the series themselves. In particular, if the vector of series has some stochastic trend, it is interesting to find linear combinations that are stable around some fixed mean. This is the idea of cointegration. Finally, multivariate model can be analyzed form the state-space approach and a review of the field is also included in this third part of the book.

The subject matter of Part 1 is distributed into chapters as follows. Chapter 2 introduces stationary time series models and presents the three most important classical approaches to analyze time series data: ARIMA models, periodicity analysis in the frequency domain, and state-space models. The chapter introduces the basic tools for each of these analysis: the autocorrelation function and the partial autocorrelation function for linear stationary process, the periodogram, and the spectrum for periodicity analysis and the dynamic linear system for the state-space representation. Chapter 3 considers the model specification strategy for univariate ARIMA models. Assuming that we have a linear process, the key problem is how to parametrize it in order to represent its structure with a small number of parameter values that can be estimated from the data. The chapter explains three statistical tools that can be used for this objective: the autocorrelation function, partial autocorrelation function, and the extended autocorrelation function. It is shown how an ARIMA model can be identified by these tools, and examples are given of the use of this methodology. Chapter 4 presents an introduction to the maximum likelihood estimation of ARMA models and discusses the diagnostic checking of the fitted model. It also includes procedures for parameter estimation using the sample spectrum and the estimation of state-space models by the Kalman filter. Chapter 5 presents the prediction problem and concentrates in the computation of ARIMA forecasts with emphasis on understanding the structure of the forecasts generated by these models. It is shown that the forecast function can be easily understood in terms of the main parts of the ARIMA model. This chapter also includes an introduction to the combination of forecast from different sources and to the problem of model selection in time series. Time series, as any kind of statistical data, are often subject to outliers, and Chapter 6 discusses outliers and influential observations in univariate time series. Different types of outliers are introduced, and a methodology is presented to identify them and estimate their effects. It is shown that outlier analysis is related to the estimation of missing values in a time series, and a brief introduction to this important practical problem is presented. Chapter 7 presents a procedure for automatic ARIMA modeling of univariate time series that is implemented in the program TRAMO. This program allows a powerful and fast application of the methodology presented in the previous chapters. Finally, Chapter 8 discusses the use of ARIMA models in the important problem of seasonal adjustment in economic time series. The chapter shows how ARIMA models provide a powerful tool for decomposing the observed time series and carrying out seasonal adjustment and discusses this topic within the broader context of signal extraction.

Part 2, on advanced topics, includes Chapters 9–13. Chapter 9 considers a particular class of nonlinear time series models with many application in finance: heteroscedastic models. The chapter concentrates in the most often used ARCH and GARCH models and presents examples of their applications. Chapter 10 considers more general nonlinear time series models. This chapter presents a general test to detect three kinds of non linearity (bilinear, exponential autoregressive, and threshold autoregressive) often found in time series and discusses in more detail the fitting of threshold models. Chapter 11 analyzes in a common framework linear and nonlinear model by using the Bayesian approach. It is shown how Markov chain Monte Carlo methods (MCMC) provides a powerful tool for the analysis of complex models within the Bayesian framework. Chapter 12 presents an alternative way to analyze time series: the nonparametric approach. In particular, it is shown how this approach can be applied to the decomposition and seasonal adjustment of economic time series. Finally, Chapter 13 includes an introduction of neural network models in time series. This method provides a simple way to generate forecast for a time series with a minimum set of assumptions about the underlying structure.

Part 3, on multiple time series analyses, includes Chapter 14–16. Chapter 14 presents a methodology for building multivariate time series ARIMA models. The three stages of identification, estimation, and diagnostic are presented, and illustrated with examples. The chapter also discusses the important problem of model simplification by different types of eigenvalue analysis. A key idea in multivariate modeling is finding simplifying structure in the vector time series, and, in particular, this includes finding linear combination of the vector time series that are more stable than the observed series. This leads to the idea of cointegration, developed in Chapter 15, in which a general methodology is presented for testing and estimating cointegration relationships. Finally, Chapter 16 presents an introduction to the multivariate analyses of linear system from the linear system approach. This methodology offers an alternative to the vector ARIMA methodology for multivariate analysis of time series.

1.3. FURTHER READING

The reader interested in a deeper analysis of the basic concepts in time series should consult the books by Abraham and Ledolter (1983), Anderson (1971), Box and Jenkins (1976), Box et al. (1994), Brockwell and Davis (1987, 1996), Gourieroux and Monfort (1997), Granger and Newbold (1977), Fuller (1976), Pandit and Wu (1983), Shumway (1988), Shumway and Stoffer (2000), and Wei (1990). The spectral approach is presented in Brillinger (1975), Granger and Hatanaka (1964), Jenkins and Watts (1968), and Priestley (1981). Harvey (1989) discusses with detail the structural approach in time series based on the state-space representation (see Anderson and Moore 1979) for economic time series. Hamilton (1994) and Enders (1995) also emphasize economic time series and econometrics. Hendry and Clements (1998) concentrates on economic forecasting.

Moving to the advanced topic section, ARCH and GARCH models are discussed in recent econometric texts, and the reader can find a deeper study in the books by Engle (1995) and Gourieroux (1997). Nonlinear models are discussed by Granger and Andersen (1980), Priestley (1988), and Tong (1990). Bayesian models are discussed by West and Harrison (1997), nonparametric regression by Hardle (1990), and neural networks by Ripley (1996).

Multivariate ARMA time series models are considered by Hannan (1970), Lutkepohl (1993), Reinsel (1993), and Reinsel and Velu (1998). Aoki (1990) and Hannan and Deistler (1988) are important references for state-space modeling of multivariate time series.

The limitation of space and time has made that many interesting development in time series have not been introduced in this text. Among them are long memory processes (Beran 1994), wavelets (Hardle et al. 1998, Morettin 1999), and discrimination and clustering in time series (Karizawa et al. 1998).

REFERENCES

Abraham, B. and Ledolter, J. (1983). *Statistical Methods for Forecasting*. Wiley, New York.

Anderson, B. O. and Moore, J. B. (1979). *Optimal Filtering*. Prentice-Hall, Englewood Cliffs, NJ.

Anderson, T. W. (1971). *The Statistical Analysis of Time Series*. Springer-Verlag, New York.

Aoki, M. (1990). *State Space Modeling of Time Series*. Springer-Verlag, New York.

Beran, J. (1994). *Statistics for Long-Memory Processes*. Cambridge Univ. Press, Cambridge, UK.

Box, G. E. P. and Jenkins, G. M. (1976). *Time Series Analysis: Forecasting and Control.* Holden-Day, San Francisco.

Box, G. E. P., Jenkins, G. M. and Reinsel, G. (1994). *Time Series Analysis: Forecasting and Control*, 3rd ed. Prentice-Hall, Englewood Cliffs, NJ.

Box, G. E. P. and Tiao, G. C. (1975). Intervention analysis with applications to economic and environmental problems. *J. Am. Stat. Assoc.* **75**, 70–79.

Box, G. E. P. and Tiao, G. C. (1977). A canonical analysis of multiple time series. *Biometirka* **64**, 355–366.

Brillinger, D. R. (1975). *Time Series: Data Analysis and Theory.* Holt, Rinehart & Winston, New York.

Brockwell, P. J. and Davis, R. A. (1987). *Time Series: Theory and Methods.* Springer-Verlag, New York.

Brockwell, P. J. and Davis, R. A. (1996). *An Introduction to Time Series: Theory and Methods.* Springer-Verlag, New York.

Coen, P. J. Gomme, E. D., and Kendall, M. G. (1969). Lagged relationships in economic forecasting (with discussion). *J. Roy. Stat. Soc. A* **132**, 133–163.

Enders W. (1995). *Applied Econometric Time Series.* Cambridge Univ. Press, Cambridge UK.

Engle, R. F. (1995). *ARCH: Selected Readings.* Oxford Univ. Press, Oxford, UK.

Engle, R. F. and Granger, C. W. J. (1986). Co-integration and error correction: Representation, estimation, and testing. *Econometrica* **55**, 251–267.

Fuller, W. A. (1976). *Introduction to Statistical Time Series.* Wiley, New York.

Gourieroux, C. (1997). *ARCH Models and Financial Applications.* Springer, Berlin.

Gourieroux, C. and Monfort, A. M.(1997). *Time Series and Dynamic Models.* Cambridge Univ. Press, Cambridge, UK.

Granger, C. W. J. and Adensen, A. P. (1980). *An Introduction to Bilinear Time Series Models.* Vandenhoeck and Ruprecht, Amsterdam.

Granger, C. W. J. and Hatanaka, M. (1964). *Spectral Analysis of Economic Time Series.* Princeton Univ. Press, Princeton, NJ.

Granger, C. W. J. and Newbold, T. (1977). *Forecasting Economic Time Series.* Academic Press, New York.

Hamilton, J. D.(1994). *Time Series Analysis,* Princeton Univ. Press, Princeton, NJ.

Hannan, E. J. (1970). *Multiple Time Series.* Wiley, New York.

Hannan, E. J. and Deistler, M. (1988). *The Statistical Theory of Linear Systems.* Wiley, New York.

Hardle W. (1990). *Applied Nonparametric Regression.* Cambridge Univ. Press, Cambridge, UK.

Hardle, W. et al. (1998). *Wavelets, Approximation and Statitical Applications.* Cambridge Univ. Press, Cambridge, UK.

Harvey, A. C. (1989). *Forecasting, Structural Time Series Models and the Kalman Filter.* Cambridge Univ. Press, Cambridge, UK.

Hendry, D. F. and Clements, M. P. (1998). *Forecasting Economic Time Series.* Cambridge Univ. Press, Cambridge, UK.

Jenkins, G. M, and Watts, D. G. (1968). *Spectral Analysis and Its Applications.* Holden-Day, San Francisco.

REFERENCES

Karizawa, Y., Shumway, R. H., and Taniguchi, M. (1998). Discrimination and clustering for multivariate time series. *J. Am. Stat. Assoc.* **93**, 328–340.

Lutkepohl, H. (1993). *Introduction to Multiple Time-Series Analysis*. Springer-Verlag, Berlin.

Morettin, P. (1999). *Ondas e Ondaletas*. Edusp, Sao-Paulo.

Pandit, D. M. and Wu, S. M. (1983). *Time Series and System Analysis with Applications*. Wiley, New York.

Priestley, M. B. (1981). *Spectral Analysis and Time Series*. Academic Press, Orlando, FL.

Priestley, M. B. (1988). *Non-linear and Non-stationary Time Series Analysis*. Academic Press, Orlando, FL.

Quenouille, M. H. (1957). *The Analysis of Multiple Time-Series*. Griffin, London.

Reinsel, G. C. (1993). *Elements of Multivariate Time-Series Analysis*. Springer-Verlag, New York.

Reinsel, G. C. and Velu, R. P. (1998). *Multivariate Reduced Rank Regression*. Springer-Verlag, New York.

Ripley B. D. (1996). *Pattern Recognition and Neural Networks*. Cambridge Univ. Press, Cambridge, UK.

Shumway, R. H. (1988). *Applied Statistical Time Series Analysis*. Prentice-Hall, Englewood Cliffs, NJ.

Shumway, R. H. and Stoffer, D. A. (2000). *Time Series Analysis and its Applications*. Springer-Verlag, New York.

Tiao, G. C. and Tsay, R. S. (1994). Some advances in non-linear and adaptive modelling in time-series. *J. Forecasting* **13**, 109–131.

Tiao, G. C., Tsay, R. S., and Wang, T. (1993). Usefulness of linear transformations in multivariate time-series analysis. *Empirical Econ.* **18**, 567–593.

Tong, H. (1990). *Nonlinear Time Series. A Dynamical System Approach*. Oxford Science Publications, New York.

Wei, W. W. S. (1990). *Time Series Analysis*. Addison-Wesley, Reading, MA.

West, M. and Harrison, J. (1997). *Bayesian Forecasting and Dynamic Models*, 2nd ed. Springer-Verlag, New York.

PART I

Basic Concepts in Univariate Time Series

CHAPTER 2

Univariate Time Series: Autocorrelation, Linear Prediction, Spectrum, and State-Space Model

G. Tunnicliffe Wilson
Lancaster University

2.1. LINEAR TIME SERIES MODELS

Figures 2.1–2.3 illustrate a variety of time series. Let us start by looking at these from the point of view of linear prediction. By *prediction* we mean estimation of one or more values of the series using previous values. A *linear prediction* is one that can be represented as a linear combination of the previous values.

The simplest linear prediction method is to extrapolate a deterministic curve fitted to the time series by linear regression. This curve may, for example, consist of polynomials for trend, sinusoids for cycles, and indicator variables for seasonality. Consider the monthly atmospheric carbon dioxide series shown in Figure 2.1a, which may be modeled, for $t = 1, 2, \ldots, n = 161$, as the response variable z_t in the linear regression:

$$z_t = c + bt + \alpha_1 I_{1,t} + \cdots + \alpha_{12} I_{12,t} + e_t \qquad (2.1)$$

where

$$I_{1,t} = \begin{cases} 1 & \text{in each January} \\ 0 & \text{in other months} \end{cases} \cdots I_{12,t} = \begin{cases} 1 & \text{in each December} \\ 0 & \text{in other months} \end{cases}. \qquad (2.2)$$

A Course in Time Series Analysis, Edited by Daniel Peña, George C. Tiao, and Ruey S. Tsay.
ISBN 0-471-36164-X. © 2001 John Wiley & Sons, Inc.

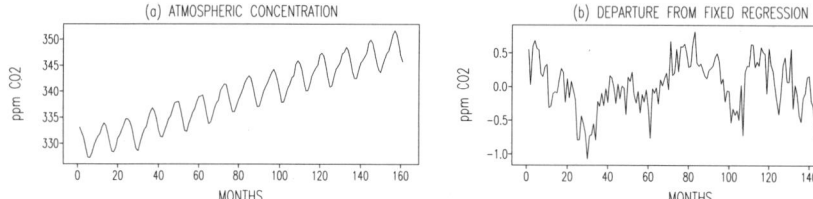

FIGURE 2.1 Atmospheric carbon dioxide concentrations from 1974 to 1986: (a) monthly values in ppm showing trend and seasonal pattern; (b) variation of values around a fixed trend and seasonality.

This represents a fixed trend and a fixed annual seasonal pattern around that trend. One of the indicator variables $I_{j,t}$, usually the first or last, must be removed to avoid collinearity.

The series in Figure 2.1a is very regular and the chosen model appears to fit very well, but two criticisms can be made of this approach. The first of these comes from considering how the prediction of the next value, obtained by extrapolating the fitted model, can be expressed as a linear combination

$$\hat{z}_{n+1} = w_1 z_1 + \cdots + w_n z_n. \qquad (2.3)$$

For this example $w_1 = -.0189$ and $w_n = .0189$. This has the undesirable property of placing comparable weight, although of opposite sign, on the earliest and latest points of the data. For prediction it is desirable that much greater weight be placed on later data, closer in time to the values that are being predicted.

Second, the regression is based on the simple statistical assumption that the errors are uncorrelated, but this is not supported by the graph of the residuals from model (2.1) shown in Figure 2.1b on a much enlarged scale compared with the series. The persistence shown in these errors, specifically, the tendency for one value to be close to the previous value, could be used to improve the prediction.

Various methods have been used to overcome these criticisms, notably the use of regression with heavier weight placed on the most recent values; the weights typically are discounted into the past. These ideas have been part of the development toward the *general linear (time series) model* (GLM), which is the basis of the *autoregressive integrated moving–average* (ARIMA) models presented in the next chapter. One form

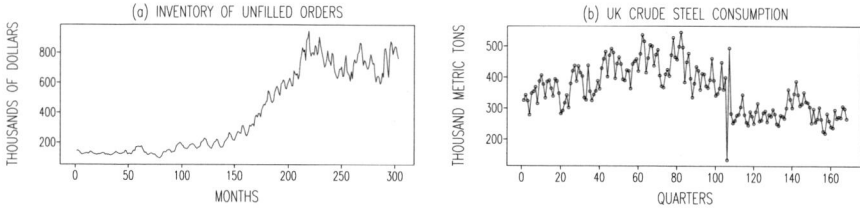

FIGURE 2.2 (a) Monthly series of unfilled orders; (b) UK quarterly consumption of crude steel.

2.1. LINEAR TIME SERIES MODELS

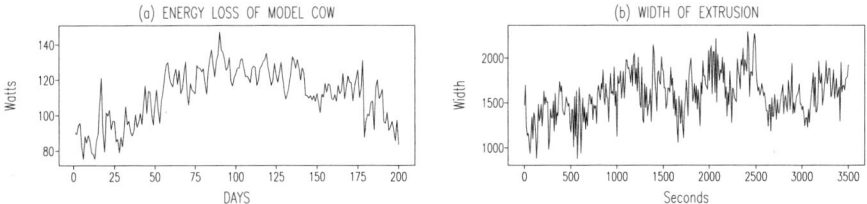

FIGURE 2.3 (a) Daily energy loss of a model cow; (b) width of extruded plastic product.

of the GLM is

$$z_t = \sum_{k=1}^{\infty} \pi_k z_{t-k} + a_t = \hat{z}_{t-1}(1) + a_t \qquad (2.4)$$

where $\hat{z}_{t-1}(1)$ is the linear combination of past values that best predicts z_t and a_t is a times series of independent prediction errors.

A particular example of this model is one for which $\hat{z}_{t-1}(1)$ is given by the widely used *exponentially weighted moving–average* (EWMA) predictor:

$$\begin{aligned}\hat{z}_{t-1}(1) &= (1-\theta)z_{t-1} + \theta(1-\theta)z_{t-2} + \theta^2(1-\theta)z_{t-3} + \cdots \\ &= (1-\theta)z_{t-1} + \theta\hat{z}_{t-2}(1).\end{aligned} \qquad (2.5)$$

Here, the coefficients π_k are discounted by the factor θ for increasing lag k and scaled by $(1-\theta)$ to sum to unity. The second equation is the recursive form used in practice. We make the following remarks about the GLM (2.4):

1. We assume the property of *time invariance*, that the coefficients π_k depend only on the lag k and not on t and that the errors or *linear innovations* a_t have constant variance besides being an independent series.
2. We consider z_t to be a stochastic process for which (2.4) is just one possible valid representation, rather than the defining equation. Another representation is to express z_t formally as a combination of present and past values of a_t:

$$z_t = a_t + \sum_{k=1}^{\infty} \psi_k a_{t-k}. \qquad (2.6)$$

This is a formal expression in the sense that for some series the sum may not converge, but it may be extended back as far as desired to some time origin and the remaining terms of the sum represented by well-defined initial conditions. For example the EWMA model above can be expressed formally as

$$z_t = a_t + \sum_{k=1}^{\infty}(1-\theta)a_{t-k} \qquad (2.7)$$

and in the well-defined form

$$z_t = a_t + \sum_{k=1}^{K}(1-\theta)a_{t-k} + [z_{t-K-1} - \theta a_{t-K-1}]. \tag{2.8}$$

3. The two forms of the GLM as expressed in (2.4) and (2.6) are known as the *infinite autoregressive* and *infinite moving–average* forms, respectively. These are taken by Box and Jenkins (1976) as the starting point for the introduction of ARIMA models presented in the next chapter. Although both (2.4) and (2.6) involve sums into the infinite past, ARIMA modeling can be applied without difficulty to finite observed data sets, usually by implicitly estimating the contribution of the series values prior to those observed.

The general linear model is capable of representing a wide variety of series. For some series, of which the carbon dioxide concentration is a good example, there may well be a fixed trend and seasonality, or other deterministic component, which is best removed before applying ARIMA modeling to the remaining variation such as that shown in Figure 2.1b. For most series, though not all, such a deterministic component consists of just a constant term, which is normally assumed as part of the ARIMA model.

For other series, such as that shown in Figure 2.2a, which is a monthly series of unfilled orders for newspapers and magazines from 1964 to 1989, the trend and seasonality are clearly changing, and the GLM can represent such stochastic variations very well through *seasonal* ARIMA modeling. The series in Figure 2.2b, which is a quarterly series of crude steel consumption from 1953 to 1995 shows, besides a clear quarterly seasonality, a cyclical variation with a period of about 4 years and all these features can again be well represented by a general linear model.

The main purpose of this chapter is to lay the foundations on which the appropriate general linear model from the ARIMA class can be determined, by describing the statistical summaries of the observed time series used in this task. The two series shown in Figure 2.3 provide further examples to illustrate the behaviour that can be represented by the linear model. Figure 2.3a shows the daily energy loss recorded for a model cow. The variations of this series about an annual cycle could be described as irregular rather than random; there is a slight smoothness to the pattern. Figure 2.3b is a series of measurements of the width of a product formed from a continuous plastics extrusion process. A slow but inconsistent drift of level can be seen in the very "noisy" measurements.

2.2. THE AUTOCORRELATION FUNCTION

Prediction and correlation go hand in hand. In simple linear regression the sample correlation coefficient provides a direct measure of predictive ability. To define a theoretical correlation coefficient we need a well-defined population from which both the explanatory variable, or predictor, and the response are jointly drawn. In a

2.2. THE AUTOCORRELATION FUNCTION

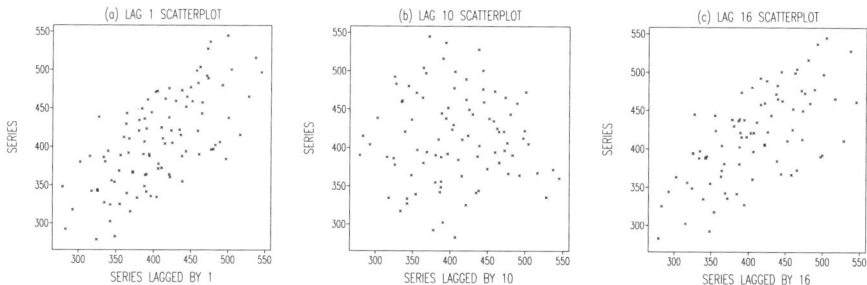

FIGURE 2.4 Scatterplots of series of steel consumption against the same series values lagged by 1, 10, and 16 quarters.

univariate time series context we consider that the pair of values z_{t-k}, z_t constitute the predictor and response. The requirement that these come from a well-defined population as t varies is satisfied by the assumption of *stationarity*. This has various forms and we state here the weak form, that

1. $E(z_t) = \mu_z$ is constant for all t.
2. $Var(z_t) = \sigma_z^2$ is constant for all t.
3. $Cov(z_{t-k}, z_t) = \gamma_{z,k}$ depends only the separation lag k and not on t.

The sequence $\gamma_{z,k}$ is the *autocovariance function* of the series and, dropping the suffix z for simplicity, $\rho_k = \gamma_k/\gamma_0$ is the *autocorrelation function*.

We illustrate the definition of the autocorrelations by considering the scatterplots in Figure 2.4 between the first 100 values of the series of steel production and the values at, respectively, lags of 1, 10 and 16 quarters. The first plot for example has the values $z_2, \ldots, z_{100}$ plotted vertically against $z_1, \ldots, z_{99}$ horizontally. Over the earlier period this series appears to be approximately stationary, although in the longer term there is certainly a change of structure. From these three scatterplots we obtain sample correlations of .673, −.018, and .660, which estimate the autocorrelations at the respective lags. It should be realized that these scatter plots differ in an essential way from the typical scatter plot of bivariate data in that the points displayed are *not* independent of one another; they come from the same record of a series of dependent values. This affects the sampling properties of the estimated autocorrelation, which at any given lag depend also on the autocorrelations at other lags. This limits the usefulness of formulas for the standard errors of these estimates developed by Bartlett (1946) and given by Box and Jenkins (1976, p. 34). Special cases, given below, are however very useful. The commonly used definitions of the sample mean, variance, autocovariances, and autocorrelations calculated from observations $z_1, z_2, \ldots, z_n$ are

$$\bar{z} = \frac{1}{n} \sum_{t=1}^{n} z_t \qquad (2.9)$$

$$s_z^2 = \frac{1}{n} \sum_{t=1}^{n} (z_t - \bar{z})^2 \qquad (2.10)$$

$$c_{z,k} = \frac{1}{n} \sum_{t=1}^{n-k} (z_t - \bar{z})(z_{t+k} - \bar{z}) \qquad (2.11)$$

$$r_{z,k} = \frac{c_{z,k}}{s_z^2}. \qquad (2.12)$$

There are differences between the sample correlations given above, of the scatterplots in Figure 2.4, and those defined by (2.12), which are .662, −.014, and .570 at the lags of $k = 1$, 10 and 16. Although small at low lags, the differences increase at higher lags. They arise partly because in (2.11) and (2.12) the series values are corrected by the *same* mean and divided by the *same* variance, calculated for the whole sample. In Figure 2.4a, for example, the correlation is calculated between the sub-samples $z_2, z_3, \ldots, z_{100}$ and $z_1, z_2, \ldots, z_{99}$, correcting them by their slightly different means, and dividing by the product of their slightly different standard deviations. At higher lags an important difference arises because the divisor of n is retained in (2.11), even though the covariance is formed from only $n - k$ lagged products. A divisor of $n - k$ in place of n will be seen in the form of (2.11) given in some older texts, but it can result in a sample autocorrelation (2.12) larger than 1 at high lags. These differences arise also in lagged regression and other areas of time series modeling, and are generally referred to as *end effects*.

The sample autocorrelation function (acf) provides a widely used statistical summary of the properties of an observed times series. Figure 2.5 contains four frames in the first of which a plot of the first 100 values of the steel consumption time series is shown, and in the second its sample autocorrelations. The remaining two frames show the sample *partial* autocorrelation function and sample *spectral density*, which will be described shortly. Note how the sample autocorrelation in this case summarizes the cyclical properties of the series with peaks at lags 4 and 16 indicating the periods of the cycles.

Although we stated earlier that this series was approximately stationary, the aspect ratio of the first frame in this figure suggests that it might contain a trend and may not therefore be stationary. This is a typical problem, that from a finite sample it may not be evident whether slow changes in level are part of long-term variation of a stationary series about a fixed mean, or are best described by a trend in level that requires correction before stationarity can be assumed. Figure 2.6 shows a similar set of graphs for the energy loss series *after* it has been corrected, by regression, for a smooth annual cycle. The regressors for this cycle are defined by $\cos(2\pi t/p)$ and $\sin(2\pi t/p)$, where t is the time index and p is the period, of 365 in this case. Figure 2.6 gives no indication of shorter-term cycles, but some autocorrelation at low lags that reflects a modest degree of persistence in successive values of the series.

In these last two figures the sample autocorrelation plots have included parallel lines equally spaced about the horizontal axis. These are shown to indicate the range of sampling variability to be expected of the (low lag) sample autocorrelations *when the series is random*, that is, when its acf is zero at all lags. They are drawn at $\pm 2/\sqrt{n}$ where n is the sample length, to give approximate two standard error limits in that case.

2.2. THE AUTOCORRELATION FUNCTION

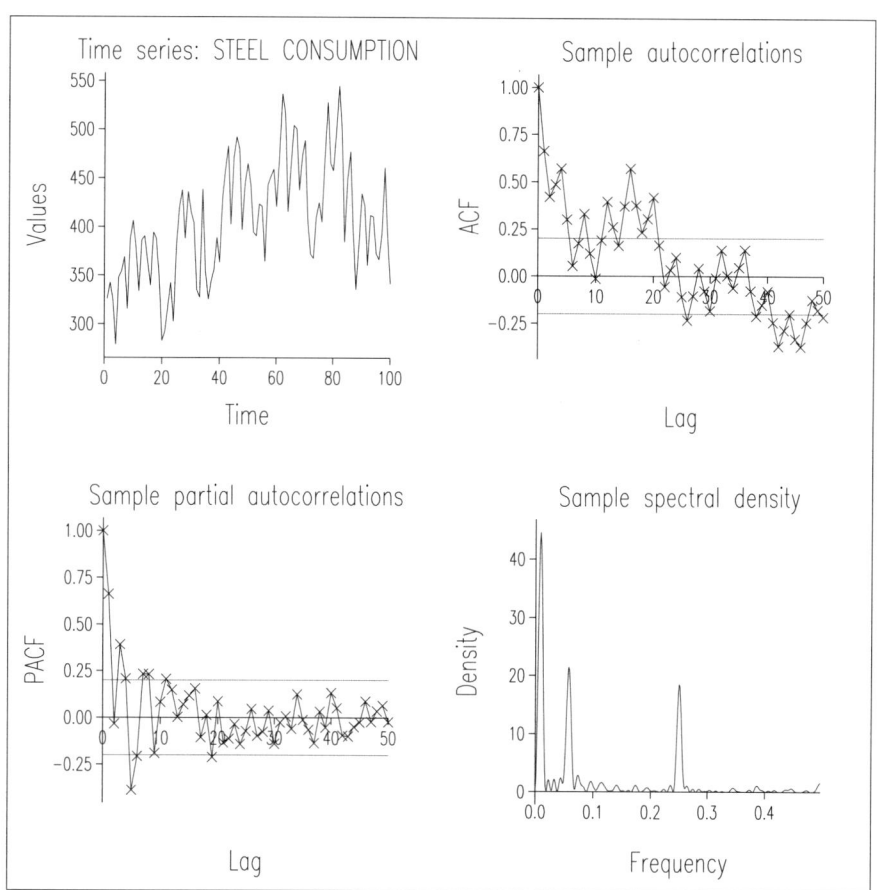

FIGURE 2.5 First 100 values of the series of steel consumption with sample statistical properties.

This is an important case because we shall often wish to check whether there is *any* evidence of autocorrelation in a series. A series for which there is no autocorrelation is known as *white noise*.

If the series is autocorrelated at *any* lags, the limits drawn in these plots *underestimate* the sampling variability even for those lags k where ρ_k *is* zero. If it supposed that the true autocorrelations are all zero beyond some lag q, then the approximate standard error of the estimated autocorrelations at those same lags is

$$\text{SE}(r_k) \simeq \sqrt{\frac{1}{n}\left(1 + 2\rho_1^2 + \cdots + 2\rho_q^2\right)}. \tag{2.13}$$

For example, in Figure 2.6, if we take the correlations at lags 1, 2, and 3 to be nonzero

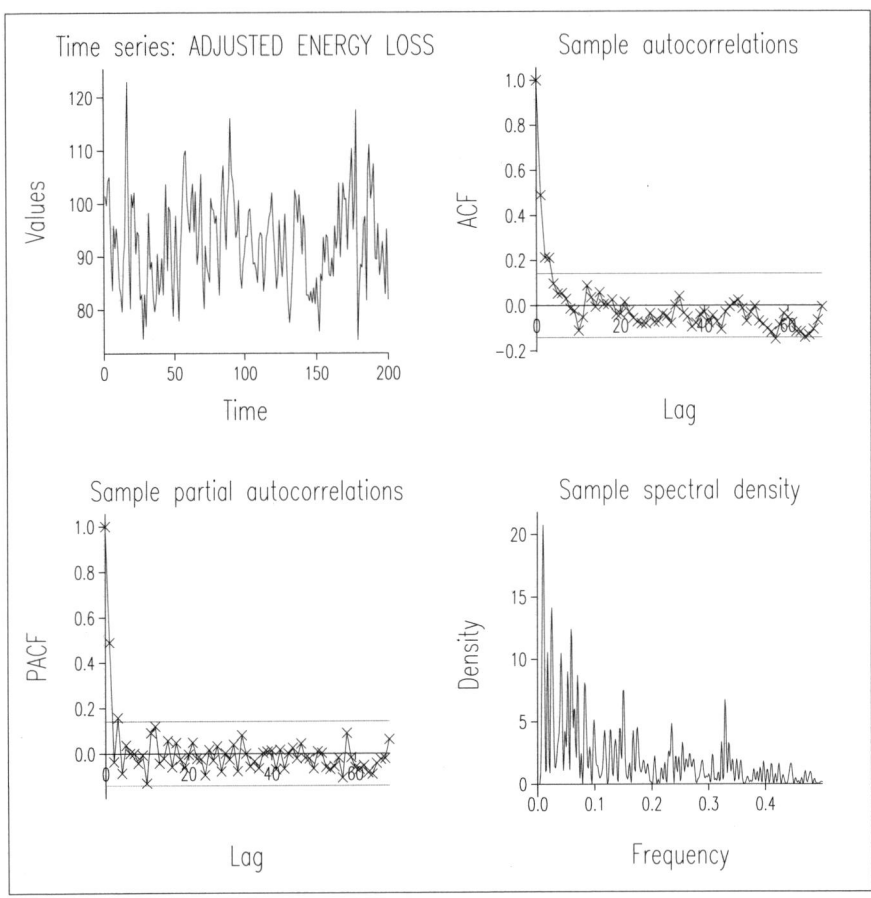

FIGURE 2.6 Model cow daily energy loss corrected for annual cycle with sample statistical properties.

and equal to their sample values, and assume all the remaining correlations to be zero, the value for the standard error of the sample correlations at lags beyond 3 is approximately

$$\sqrt{\frac{1}{200}[1 + 2(0.5^2 + 0.2^2 + 0.2^2)]} = .091. \tag{2.14}$$

The limits of plus or minus two standard errors should therefore be taken as appreciably wider than those shown. The sample values close to lag 60, which lie just outside the limits as shown, would not then be considered extreme using the revised limits.

2.3. LAGGED PREDICTION AND THE PARTIAL AUTOCORRELATION FUNCTION

If a regression line is fitted to the scatterplot in Figure 2.4a, the resulting prediction equation is

$$z_t = c + 0.67 z_{t-1} + a_t \tag{2.15}$$

which accounts for 44.7% of the variance of z_t. This provides a simple one–step ahead predictor for the series. The regression coefficient is very close to the first lag sample autocorrelation as defined in (2.12), which is .66. To explain this, consider the usual relationship $\hat{\phi} = r s_y/s_x$ between the estimated coefficient $\hat{\phi}$, say, in a regression of y on x and the sample correlation coefficient r. Here s_y and s_x are the sample standard deviations of y and x, and in the time series case these are almost the same, differing only in the end effects described in the previous section.

The regression of z_t on a range of lagged values $z_{t-1}, \ldots, z_{t-k}$ can be expected to improve prediction. In a similar manner the least-squares equations of this regression can be approximated in terms of the sample autocorrelations up to lag k. Consider the example with $k = 5$. The lagged regression equations for the model

$$z_t = c + \phi_1 z_{t-1} + \phi_2 z_{t-2} + \phi_3 z_{t-3} + \phi_4 z_{t-4} + \phi_5 z_{t-5} + a_t \tag{2.16}$$

may be written out for $t = 6, \ldots, n$ in the standard form $Y = c + X\phi + E$ with

$$Y = \begin{pmatrix} z_6 \\ z_7 \\ \vdots \\ z_n \end{pmatrix}, \quad X = \begin{pmatrix} z_5 & z_4 & \cdots & z_1 \\ z_6 & z_5 & \cdots & z_2 \\ \vdots & \vdots & \vdots & \vdots \\ z_{n-1} & z_{n-2} & \cdots & z_{n-5} \end{pmatrix}. \tag{2.17}$$

After correcting each column of Y and X for its mean, the least-squares equations $X'X\hat{\phi} = X'Y$, when divided by $n s_z^2$, are in large samples close to the equations

$$\begin{pmatrix} 1 & r_1 & r_2 & r_3 & r_4 \\ r_1 & 1 & r_1 & r_2 & r_3 \\ r_2 & r_1 & 1 & r_1 & r_2 \\ r_3 & r_2 & r_1 & 1 & r_1 \\ r_4 & r_3 & r_2 & r_1 & 1 \end{pmatrix} \begin{pmatrix} \hat{\phi}_1 \\ \hat{\phi}_2 \\ \hat{\phi}_3 \\ \hat{\phi}_4 \\ \hat{\phi}_5 \end{pmatrix} = \begin{pmatrix} r_1 \\ r_2 \\ r_3 \\ r_4 \\ r_5 \end{pmatrix} \tag{2.18}$$

The differences again are due to the end effects in forming the sums of products of lagged values.

TABLE 2.1. Sample Autocorrelations and Prediction Coefficients

Lag	Sample Acf	Regression Coefficients	Yule–Walker Coefficients	Sample Pacf	$1 - R^2$
1	.662	.709	.697	.662	.562
2	.420	−.152	−.143	.033	.561
3	.486	.116	.154	.392	.475
4	.570	.538	.447	.208	.454
5	.300	−.445	−.389	−.389	.385

The residual sum of squares from the regression, $Y'Y - \hat{\phi}'X'Y$, expressed as a fraction of the raw variance $Y'Y$, is similarly approximated by

$$1 - R^2 = 1 - \hat{\phi}_1 r_1 - \hat{\phi}_2 r_2 - \hat{\phi}_3 r_3 - \hat{\phi}_4 r_4 - \hat{\phi}_5 r_5. \tag{2.19}$$

Equations (2.18) are known as the *Yule–Walker* equations. With the sample acf values r_k replaced by the population quantities ρ_k and correspondingly $\hat{\phi}_k$ replaced by ϕ_k, they become equations for the *minimum mean-square error* lagged prediction coefficients. Table 2.1 shows the sample autocorrelations up to lag 5, the coefficients which result from carrying out the regression up to lag 5 and for comparison with those, the coefficients obtained from solving the Yule–Walker equations. Noticeable differences arise from the end effects.

It should be appreciated that as more lags are included in this regression, the coefficient of a particular lag, say, the first lag, will change. Thus each coefficient should be doubly indexed, as $\phi_{k,j}$, by the *order*, or maximum lag, k of the regression besides the lag j with which it is associated.

Table 2.1 also shows the coefficients $\hat{\phi}_{k,k}$ up to lag 5. These play a valuable role because, just as the coefficient $\phi_{1,1}$ is identical to the first lag acf ρ_1, so the coefficients $\phi_{k,k}$ may be interpreted as the *partial* autocorrelations associated with lags k. These measure the *improvement* in prediction gained from progression to the order k regression from the order $k-1$. The value of $1 - R^2$ for the order 5 regression is obtained from that for the order 4 regression by applying the factor $1 - \hat{\phi}_{5,5}^2$. Progressing down to the order 1 then gives

$$1 - R^2 = \left(1 - \hat{\phi}_{5,5}^2\right)\left(1 - \hat{\phi}_{4,4}^2\right)\left(1 - \hat{\phi}_{3,3}^2\right)\left(1 - \hat{\phi}_{2,2}^2\right)\left(1 - \hat{\phi}_{1,1}^2\right). \tag{2.20}$$

The reducing sequence of values of $1 - R^2$ is shown as the last column in Table 2.1. We illustrate the direct calculation of the partial autocorrelation at lag 2. The first step is to *correct* both the regressor z_t and the *new* predictor z_{t-2} for the effects of the *current* predictor z_{t-1} by forming $z_t - \phi_{1,1} z_{t-1}$ and $z_{t-2} - \phi_{1,1} z_{t-1}$. The coefficient $\phi_{1,1}$ is the same in both of these because both z_t and z_{t-2} have the same autocorrelation with, and hence the same dependence upon, z_{t-1}. The partial autocorrelation at lag 2 is just the correlation between these corrected terms. Each has variance $\sigma_z^2 (1 - \phi_{1,1}^2)$ and the covariance between them can be simplified to that between the first, $z_t - \phi_{1,1} z_{t-1}$,

2.4. TRANSFORMATIONS TO STATIONARITY

and just z_{t-2}, giving $\sigma_z^2(\rho_2 - \phi_{1,1}\rho_1)$. The required partial autocorrelation is then the ratio

$$\phi_{2,2} = \frac{\rho_2 - \phi_{1,1}\rho_1}{1 - \phi_{1,1}^2}. \tag{2.21}$$

The lag 2 predictor may be expressed by *updating* the lag 1 predictor:

$$\phi_{2,1}z_{t-1} + \phi_{2,2}z_{t-2} = \phi_{1,1}z_{t-1} + \phi_{2,2}(z_{t-2} - \phi_{1,1}z_{t-1}) \tag{2.22}$$

from which, by comparing coefficients, we see that $\phi_{2,1} = \phi_{1,1} - \phi_{2,2}\phi_{1,1}$. This computation generalizes to provide an efficient means of solving the Yule–Walker equations which has been historically important. Display of the sample partial autocorrelation coefficients (pacf) $\hat{\phi}_{k,k}$ provides a valuable further insight into the lagged dependency of a time series, although the information it contains is in fact equivalent to that in the acf. In Figures 2.5 and 2.6 the third frame shows plots of these values for the corresponding series. The partial acf generally dies out faster than the acf because a limit to the predictability of the series, as measured by $1 - R^2$, is usually reached at a fairly low lag. From Figure 2.5 the gains of lagged prediction beyond lag 5 appear to be modest.

As for the sample acf, approximate standard error lines are drawn for the sample pacf. These are appropriate for judging the lag beyond which one might assume that all values of the pacf are negligible.

Lagged prediction provides a justification for the GLM representation (2.4) of a stationary time series. Provided the order is sufficiently high, a minimum mean square error lagged predictor will approximate (2.4) with arbitrary precision. The error will necessarily be close to white noise and because of its derivation is known also as the *linear innovation* of the series.

2.4. TRANSFORMATIONS TO STATIONARITY

We have already mentioned that, for some series, correcting for a deterministic component by regression may result in a series that appears stationary. This is illustrated by the examples of the atmospheric carbon dioxide series in Figures 2.1a and 2.1b and the energy loss series in Figures 2.3a and 2.6. Such series are called *trend stationary*; the only feature that is nonstationary is the variation in mean level, which is accounted for by the regression components. If the corrected series, namely, the errors in the regression, are autocorrelated, then this fact ought to be taken into account when fitting the regression. When trends, seasonality, and cycles are being fitted, the results of ordinary least-squares regression are usually quite adequate. However, it is quite possible to fit the regression with autocorrelated errors and, in particular, errors which follow the ARIMA models of the next chapter.

It is sometimes appropriate to take logarithms of the values of a time series before embarking on any other analysis. This can help to improve the fit of a linear model before regression or other modeling of the series. Other such *instantaneous nonlinear transformations*, such as taking the square root, can be useful for improving the linear modeling of stationary series.

Differencing of a time series is a simple operation that can often transform a nonstationary time series to a stationary series. The simplest example arises when the series is a *random-walk* z_t. This is defined by the property that its first difference ∇z_t is white noise:

$$\nabla z_t = z_t - z_{t-1} = a_t. \tag{2.23}$$

Such a process appears to be a good description of many financial time series such as stock prices. The series itself is then made up of the cumulative sum of independent successive steps:

$$z_1 = z_0 + a_1 \tag{2.24}$$

$$z_2 = z_1 + a_2 = z_0 + a_1 + a_2$$

$$\vdots$$

$$z_t = z_{t-1} + a_t = z_0 + a_1 + a_2 + \cdots + a_t. \tag{2.25}$$

Figure 2.7a shows an example of a process simulated in this way, and Figure 2.7b shows a series of daily dollar term interest rates that appears to follow such a process. The random walk is *not* stationary because the variance increases with time. It may however, quite simply, be represented by the GLM (2.4) with $\pi_1 = 1$ and $\pi_2 = \pi_3 = \cdots = 0$.

Such processes, which are formed, as shown in (2.25), by the summation of a stationary process, are known as *integrated processes*. They can show features of trend, seasonality, and cycles which are similar to those of trend stationary processes; the main distinction is that for integrated processes these features are *evolving*. For example, the trend slope may be slowly changing. These processes are also known as *difference stationary* processes.

An integrated process that is similar to the random walk is that which is predicted by the EWMA (2.5). It is a short exercise to show that for this process the first

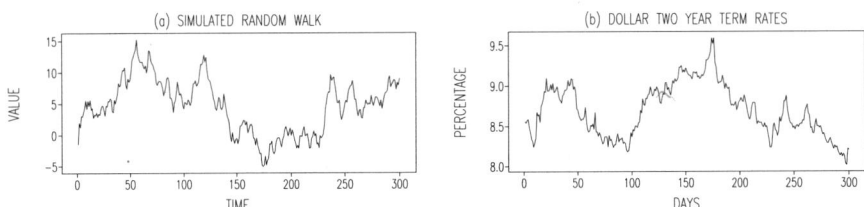

FIGURE 2.7 (a) A simulated random-walk process; (b) series of daily 2-year term interest rates.

difference is

$$\nabla z_t = a_t - \theta a_{t-1} \qquad (2.26)$$

which is a simple stationary series.

Of course, differencing a stationary process gives another stationary process; one whose autocorrelation is usually complicated by this unnecessary differencing. Such inappropriate differencing of an already stationary series is undesirable. There are, however, series for which it is not clear from a finite sample whether they are stationary or whether they have some features of an integrated process. It is then usual to consider modeling both the series and its differences, starting with inspection of the sample acf and pacf of both.

Another common form of differencing is *seasonal differencing*. As an example, consider the first seasonal difference of the atmospheric carbon dioxide series defined as

$$\nabla_{12} z_t = z_t - z_{t-12}. \qquad (2.27)$$

Figure 2.8 shows this series together with its sample acf and pacf. This is just the series of annual changes (increases in this case) computed month by month. The effect of any fixed seasonal pattern in the series, as modeled by the indicator variables in (2.1), is canceled out by this difference, which generally appears to simplify such a series. However, just as with ordinary differencing, it is possible that seasonal differencing is *inappropriate*, and this is the case if the series is, in reality, trend stationary. The seasonally differenced series may then be unnecessarily complicated in structure, compared with the trend-corrected series.

Although the theoretical acf ρ_k is defined only for stationary series, the sample acf (and pacf) can, of course, be computed and displayed for any time series record. It will *not* have well-defined properties, such as useful large sample limiting values, but can generally be useful to indicate that some kind of transformation is necessary, whether trend correction or differencing, to produce a stationary series. For example, in Figure 2.5 the peaks in the sample acf at lags 4, 8, 12, ... draw attention to the quarterly seasonality in the series which requires modeling by means of some such transformation.

2.5. CYCLES AND THE PERIODOGRAM

Periodic or cyclic behaviour is a feature of many time series. In this section we define first the *periodogram*, which is a general tool for revealing unknown periodicities as well as confirming known periodicities in an observed series. The periodogram provides another statistical summary of an observed time series, just like the sample acf and pacf. It is in fact equivalent to these in its information content but complements them by what it can reveal as to the nature of the series. After defining the periodogram, we consider its properties under various modeling assumptions, and this leads us to the spectral representation of stationary time series.

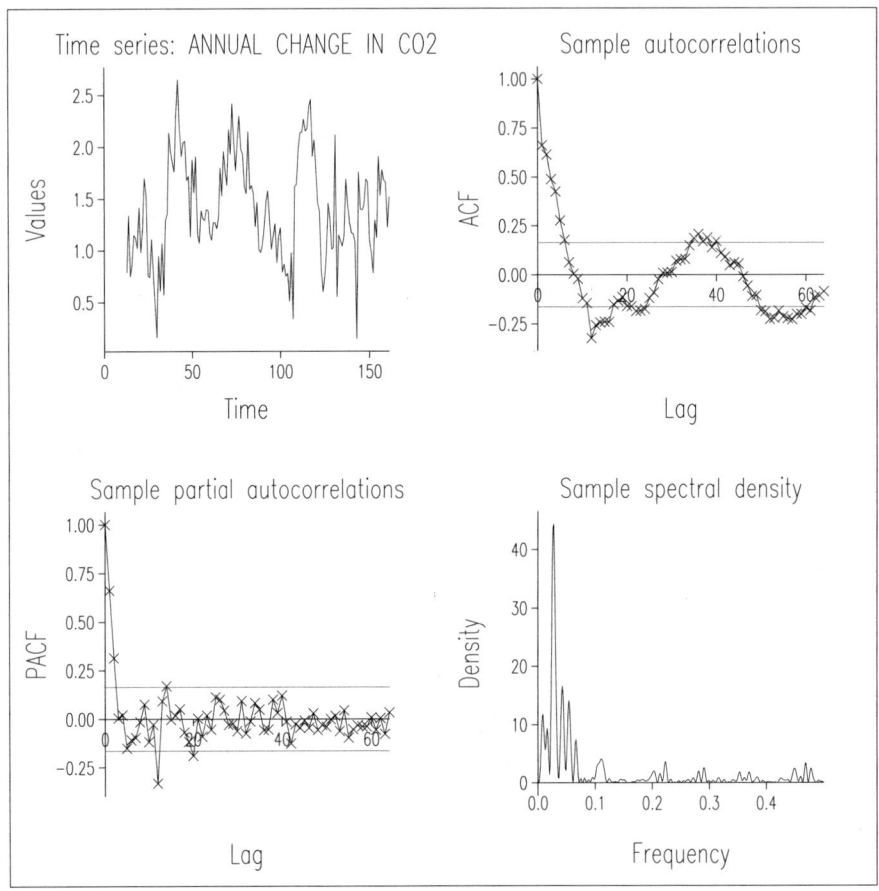

FIGURE 2.8 Series of annual changes of monthly atmospheric CO_2 levels with sample statistical properties.

We motivate the periodogram by considering the estimation of a deterministic cyclical component in a series z_t. The linear model used to fit such a cycle is

$$z_t = R \cos 2\pi(ft + g) + e_t = A \cos 2\pi ft + B \sin 2\pi ft + e_t. \qquad (2.28)$$

Here R is the *amplitude* of the cycle and f the *frequency* in cycles per sampling point so that the *period* of the cycle is $p = 1/f$. The *phase*, or fraction of a cycle completed at time $t = 0$, is g. If f is taken as known, the model is linear in A and B, from which the amplitude and phase can be derived from $A = R \cos g$ and $B = -R \sin g$. In particular, $R^2 = A^2 + B^2$.

Because the series is sampled at regular intervals, it is necessary to restrict the range of frequencies to $0 \leq f \leq 0.5$; the upper frequency limit of one half a cycle

2.5. CYCLES AND THE PERIODOGRAM

per sampling interval is known as the *Nyquist frequency*. This is because any cycle with a frequency greater than this, or equivalently of period less than 2, cannot be distinguished from a cycle with a frequency within this range.

The least-squares equations for $\hat{A}$ and $\hat{B}$ are

$$\begin{pmatrix} \sum_1^n (\cos 2\pi f t)^2 & \sum_1^n \cos 2\pi f t \sin 2\pi f t \\ \sum_1^n \sin 2\pi f t \cos 2\pi f t & \sum_1^n (\sin 2\pi f t)^2 \end{pmatrix} \begin{pmatrix} \hat{A} \\ \hat{B} \end{pmatrix} = \begin{pmatrix} \sum_1^n z_t \cos 2\pi f t \\ \sum_1^n z_t \sin 2\pi f t \end{pmatrix}. \tag{2.29}$$

We can approximate, provided that $f \neq 0, \frac{1}{2}$, $\sum_1^n (\cos 2\pi f t)^2$ and $\sum_1^n (\sin 2\pi f t)^2$ by $n/2$ and $\sum_1^n \sin 2\pi f t \cos 2\pi f t$ by 0 so that these equations may be approximated as

$$\begin{pmatrix} \frac{1}{2}n & 0 \\ 0 & \frac{1}{2}n \end{pmatrix} \begin{pmatrix} \hat{A} \\ \hat{B} \end{pmatrix} = \begin{pmatrix} \sum_1^n z_t \cos 2\pi f t \\ \sum_1^n z_t \sin 2\pi f t \end{pmatrix} \tag{2.30}$$

giving

$$\hat{A} \approx \frac{2}{n} \sum_1^n z_t \cos 2\pi f t, \qquad \hat{B} \approx \frac{2}{n} \sum_1^n z_t \sin 2\pi f t. \tag{2.31}$$

The periodogram is then defined as

$$I(f) = \tfrac{1}{2} n \left(\hat{A}^2 + \hat{B}^2 \right) = \tfrac{1}{2} n \hat{R}^2. \tag{2.32}$$

There are efficient methods of calculating the periodogram over a fine grid of frequencies spanning $0 \leq f \leq \frac{1}{2}$, and we shall graph it over this range. The scaling factor in the definition is chosen so that the area under the graph of $I(f)$ is the mean square value of the series:

$$\int_{f=0}^{\frac{1}{2}} I(f) \, df = \frac{1}{n} \sum_{t=1}^n z_t^2$$

Usually the series is mean corrected before calculating the periodogram, so this becomes the sample variance. The periodogram then describes the distribution or analysis of the sample variance of the series over the frequency range. It is also common practice to scale the periodogram, dividing by the series variance so that the total area is unity. We shall choose a frequency grid with divisions of approximate width $1/4n$ so that the periodogram appears continuous for most of our examples. Historically the frequency values were chosen to be the harmonics of the fundamental frequency of one cycle in the sample length, specifically, $1/n, 2/n, 3/n$, and so on.

The periodogram can be considered as an investigative tool, a transformation of the data designed to reveal cyclical behaviour. We first consider its properties

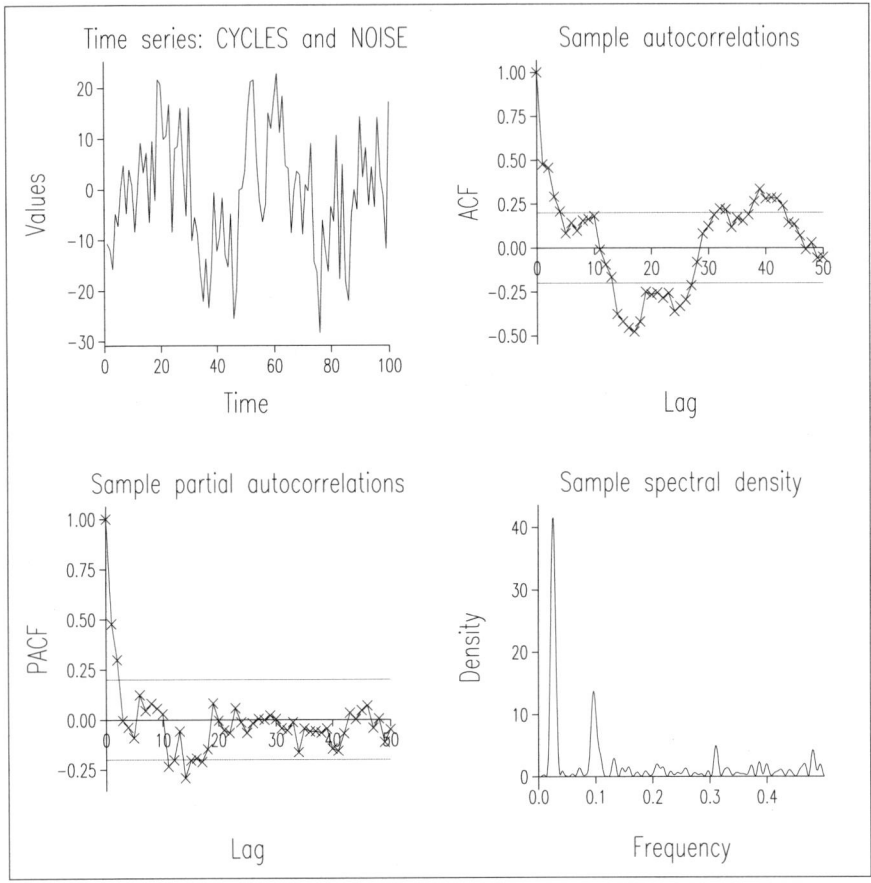

FIGURE 2.9 Constructed series of two cycles and a random error component with sample statistical properties.

assuming that the data does consist of a combination of deterministic cycles of the form assumed in the motivating regression. We then introduce the spectrum and spectral representation by considering its properties when the series is stationary and contains no deterministic components. In this context the periodogram is called the *sample spectral density*, and this is how it is described in the figures.

Figure 2.9 shows the analysis of a series generated as

$$z_t = R_1 \cos 2\pi(f_1 t + g_1) + R_2 \cos 2\pi(f_2 t + g_2) + a_t \quad (2.33)$$

where the amplitudes of the cycles are $R_1 = 10$ and $R_2 = 7$, the periods are $1/f_1 = 40$ and $1/f_2 = 10$, the phases are $g_1 = 0.5$ and $g_2 = 0.2$, and the error standard deviation is $\sigma_a = 8$.

2.5. CYCLES AND THE PERIODOGRAM

The graph of the series in the first frame shows some cyclical pattern but it is quite well masked by the white-noise errors whose standard deviation is comparable with the amplitudes of the cycles. The scaled periodogram in the fourth frame clearly reveals the evidence for the cycles with a peak at each of the two frequencies .025 and .1. There are, however, smaller peaks which can only be ascribed to the white-noise term in the series. To understand these features, we use the approximate properties of the estimates, that

$\hat{A}$ and $\hat{B}$ are independent *normal* with means A and B and variance $(2/n)\sigma_a^2$.

(2.34)

From this the expected value of the periodogram is

$$E[I(f)] = E\left[\frac{n}{2}(\hat{A}^2 + \hat{B}^2)\right] \simeq \frac{n}{2}(A^2 + B^2) + 2\sigma_e^2 = \tfrac{1}{2}nR^2 + 2\sigma_e^2. \quad (2.35)$$

Evidence for frequencies that are present in the data, that is, for which $R > 0$, therefore appear as peaks of height proportional to n, which will become prominent as the series length increases.

At frequencies for which no cycle is present (i.e., $R = 0$), the expected value of the periodogram is just $2\sigma_a^2$ whatever the length of the series; the peaks due to the cycles become prominent because this remains low in comparison. Furthermore, because $A = B = 0$, the distribution of the periodogram at such frequencies is a scaled chi-squared variable on 2 degrees of freedom, also known as an *exponential distribution*:

$$I(f) = \frac{n}{2}(\hat{A}^2 + \hat{B}^2) \sim \sigma_e^2 \chi_2^2 = \text{exponential}(2\sigma_e^2). \quad (2.36)$$

These properties are exact if the frequencies of the cycles are exact multiples j of the harmonic frequency $1/n$, that is, if they have an exact number j of cycles in the data length. The cyclical regressors are then mutually orthogonal, and the estimates at these frequencies are independent. This would be unusual in practice, but the properties are in fact very close to this. The estimates are approximately independent for frequencies separated by an interval greater than $1/n$, and the width of the peak associated with a cycle in the series is approximately $1/n$.

The statistical problem of detecting cycles in the presence of white-noise observational error is illustrated by the example in Figure 2.9. We know that the two largest peaks are due to real cyclical components. There are other smaller peaks that we also know are due only to the observational noise. These are the largest of the independent exponentially distributed periodogram values at the other frequencies. Because the central 90% of this distribution ranges from about $\frac{1}{20}$th to 3 times its mean, the general picture of the periodogram at these other frequencies is very variable with many "peaks." These are due simply to natural statistical fluctuation but are easily misinterpreted as indicating regular cycles in the data. If we could increase the observed series length, we could eventually discriminate with certainty between such spurious peaks and peaks due to real cycles. In practice there is always the scope for some misinterpretation, as with all statistical modeling.

Turning now to a real example, reconsider Figure 2.5. The scaled periodogram here shows peaks at frequencies .01, .06, and .25, associated with periods of 100, 16, and 4 quarters. The first peak is typical of series that show some evidence of trendlike behaviour. This may be due to a real trend or a cycle with period much greater than the series length, which appears as a low-frequency component of the periodogram. It is typically accentuated as a peak centered on a frequency of $1/n$ because mean correction of the data means that the periodogram always falls to zero at zero frequency. The other peaks certainly appear to be associated with real cycles in the data; one with frequency .06 (period 16) may be due to a business cycle and one with frequency .25 (period 4) is certainly associated with the annual cycle.

This is a rather special example. Peaks in the periodogram are commonly much wider than $1/n$, and there are various explanations for this. It means essentially that the cycle associated with this peak is irregular in some way, so that its amplitude, frequency, or phase may be varying. This is sometimes described as modulation of the cycle. There are many possible mechanisms for this. The linear autoregressive models described in the next chapter were proposed by Yule (1927) as one such simple linear mechanism.

2.6. THE SPECTRUM

This is a natural stage at which to put aside the deterministic sinusoidal models for cyclical behaviour and consider the cyclical properties of stationary time series. The point is that a stationary time series has a natural expression in terms of cyclical components *but* the coefficients of these components are now random quantities. We support this statement by showing that the periodogram defined in (2.32) can be expressed as a combination of the sample autocovariances c_k of z_t:

$$I(f) = S^*(f) = 2\left(c_0 + 2\sum_{k=1}^{n-1} c_k \cos 2\pi k f\right). \quad (2.37)$$

The derivation is as follows:

$$I(f) = \frac{2}{n}\left[\left(\sum_{t=1}^{n} z_t \cos 2\pi f t\right)^2 + \left(\sum_{t=1}^{n} z_t \sin 2\pi f t\right)^2\right]$$

$$= \frac{2}{n}\sum_{t=1}^{n}\sum_{s=1}^{n}\{z_t z_s [\cos(2\pi f t)\cos(2\pi f s) + \sin(2\pi f t)\sin(2\pi f s)]\} \quad (2.38)$$

$$= \frac{2}{n}\sum_{t=1}^{n}\sum_{s=1}^{n}\{z_t z_s \cos[2\pi f(s-t)]\}.$$

2.6. THE SPECTRUM

Now setting $s = t + k$ and rearranging the sum

$$I(f) = 2 \sum_{|k|<n} \left[\frac{1}{n} \left(\sum_{t-s=k} z_t z_s \right) \cos(2\pi f k) \right] \qquad (2.39)$$

$$= 2 \left(c_0 + 2 \sum_{k=1}^{n-1} c_k \cos 2\pi k f \right).$$

Provided z_t is a stationary time series, we will refer to this as the *sample spectrum* $S^*(f)$ because it is the sample value of the *spectrum* obtained by replacing the sample values c_k by γ_k:

$$S(f) = 2 \left(\gamma_0 + 2 \sum_{k=1}^{\infty} \gamma_k \cos 2\pi k f \right). \qquad (2.40)$$

This is a well-defined quantity provided the autocovariances γ_k decay sufficiently as k increases. By analogy with the periodogram, it shows how the variance of z_t is distributed over the range of frequencies:

$$\sigma_z^2 = \int_0^{\frac{1}{2}} S(f) df \qquad (2.41)$$

and more generally all the autocovariances can be derived from $S(f)$:

$$\gamma_k = \int_0^{\frac{1}{2}} S(f) \cos 2\pi k f df. \qquad (2.42)$$

Figure 2.10 shows the sample spectrum of the corrected energy-loss series with a superimposed smooth solid line, which is an estimate of the spectrum. This illustrates the property that, however large the sample size, the sample spectrum has an exponential distribution about the spectrum:

$$S^*(f) \simeq \text{exponential}[S(f)]. \qquad (2.43)$$

The values are also independent at frequencies separated by more than $1/n$, resulting in the rapidly fluctuating values seen in the figure. The appearance is therefore similar to that part of the sample spectrum in Figure 2.9 that arises from the white-noise error, but the expected level changes with frequency.

This distributional property follows from evaluating the mean and variance of the coefficients $\hat{A}$ and $\hat{B}$ used to form the periodogram:

$$\hat{A} \sim \text{normal}\,(0, \tfrac{1}{n} S(f)) \qquad \hat{B} \sim \text{normal}\,(0, \tfrac{1}{n} S(f)). \qquad (2.44)$$

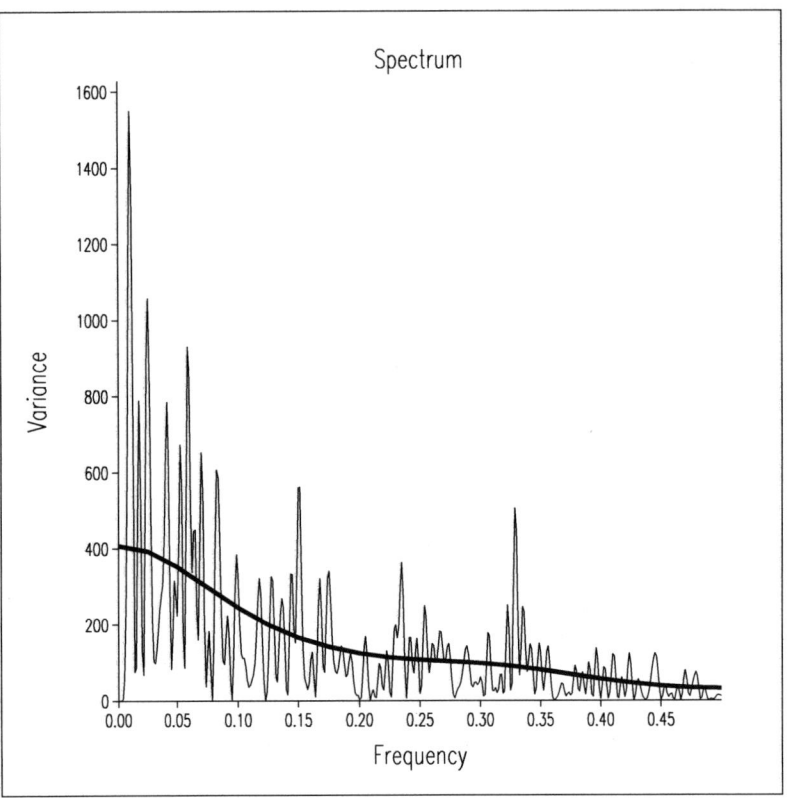

FIGURE 2.10 Spectrum and sample spectrum for the corrected energy loss series.

Then

$$S^*(f) = \frac{n}{2}\left(\hat{A}^2 + \hat{B}^2\right) \sim S(f)\tfrac{1}{2}\chi_2^2 = \text{exponential}[S(f)]. \quad (2.45)$$

The spectrum is well defined for a stationary series represented by the linear model (2.6) as

$$S_z(f) = \left|1 + \sum_{k=1}^{\infty} \psi_k \exp(i2\pi k f)\right|^2 2\sigma_a^2. \quad (2.46)$$

In this expression the constant, or *uniform*, spectrum $S_a(f) = 2\sigma_a^2$ of the white-noise process a_t, is multiplied by a factor that is a function of f. This is the *squared gain* of the linear relationship between the series z_t and a_t. It is possible to interpet z_t as being obtained by multiplying the cyclical component of a_t at each frequency

2.6. THE SPECTRUM

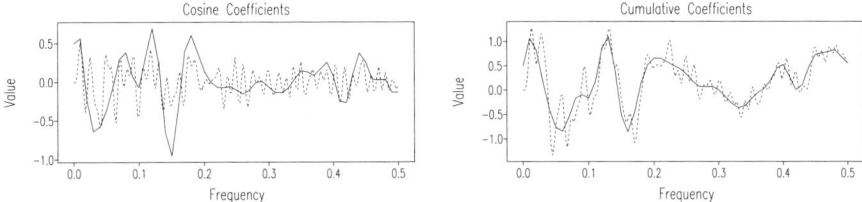

FIGURE 2.11 (a) Superimposed sample Fourier cosine coefficients from increasing lengths of energy loss series; (b) superimposed *cumulative* sample Fourier coefficients of the same series.

by the corresponding gain, thus transforming the uniform spectrum into one that is proportional to the squared gain.

We now consider more carefully the sense in which a stationary time series can be considered to have well-defined cyclical components with random coefficients. Figure 2.11a shows the coefficients $\hat{A}_j$ calculated at the frequency intervals $f_j = j/N$ where $N = 2n$ for, respectively, 50 and 200 central values of the energy-loss series (i.e., taking the 101st point to be $t = 1$). Figure 2.11b shows the cumulative values of these coefficients:

$$\sum_{f_j \leq f} \hat{A}_j. \tag{2.47}$$

As the sample size increases fourfold, so does the number of individual coefficients. The coefficient at any individual frequency splits approximately into four new independent coefficients. As the sample size increases these do not converge because finer structure emerges in the graph. The cumulative coefficients do, however converge, but to a random function of frequency. Finer detail emerges in the cumulative plot as the sample size increases, but the perturbations become smaller as the graph converges. This behaviour illuminates the following formal spectral representation of a stationary time series:

- The autocovariances may be represented as

$$\gamma_k = \int_0^{\frac{1}{2}} \cos 2\pi k f \, dF(f) \tag{2.48}$$

where $F(f)$ is the nondecreasing spectral distribution function, which, in the case of a continuous spectrum, can be expressed as the integral of the spectral density:

$$F(f) = \int_0^f S(f') \, df'. \tag{2.49}$$

- The series values may be represented as

$$z_t = \int_0^{\frac{1}{2}} \cos 2\pi f t \, d\alpha(f) + \int_0^{\frac{1}{2}} \sin 2\pi f t \, d\beta(f) \qquad (2.50)$$

where $\alpha(f)$ and $\beta(f)$ are uncorrelated processes, each of which is a process with uncorrelated (usually termed orthogonal) increments having variances

$$\text{Var}\,[\alpha(f)] = \text{Var}\,[\beta(f)] = F(f). \qquad (2.51)$$

For a Gaussian process the increments are independent. In practice this means that, using a sufficiently fine grid of frequencies $f_j = j/N$, a finite set of values $z_1, z_2, \ldots, z_n$ of a stationary Gaussian time series z_t with mean μ and spectrum $S(f)$ may be well approximated by the representation

$$z_t = \mu + \sum_j (A_j \cos 2\pi f_j t + B_j \sin 2\pi f_j t). \qquad (2.52)$$

Here, A_j and B_j are zero-mean normal random variables that are independent of each other and for different j, and that both have variance $S(f_j)/N$.

An important point is the independence between *all* frequencies. The frequency division $1/N$ can be as fine as desired while retaining this independence. It is for the *estimates* of these coefficients that a frequency separation of at least $1/n$ is required to ensure independence when a sample of only length n is available. A comprehensive account of the subject of spectral analysis is given by Priestley (1981).

2.7. FURTHER INTERPRETATION OF TIME SERIES ACF, PACF, AND SPECTRUM

At the end of Section 2.4 we drew attention to the fact that the sample acf and pacf can be calculated for series that are not stationary and that these may then give some useful indication of possible transformation to stationarity, by correcting for regression components or by differencing. The sample spectrum may be of similar value. A trend component of a series will lead to very high values of the sample spectrum at low frequencies. This will often mask the other features of the spectrum; the values at higher frequencies are barely discernible. Sometimes the logarithms of the spectrum are plotted to alleviate this problem. The best remedy is to correct for the trend or to difference the series.

Deterministic seasonality of large amplitude has a similar but less drastic effect, introducing sharp peaks in the sample spectrum at the seasonal frequency and its multiples. This is evident in Figure 2.5. When such features are seen, they can be removed by seasonal regression or seasonal differencing.

The sample spectrum has some advantage over the sample acf in that the anomalies arising from certain types of nonstationary behaviour are often isolated to particular frequencies whereas they affect all the values of the sample acf. Thus, for the

2.7. FURTHER INTERPRETATION OF TIME SERIES ACF, PACF, AND SPECTRUM

random-walk model, which is *not* stationary, a sensible extension of the definition of the spectrum, called the *pseudospectrum* may be defined at all frequencies other than frequency zero, where an infinite peak occurs. The sample spectrum of a random walk is able to reflect the pseudospectrum values at other frequencies. The sample acf can be most uninformative in this case, with values close to one appearing at all lags.

Both the sample acf and the sample spectrum are useful for revealing the possibility that a series consists of independent components that are not themselves directly observed. If series u_t and v_t are both stationary and independent of each other with respective acfs $\gamma_{u,k}$ and $\gamma_{v,k}$ and spectra $S_u(f)$ and $S_v(f)$, then their sum $z_t = u_t + v_t$ has acf $\gamma_{u,k} + \gamma_{v,k}$ and spectra $S_u(f) + S_v(f)$. Discovering such a structure may provide a useful simplification of, or interpretation of, the observed series.

A simple example is that of "signal" plus "noise." Figure 2.12 shows the analysis of the plastic extrusion width measurement series. This is typical of a series consisting

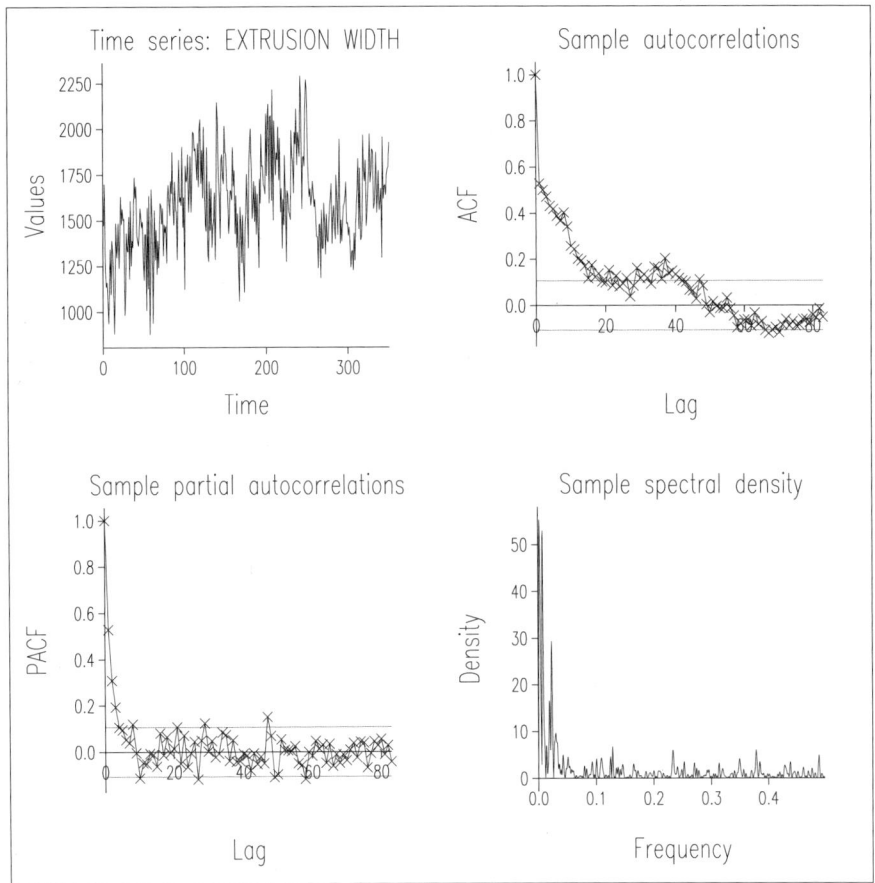

FIGURE 2.12 Analysis of the plastic extrusion width measurement.

in almost equal measure (over the sample length) of a random-walk component and a white-noise component. The first value of the sample acf is about .5, after which the values reduce slowly. This is perhaps an appropriate point to make a remark about the fact that the sample acf follows a typical pattern for such series, of drifting down to quite a long sequence of negative values. This can be explained as a consequence of mean correcting the series, which causes the sum of the sample acf values from lags 1 to $n-1$ to be $-\frac{1}{2}$. This is just the fact that the sample spectrum is zero at frequency zero. Large values of the sample acf at lower lags must therefore be compensated by negative values at a range of higher lags.

To return to the main point, the observed sample acf looks like the mean of two sample acfs; one is that of a white noise series and the other, that of a random walk. These together account for the sudden fall from $r_0 = 1$ to $r_0 = 0.5$ and the slow decay thereafter. The sample spectrum reflects the white-noise component in the fairly uniform values at frequencies away from zero, and the random-walk component in the peak close to frequency zero.

2.8. STATE-SPACE MODELS AND THE KALMAN FILTER

In the last section of this chapter we use the simple signal–noise model of the previous section as an introduction to linear state-space models. Such models are also called *structural models* if they are believed to explain the statistical behaviour of the series in terms of more simple components, and a further example, of a seasonal structural model, is presented to illustrate this. These models contrast with the ARIMA models of the next chapter, which are essentially empirical, having the parametric flexibility to represent statistically a wide range of time series behavior.

Applications of state-space models usually exploit the *Kalman filter*, which is a means of estimating the unobserved states or components and also, most importantly, of constructing predictions. Because these predictions are linear functions of past observations of the series, state-space models are seen to belong to the class of general linear models. The prediction errors or innovations a_t constructed by the Kalman filter play a central rôle in the estimation of the states.

Consider now the simple state–space model:

$$z_t = x_t + \epsilon_t \qquad (2.53)$$

where

$$x_t = x_{t-1} + \alpha_t. \qquad (2.54)$$

To accord with common practice, we use x_t here as the state variable, which is typically an unobserved signal. The observed series z_t is a measurement of the signal with observation error ϵ_t and (2.53) is known as the *observation equation*. The structure of the signal is described by the *state transition equation* (2.54), which in this example is a random walk. Both ϵ_t and α_t are assumed to be white-noise series

2.8. STATE-SPACE MODELS AND THE KALMAN FILTER

and independent of each other. Their variances are parameters of this model, and their estimation is considered in Chapter 4.

For the general discrete-time state-space model the corresponding equations are

$$z_t = Hx_t + \epsilon_t \tag{2.55}$$

and

$$x_t = Ax_{t-1} + \alpha_t \tag{2.56}$$

where now both z_t and x_t may be vector-valued so that both H and A are matrices. The transition matrix A captures the dynamic structure of the model, and the observation matrix H expresses how these are combined in the observations. The observation vector z_t typically has a small number of elements compared with the state vector x_t. The essential property of a state model may be summarized as

> Conditional upon the state x_t, the distribution of the observation z_t and of any future observations and states, does not depend on the value of any past observations and states.

This property allows estimation of the state x_t from present and past observations, z_t, $z_{t-1}, \ldots$, by the Kalman filter. We describe this filter now for the simple model to emphasize the principles. The filter equations for the general model are presented in Chapter 4.

Assume that we have the estimate $\hat{x}_{t-1,t-1}$ of the state x_{t-1} at time $t-1$ using all the observations up to time $t-1$. The first step is *prediction* of the state x_t for which we use the transition equation (2.54) to obtain simply

$$\hat{x}_{t,t-1} = \hat{x}_{t-1,t-1}. \tag{2.57}$$

Note that the second index always indicates the range of past observations that have been used to form the estimate.

From the observation equation (2.53) this estimate is also the predictor of z_t from past values:

$$\hat{z}_{t,t-1} = \hat{x}_{t,t-1} \tag{2.58}$$

When z_t is observed, the prediction error or innovation $a_{t,t-1} = z_t - \hat{z}_{t,t-1}$ is known and may be used in the second, or *correction*, step, to update the estimate of the state x_t. To see how this is done, we follow through the errors of the estimates of these states and observations, starting with that of the state x_{t-1} whose error variance we assume known:

$$e_{t-1,t-1} = x_{t-1} - \hat{x}_{t-1,t-1}; \quad \text{Var } e_{t-1,t-1} = p_{t-1,t-1}. \tag{2.59}$$

Then the error $x_t - \hat{x}_{t,t-1}$ of the subsequent state is

$$e_{t,t-1} = e_{t-1,t-1} + \alpha_t; \quad \text{Var } e_{t,t-1} = p_{t,t-1} = p_{t-1,t-1} + \sigma_\alpha^2. \tag{2.60}$$

Finally the observation prediction error, or innovation $z_t - \hat{z}_{t,t-1}$, is

$$a_{t,t-1} = e_{t,t-1} + \epsilon_t; \quad \text{Var } a_{t,t-1} = p_{t,t-1} + \sigma_\epsilon^2. \tag{2.61}$$

The regression of the state error $e_{t,t-1}$ on the known innovation then gives the correction to the state estimate as

$$\hat{x}_{t,t} = \hat{x}_{t,t-1} + K_t a_{t,t-1}. \tag{2.62}$$

The regression coefficient K_t, known as the *Kalman gain*, is just the ratio of the covariance between $e_{t,t-1}$ and $a_{t,t-1}$ to the variance of the latter:

$$K_t = \frac{p_{t,t-1}}{p_{t,t-1} + \sigma_\epsilon^2}. \tag{2.63}$$

The reduction in the error variance that follows from this regression gives

$$p_{t,t} = p_{t,t-1} - K_t p_{t,t-1}. \tag{2.64}$$

Equations (60)–(64) provide the complete cycle of updating the estimates from time $t - 1$ to time t. Starting from the beginning of the series, taking z_1 as the first state estimate $\hat{x}_{1,1}$ with error variance $p_{1,1} = \sigma_\epsilon^2$, the Kalman gain K_t soon converges to a constant K. In the case of this simple model the updating equation for the prediction can be reduced to

$$\hat{z}_{t+1,t} = \hat{z}_{t,t-1} + K(z_t - \hat{z}_{t,t-1}). \tag{2.65}$$

This is of the same form as the EWMA predictor presented in (2.5). The simple state-space model therefore provides a structural explanation for this very widely used predictor.

The seasonal structural model that we use as a further illustration is one considered by Harvey (1990). We apply it to the series of atmospheric carbon dioxide concentrations shown in Figure 2.1. Besides the observation error, this model consists of two independent components to model first the trend and then the seasonality. The attraction of such a model is that it generalizes the regression model in which the trend is represented by a fixed straight line, and the seasonality by fixed seasonal effects using indicator variables. It replaces these by stochastic components that permit the trend and seasonality to vary through time. It is possible to recognize these components in the sample spectrum of the series in a similar manner to those of signal and noise in Figure 2.12.

2.8. STATE-SPACE MODELS AND THE KALMAN FILTER

The model for the trend involves two state variables

$$\begin{aligned} \mu_t &= \mu_{t-1} + \beta_{t-1} + \eta_t \\ \beta_t &= \beta_{t-1} + \xi_t \end{aligned} \quad (2.66)$$

where two free parameters, σ_η^2 and σ_ξ^2, are the variances of the white-noise *disturbance* terms η_t and ξ_t. If these variances are both zero, that is, if the terms η_t and ξ_t are zero in these equations, it is seen that β_t then retains a constant value, β, and μ_t increases by this constant amount at each timepoint, and thus follows a straight-line trend. When these variances are positive, the trend will have varying level and slope.

The model for the seasonal component s_t is

$$s_t = -(s_{t-1} + s_{t-2} + \cdots + s_{t-11}) + \omega_t \quad (2.67)$$

where ω_t is white-noise disturbance with variance σ_ω^2. If this is zero, the model specifies that any 12 successive values sum to zero, so that the next value s_t must be equal to the value s_{t-12} just one year previously, resulting in a fixed seasonal pattern about an average of zero. The disturbance ω_t allows some variation of this pattern when its variance is positive.

Together, therefore, these models encompass the regression model (2.1) for fixed trend and seasonality, but also extend it to allow variations in these over time. There are altogether, including the observation error, four variance parameters in this model corresponding to four different white-noise sources of variability.

To define the state-space model, the first two states are taken to be $x_{1,t} = \mu_t$ and $x_{2,t} = \beta_t$. The remaining states are taken as $x_{3,t} = s_t, x_{4,t} = s_{t-1}, \ldots, x_{13,t} = s_{t-10}$. The state transition equation then becomes

$$x_t = \begin{pmatrix} 1 & 1 & 0 & 0 & \cdots & 0 & 0 \\ 0 & 1 & 0 & 0 & \cdots & 0 & 0 \\ 0 & 0 & -1 & -1 & \cdots & -1 & -1 \\ 0 & 0 & 1 & 0 & \cdots & 0 & 0 \\ 0 & 0 & 0 & 1 & \cdots & 0 & 0 \\ \vdots & \vdots & \vdots & \vdots & \ddots & \vdots & \vdots \\ 0 & 0 & 0 & 0 & \cdots & 1 & 0 \end{pmatrix} x_{t-1} + \begin{pmatrix} \eta_t \\ \xi_t \\ \omega_t \\ 0 \\ 0 \\ \vdots \\ 0 \end{pmatrix} \quad (2.68)$$

where the last 10 lines are merely a device for retaining the latest 10 values of s_t in the state vector.

The Kalman filter can again be applied to predict this series and estimate the states. Besides constructing the *filtered* state estimates from present and past observations, it is also possible to construct what are known as their *smoothed* estimates based on all available past, present, and future values. These estimates are shown in Figure 2.13a for the trend and Figure 2.13b for the seasonal component. Both of these are fairly regular for this model. The general equations for constructing such filtered and smoothed state estimates are presented in Chapter 4. Other methods of obtaining

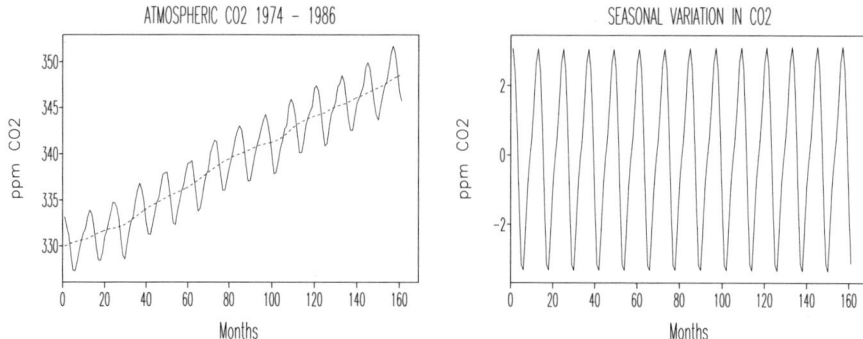

FIGURE 2.13 (a) Monthly CO_2 with smoothed trend line and (b) the estimated seasonal component.

decompositions of such series into trend and seasonal components are described in Chapter 7.

REFERENCES

Bartlett, M. S. (1946). On the theoretical specification of sampling properties of autocorrelated time series. *J. Royal Stat. Soc. B* **8**, 27.

Box, G. E. P. and Jenkins, G. M. (1976). *Time Series Analysis, Forecasting and Control*, revised ed. Holden-Day, San Francisco.

Harvey, A. C. (1990). *Forecasting, Structural Time Series Model and the Kalman Filter.* Cambridge Univ. Press, Cambridge.

Priestley, M. B. (1981). *Spectral Analysis and Time Series*. Academic Press, London.

Yule, G. U. (1927). On a method of investigating periodicities in disturbed series, with special reference to Wölfer's sunspot numbers. *Phil. Trans.* **A226**, 267.

CHAPTER 3

Univariate Autoregressive Moving-Average Models

George C. Tiao
University of Chicago

3.1. INTRODUCTION

Economic, engineering, environmental, and other scientific data are often taken in roughly equally spaced time intervals, for example, hour, day, month, quarter or year. Because of the inertia of the system, such time series are frequently serially dependent. For instances, the morning temperature of a given day tends to be correlated with the evening temperature of the previous day, the air pollution index at noon may be highly influenced by the weather conditions and traffic level in the morning, and an individual's spending of a given month may be highly correlated with his income and spending of the previous month. Dynamic relationships in time series data can often be represented by a linear transfer function model. Consider the simple case in which y_t is the output and x_t the input of a system at time t. A linear dynamic model takes the form

$$
\begin{aligned}
y_t &= v_0 x_t + v_1 x_{t-1} + \cdots + v_j x_{t-j} + \cdots \\
&= v(B) x_t
\end{aligned}
\tag{3.1}
$$

where B is the backshift operator such that $Bx_t = x_{t-1}$, and the quantity v_j measures the effect of x_{t-j} on y_t. The v_j's are known as the impulse responses and $v(B)$, the transfer function. Rather than dealing with a possibly infinite number of the v_j's, in practice, a parsimonious representation takes the rational polynomial form

$$
v(B) = \frac{\omega(B)}{\delta(B)} B^b
\tag{3.2}
$$

A Course in Time Series Analysis, Edited by Daniel Peña, George C. Tiao, and Ruey S. Tsay.
ISBN 0-471-36164-X. © 2001 John Wiley & Sons, Inc.

where $\omega(B) = \omega_0 - \omega_1 B - \cdots - \omega_s B^s$ and $\delta(B) = \delta_0 - \delta_1 B - \cdots - \delta_s B^r$ are polynomials in B of degrees s and r, respectively, and b measures the delay or dead time of the system.

3.1.1. Univariate ARMA models

In practice, interest often centers on modeling the temporal dependence of an observable time series for forecasting future observations. Since the 1970s, primarily due to the work of Box and Jenkins (1976), a class of mixed autoregressive moving-average (ARMA) models of the form

$$\varphi(B)z_t = c + \theta(B)a_t \tag{3.3}$$

originally proposed by Yule (1927) and Slutsky (1937), has been found useful to represent the serially dependent relationship of many time series encountered in practice. In (3.3), $\{z_t\}$ is the observable time series, $\{a_t\}$ is a sequence of white noise, identically and independently distributed as $N(0, \sigma_a^2)$, c is a constant, $\varphi(B) = 1 - \varphi_1 B - \cdots - \varphi_p B^p$ is the autoregressive polynomial, $\theta(B) = 1 - \theta_1 B - \cdots - \theta_q B^q$ is the moving-average polynomial, and $\varphi(B)$ and $\theta(B)$ are assumed to have no common factor. The model (3.3) is of the form (3.1) and (3.2) where $y_t = z_t$ and the input x_t is now the white noise a_t. It will be denoted as the ARMA(p,q) model.

The model (3.3) will be stationary if all the zeros of $\varphi(B)$ are restricted to lie outside the unit circle and in this case $c = (1 - \varphi_1 - \cdots - \varphi_p)\mu$ where μ is the mean of the series. Theoretically, *stationarity* means that the probability density functions of $(z_{t_1}, \ldots, z_{t_1+k})$ and $(z_{t_2}, \ldots, z_{t_2+k})$ are of identical forms for any arbitrary choice of the integers (t_1, t_2, k). In practice, this is saying that the overall behavior of the series remains the same over time.

Now, real-world time series data often exhibit a drifting behavior. Such nonstationary series can be modeled by allowing some of the zeroes of $\varphi(B)$ to be equal to one. Thus, writing $\varphi(B) = \phi(B)(1 - B)^d$ we have from (3.3) that

$$\phi(B)(1 - B)^d z_t = c + \theta(B)a_t \tag{3.4}$$

where the zeros of $\phi(B) = 1 - \phi_1 B - \cdots - \phi_{p_*} B^{p_*}$, $p_* = p - d$, are all lying outside the unit circle. The model (3.4) is known as the *autoregressive integrated moving-average model* of order (p_*,d,q), or ARIMA(p_*,d,q). Furthermore, economic and environmental data often exhibit an evolving cyclical or seasonal behavior, and this type of pattern can be modeled by letting $\varphi(B)$ to have zeros on the unit circle. Thus, more generally, we can write (3.3) in the form

$$\phi(B)u(B)z_t = c + \theta(B)a_t \tag{3.5}$$

where $u(B) = 1 - u_1 B - \cdots - u_d B^d$ has all its zeroes on the unit circle. The special form

$$u(B) = (1 - B)^d (1 - B^s)^{d_s} \tag{3.6}$$

3.2. SOME BASIC PROPERTIES OF UNIVARIATE ARMA MODELS

where s is a positive integer representing the seasonal period, has been widely used in practice to model seasonal time series ($s = 12$ for monthly data and $s = 4$ for quarterly data).

In what follows we list some special cases of the class in (3.5) that have been widely used in practice:

$$(1 - \varphi B)z_t = c + a_t \tag{3.7}$$

$$z_t = c + (1 - \theta B)a_t \tag{3.8}$$

$$(1 - B)z_t = (1 - \theta B)a_t \tag{3.9}$$

$$(1 - B^s)z_t = (1 - \theta_1 B)(1 - \theta_s B^s)a_t \tag{3.10}$$

$$(1 - B)(1 - B^s)z_t = (1 - \theta_1 B)(1 - \theta_s B^s)a_t. \tag{3.11}$$

The model (3.7) is an ARMA(1,0) or AR(1) model and is stationary when $-1 < \varphi < 1$. The second model, (3.8), is a stationary ARMA(0,1) or MA(1) model. The third, (3.9), is nonstationary and for $-1 < \theta < 1$, can be written in the alternative form (see discussion of the π form below)

$$z_t = (1 - \theta)z_{t-1} + \theta(1 - \theta)z_{t-2} + \theta^2(1 - \theta)z_{t-3} + \cdots + a_t, \tag{3.12}$$

showing that the dependence of z_t on past values $z_{t-1}, z_{t-2}, \ldots$ decreases exponentially as we stretch into the past. This model is commonly called the *exponential smoothing model* in the forecasting literature. The models (3.7)–(3.9) will be illustrated in Section 3.4 on examples later. Finally, (3.10) and (3.11) are two models most commonly used for seasonal time series data.

3.1.2. Outline of the chapter

The main purpose of this chapter is to provide the reader with an expository account of the class of univariate ARMA(p,q) models in (3.3), and some useful modeling specification techniques for applying these models to real data. Section 3.2 considers some basic properties of the models with special emphasis on autocorrelation function, partial autocorrelation function, and extended autocorrelation function; Section 3.3 discusses model specification strategy, and Section 3.4 presents several illustrative examples.

3.2. SOME BASIC PROPERTIES OF UNIVARIATE ARMA MODELS

For simplicity in presenting the main results, we shall, until further notice, assume that $c = 0$. To discuss the properties of (3.3), we suppose that the series starts at some timepoint m. This is a realistic assumption since any real-world time series must begin at a fixed time origin. By making this assumption, we will obtain some general results covering both the stationary and the nonstationary cases. From (3.3) we can then write

$$\mathbf{D}_\varphi \mathbf{z} = \mathbf{D}_\theta \mathbf{a} + \mathbf{w} \qquad (3.13)$$

where

$$D_\varphi = \begin{bmatrix} 1 & & & & & \\ -\varphi_1 & \ddots & & & & \\ \vdots & & \ddots & & & \\ -\varphi_p & & & \ddots & & \\ & \ddots & & & \ddots & \\ & & -\varphi_p & \cdots & -\varphi_1 & 1 \end{bmatrix} \quad \text{and}$$

$$D_\theta = \begin{bmatrix} 1 & & & & & \\ -\theta_1 & \ddots & & & & \\ \vdots & & \ddots & & & \\ -\theta_q & & & \ddots & & \\ & \ddots & & & \ddots & \\ & & -\theta_q & \cdots & -\theta_1 & 1 \end{bmatrix}$$

are $(t - m + 1) \times (t - m + 1)$ matrices, $\mathbf{z} = (z_m, \ldots, z_t)'$, $\mathbf{a} = (a_m, \ldots, a_t)'$, $\mathbf{w} = (w_m, \ldots, w_{m+r-1}, 0, \ldots, 0)'$, and $w_m, \ldots, w_{m+r-1}$ are $r = \max(p,q)$ initial values. The w's can be deterministic or stochastic, and in this chapter we shall suppose they are normally distributed and are independent of the a_t's.

3.2.1. The ψ and π weights

From (3.13), we have that

$$\mathbf{z} = \mathbf{D}_\varphi^{-1} \mathbf{D}_\theta \mathbf{a} + \mathbf{D}_\varphi^{-1} \mathbf{w} \qquad (3.14)$$

where

$$D_\varphi^{-1} D_\theta = \begin{bmatrix} 1 & & & \\ \psi_1 & \ddots & & \\ \vdots & & \ddots & \\ \psi_{t-m} & \cdots & \psi_1 & 1 \end{bmatrix}; \quad \text{and} \quad D_\varphi^{-1} = \begin{bmatrix} 1 & & & \\ \psi_1^* & \ddots & & \\ \vdots & & \ddots & \\ \psi_{t-m}^* & \cdots & \psi_1^* & 1 \end{bmatrix}$$

and the ψ's can be obtained by equating coefficients of powers of B from the relations

$$\varphi(B)\psi(B) = \theta(B) \qquad (3.15)$$

where $\psi(B) = 1 + \psi_1 B + \psi_2 B^2 + \cdots$. It is easy to verify that, for $\ell \geq 0$, the ψ's

3.2. SOME BASIC PROPERTIES OF UNIVARIATE ARMA MODELS

satisfy the difference equation

$$\psi_\ell = \varphi_1 \psi_{\ell-1} + \cdots + \varphi_p \psi_{\ell-p} - \theta_\ell \tag{3.16}$$

where $\psi_0 = 1$, $\psi_j = 0$ for $j < 0$ and $\theta_\ell = 0$ for $\ell > q$. Thus for $\ell \geq r$, the ψ's are given by

$$\psi_\ell = A_1 \alpha_1^\ell + \cdots + A_{p_0} \alpha_{p_0}^\ell \tag{3.17}$$

where $p_0 \leq p$, $A_1, \ldots, A_{p_0}$ are polynomials in ℓ and $\alpha_1^{-1}, \ldots, \alpha_{p_0}^{-1}$ are the p_0 distinct zeros of $\varphi(B)$. Analogous expressions can be obtained for the ψ^*'s.

From (3.14), we can write z_t in the ψ form as

$$z_t = a_t + \sum_{h=1}^{t-m} \psi_h a_{t-h} + \sum_{h=t-(m+r)+1}^{t-m} \psi_h^* w_{t-h} \tag{3.18}$$

In a similar way, we have from (3.13) that

$$\mathbf{D}_\theta^{-1} \mathbf{D}_\varphi \mathbf{z} - \mathbf{D}_\theta^{-1} \mathbf{w} = \mathbf{a}, \tag{3.19}$$

where

$$\mathbf{D}_\theta^{-1} \mathbf{D}_\varphi = \begin{bmatrix} 1 & & & \\ -\pi_1 & \ddots & & \\ \vdots & \ddots & \ddots & \\ -\pi_{t-m} & \cdots & -\pi_1 & 1 \end{bmatrix} \quad \text{and} \quad \mathbf{D}_\theta^{-1} = \begin{bmatrix} 1 & & & \\ -\pi_1^* & \ddots & & \\ \vdots & \ddots & \ddots & \\ -\pi_{t-m}^* & \cdots & -\pi_1^* & 1 \end{bmatrix}$$

and the π's can be obtained from the relation $\theta(B)\pi(B) = \varphi(B)$, where $\pi(B) = 1 - \pi_1 B - \pi_2 B^2 - \cdots$, so that the π's satisfy the difference equation

$$\pi_\ell = \theta_1 \pi_{\ell-1} + \cdots + \theta_q \pi_{\ell-q} + \varphi_\ell \tag{3.20}$$

where $\pi_0 = -1$, $\pi_j = 0$ for $j < 0$ and $\varphi_\ell = 0$ for $\ell > p$. Thus for $\ell \geq r$, the π's can be written

$$\pi_\ell = G_1 \beta_1^\ell + \cdots + G_{q_0} \beta_{q_0}^\ell \tag{3.21}$$

where $q_0 \leq q$, $G_1, \ldots, G_{q_0}$ are polynomials in ℓ and $\beta_1^{-1}, \ldots, \beta_{q_0}^{-1}$ are the q_0 distinct zeros of $\pi(B)$. Similar expressions can be obtained for the π^*'s.

From (3.19), we can write z_t in the π form as

$$z_t = \sum_{h=1}^{t-m} \pi_h z_{t-h} + \sum_{h=t-(m+r)+1}^{t-m} \pi_h^* w_{t-h} + a_t. \tag{3.22}$$

Note that by supposing the series to start at a fixed timepoint m and introducing the initial values $w_m, \ldots, w_{m+r-1}$, we have obtained *two alternative* forms of the ARMA model (3.3). The ψ form (3.18) shows how the observation z_t is affected by current and past shocks or innovations a_{t-j}'s, and the π form (3.22) indicates how z_t is related to its own past values z_{t-j}'s. These two expressions are of fundamental importance in understanding the nature of the model. In obtaining (3.18) and (3.22), no restriction is made on $\varphi(B)$ and $\theta(B)$.

For illustrations, consider the models in (3.7)–(3.9). Ignoring the initial values and the constant c, we have that, for the AR(1) model in (3.7)

$$\begin{aligned} \psi \ form : \ & z_t = a_t + \varphi a_{t-1} + \varphi^2 a_{t-2} + \cdots + \varphi^j a_{t-j} + \cdots \\ \pi \ form : \ & z_t = \varphi z_{t-1} + a_t \end{aligned} \quad (3.23)$$

the ψ form shows that the effect of the innovations a_{t-j}'s on z_t decreases exponentially as j is increased, and the π form shows that z_t depends only on the last value z_{t-1} apart from a_t. For the MA(1) model (3.8),

$$\begin{aligned} \psi \ form : \ & z_t = a_t - \theta a_{t-1} \\ \pi \ form : \ & z_t = -\theta z_{t-1} - \cdots - \theta^j z_{t-1} - \cdots + a_t \quad (\text{for } -1 < \theta < 1) \end{aligned} \quad (3.24)$$

the ψ form shows that the effect of the innovations a_{t-j}'s cuts off after the first lag, and the π form shows that z_t depends on all of its past values but with exponentially decreasing weight. Note that for both models, the effects of the innovations eventually disappear and hence the observations will tend to fluctuate about a fixed mean level in a stationary manner. Finally for the ARMA(1,1) model with $\varphi = 1$ in (3.9)

$$\begin{aligned} \psi \ form : \ & z_t = a_t + (1 - \theta)(a_{t-1} + a_{t-2} + \cdots) \\ \pi \ form : \ & z_t = (1 - \theta)z_{t-1} + \theta(1 - \theta)z_{t-2} + \theta^2(1 - \theta)z_{t-3} + \cdots + a_t \end{aligned} \quad (3.25)$$

the ψ form shows that the effects of the innovations a_{t-j}'s stay permanently in the system, and the π form gives rise to what is called 'exponential smoothing' in forecasting future z_t's as mentioned earlier. The permanency of the effects of the a_{t-j}'s means that the level of the series is changing from one time period to the next, underlining the nonstationarity of the series.

3.2.2. Stationarity condition and autocovariance structure of z_t

In (3.18), since the initial values w_t and the innovations a_t are assumed normally distributed, it follows that the observations z_t are also normally distributed. It is readily seen that, if in (3.17)

$$|\alpha_j| < 1, \quad j = 1, \ldots, p_0 \quad (3.26)$$

3.2. SOME BASIC PROPERTIES OF UNIVARIATE ARMA MODELS

then as $t - m \to \infty$, we have that

$$E(z_t) \to 0, \quad \text{cov}(z_t, z_{t+\ell}) \to \sigma_a^2 \left(\sum_{h=0}^{\infty} \psi_h \psi_{h+\ell} \right) \tag{3.27}$$

so that z_t will be stationary in this asymptotic sense. In what follows, we shall refer to (3.26) as the *stationarity condition* of the ARMA model, which is equivalent to requiring that the zeros of $\varphi(B)$ are lying outside of the unit circle.

Let us denote

$$\gamma(\ell) = \text{cov}(z_t, z_{t+\ell}) = \text{cov}(z_t, z_{t-\ell}) = \gamma(-\ell) \tag{3.28}$$

as the lag ℓ autocovariance of a stationary series. For an alternative expression of $\gamma(\ell)$ in terms of the parameters of the ARMA models, we proceed as follows. From (3.3), we have that

$$z_{t-\ell}(z_t - \varphi_1 z_{t-1} - \cdots - \varphi_p z_{t-p}) = z_{t-\ell}(a_t - \theta_1 a_{t-1} - \cdots - \theta_q a_{t-q}).$$

By taking expectation on both sides of this equation and making use of (3.18), we obtain, for $\ell \geq 0$,

$$\gamma(\ell) = \sum_{h=1}^{p} \varphi_h \gamma(\ell - h) + g_\ell \tag{3.29}$$

where

$$g_\ell = \begin{cases} -\sigma_a^2 \sum_{h=0}^{q-\ell} \psi_h \theta_{h+\ell}, & \ell = 0, \ldots, q \\ 0, & \ell > q \end{cases}$$

and $\theta_0 = -1$. By comparing the expression for ψ_ℓ in (3.16) with that for $\gamma(\ell)$ in (3.29), it is clear that there is a one to one correspondence between the ψ_ℓ's and the $\gamma(\ell)$'s and their behavior closely resembles each other.

3.2.3. The autocorrelation function

The autocorrelation function is defined as

$$\rho(\ell) = \frac{\gamma(\ell)}{\gamma(0)}, \quad \ell = 0, \pm 1, \pm 2, \ldots \tag{3.30}$$

From (3.29), we have that

$$\rho(\ell) = \sum_{h=1}^{p} \varphi_h \rho(\ell - h) + \frac{g_\ell}{\gamma(0)} \tag{3.31}$$

As in the case of the autocovariances $\gamma(\ell)$'s, the behavior of the $\rho(\ell)$'s closely resembles that of the ψ_ℓ's. In particular, for $\ell > q$, the $\rho(\ell)$'s satisfy the homogeneous difference equation

$$\rho(\ell) = \sum_{h=1}^{p} \varphi_h \rho(\ell - h)$$

the solution of which takes the form

$$\rho(\ell) = A_1^* \alpha_1^\ell + \cdots + A_{p_0}^* \alpha_{p_0}^\ell \tag{3.32}$$

where $p_0 \leq p$, $A_1^*, \ldots, A_{p_0}^*$ are polynomials in ℓ and, as before, $\alpha_1^{-1}, \ldots, \alpha_{p_0}^{-1}$ are the p_0 distinct zeros of $\varphi(B)$. Thus, for $\varphi(B) \neq 1$, the $\rho(\ell)$'s are mixtures of exponentials, polynomials and damped sine functions of ℓ, and converge to 0 as ℓ increases.

When $\varphi(B) = 1$, that is, when (3.3) is a moving-average model of order q, MA(q), then

$$\rho(\ell) = \begin{cases} -\theta_q \left(1 + \theta_1^2 + \cdots + \theta_q^2\right)^{-1}, & \ell = q \\ 0, & \ell > q. \end{cases} \tag{3.33}$$

In other words, for a MA(q) model, the autocorrelation function cuts off after lag q. This is an important property that will prove useful in the model building process.

3.2.4. The partial autocorrelation function

Consider first an stationary AR(p) model [i.e., when $\theta(B) = 1$]. From the correlation structure in (3.31), we see that for $\ell = 1, \ldots, p$, the p autoregressive coefficients in the vector $\Phi_{(p)} = (\varphi_1, \ldots, \varphi_p)'$ are related to the p autocorrelations in the vector $\boldsymbol{\rho}_{(p)} = [\rho(1), \ldots, \rho(p)]'$ by the system of equations

$$\mathbf{G}_{(p)} \Phi_{(p)} = \boldsymbol{\rho}_{(p)} \tag{3.34}$$

where $\mathbf{G}_{(p)}$ is the $p \times p$ matrix

$$\mathbf{G}_{(p)} = \begin{bmatrix} 1 & \rho(-1) & \cdots & \rho(-p+2) & \rho(-p+1) \\ \rho(1) & 1 & & \vdots & \rho(-p+2) \\ \vdots & & \ddots & & \vdots \\ \rho(p-2) & & & 1 & \rho(-1) \\ \rho(p-1) & \rho(p-2) & \cdots & \rho(1) & 1 \end{bmatrix}$$

Regarding this as a system of p equations in the p unknowns $(\varphi_1, \ldots, \varphi_p)$, the solution

3.2. SOME BASIC PROPERTIES OF UNIVARIATE ARMA MODELS

of φ_p is, for $p > 1$, the ratio of two determinants

$$\varphi_p = |\mathbf{H}_{(p)}|/|\mathbf{G}_{(p)}| \tag{3.35}$$

where

$$\mathbf{H}_{(p)} = \begin{bmatrix} 1 & \rho(-1) & \cdots & \rho(-p+2) & \rho(1) \\ \rho(1) & 1 & & \vdots & \rho(2) \\ \vdots & & \ddots & & \vdots \\ \rho(p-2) & & & 1 & \rho(p-1) \\ \rho(p-1) & \rho(p-2) & \cdots & \rho(1) & \rho(p) \end{bmatrix}$$

is a $p \times p$ matrix, and of course $\varphi_p = \rho(1)$ for $p = 1$. This result then leads to defining the following function of $\rho(1), \ldots, \rho(\ell)$ for any stationary model:

$$\wp(\ell) = \begin{cases} \rho(\ell), & \ell = 1 \\ |\mathbf{H}_{(\ell)}|/|\mathbf{G}_{(\ell)}|, & \ell > 1 \end{cases} \tag{3.36}$$

which is known as the *partial autocorrelation function*. It has the property that, for a stationary AR(p) model

$$\wp(\ell) = \begin{cases} \varphi_\ell, & \ell = p \\ 0, & \ell > p \end{cases} \tag{3.37}$$

in other words, $\wp(\ell)$ vanishes for $\ell > p$ when the model is AR(p). This is akin to the property of the autocorrelation function $\rho(\ell)$ with respect to the MA(q) model, and will prove to be an useful tool in model building.

3.2.5. The extended autocorrelation function

The "cutting off" property of the autocorrelation function $\rho(\ell)$ with respect to MA(q) models and the partial autocorrelation function $\wp(\ell)$ with respect to AR(p) models will no longer hold for mixed ARMA(p,q) models. We now introduce the extended autocorrelation function (EACF), which does possess a similar property for the mixed models.

Consider first the simple case of the ARMA(2,1) model. From (3.31), we see that

$$\begin{bmatrix} \rho(\ell) & \rho(\ell-1) \\ \rho(\ell+1) & \rho(\ell) \end{bmatrix} \begin{bmatrix} \varphi_1 \\ \varphi_2 \end{bmatrix} \neq \begin{bmatrix} \rho(\ell+1) \\ \rho(\ell+2) \end{bmatrix}, \quad \ell = 0$$

$$\begin{bmatrix} \rho(\ell) & \rho(\ell-1) \\ \rho(\ell+1) & \rho(\ell) \end{bmatrix} \begin{bmatrix} \varphi_1 \\ \varphi_2 \end{bmatrix} = \begin{bmatrix} \rho(\ell+1) \\ \rho(\ell+2) \end{bmatrix}, \quad \ell \geq 1$$

where it is noted that $\rho(j) = \rho(-j)$ and $\rho(0) = 1$. For a given ℓ, let the 2×1 vector

$(\varphi_1^{(\ell)}, \varphi_2^{(\ell)})'$ satisfies the equations

$$\begin{bmatrix} \rho(\ell) & \rho(\ell-1) \\ \rho(\ell+1) & \rho(\ell) \end{bmatrix} \begin{bmatrix} \varphi_1^{(\ell)} \\ \varphi_2^{(\ell)} \end{bmatrix} = \begin{bmatrix} \rho(\ell+1) \\ \rho(\ell+2) \end{bmatrix}.$$

Clearly, $(\varphi_1^{(\ell)}, \varphi_2^{(\ell)})' \neq (\varphi_1, \varphi_2)'$ for $\ell = 0$ and $(\varphi_1^{(\ell)}, \varphi_2^{(\ell)})' = (\varphi_1, \varphi_2)'$ for $\ell \geq 1$. Letting

$$w_t^{(\ell)} = \left(1 - \varphi_1^{(\ell)} B - \varphi_2^{(\ell)} B^2\right) z_t$$

it then follows that for $\ell \geq 1$, the transformed series $\{w_t^{(\ell)}\}$ follows a MA(1) process and its autocorrelation function will have the "cutting off" property. That is, the lag 1 autocorrelation of $w_t^{(\ell)}$ will be nonzero, but autocorrelations of higher lags will all be zero.

In general, for $k = 1, 2, 3, \ldots$ and $\ell = 0, 1, 2, 3, \ldots$, let the $k \times 1$ vector

$$\Phi_{(k)}^{(\ell)} = \left(\varphi_1^{(\ell)}, \ldots, \varphi_k^{(\ell)}\right)'$$

satisfy the system of equations

$$\mathbf{G}_{(k)}^{(\ell)} \Phi_{(k)}^{(\ell)} = \mathbf{P}_{(k)}^{(\ell)} \tag{3.38}$$

where

$$\mathbf{G}_{(k)}^{(\ell)} = \begin{bmatrix} \rho(\ell) & \rho(\ell-1) & \cdots & & \rho(\ell-k+1) \\ \rho(\ell+1) & \rho(\ell) & & & \vdots \\ \vdots & & \ddots & & \vdots \\ & & & \rho(\ell) & \rho(\ell-1) \\ \rho(\ell+k-1) & \cdots & \cdots & \rho(\ell+1) & \rho(\ell) \end{bmatrix}$$

and

$$\mathbf{P}_{(k)}^{(\ell)} = (\rho(\ell+1), \ldots, \rho(\ell+k))',$$

and let $\rho(k, \ell)$ be the lag ℓ autocorrelation of the transformed process $\{w_{k,t}^{(\ell)}\}$, where $w_{k,t}^{(\ell)} = (1 - \varphi_1^{(\ell)} B - \cdots - \varphi_k^{(\ell)} B^k) z_t$. It is readily shown that $\rho(k, \ell)$ is a function of the autocorrelations $\rho(1), \ldots, \rho(k+\ell)$. Specifically

$$\rho(k, \ell) = \frac{\mathbf{b}' \mathbf{G}_{(k+1)}^{(\ell)} \mathbf{b}}{\mathbf{b}' \mathbf{G}_{(k+1)}^{(0)} \mathbf{b}} \tag{3.39}$$

where $\mathbf{b}' = (1, \Phi_{(k)}^{(\ell)'})$.

3.3. MODEL SPECIFICATION STRATEGY

Now, for an ARMA(p,q) model, $\rho(k, \ell)$ has the "cutting off" property such that for $k = p$,

$$\rho(k, \ell) = \begin{cases} -\theta_q\left(1 + \theta_1^2 + \cdots + \theta_q^2\right)^{-1}, & \ell = q, \\ 0, & \ell > q, \end{cases} \quad (3.40)$$

which is analogous to the property of $\rho(\ell)$ in (3.33) for the MA(q) model. Following the work of Tsay and Tiao (1984), we call $\rho(k, \ell)$ the kth *extended autocorrelation* of lag ℓ for z_t. We shall also denote $\rho(\ell) = \rho(0, \ell)$ so that $\rho(k, \ell)$, will be defined for $k \geq 0$, and $\ell \geq 1$. It can be readily shown that for a stationary ARMA(p,q) model, when $k \geq p$

$$\rho(k, \ell) = \begin{cases} c, & \ell = q + k - p \\ 0, & \ell > q + k - p \end{cases} \quad (3.41)$$

where $-1 < c < 1$. This property for $\rho(k, \ell)$ will be exploited later in the model building process.

3.3. MODEL SPECIFICATION STRATEGY

The class of ARMA(p,q) models in (3.3) is extensive. In practice, guidelines are needed in selecting a member of the class to represent the time series data at hand. Box and Jenkins (1976) have proposed an iterative model building strategy that has been widely adopted by practitioners. The strategy consists of three main phases:

- Tentative specification or identification of a model
- Efficient estimation of model parameters
- Diagnostic checking of fitted model for further improvement

In the remainder of this chapter, we shall focus on tentative specification.

3.3.1. Tentative specification

The aim here is to employ statistics that (1) can be readily calculated from the data and (2) allow the user to tentatively select a model, that is, determine (p,q) in (3.3) or (p-d, d, q) in (3.4). We shall discuss three methods:

1. The sample autocorrelation function (SACF)
2. The sample partial autocorrelation function (SPACF)
3. The sample extended sample autocorrelation function (SEACF)

SACF

The sample autocorrelations of z_t are defined as

$$\hat{\rho}(\ell) = C(\ell)/C(0), \quad \ell = 1, 2, \ldots \quad (3.42)$$

where $C(j) = \sum_{t=1}^{n-j}(z_t - \bar{z})(z_{t+j} - \bar{z})$ and $\bar{z}$ is the sample mean of the n available observations $z_1, \ldots, z_n$. It is well known that, as $n \to \infty$, for stationary models,

$$\hat{\rho}(\ell) \xrightarrow{P} \rho(\ell). \quad (3.43)$$

Also, if $\varphi(B)$ is of the form in (3.4), then

$$\hat{\rho}(\ell) \xrightarrow{P} 1. \quad (3.44)$$

Thus, from (3.44), if the SACF $\hat{\rho}(\ell)$ (as a function of the lag ℓ) of the original series z_t is persistently close to 1 as ℓ increases, one then forms the first difference series $w_t = (1 - B)z_t$ and studies its SACF to determine whether further differencing is called for. Once stationarity is achieved, from (3.33) and (3.43), a "cutting off" pattern after, say, lag q, in the SACF will then lead to tentative specification of a MA(q) model.

For stationary models, the $\hat{\rho}(\ell)$'s are asymptotically normally distributed, but their covariance structure is rather complex in general (see Bartlett 1946). For the MA(q) model and $\ell > q$, the asymptotic variance of $\hat{\rho}(\ell)$ is

$$\text{Var}(\hat{\rho}(\ell)) \cong n^{-1}\left[1 + 2\sum_{j=1}^{q}\rho^2(j)\right]. \quad (3.45)$$

By substituting $\hat{\rho}(\ell)$ for the unknown $\rho(\ell)$ in (3.45), the estimated variances of the $\hat{\rho}(\ell)$'s are often used to help specify the order q of an MA model.

SPACF

The sample partial autocorrelations

$$\hat{\wp}(\ell), \quad \ell = 1, 2, \ldots \quad (3.46)$$

of z_t are obtained by replacing the $\rho(\ell)$'s in (3.36) by their sample estimates $\hat{\rho}(\ell)$'s. For stationary models

$$\hat{\wp}(\ell) \xrightarrow{P} \wp(\ell) \quad (3.47)$$

and the $\hat{\wp}(\ell)$'s are asymptotically normally distributed. Also, for a stationary AR(p) model,

$$\text{Var}(\hat{\wp}(\ell)) \cong n^{-1}, \quad \ell > p. \quad (3.48)$$

3.3. MODEL SPECIFICATION STRATEGY

The properties in (3.47) and (3.48) make SPACF a convenient tool for specifying the order p of a stationary AR model in practice. For nonstationary models, where $\varphi(B)$ contains the factor $u(B)$ in (3.5), the asymptotic property of $\hat{\wp}(\ell)$ is rather complex, however.

In the past, the SACF and SPACF have been the most commonly used statistical tools for tentative model specification. Specifically, a persistently high SACF signals the need for differencing; a low-order moving-average model is suggested by SACF exhibiting a small number of large values at low lags; and a low-order autoregressive model, by SPACF showing a similar "cutting off" pattern. Also, for series exhibiting a strong cyclical behavior of period s, persistent high SACF at lags that are multiples of s signals, there is the need to apply the "seasonal differencing" operator $(1 - B^s)$ to the data, and so on. Illustrative examples will be given in the next section.

It should be noted that the weaknesses of the SACF and SPACF in model specification are

1. Subjective judgment is often required to decide on the order of differencing.
2. For stationary mixed ARMA models, both SACF and SPACF tend to exhibit a gradual "tapering off" behavior, making specification of the autoregressive and the moving average parts difficult.

SEACF

Several approaches have been proposed to handle the mixed ARMA model specification problem. These include the *R*- and *S*-array methods of Gray et al. (1978) and the generalized partial autocorrelations by Woodward and Gray (1981). In what follows, we discuss the sample extended autocorrelation function (SEACF) approach proposed by Tsay and Tiao (1984) for tentative specification of the orders (p,q) for the general nonstationary and stationary ARMA mode (3.3). The proposed procedure eliminates the need to difference or transform the series to achieve stationarity and directly specifies the values of p and q.

For stationary ARMA models, the estimates $\hat{\rho}(k, \ell)$'s of the EACF $\rho(k, \ell)$'s as defined in (3.39) can be obtained on replacing the $\rho(\ell)$'s in the expression by their sample counterparts $\hat{\rho}(\ell)$'s. In this case, the estimated $\hat{\rho}(k, \ell)$'s will be consistent for the $\rho(k, \ell)$'s and hence the property (3.41) can be exploited for model specification. However, for nonstationary models, the $\hat{\rho}(k, \ell)$'s will not have the asymptotic property given by the right-hand side (RHS) of (3.41) in general.

Now for ARMA(p,q) models, one can view the SEACF approach as consisting of the following two steps:

1. Find consistent estimates of the autoregressive parameters φ_j's in order to transform z_t into a moving average process.
2. Make use of the "cutting off" property of the autocorrelation function of the transformed series for model identification.

For estimating the φ_j's, the following iterated regression approach has been adopted. First, let $\hat{\varphi}_{1(k)}^{(0)}, \ldots, \hat{\varphi}_{k(k)}^{(0)}$ be the ordinary least squares (OLS) estimates from fitting the

AR(k) regression to the data

$$z_t = \hat{\varphi}_{1(k)}^{(0)} z_{t-1} + \cdots + \hat{\varphi}_{k(k)}^{(0)} z_{t-k} + e_{k,t}^{(0)} \qquad (3.49)$$

where $e_{k,t}^{(0)}$ denotes the error term. The first iterated AR(k) regression is given by

$$z_t = \hat{\varphi}_{1(k)}^{(1)} z_{t-1} + \cdots + \hat{\varphi}_{k(k)}^{(1)} z_{t-k} + \hat{\beta}_{1(k)}^{(1)} \hat{e}_{k,t-1}^{(0)} + e_{k,t}^{(1)} \qquad (3.50)$$

where $\hat{e}_{k,t}^{(0)} = z_t - \hat{\varphi}_{1(k)}^{(0)} z_{t-1} - \cdots - \hat{\varphi}_{k(k)}^{(0)} z_{t-k}$ is the residual from (3.49) and $e_{k,t}^{(1)}$ denotes the error term. This yields a new set of OLS estimates $\hat{\varphi}_{1(k)}^{(1)}, \ldots, \hat{\varphi}_{k(k)}^{(1)}$. In general, for $\ell = 1, 2, \ldots$ the estimates $\hat{\varphi}_{1(k)}^{(\ell)}, \ldots, \hat{\varphi}_{k(k)}^{(\ell)}$ are obtained from the ℓth iterated AR(k) regression

$$z_t = \hat{\varphi}_{1(k)}^{(\ell)} z_{t-1} + \cdots + \hat{\varphi}_{k(k)}^{(\ell)} z_{t-k} + \hat{\beta}_{1(k)}^{(\ell)} \hat{e}_{k,t-1}^{(\ell-1)} + \cdots + \hat{\beta}_{\ell(k)}^{(\ell)} \hat{e}_{k,t,-\ell}^{(0)} + e_{k,t}^{(\ell)} \qquad (3.51)$$

where $\hat{e}_{k,t}^{(i)} = z_t - \hat{\varphi}_{1(k)}^{(i)} z_{t-1} - \cdots - \hat{\varphi}_{k(k)}^{(i)} z_{t-k} - \sum_{h=1}^{i} \hat{\beta}_{h(k)}^{(i)} \hat{e}_{k,t-h}^{i-h}$ (i.e. the residual from the ith iterated regression) and $e_{k,t}^{(\ell)}$ is the error term. In practice, these iterated estimates $\hat{\varphi}_{j(k)}^{(\ell)}$'s can be obtained from OLS estimates of the autoregressive coefficients by fitting AR(k), ..., AR($k+\ell$) to z_t, using the recursion

$$\hat{\varphi}_{j(k)}^{(\ell)} = \hat{\varphi}_{j(k+1)}^{(\ell-1)} - \frac{\hat{\varphi}_{j-1,(k)}^{(\ell-1)} \hat{\varphi}_{k+1,(k+1)}^{(\ell-1)}}{\hat{\varphi}_{k(k)}^{(\ell-1)}} \qquad (3.52)$$

where $\hat{\varphi}_{0(k)}^{(\ell)} = -1$, $j = 1, \ldots, k$, $k \geq 1$, and $\ell \geq 1$.

On the basis of some consistency results of OLS estimates of autoregressive parameters for nonstationary and stationary ARMA(p,q) models in Tiao and Tsay (1983), these authors show that for $k = p$

$$\hat{\Phi}^{(\ell)}(p) \xrightarrow{P} \Phi(p), \quad \ell \geq q \qquad (3.53)$$

where

$$\hat{\Phi}^{(\ell)}(p) = \left(\hat{\varphi}_{1(p)}^{(\ell)}, \ldots, \hat{\varphi}_{p(p)}^{(\ell)} \right)'.$$

Now, analogously to (3.39), the SEACF $\hat{\rho}(k, \ell)$ is defined as

$$\hat{\rho}(k, \ell) = \hat{\rho}_{(\ell)}\left(\hat{w}_{k,t}^{(\ell)} \right) \qquad (3.54)$$

where $\hat{\rho}_{(\ell)}(\hat{w}_{k,t}^{(\ell)})$ is the lag ℓ sample autocorrelation of the transformed series $\hat{w}_{k,t}^{(\ell)} = (1 - \hat{\varphi}_{1(k)}^{(\ell)} B - \cdots - \hat{\varphi}_{k(k)}^{(\ell)} B^k) z_t$. Also, we may denote $\hat{\rho}(0, \ell) = \hat{\rho}(\ell)$ for the ordinary sample autocorrelations, and shall call $\hat{\rho}(k, \ell)$ the kth sample extended autocorrelation

3.3. MODEL SPECIFICATION STRATEGY

TABLE 3.1. The SEACF Table

	0	1	2	3	4	...
0	$\hat{\rho}(0,1)$	$\hat{\rho}(0,2)$	$\hat{\rho}(0,3)$	$\hat{\rho}(0,4)$	$\hat{\rho}(0,5)$	...
1	$\hat{\rho}(1,1)$	$\hat{\rho}(1,2)$	$\hat{\rho}(1,3)$	$\hat{\rho}(1,4)$	$\hat{\rho}(1,5)$	...
2	$\hat{\rho}(2,1)$	$\hat{\rho}(2,2)$	$\hat{\rho}(2,3)$	$\hat{\rho}(2,4)$	$\hat{\rho}(2,5)$	...
3	$\hat{\rho}(3,1)$	$\hat{\rho}(3,2)$	$\hat{\rho}(3,3)$	$\hat{\rho}(3,4)$	$\hat{\rho}(3,5)$	...
4	$\hat{\rho}(4,1)$	$\hat{\rho}(4,2)$	$\hat{\rho}(4,3)$	$\hat{\rho}(4,4)$	$\hat{\rho}(4,5)$	...
⋮	⋮	⋮	⋮	⋮	⋮	⋱

of lag ℓ. Tsay and Tiao show that for the general ARMA(p,q) model in (3.3), stationary or nonstationary, when $k > p$

$$\hat{\rho}(k, \ell) \xrightarrow{p} \begin{cases} c, & \ell = q + k - p \\ 0, & \ell > q + k - p \end{cases} \quad (3.55)$$

where $|c| < 1$.

3.3.2. Tentative model specification via SEACF

The asymptotic property of the SEACF $\hat{\rho}(k, \ell)$ given by (3.55) can now be exploited to help tentatively identify ARMA(p,q) models in practice. For this purpose, it is useful to arrange the $\hat{\rho}(k, \ell)$'s in a two-way table as shown in Table 3.1, in which the first row gives the SACF, the second row gives the first SEACF, and so on. The rows are numbered 0, 1, 2, ... to signify the AR order and the columns in a similar way for the MA order. To illustrate the use of the table, suppose that the true model is an ARMA(1,2). For the SACF, it is well known that asymptotically $\hat{\rho}(0, \ell) \neq 0$ for $\ell \geq 2$. Now from (3.55) with $p = 1$ and $q = 2$, we see that

1. When $k = 1$, $\hat{\rho}(1, \ell) \cong 0$ for $\ell \geq 3$
2. When $k = 2$, $\hat{\rho}(2, \ell) \cong 0$ for $\ell \geq 4$

and so on. The full situation is shown in Table 3.2, where × denotes a nonzero value, 0 is zero and ∗ means a value between −1 and 1. The 0's are seen to form a triangle with boundaries given by the two lines $k = 1$ and $\ell - k = 2$. The row and column coordinates of the vertex correspond precisely to the AR and MA order, respectively.

In general, we are thus led to search from the SEACF table the vertex of a triangle of asymptotic 0's having boundary lines $k = c_1 > 0$ and $\ell - k = c_2 > 0$, and tentatively identify $p = c_1$ and $q = c_2$ as the order of the ARMA model. In practice, for finite samples, the $\hat{\rho}(k, \ell)$'s will not be zero. The asymptotic variance of the $\hat{\rho}(k, \ell)$'s can be approximately obtained by using Bartlett's formula. As a crude but

TABLE 3.2. Asymptotic SEACF Table for an ARMA(1,2) Model[a]

	0	1	2	3	4	...
0	*	×	×	×	×	×
1	*	×	0	0	0	0
2	*	×	×	0	0	0
3	*	×	×	×	0	0
4	*	×	×	×	×	0

[a] Multiplication sign (×) denotes a nonzero value and asterisk (*), a value between −1 and 1.

simple approximation, we may use the value $(n - k - \ell)^{-1}$ on the hypothesis that the transformed series $\hat{w}_{k,t}^{(\ell)}$ is white noise to estimate the variance of $\hat{\rho}(k, \ell)$. Of course, it is understood that this simple approximation might underestimate the variance of $\hat{\rho}(k, \ell)$ and further study is needed. As a preliminary but informative guide for model specification, the SEACF table may be supplemented by an analogous table consisting of indicator symbols × denoting values beyond ±2 standard deviations and 0 for in between values. This is shown in Table 3.2. Illustrative examples will be given in the next section.

3.4. EXAMPLES

Example 3.1: A Generated MA(1) Example. The following data are generated from the MA(1) model:

$$Z_t = 5 + a_t - .7a_{t-1},$$

where the a_t's are iid $N(0,1)$.

5.212	5.734	3.822	6.633	4.258	4.355	6.173
4.337	5.473	4.736	3.827	8.644	1.854	5.109
5.318	5.293	5.331	4.462	5.437	6.309	5.149
3.566	6.139	6.391	4.263	4.031	5.474	4.315
5.701	2.572	4.280	5.903	5.964	5.126	5.512
5.475	3.507	5.914	6.951	4.435	5.604	6.113
3.568	5.885	3.148	5.054	5.783	6.228	4.621
5.139	5.947	2.685	7.020	3.625	5.206	6.000
5.149	3.134	5.666	3.711	5.812	5.673	1.948
6.174	5.230	5.815	5.465	2.898	6.592	5.704
4.472	6.062	5.690	1.988	5.424	5.713	5.610
4.681	5.938	5.208	4.737	4.615	5.337	5.550
2.866	5.839	4.945	4.019	6.561	4.140	5.615
3.983	5.618	4.356	6.710	3.335	5.379	6.604
2.209	6.389					

3.4. EXAMPLES

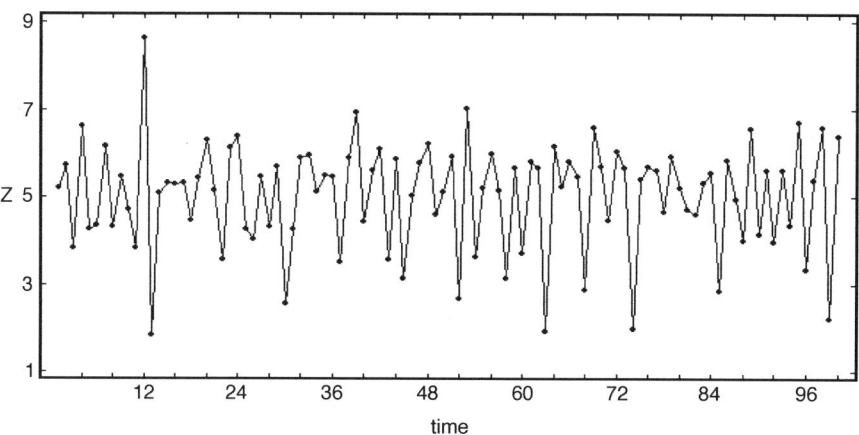

FIGURE 3.1 Generated MA(1) series.

A time series plot of the data is shown in Figure 3.1. It is clear that the series is stationary, with the observations fluctuating tightly about the mean level of the series.

In what follows we give three methods—SACF, SPACF, and SEACF—for tentative specification of a model for this data set.

1. The sample autocorrelation function (SACF):

```
AUTOCORRELATIONS

1–12   –.47  –.05   .16  –.17   .11  –.04  –.08   .08   .06  –.23   .17  –.04
ST.E.   .10   .12   .12   .12   .12   .13   .13   .13   .13   .13   .13   .13
Q      22.5  22.7  25.6  28.6  29.9  30.1  30.7  31.5  31.9  37.8  41.0  41.2

        –1.0   –.8   –.6   –.4   –.2    .0    .2    .4    .6    .8   1.0
         +----+----+----+----+----+----+----+----+----+----+
                                       I
 1   –.47          XXXXXXX+XXXXI         +
 2   –.05                     + XI       +
 3    .16                     +    IXXXX +
 4   –.17                     + XXXXI    +
 5    .11                     +    IXXX  +
 6   –.04                     + XI       +
 7   –.08                     + XXI      +
 8    .08                     +    IXX   +
 9    .06                     +    IX    +
10   –.23                XXXXXXI         +
11    .17                     +    IXXXX +
12   –.04                     + XI       +
```

2. The sample partial autocorrelation function (SPACF):

PARTIAL AUTOCORRELATIONS

1–12	−.47	−.35	−.04	−.15	−.00	−.03	−.10	−.05	.10	−.19	−.06	−.07
ST.E.	.10	.10	.10	.10	.10	.10	.10	.10	.10	.10	.10	.10

```
            -1.0   -.8   -.6   -.4   -.2    .0    .2    .4    .6    .8   1.0
            +----+----+----+----+----+----+----+----+----+----+
                                     I
 1   -.47                  XXXXXXX+XXXXI         +
 2   -.35                    XXXX+XXXXI          +
 3   -.04                        +   XI          +
 4   -.15                       +XXXXI           +
 5   -.00                        +    I          +
 6   -.03                        +   XI          +
 7   -.10                        + XXXI          +
 8   -.05                        +   XI          +
 9    .10                        +    IXX        +
10   -.19                      XXXXXXI           +
11   -.06                        +  XXI          +
12   -.07                        +  XXI          +
```

3. The sample extended autocorrelation function (SEACF):

THE EXTENDED ACF TABLE

(Q-->)	0	1	2	3	4	5	6	7	8	9	10	11	12
(P= 0)	−.47	−.05	.16	−.17	.11	−.04	−.08	.08	.06	−.23	.17	−.04	−.10
(P= 1)	−.52	−.20	.08	−.06	.08	−.03	−.10	.07	.03	−.22	.13	−.04	−.12
(P= 2)	−.10	−.34	−.18	.01	−.01	−.09	−.01	−.02	−.02	−.21	−.10	−.10	.02
(P= 3)	−.20	−.38	−.21	−.03	.00	−.07	−.05	−.02	−.03	−.18	.08	−.12	.04
(P= 4)	−.00	−.18	−.14	.07	−.08	−.06	−.01	.02	−.01	−.15	−.09	−.15	.04
(P= 5)	−.00	−.01	−.40	−.16	.09	−.05	−.01	.02	−.02	−.14	.06	−.01	−.06
(P= 6)	−.24	.08	−.15	−.27	.04	.04	.15	−.00	−.05	−.13	.06	−.02	−.04

SIMPLIFIED EXTENDED ACF TABLE (5% LEVEL)

(Q-->)	0	1	2	3	4	5	6	7	8	9	10	11	12
(P= 0)	X	O	O	O	O	O	O	O	O	O	O	O	O
(P= 1)	X	O	O	O	O	O	O	O	O	O	O	O	O
(P= 2)	O	X	O	O	O	O	O	O	O	O	O	O	O
(P= 3)	O	X	O	O	O	O	O	O	O	O	O	O	O
(P= 4)	O	O	O	O	O	O	O	O	O	O	O	O	O
(P= 5)	O	O	X	O	O	O	O	O	O	O	O	O	O
(P= 6)	X	O	O	O	O	O	O	O	O	O	O	O	O

The SACF and SEACF clearly suggest an MA(1) model, and the SPACF indicates that an AR model of at least 2 would be needed. Thus, one would be led to tentatively specify an MA(1) model for this data set.

3.4. EXAMPLES

Example 3.2: A Generated AR(1) Example. The following data are generated from the AR(1) model:

$$Z_t = 1.5 + .7Z_{t-1} + a_t,$$

or

$$Z_t - 5 = .7(Z_{t-1} - 5) + a_t,$$

where the a_t's are iid $N(0,1)$.

5.449	6.198	5.279	6.436	6.132	5.237	5.950
5.550	5.778	5.555	4.223	7.284	5.433	4.595
4.540	4.847	5.343	5.017	5.292	6.710	7.400
6.088	6.485	7.938	7.648	6.299	5.994	5.071
5.313	2.976	1.292	1.704	3.166	4.174	5.255
6.236	5.112	5.466	7.548	7.773	8.238	9.288
7.985	7.962	5.833	4.768	5.050	6.412	6.573
6.649	7.485	5.356	6.301	5.272	4.949	5.796
6.288	4.547	4.401	3.094	3.438	4.419	1.901
2.120	2.716	4.029	5.225	3.689	4.646	5.850
5.836	6.815	7.822	5.049	4.109	4.442	5.265
5.326	6.264	6.818	6.663	6.053	5.996	6.428
4.377	4.267	4.224	3.292	4.549	4.346	4.920
4.191	4.525	4.087	5.665	4.713	4.652	6.257
4.140	4.569					

A time series plot of the data is shown in Figure 3.2.

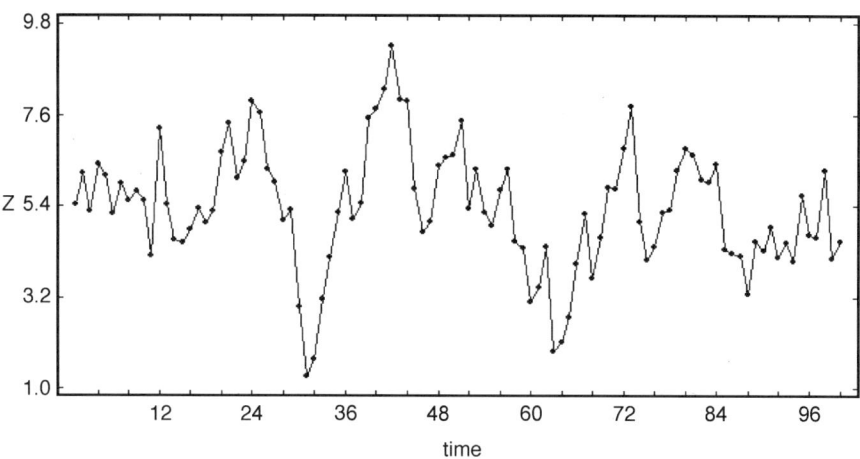

FIGURE 3.2 Generated AR(1) series.

Compared with the generated MA(1) series, we see that this series exhibits more momentum over time.

The following are the three methods to help specify a model for this series: SACF, SPACF, and SEACF.

1. **SACF:**

   AUTOCORRELATIONS

1–12	.72	.51	.37	.17	.06	–.06	–.16	–.19	–.26	–.31	–.24	–.21
ST.E.	.10	.14	.16	.17	.17	.17	.17	.17	.17	.18	.18	.19
Q	53.1	80.5	94.6	97.8	98.2	98.6	101	105	113	124	131	136

   ```
                    -1.0  -.8   -.6   -.4   -.2    .0    .2    .4    .6    .8   1.0
                    +----+----+----+----+----+----+----+----+----+----+
                                              I
    1   .72                             +     IXXXX+XXXXXXXXXXXXX
    2   .51                            +      IXXXXXX+XXXXXX
    3   .37                           +       IXXXXXXX+X
    4   .17                           +       IXXXX    +
    5   .06                           +       IXX      +
    6  -.06                           +      XXI       +
    7  -.16                           +     XXXXI      +
    8  -.19                           +     XXXXXI     +
    9  -.26                          +    XXXXXXI       +
   10  -.31                          +XXXXXXXXI         +
   11  -.24                           +    XXXXXXI      +
   12  -.21                           +     XXXXXI      +
   ```

2. **SPACF:**

   PARTIAL AUTOCORRELATIONS

1–12	–.72	–.00	–.00	–.18	.01	–.14	–.04	–.03	–.11	–.12	.14	–.06
ST.E.	.10	.10	.10	.10	.10	.10	.10	.10	.10	.10	.10	.10

   ```
                    -1.0  -.8   -.6   -.4   -.2    .0    .2    .4    .6    .8   1.0
                    +----+----+----+----+----+----+----+----+----+----+
                                              I
    1   .72                             +     IXXXX+XXXXXXXXXXXXX
    2   .00                             +     I     +
    3   .00                             +     I     +
    4  -.18                          XXXXXI          +
    5   .01                             +     I     +
    6  -.14                            + XXXI       +
    7  -.04                            +   XI       +
    8  -.03                            +   XI       +
    9  -.11                            + XXXI       +
   10  -.12                            + XXXI       +
   11   .14                            +     IXXXX+
   12  -.06                            +    XI     +
   ```

3.4. EXAMPLES

3. SEACF:

THE EXTENDED ACF TABLE

(Q-->)	0	1	2	3	4	5	6	7	8	9	10	11	12
(P= 0)	.72	.51	.37	.17	.06	−.06	−.16	−.19	−.26	−.31	−.24	−.21	−.16
(P= 1)	.00	.01	.22	.06	.07	−.07	−.10	.01	−.03	−.24	.06	−.06	−.13
(P= 2)	−.03	−.00	.23	−.04	.15	−.02	−.12	−.05	−.00	−.25	−.10	−.03	−.13
(P= 3)	−.06	−.03	.12	−.06	.14	−.01	−.12	.01	−.02	−.25	.05	−.13	−.10
(P= 4)	.08	−.48	−.20	.06	.03	−.08	.02	.00	.04	−.15	−.02	−.16	−.14
(P= 5)	.10	−.45	−.25	.19	.03	−.02	.01	.05	.06	−.16	−.00	−.11	.04
(P= 6)	−.24	−.05	−.37	−.13	−.05	.00	−.01	.07	.08	−.13	.02	−.04	−.08

SIMPLIFIED EXTENDED ACF TABLE (5% LEVEL)

(Q-->)	0	1	2	3	4	5	6	7	8	9	10	11	12
(P= 0)	X	X	X	O	O	O	O	O	O	O	O	O	O
(P= 1)	O	O	O	O	O	O	O	O	O	X	O	O	O
(P= 2)	O	O	X	O	O	O	O	O	O	X	O	O	O
(P= 3)	O	O	O	O	O	O	O	O	O	X	O	O	O
(P= 4)	O	X	O	O	O	O	O	O	O	O	O	O	O
(P= 5)	O	X	O	O	O	O	O	O	O	O	O	O	O
(P= 6)	X	O	X	O	O	O	O	O	O	O	O	O	O

The SACF shows that a low-order MA model will not be appropriate, and the SPACF strongly suggests an AR(1) model. This tentative specification is also supported by the SEACF.

Example 3.3: A Generated ARIMA(0,1,1) Example. Here we consider 150 observations generated from the following nonstationary model:

$$Z_t - Z_{t-1} = a_t - .4\,a_{t-1}$$

where the a_t's are iid $N(0,1)$. The data are shown in Figure 3.3

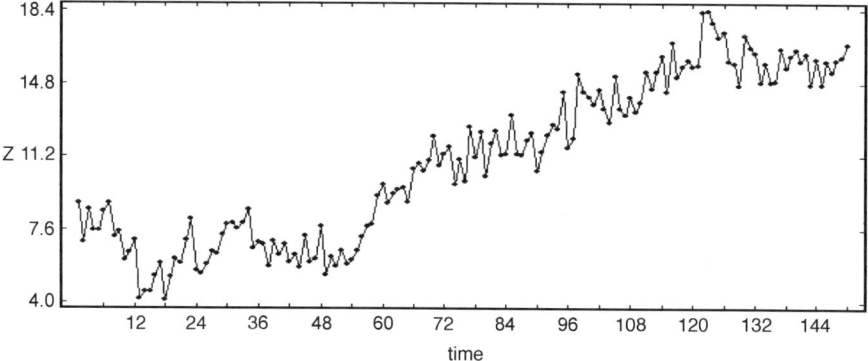

FIGURE 3.3 Generated (0,1,1) series.

UNIVARIATE AUTOREGRESSIVE MOVING-AVERAGE MODELS

It is clear that the series exhibits a drifting behavior. The observations do not cling to any stable mean level. For tentative specification of a model, we first apply SACF and SPACF:

AUTOCORRELATIONS

1–12	.95	.94	.93	.91	.89	.89	.87	.86	.84	.83	.81	.79
ST.E.	.08	.14	.17	.20	.23	.25	.27	.29	.31	.32	.34	.35
Q	138	274	408	537	662	787	908	1026	1140	1251	1359	1461

```
        -1.0   -.8   -.6   -.4   -.2    .0    .2    .4    .6    .8   1.0
        +----+----+----+----+----+----+----+----+----+----+----+
                                          I
 1  .95                              +    IXXX+XXXXXXXXXXXXXXXXXXXX
 2  .94                             +     IXXXXXX+XXXXXXXXXXXXXXXX
 3  .93                            +      IXXXXXXXX+XXXXXXXXXXXXXX
 4  .91                           +       IXXXXXXXXX+XXXXXXXXXXXXX
 5  .89                          +        IXXXXXXXXXX+XXXXXXXXXXX
 6  .89                         +         IXXXXXXXXXXX+XXXXXXXXXX
 7  .87                        +          IXXXXXXXXXXXX+XXXXXXXXX
 8  .86                       +           IXXXXXXXXXXXXX+XXXXXXX
 9  .84                      +            IXXXXXXXXXXXXXX+XXXXXX
10  .83                     +             IXXXXXXXXXXXXXXX+XXXXX
11  .81                    +              IXXXXXXXXXXXXXXX+XXXX
12  .79                   +               IXXXXXXXXXXXXXXXX+XXX
```

PARTIAL AUTOCORRELATIONS

1–12	.95	.37	.17	.00	–.07	.12	.01	–.04	–.08	–.02	–.01	–.09
ST.E.	.08	.08	.08	.08	.08	.08	.08	.08	.08	.08	.08	.08

```
        -1.0   -.8   -.6   -.4   -.2    .0    .2    .4    .6    .8   1.0
        +----+----+----+----+----+----+----+----+----+----+----+
                                          I
 1   .95                              +    IXXX+XXXXXXXXXXXXXXXXXXXX
 2   .37                              +    IXXX+XXXXX
 3   .17                              +    IXXXX
 4   .00                              +    I    +
 5  -.07                              +   XXI   +
 6   .12                              +    IXXX+
 7   .01                              +    I    +
 8  -.04                              +    XI   +
 9  -.08                              +   XXI   +
10  -.02                              +    I    +
11  -.01                              +    I    +
12  -.09                              +   XXI   +
```

3.4. EXAMPLES

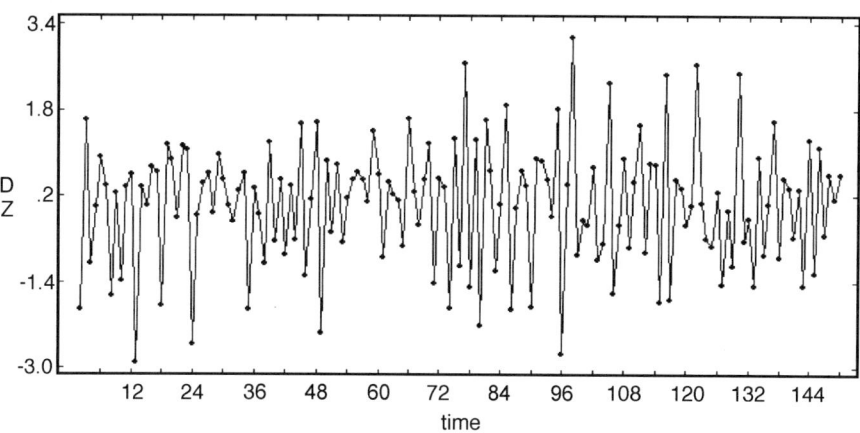

FIGURE 3.4 First difference of the generated (0,1,1) series.

We see that the sample autocorrelations stay persistently high and close to one. This type of behavior suggests that the series is nonstationary, calling for differencing the data.

Figure 3.4 shows the differenced series and stationarity is clearly seen. The SACF and SPACF of the differenced series below indicate that an MA(1) model would be adequate so that an ARIMA(0,1,1) model would be specified for the original series:

```
AUTOCORRELATIONS
1–12      –.47   .02   .08   .01  –.14   .08   .03  –.03  –.04   .03   .07  –.03
ST.E.      .08   .10   .10   .10   .10   .10   .10   .10   .10   .10   .10   .10
Q         33.2  33.2  34.3  34.3  37.4  38.5  38.6  38.7  39.0  39.1  39.9  40.0

          –1.0  –.8   –.6   –.4   –.2   .0    .2    .4    .6    .8   1.0
          +----+----+----+----+----+----+----+----+----+----+
                                        I
  1  –.47              XXXXXXXX+XXXI         +
  2   .02                          +   I     +
  3   .08                          +   IXX   +
  4   .01                          +   I     +
  5  –.14                         +XXXXI     +
  6   .08                          +   IXX   +
  7   .03                          +   IX    +
  8  –.03                          +   XI    +
  9  –.04                          +   XI    +
 10   .03                          +   IX    +
 11   .07                          +   IXX   +
 12  –.03                          +   XI    +
```

PARTIAL AUTOCORRELATIONS

1–12	−.47	−.26	−.04	.06	−.11	−.06	.03	.05	−.03	−.06	.07	.09
ST.E.	.08	.08	.08	.08	.08	.08	.08	.08	.08	.08	.08	.08

```
            −1.0   −.8    −.6    −.4    −.2    .0     .2     .4     .6     .8    1.0
            +----+----+----+----+----+----+----+----+----+----+
                                         I
 1   −.47                    XXXXXXXX+XXXI    +
 2   −.26                         XXX+XXXI    +
 3   −.04                            +   XI   +
 4    .06                            +  IXX   +
 5   −.11                           +XXXI     +
 6   −.06                            + XXI    +
 7    .03                            +   IX   +
 8    .05                            +   IX   +
 9   −.03                            +   XI   +
10   −.06                            +   XI   +
11    .07                            +  IXX   +
12    .09                            +  IXX   +
```

Alternatively, consider the SEACF of the original series:

THE EXTENDED ACF TABLE

(Q-->)	0	1	2	3	4	5	6	7	8	9	10	11	12
(P= 0)	.95	.94	.93	.91	.89	.89	.87	.86	.84	.83	.81	.79	.77
(P= 1)	−.47	.01	.08	.01	−.14	.08	.03	−.03	−.04	.03	.07	−.03	−.05
(P= 2)	−.44	.07	.07	.00	−.16	.08	.09	−.02	−.08	.01	.06	−.02	−.03
(P= 3)	−.14	.22	−.13	.04	−.09	.07	−.02	.01	.01	.00	.01	.03	−.04
(P= 4)	.36	.01	−.22	−.23	−.04	.01	−.02	.00	.00	.01	−.01	.01	−.05
(P= 5)	.46	.09	−.36	−.24	−.12	.00	−.02	.01	.03	−.01	.00	.01	−.05
(P= 6)	−.37	.34	.16	−.08	−.02	.01	−.01	.00	.00	−.00	−.00	.01	−.05

SIMPLIFIED EXTENDED ACF TABLE (5% LEVEL)

(Q-->)	0	1	2	3	4	5	6	7	8	9	10	11	12
(P= 0)	X	X	X	X	X	X	X	X	X	X	X	X	X
(P= 1)	X	O	O	O	O	O	O	O	O	O	O	O	O
(P= 2)	X	O	O	O	O	O	O	O	O	O	O	O	O
(P= 3)	O	O	O	O	O	O	O	O	O	O	O	O	O
(P= 4)	X	O	X	X	O	O	O	O	O	O	O	O	O
(P= 5)	X	O	X	X	O	O	O	O	O	O	O	O	O
(P= 6)	X	X	O	O	O	O	O	O	O	O	O	O	O

The SEACF table suggests a mixed ARMA(1,1) model for the original series.

3.4. EXAMPLES

From the preceding discussion, we are led to consider two tentative models: an ARIMA(0,1,1) and an ARMA(1,1).

$$\text{ARIMA}(0,1,1): \quad (1-B)Z_t = (1-\theta B)a_t,$$
$$\text{ARMA}(1,1): \quad (1-\phi B)Z_t = (1-\theta B)a_t,$$

Details of the parameter estimation process will be given in the next chapter. The results are:

$$\text{ARIMA}(0,1,1): \quad \hat{\theta} = 0.5376(.0690^*); \quad \hat{\sigma}_a = .9938$$
$$\text{ARMA}(1,1): \quad \hat{\phi} = 1.0043(.0031),$$
$$\hat{\theta} = 0.5582(.0690); \quad \hat{\sigma}_a = .9878$$

where * denotes estimated standard error. The fits are almost identical. In particular, for the ARMA(1,1) model the estimate $\hat{\phi}$ is 1.0043, which is very close to $\phi = 1$ corresponding to differencing the data. Thus, the use of SEACF eliminates the need to make decision on differencing the series at the model specification stage.

Example 3.4: Series A of Box, Jenkins, and Reinsel. We consider here series A of Box et al. (1994), which consists of 197 two-hour concentration readings of a chemical process. A time series plot is shown in Figure 3.5.

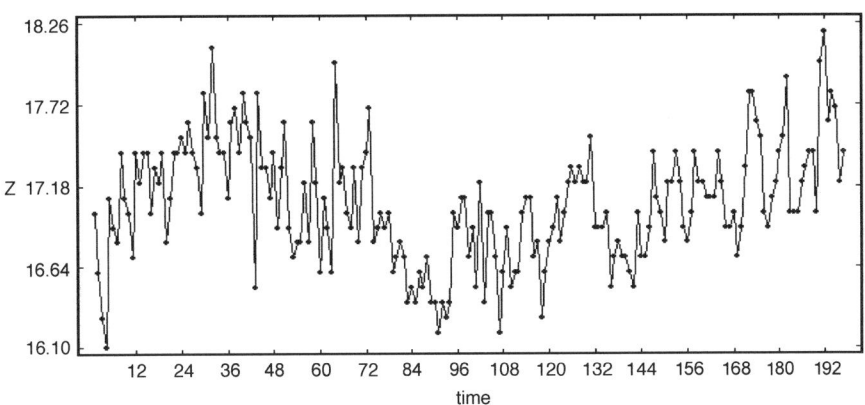

FIGURE 3.5 Series A: 2-h concentration readings of a chemical process.

For tentative specification, we turn to SACF, SPACF, and SEACF:

AUTOCORRELATIONS

1–12	.57	.50	.40	.36	.33	.35	.39	.32	.30	.25	.19	.16
ST.E.	.07	.09	.10	.11	.12	.12	.13	.13	.14	.14	.14	.14
Q	65.0	114	146	172	194	219	251	272	291	305	312	318

```
             –1.0  –.8  –.6  –.4  –.2   .0   .2   .4   .6   .8  1.0
             +----+----+----+----+----+----+----+----+----+----+
                                     I
  1   .57                            +   IXX+XXXXXXXXXXX
  2   .50                            +   IXXX+XXXXXXXX
  3   .40                          +     IXXXX+XXXXX
  4   .36                          +     IXXXX+XXXX
  5   .33                        +       IXXXXX+XX
  6   .35                        +       IXXXXX+XXX
  7   .39                        +       IXXXXX+XXXX
  8   .32                      +         IXXXXXX+X
  9   .30                      +         IXXXXXX+X
 10   .25                      +         IXXXXXXX+
 11   .19                      +         IXXXXX  +
 12   .16                      +         IXXXX   +
```

PARTIAL AUTOCORRELATIONS

1–12	.57	.25	.07	.07	.07	.12	.16	–.03	.01	–.02	–.07	–.02
ST.E.	.07	.07	.07	.07	.07	.07	.07	.07	.07	.07	.07	.07

```
             –1.0  –.8  –.6  –.4  –.2   .0   .2   .4   .6   .8  1.0
             +----+----+----+----+----+----+----+----+----+----+
                                     I
  1   .57                            +   IXX+XXXXXXXXXXX

  2   .25                            +   IXX+XXX
  3   .07                            +   IXX+
  4   .07                            +   IXX+
  5   .07                            +   IXX+
  6   .12                            +   IXXX
  7   .16                            +   IXX+X
  8  –.03                          + XI  +
  9   .01                          + I   +
 10  –.02                          + I   +
 11  –.07                         +XXI   +
 12  –.02                          + XI  +
```

3.4. EXAMPLES

THE EXTENDED ACF TABLE

(Q-->)	0	1	2	3	4	5	6	7	8	9	10	11	12
(P= 0)	.57	.50	.40	.36	.33	.35	.39	.32	.30	.25	.19	.16	.19
(P= 1)	−.39	.04	−.06	−.01	−.07	−.01	.16	−.07	.04	.04	−.04	−.06	−.00
(P= 2)	−.29	−.27	−.04	.01	−.05	−.01	.17	.03	.04	.07	−.02	−.05	−.00
(P= 3)	−.50	−.01	.09	−.01	−.01	−.03	.16	−.03	.11	−.02	−.01	.01	−.06
(P= 4)	−.48	−.02	.08	−.02	−.01	−.04	.14	.03	.09	−.03	−.02	.00	−.08
(P= 5)	−.39	−.41	−.17	.01	−.17	−.02	.10	−.01	.06	.07	−.01	.01	−.06
(P= 6)	−.49	.15	−.18	−.00	−.26	−.06	.09	−.10	.05	.02	−.02	−.03	−.05

SIMPLIFIED EXTENDED ACF TABLE (5% LEVEL)

(Q-->)	0	1	2	3	4	5	6	7	8	9	10	11	12
(P= 0)	X	X	X	X	X	X	X	X	X	O	O	O	O
(P= 1)	X	O	O	O	O	O	O	O	O	O	O	O	O
(P= 2)	X	X	O	O	O	O	X	O	O	O	O	O	O
(P= 3)	X	O	O	O	O	O	X	O	O	O	O	O	O
(P= 4)	X	O	O	O	O	O	O	O	O	O	O	O	O
(P= 5)	X	X	O	O	O	O	O	O	O	O	O	O	O
(P= 6)	X	O	O	O	X	O	O	O	O	O	O	O	O

We see from SACF and SPACF that low-order MA or AR models are not likely. On the other hand, the SEACF pattern suggests that an ARMA(1,1) might be appropriate. Alternatively, the apparent persistence of the sample autocorrelations might suggest differencing the data. The SACF and SPACF of $(1-B)Z_t$ are shown below:

AUTOCORRELATIONS

1–12	−.41	.02	−.07	−.01	−.07	−.02	.15	−.07	.04	.02	−.05	−.06
ST.E.	.07	.08	.08	.08	.08	.08	.08	.08	.08	.08	.08	.09
Q	33.9	34.0	34.9	34.9	35.9	35.9	40.3	41.2	41.5	41.6	42.1	42.9

```
            −1.0  −.8  −.6  −.4  −.2  .0  .2  .4  .6  .8  1.0
            +----+----+----+----+----+----+----+----+----+----+
                                     I
 1   −.41                  XXXXXX+XXXI       +
 2    .02                         +   I      +
 3   −.07                         + XXI      +
 4   −.01                         +   I      +
 5   −.07                         + XXI      +
 6   −.02                         +  XI      +
 7    .15                         +   IXXXX
 8   −.07                         + XXI      +
 9    .04                         +   IX     +
10    .02                         +   IX     +
11   −.05                         +  XI      +
12   −.06                         + XXI      +
```

PARTIAL AUTOCORRELATIONS

1–12	−.41	−.18	−.17	−.14	−.19	−.21	−.00	−.05	−.02	.04	−.01	−.08
ST.E.	.07	.07	.07	.07	.07	.07	.07	.07	.07	.07	.07	.07

```
                  -1.0  -.8   -.6   -.4   -.2    .0    .2    .4    .6    .8   1.0
                   +----+----+----+----+----+----+----+----+----+----+----+
                                              I
   1   -.41                            XXXXXX+XXXI    +
   2   -.18                                 X+XXXI    +
   3   -.17                                  XXXXI    +
   4   -.14                                  +XXXI    +
   5   -.19                                 X+XXXI    +
   6   -.21                                 X+XXXI    +
   7    .00                                     +  I  +
   8   -.05                                     +  XI +
   9   -.02                                     +  I  +
  10    .04                                     +  IX +
  11   -.01                                     +  I  +
  12   -.08                                     + XXI +
```

It is clear that an MA(1) model for the differenced data might be appropriate. Thus, we have two choices: an ARMA(1,1) or an ARIMA(0,1,1) model for Z_t.

The two fitted models are

1. $Z_t - .91 Z_{t-1} = 1.496 + a_t - .59 a_{t-1}$, $\hat{\sigma}_a = .312$ (stationary)

2. $Z_t - Z_{t-1} = a_t - .70 a_{t-1}$, $\hat{\sigma}_a = .317$ (nonstationary)

The fits are very close. It has been shown in the literature that for long horizon forecasts, the stationary model does a better job for this series.

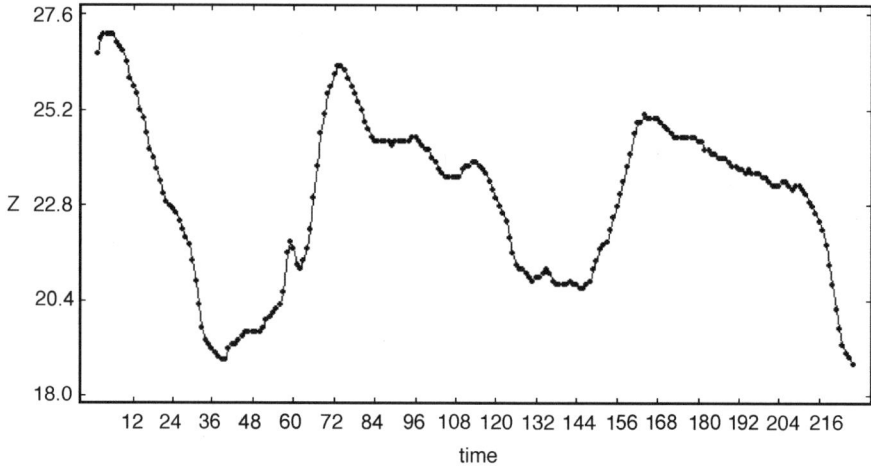

FIGURE 3.6 Series C of BJR: temperature readings.

3.4. EXAMPLES

Example 3.5: Series C of Box, Jenkins, and Reinsel. This series consists of 226 temperature readings. A time series plot is shown in Figure 3.6, exhibiting apparent nonstationary behavior.

We now apply the tentative specification tools:

AUTOCORRELATIONS – Z_t

1–12	.98	.94	.90	.85	.80	.75	.69	.64	.58	.52	.47	.41
ST.E.	.07	.11	.14	.17	.19	.20	.21	.22	.23	.24	.24	.25
Q	219	424	612	781	931	1062	1175	1270	1350	1415	1468	1509

```
              -1.0  -.8  -.6  -.4  -.2   .0   .2   .4   .6   .8  1.0
              +----+----+----+----+----+----+----+----+----+----+
                                         I
   1   .98                               +   IXX+XXXXXXXXXXXXXXXXXXXX
   2   .94                            +      IXXXXX+XXXXXXXXXXXXXXXX
   3   .90                          +        IXXXXXX+XXXXXXXXXXXXXXX
   4   .85                        +          IXXXXXXX+XXXXXXXXXXXXX
   5   .80                      +            IXXXXXXXX+XXXXXXXXXXXX
   6   .75                    +              IXXXXXXXXX+XXXXXXXXX
   7   .69                   +               IXXXXXXXXX+XXXXXXX
   8   .64                 +                 IXXXXXXXXXX+XXXXX
   9   .58               +                   IXXXXXXXXXX+XXX
  10   .52             +                     IXXXXXXXXXXX+X
  11   .47           +                       IXXXXXXXXXXXX
  12   .41          +                        IXXXXXXXXXX  +
```

PARTIAL AUTOCORRELATIONS – Z_t

1–12	.98	–.26	–.16	–.09	–.06	–.05	–.01	–.04	–.02	–.01	–.04	–.04
ST.E.	.07	.07	.07	.07	.07	.07	.07	.07	.07	.07	.07	.07

```
              -1.0  -.8  -.6  -.4  -.2   .0   .2   .4   .6   .8  1.0
              +----+----+----+----+----+----+----+----+----+----+
                                         I
   1   .98                                +   IXX+XXXXXXXXXXXXXXXXXXXX
   2  -.26                       XXXX+XXI    +
   3  -.16                           X+XXI   +
   4  -.09                             +XXI  +
   5  -.06                              + XI +
   6  -.05                              + XI +
   7  -.01                              + I  +
   8  -.04                              + XI +
   9  -.02                              + XI +
  10  -.01                              + I  +
  11  -.04                              + XI +
  12  -.04                              + XI +
```

AUTOCORRELATIONS $- (1 - B)Z_t$

1–12	.81	.65	.53	.44	.38	.32	.26	.19	.14	.14	.10	.09
ST.E.	.07	.10	.12	.13	.13	.14	.14	.14	.15	.15	.15	.15
Q	148	245	309	354	388	411	427	436	440	445	447	449

```
              -1.0  -.8  -.6  -.4  -.2   .0   .2   .4   .6   .8  1.0
              +----+----+----+----+----+----+----+----+----+----+
                                         I
 1    .81                                +   IXX+XXXXXXXXXXXXXXXXX
 2    .65                                +   IXXXX+XXXXXXXXXXX
 3    .53                              +     IXXXXX+XXXXXXX
 4    .44                              +     IXXXXX+XXXXX
 5    .38                          +         IXXXXXX+XX
 6    .32                          +         IXXXXXX+X
 7    .26                          +         IXXXXXXX
 8    .19                          +         IXXXXX  +
 9    .14                          +         IXXX    +
10    .14                          +         IXXXX   +
11    .10                          +         IXX     +
12    .09                          +         IXX     +
```

PARTIAL AUTOCORRELATIONS $- (1-B)Z_t$

1–12	.81	.01	–.01	.05	.03	–.02	–.01	–.08	.02	.12	–.14	.09
ST.E.	.07	.07	.07	.07	.07	.07	.07	.07	.07	.07	.07	.07

```
              -1.0  -.8  -.6  -.4  -.2   .0   .2   .4   .6   .8  1.0
              +----+----+----+----+----+----+----+----+----+----+
                                         I
 1    .81                                +   IXX+XXXXXXXXXXXXXXXXX
 2    .01                                +   I   +
 3   -.01                                +   I   +
 4    .05                                +   IX  +
 5    .03                                +   IX  +
 6   -.02                                +   I   +
 7   -.01                                +   I   +
 8   -.08                               +XXI  +
 9    .02                                +   IX  +
10    .12                                +   IXXX
11   -.14                              XXXI  +
12    .09                                +   IXX+
```

3.4. EXAMPLES

THE EXTENDED ACF TABLE – Z_t

(Q-->)	0	1	2	3	4	5	6	7	8	9	10	11	12
(P= 0)	.98	.94	.90	.85	.80	.75	.69	.64	.58	.52	.47	.41	.36
(P= 1)	.81	.66	.55	.48	.43	.38	.34	.28	.25	.25	.22	.22	.20
(P= 2)	−.04	−.03	−.12	−.06	.02	−.01	.07	−.04	−.12	.13	−.12	.08	−.08
(P= 3)	−.50	.01	−.07	−.11	−.01	−.00	.03	−.03	−.10	.01	−.05	−.03	−.06
(P= 4)	−.25	−.27	−.05	−.11	−.01	.03	.00	−.02	−.09	−.01	−.04	.02	−.06
(P= 5)	−.48	.28	−.29	−.07	.04	−.05	−.00	−.01	−.08	.07	.00	−.01	−.00
(P= 6)	−.08	−.32	.14	.04	−.03	−.04	−.01	−.03	−.08	.07	.00	−.02	.00

SIMPLIFIED EXTENDED ACF TABLE (5% LEVEL)

(Q--->)	0	1	2	3	4	5	6	7	8	9	10	11	12
(P= 0)	X	X	X	X	X	X	X	X	X	X	X	X	X
(P= 1)	X	X	X	X	X	X	X	O	O	O	O	O	O
(P= 2)	O	O	O	O	O	O	O	O	O	O	O	O	O
(P= 3)	X	O	O	O	O	O	O	O	O	O	O	O	O
(P= 4)	X	X	O	O	O	O	O	O	O	O	O	O	O
(P= 5)	X	X	X	O	O	O	O	O	O	O	O	O	O
(P= 6)	O	X	O	O	O	O	O	O	O	O	O	O	O

From SACF and SPACF of Z_t and those of $(1-B)Z_t$, an ARIMA(1,1,0) model is suggested. On the other hand, from the SEACF pattern an ARMA(2,0) model seems appropriate. The fitting results are practically identical.

The last three examples have demonstrated the usefulness of SEACF in tentative model specification. Instead of judgmentally deciding whether to difference a series, SEACF can help specify mixed ARMA models irrespective of whether the AR polynomial has zeros on the unit circle. It simplifies the job of tentative specification, leaving the question of differencing to the estimation stage in model building.

Example 3.6: Monthly Changes in 3-Month T-Bill, 1/83–12/93. As a further example, we consider the series of monthly changes of the 90-day T-bill rates over the period 1/83–12/93. The data are plotted in Figure 3.7.

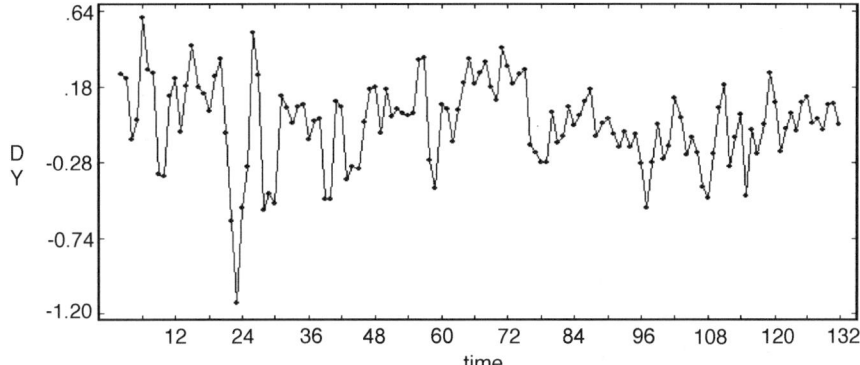

FIGURE 3.7   Monthly changes in 90-day T-bill rates 1/83–12/93.

Except for a substantial dip near the end of 1984, the series appears to be stationary. For tentative specification of a model, we again apply SACF, SPACF, and SEACF to the data:

AUTOCORRELATIONS

```
1-12      .49   .05  -.01   .04   .13   .15   .08   .01   .03  -.02  -.01   .02
ST.E.     .09   .11   .11   .11   .11   .11   .11   .11   .11   .11   .11   .11
Q        32.4  32.7  32.7  32.9  35.2  38.4  39.2  39.2  39.4  39.4  39.4  39.5

             -1.0  -.8  -.6  -.4  -.2   .0   .2   .4   .6   .8  1.0
             +----+----+----+----+----+----+----+----+----+----+
                                      I
  1   .49                              +   IXXX+XXXXXXXX
  2   .05                              +   IX    +
  3  -.01                              +   I     +
  4   .04                              +   IX    +
  5   .13                              +   IXXX  +
  6   .15                              +   IXXXX+
  7   .08                              +   IXX   +
  8   .01                              +   I     +
  9   .03                              +   IX    +
 10  -.02                              +   I     +
 11  -.01                              +   I     +
 12   .02                              +   IX    +
```

PARTIAL AUTOCORRELATIONS

```
1-12      .49  -.25   .11   .01   .13   .03  -.01  -.00   .06  -.11   .07  -.03
ST.E.     .09   .09   .09   .09   .09   .09   .09   .09   .09   .09   .09   .09

             -1.0  -.8  -.6  -.4  -.2   .0   .2   .4   .6   .8  1.0
             +----+----+----+----+----+----+----+----+----+----+
                                      I
  1   .49                              +   IXXX+XXXXXXXX
  2  -.25                          XX+XXXI     +
  3   .11                              +   IXXX+
  4   .01                              +   I    +
  5   .13                              +   IXXX+
  6   .03                              +   IX   +
  7  -.01                              +   I    +
  8   .00                              +   I    +
  9   .06                              +   IX   +
 10  -.11                             +XXXI    +
 11   .07                              +   IXX  +
 12  -.03                              +   XI   +
```

THE EXTENDED ACF TABLE

(Q-->)	0	1	2	3	4	5	6	7	8	9	10	11	12
(P= 0)	.49	.05	−.01	.04	.13	.15	.08	.01	.03	−.02	−.01	.02	−.02
(P= 1)	.43	.14	.01	.01	.08	.11	.08	−.02	.02	−.04	−.01	.00	.00
(P= 2)	.37	.13	−.04	.02	.05	.06	.07	.00	.03	.00	−.01	.00	−.00
(P= 3)	−.10	−.14	.13	−.20	.00	−.01	.07	−.00	.03	.01	−.01	−.00	.00
(P= 4)	−.08	−.48	.32	−.13	−.03	−.00	.08	.02	.02	.01	.00	−.00	.00
(P= 5)	−.19	.05	.08	−.13	−.02	−.01	.07	−.07	−.02	.01	.01	−.02	.00
(P= 6)	.07	.05	.01	−.06	.16	.03	.05	−.03	−.01	.01	.01	−.01	.00

SIMPLIFIED EXTENDED ACF TABLE (5% LEVEL)

(Q-->)	0	1	2	3	4	5	6	7	8	9	10	11	12
(P= 0)	X	O	O	O	O	O	O	O	O	O	O	O	O
(P= 1)	X	O	O	O	O	O	O	O	O	O	O	O	O
(P= 2)	X	O	O	O	O	O	O	O	O	O	O	O	O
(P= 3)	O	O	O	O	O	O	O	O	O	O	O	O	O
(P= 4)	O	X	X	O	O	O	O	O	O	O	O	O	O
(P= 5)	X	O	O	O	O	O	O	O	O	O	O	O	O
(P= 6)	O	O	O	O	O	O	O	O	O	O	O	O	O

Both the SACF and SEACF suggest that an MA(1) model might be adequate for the series.

REFERENCES

Bartlett, M. S. (1946). On the theoretical specification of sampling properties of autocorrelated time series. *J. Roy. Stat. Soc.* **8**, 27–41.

Box, G. E. P. and Jenkins, G. M. (1976). *Time Series Analysis, Forecasting and Control.* Holden-Day, San Francisco.

Box, G. E. P., Jenkins, G. M., and Reinsel, G. C. (1994). *Time Series Analysis, Forecasting and Control.* 3rd Edition. Prentice-Hall, London.

Box, G. E. P. and Tiao, G. C. (1975). Intervention analysis with application to economic and environmental problems. *J. Am. Stat. Assoc.* **70**, 70–79.

Gray, H. L., Kelly, G. D., and McIntire, D. D. (1978). A new approach to ARMA modeling. *Commun. Stat.* **B7**, 1–77.

Slutsky, E. (1937). The summation of random causes as the source of cyclic processes. *Econometrica* **5**, 105–146.

Tiao, G. C. and Tsay, R. S. (1983). Consistency properties of least squares; estimates of autoregressive parameters in ARMA models. *Ann. Stat.* **11**, 856–871.

Tsay, R. S. and Tiao, G. C. (1984). Consistent estimates of autoregressive parameters and extended sample autocorrelation function for stationary and nonstationary ARMA models. *J. Am. Stat. Assoc.* **79**, 84–96.

Woodward, W. A. and Gray, H. L. (1981). On the relationship between the S-Array and the Box-Jenkins method of ARMA model identification. *J. Am. Stat. Assoc.* **79**, 579–587.

Yule, G. U. (1927). On a method of investigating periodicities in disturbed series with special reference to Wolfer's Sunspot numbers. *Phil. Trans. Roy. Soc. Lond.* **A226**, 267–298.

CHAPTER 4

Model Fitting and Checking, and the Kalman Filter

G. Tunnicliffe Wilson
Lancaster University

4.1. PREDICTION ERROR AND THE ESTIMATION CRITERION

The estimation of the parameters of the time series models introduced in the previous chapter could be considered to be just a technical matter carried out by the computer. The aim of this chapter is to explain the criteria and methods by which parameter estimates are obtained. Although some of the explanation is technical, the intention is that it should have practical value through enabling the reader to interpret and use the results of estimation software intelligently.

It is true that the more important tasks to be carried out by the modeler, which require understanding of the models and data, are model selection (identification) and checking. However, it is also important to understand

- The model estimation criterion
- What features of the data it captures
- Whether the fitted model has those properties considered important in the identification stage

This last point is important, firstly because model identification is tentative and the chosen model may or may not have the capability of representing the data well. Moreover, the estimation method is effectively one of nonlinear least squares requiring iterative steps. As with all such methods parameter estimation may fail to provide good estimates, even though the model is appropriate for the data. Such failure should be recognized from the results of the estimation. It can usually be avoided by providing

---

A Course in Time Series Analysis, Edited by Daniel Peña, George C. Tiao, and Ruey S. Tsay.
ISBN 0-471-36164-X. © 2001 John Wiley & Sons, Inc.

4.1. PREDICTION ERROR AND THE ESTIMATION CRITERION 87

initial estimates determined by some simple and reliable scheme. The model selection methods described in Chapters 3, 5, and 8 generally involve methods of obtaining preliminary estimates, so this chapter will deal only briefly with this aspect.

Model estimation, which is efficient in the statistical sense of making best use of the information in the data, is based on assumptions about the distributional properties of the data and makes use of standard statistical inference procedures; the important ones are Bayes and likelihood inference. The practical results of using either of these two procedures are similar and effectively lead to the following scheme:

- Apply the model to predicting successive values of the recorded time series data.
- Choose the parameters that minimize the sum of squares of the resulting *one–step-ahead* prediction errors.

The question of whether *multi-step-ahead* errors would be better used in the sum-of-squares criterion has been well studied. Simple examples show that the answer may be *yes*, if the model does not perfectly describe the data and multistep prediction is the object of the modeling. Assuming that the data *do* arise from the chosen model, one-step-ahead errors are best used *even* if multi-step prediction *is* the object of the modeling.

The models we consider are all members of the class of general linear models described by (2.4) (of Chapter 2). The prediction errors we use in the sum of squares would then be the innovations a_t of that model except that not all past values are known because of the finite length of the observed time series data. This *end effect* is generally handled in one of two ways:

- Estimation of series values previous to the observed data
- Use of errors of predictions made using *only* previous observed data

When properly computed, that is, without further approximations, the likelihoods calculated from these two approaches are identical, although there will be a *transient* discrepancy between the estimated errors for the early part of the data. Because different software may use different methods, it may be important to understand how and why the results from one package may differ from another.

Chapter 7 of Box and Jenkins (1976) covers most aspects of the first approach. It provides a detailed description of the *backforecasting* scheme that they devised to estimate the series values previous to the observed data. That scheme was important to the early widespread application of ARIMA models and is still widely used. In the appendix to that chapter they also present methods for directly calculating the exact likelihood function and extensions of these methods are used increasingly. The following presentation concentrates on these exact methods.

The simplest distributional assumption to make, and the one that underlies all the estimation methods described in this chapter, is that the series being modeled is Gaussian, that is, the joint distribution of any sample is multivariate normal. Equivalently, and more usefully, the errors from the linear prediction of each term on previous terms are independent normal. Estimation methods based on this assumption have good properties even if the assumption is not a perfect description of the data. Provided the errors are independent with zero mean, zero skewness and finite fourth moment it is safe to rely on the "asymptotic" standard errors provided by estimation packages.

Before estimation we assume that any transformation, differencing or trend correction, including correction for a constant mean, has been carried out so that the resulting series w_t is stationary with zero mean. The observed sample of this series is assumed to be from a multivariate normal distribution whose covariance structure is specified by the autocovariances implied by the model. Placing the observations in a column vector $(w_1, w_2, \ldots, w_n)'$, the covariance structure of this sample is described by the symmetric $n \times n$ matrix V_n with elements

$$V_{i,j} = \text{Cov}(w_i, w_j) = \gamma_{i-j} \tag{4.1}$$

The likelihood of the observations is then derived from the joint pdf (probability density function), which apart from a constant factor is

$$f(w_1, w_2, \ldots, w_n) = |V_n|^{-(1/2)} \exp\{-\tfrac{1}{2} w' V_n^{-1} w\} \tag{4.2}$$

where $|V_n|$ is the determinant of V_n. For ARMA models the innovation variance σ_a^2 is a natural scale parameter for V_n; thus we write

$$V_n = \sigma_a^2 M_n \tag{4.3}$$

where M_n depends only on the ARMA model parameters $\beta = (\phi_1, \phi_2, \ldots, \phi_p, \theta_1, \theta_2, \ldots, \theta_q)$. Then the log likelihood is

$$-\tfrac{1}{2}\left\{\log|M_n| + n \log \sigma_a^2 + \frac{S}{\sigma_a^2}\right\} \tag{4.4}$$

where we have replaced the quadratic form $w' M_n^{-1} w$ by S in recognition of the fact that, as we show later, it can be expressed as a sum of squares of prediction errors, although possibly with a small correction for the end effect.

The value of this is that we can "concentrate out" the scale parameter σ_a^2, specifically, maximize the log likelihood with respect to σ_a^2. This is done by setting $\sigma_a^2 = \hat{\sigma}_a^2 = S/n$. A similar result is obtained using Bayes procedures by integrating out σ^2 in the posterior density for the model parameters. Omitting additive constants involving n and the factor $-\tfrac{1}{2}$, we obtain the conventional criterion, minus twice the concentrated likelihood:

$$-2L(\beta) = n \log\{|M_n|^{(1/n)} S\}. \tag{4.5}$$

Maximizing the likelihood with respect to the remaining parameters β is therefore equivalent to minimizing either this quantity or, more simply, $|M_n|^{(1/n)} S$. The factor

4.1. PREDICTION ERROR AND THE ESTIMATION CRITERION

$|M_n|^{(1/n)}$ is associated with the end effect of estimating series values previous to the observed data, as examples will illustrate. As the series length n increases, this effect becomes relatively small and $|M_n|^{(1/n)}$ in fact tends to 1 for all values of the parameters as n increases. In large samples it could therefore be omitted, but experience suggests that it is best retained in the criterion. We shall show in examples how, together with S, it is calculated with numerical efficiency, without explicitly inverting V_n or evaluating its determinant.

After substituting the parameter estimates, $-2L(\hat{\beta})$ is useful as a criterion for comparing different models by methods considered in the next section. The inverse Hessian of $-L(\beta)$ at $\hat{\beta}$ provides asymptotic standard errors of $\hat{\beta}$. For a pair of nested models the difference in $-2L$ may be used as a statistic to test the null hypothesis that the smaller model is adequate. The statistic is referred to its null chi-squared distribution with degrees of freedom equal to the difference in the number of parameters.

To illustrate these points, there follows an extract of the results of fitting an ARMA(1,1) model to the time series of extruded plastic width measurements illustrated in Figure 2.12 (of Chapter 2). This was produced by the GENSTAT package (Payne 1993). The quantity called the *deviance* is $|M_n|^{(1/n)}S$; it is this to which the nonlinear least-squares procedure is actually applied.

Convergence Monitoring

Cycle	Deviance	Current Parameters		
1	910763815	0.	0.	0.
2	96815440.	806.00	0.24723	−0.24723
3	15775966.	1379.0	0.83408	0.38535
4	13882433.	1550.7	0.95365	0.69431
5	13860649.	1594.4	0.94433	0.67700

Convergence at cycle 5:

Parameter Estimates

Model	Seasonal Period	Difference Order	Delay	Parameter	Lag	Ref	Estimate	SE	t
Noise	1	0	—	Constant	—	1	1590.5	58.6	27.14
				Phi (AR)	1	2	0.9434	0.0243	38.90
				Theta (MA)	1	3	0.6759	0.0536	12.61

Correlations

1	1.000		
2	0.025	1.000	
3	0.008	0.674	1.000
	1	2	3

When an AR(1) model was fitted to the same series, the deviance value was 16373064. The series length is $n = 351$, so the test statistic for comparing these nested models is

$$351 \log \left(\frac{16373064}{13860649} \right) = 58.47.$$

There is no doubt about the significance of this statistic referred to the χ_1^2 distribution. The t value of 12.61 for the moving-average parameter also gives a clear indication of this, but the likelihood ratio test is held to have more reliable properties. The standard error of the parameter can be used to supply approximate confidence limits.

In this example the initial parameter values were zero, the usual values when no attempt is made to supply helpful initial estimates. At this point the first derivatives of the deviance with respect to the autoregressive parameter ϕ and the moving-average parameter θ are of opposite sign and the approximation used for the Hessian is singular. If one or the other parameter alone were to be estimated, the results after one step would be approximately $\phi = r_1$ or $\theta = -r_1$, where r_1 is the first lag sample acf (autocorrelation function). When both parameters are included, the second derivative approximation is singular at this point and the nonlinear least-squares procedure generates a constrained step to the point $\phi_1 = \frac{1}{2}r_1$ and $\theta = -\frac{1}{2}r_1$. From that point both parameters increase toward their final values, except for a slight overshoot toward the end.

This is discussed to illustrate some of the difficulties that arise particularly in the estimation of models with both autoregressive and moving-average terms. Convergence may fail in the sense that the iterates reach a poor local optimum rather than the best overall parameter estimates. Note also the fairly high correlation between the ARIMA parameter estimates. This is another feature of time series model estimation; the regression "design" is generated by the model itself. High correlations between parameter estimates are often unavoidable and should not necessarily be taken as evidence of model overparameterization, although when that does occur, one expects to find the same symptoms as of multicollinearity in linear regression. Then at least one of the model orders p or q should be reduced.

4.2. THE LIKELIHOOD OF ARMA MODELS

We use examples to illustrate the various aspects of ARMA model parameter estimation of which it is useful to have some understanding. The emphasis is on the calculation of the sum of squares S and the determinant $|M_n|$, which appears in the criterion (4.5), with a brief outline of how the criterion may be minimized.

We start with the simple AR(1) model:

$$w_t = \phi w_{t-1} + a_t. \tag{4.6}$$

In this case we can calculate the prediction errors a_t for $t = 2, 3, \ldots, n$ from the

4.2. THE LIKELIHOOD OF ARMA MODELS

data w_t as

$$a_t = w_t - \phi w_{t-1}. \tag{4.7}$$

Because a_t, for $t = 2, 3, \ldots, n$, are independent of each other, and of w_1, we can use this relationship to obtain the pdf of the data as

$$f(w_1, w_2, \ldots, w_n) = f(w_1) f(a_2) f(a_3) \cdots f(a_n)$$

$$\propto f(w_1) \sigma_a^{-(n-1)} \exp - \frac{1}{2\sigma_a^2} \left\{ \sum_{t=2}^{n} a_t^2 \right\}. \tag{4.8}$$

It is possible to consider w_1 as a fixed quantity that, considered alone, does not contribute to the information needed to estimate ϕ. This is to *condition* on the value w_1, which is appropriate if it is possible that w_1 is not a typical value from the stationary distribution of the series. Then we obtain the concentrated likelihood as $-2L(\phi) = (n-1)\log(S)$ where

$$S = \sum_{t=2}^{n} a_t^2. \tag{4.9}$$

Minimizing S is then the standard least-squares problem of regressing $w_2, w_3, \ldots, w_n$ on $w_1, w_2, \ldots, w_{n-1}$. This *lagged* regression is a rather obvious way to estimate autoregressive models of all orders.

In order to obtain the likelihood exactly as defined by (4.8), we need to take into account the information from w_1, which has the variance $\sigma_a^2/(1-\phi^2)$ of the stationary series. Then, including the term

$$f(w_1) \propto \left\{ \frac{(1-\phi^2)}{\sigma_a^2} \right\}^{1/2} \exp - \frac{(1-\phi^2)}{2\sigma_a^2} w_1^2 \tag{4.10}$$

in the likelihood (4.8), and writing $a_t = w_t - \phi w_{t-1}$, we obtain the expression in (4.5) as

$$|M_n|^{1/n} S = (1-\phi^2)^{-(1/n)} \left\{ (1-\phi^2) w_1^2 + \sum_{t=2}^{n} (w_t - \phi w_{t-1})^2 \right\}. \tag{4.11}$$

This requires minimization by a nonlinear least-squares procedure, but the departure from linear least squares is small and convergence is usually rapid. It provides an estimate of ϕ that necessarily satisfies the stationarity requirement. The method readily generalizes to the AR(p) model.

We next consider the simple MA(1) model:

$$w_t = a_t - \theta a_{t-1}. \tag{4.12}$$

To put the estimation of θ into context, consider a series of observations $z_1, z_2, \ldots, z_n$ for which the EWMA predictor is appropriate. The first differences $w_2 = z_2 - z_1$, $w_3 = z_3 - z_2, \ldots, w_n = z_n - z_{n-1}$ then follow the MA(1) model above. The parameter θ is the smoothing parameter of the predictor (2.5) (in Chapter 2). The situation is similar to that of the AR(1) model in that we can calculate the prediction errors a_t from the data w_t, for $t = 2, 3, \ldots, n$, but only by assuming a value for a_1. The calculation is recursive:

$$a_t = w_t + \theta a_{t-1} \quad \text{for } t = 2, 3, \ldots, n. \tag{4.13}$$

The pdf of the data *together with* the assumed value of a_1 is

$$f(a_1, w_2, w_3, \ldots, w_n) = f(a_1)f(a_2)f(a_3)f(a_4)\ldots f(a_n)$$

$$\propto \sigma_a^{-n} \exp -\frac{1}{2\sigma_a^2} S \tag{4.14}$$

where

$$S = \left\{ \sum_{t=1}^{n} a_t^2 \right\}. \tag{4.15}$$

A simple strategy for dealing with the fact that a_1 is unknown is to set it to zero. This is equivalent to starting off the EWMA by using z_1 as the predictor for z_2, quite a common practice. If, however, the series has a slowly varying level with a lot of scatter about it, this strategy could distort the predictions for several early timepoints. As soon as a few series values were observed, it would make sense to go back and use a better "predictor" in place of z_1, constructed as an EWMA of these early series values, discounting into the *future*.

This is an application of the *backforecasting* method, which in this case leads to the estimate

$$\hat{a}_1 = \hat{w}_1 = -(\theta w_2 + \theta^2 w_3 + \theta^3 w_4 + \cdots). \tag{4.16}$$

This method should take account of the finite range of future series values, and Box and Jenkins (1976) do this by suggesting a repeated process of forecasting and backforecasting. A direct approach is to construct $\hat{a}_1$ as a least-squares estimate, by minimizing S above wrt a_1. The other terms $a_2, a_3, \ldots a_n$ all depend linearly on a_1 so that

$$\hat{a}_1 = -\frac{(\theta a_2 + \theta^2 a_3 + \theta^3 a_4 + \cdots + \theta^{n-1} a_n)}{K} \tag{4.17}$$

where $a_2, a_3, \ldots$ in this expression are the values obtained by the strategy of setting

4.2. THE LIKELIHOOD OF ARMA MODELS

$a_1 = 0$, and

$$K = 1 + \theta^2 + \theta^4 + \cdots + \theta^{2(n-1)}. \tag{4.18}$$

The sum of squares can be decomposed as $S = K(a_1 - \hat{a}_1)^2 + \hat{S}$ where $\hat{S}$ is the minimum value of S got by using $\hat{a}_1$ to start up the calculations for $a_2, a_3, \ldots$ in (4.13). What this exercise actually gives us is the conditional mean $\hat{a}_1$ and conditional variance σ_a^2/K of a_1 given $w_2, w_3, \ldots, w_n$. Then the pdf (4.14) can be factored:

$$f(a_1 \mid w_2, w_3, \ldots, w_n) f(w_2, w_3, \ldots, w_n) \tag{4.19}$$

where

$$f(a_1 \mid w_2, w_3, \ldots, w_n) \propto \frac{K^{1/2}}{\sigma_a} \exp\left\{ -\frac{K}{2\sigma_a^2}(a_1 - \hat{a}_1)^2 \right\} \tag{4.20}$$

and consequently

$$f(w_2, w_3, \ldots, w_n) \propto K^{-(1/2)} \sigma_a^{-(n-1)} \exp\left\{ -\frac{1}{2\sigma_a^2} \hat{S} \right\}. \tag{4.21}$$

Comparing this with the likelihood expression (4.5), we can now identify the required values of $|M_n|$ with K and S with $\hat{S}$.

The terms a_t that contribute to S in (4.15) do *not* depend linearly on θ, so iterative nonlinear least-squares methods must be used to obtain the maximum likelihood estimates.

To complete our illustrations, we now derive the likelihood for the ARMA(1,1) model:

$$w_t = \phi w_{t-1} + a_t - \theta a_{t-1}. \tag{4.22}$$

Similar principles are applied. The residuals are regenerated, for $t = 2, 3, \ldots, n$ using

$$a_t = w_t - \phi w_{t-1} + \theta a_{t-1} \tag{4.23}$$

using an initial value for a_1. We shall need to use $\mathrm{E}(w_1 \mid a_1) = a_1$ and $\mathrm{Var}(w_1 \mid a_1) = \sigma_a^2/\delta^2$ where $\delta^2 = (1 - \phi^2)/(\phi - \theta)^2$. These are obtained by expressing $w_t = a_t + (\phi w_{t-1} - \theta a_{t-1})$. The pdf of the data *together with* the assumed value of a_1 may then be expressed:

$$f(a_1, w_1, w_2, \ldots, w_n) = f(w_1 \mid a_1) f(a_1) f(a_2) f(a_3) \cdots f(a_n)$$

$$\propto \delta \sigma_a^{-n} \exp\left\{ -\frac{1}{2\sigma_a^2} S \right\} \tag{4.24}$$

where

$$S = \delta^2(w_1 - a_1)^2 + \left\{\sum_{t=1}^{n} a_t^2\right\}. \quad (4.25)$$

Minimizing S with respect to a_1 now gives

$$\hat{a}_1 = \frac{\delta^2 w_1 - \theta a_2 - \theta^2 a_3 - \theta^3 a_4 - \cdots - \theta^{n-1} a_n}{K} \quad (4.26)$$

where $a_2, a_3, \ldots$ in this expression are the values obtained using the initial setting of $a_1 = 0$, and

$$K = \delta^2 + 1 + \theta^2 + \theta^4 + \cdots + \theta^{2(n-1)}. \quad (4.27)$$

Using a similar argument as for the MA(1) case we therefore identify, in the likelihood expression (4.5), $|M_n|$ with K/δ^2 and S with $\hat{S}$ calculated using $\hat{a}_1$. A small technical point is that in the case when $\phi = \theta$, the ratios shown above can be rearranged to avoid division by $\phi - \theta$.

The backforecasting method mentioned in connection with the MA(1) model can be extended to calculate S for all ARMA models and is still widely used. However, it does in theory require convergence of cycles of forecasting and backforecasting to obtain correct results. It will not give the "correct" value of S as in the procedures discussed above if the series is short and the moving-average parameter is close to 1. This problem is more severe for seasonal models.

4.3. LIKELIHOODS CALCULATED USING ORTHOGONAL ERRORS

The methods of the previous chapter were based on the calculation of estimates of the innovations a_t in the ARMA model. These are the errors of prediction using *all* past values. We overcame the problem that we *do not have* all these values by estimating a_t using the series values actually observed.

Another approach is based on the errors of prediction using *only* the available past values. These are called *orthogonal errors*. The first step is to express the joint pdf of the observed series as

$$f(w_1, w_2, \ldots, w_n) = f(w_1) f(w_2 | w_1) f(w_3 | w_2, w_1) \ldots f(w_n | w_1, w_2, \ldots, w_{n-1}). \quad (4.28)$$

Let the error of prediction be

$$w_t - E(w_t | w_1, w_2, \ldots, w_{t-1}) = a_{t,t-1}, \quad (4.29)$$

4.3. LIKELIHOODS CALCULATED USING ORTHOGONAL ERRORS

indicating that it differs from a_t by using only the previous $t-1$ values. Letting its variance be σ_t^2, the joint pdf is proportional to

$$\prod_{t=1}^{n} \frac{1}{\sigma_t} \exp\left\{-\tfrac{1}{2} \frac{a_{t,t-1}^2}{\sigma_t^2}\right\}. \tag{4.30}$$

As t increases, so that more past information is available, $a_{t,t-1} \to a_t$ and $\sigma_t^2 \to \sigma_a^2$. As before, take σ_a^2 as a scale factor, writing $\sigma_t^2 = h_t^2 \sigma_a^2$. The factor h_t^2 is a function of the ARMA parameters $\beta = (\phi, \theta)$ only, and $h_t \to 1$ as t increases. The likelihood expression (4.5) may then be calculated by identifying $|M_n|$ with $\prod h_t^2$ and S with the weighted sum of squares $\sum a_{t,t-1}^2 / h_t^2$.

For an AR(p) model the innovations and orthogonal errors are the same provided $t > p$, so that $h_t = 1$ and

$$a_{t,t-1} = a_t = w_t - \phi_1 w_{t-1} - \phi_2 w_{t-2} - \cdots - \phi_p w_{t-p}. \tag{4.31}$$

For $t = 1, 2, \ldots, p$ formulas for $a_{t,t-1}$ are readily obtained in terms of the partial autocorrelations of the model. For example, if $p = 2$ with AR model parameters ϕ_1 and ϕ_2, we need only the values $\phi_{2,2} = \phi_2$ and $\phi_{1,1} = \rho_1 = \phi_1/(1-\phi_2)$ to derive

$$\begin{aligned} a_{1,0} &= w_1; \quad \text{Var}\, a_{1,0} = \text{Var}\, w_1 = \frac{\sigma_a^2}{(1-\phi_{1,1}^2)(1-\phi_{2,2}^2)} \\ a_{2,1} &= w_2 - \phi_{1,1} w_1; \quad \text{Var}\, a_{2,1} = \frac{\sigma_a^2}{1-\phi_{2,2}^2}. \end{aligned} \tag{4.32}$$

For the general ARMA model two methods of generating orthogonal errors are widely used. We illustrate these for the ARMA(1,1) model.

The first uses Choleski factorization of a band matrix. Orthogonalized residuals can be obtained by Choleski factorization of the covariance matrix V_n of the data, or equivalently the scaled matrix M_n, but in general this requires extensive computation. This is reduced for an MA(q) model because then M_n is zero except for the diagonal and the adjacent q diagonals above and below. For an ARMA(p,q) model it is a simple matter to transform to data with a similar covariance structure. For the ARMA(1,1) case, let $u_t = w_t - \phi w_{t-1}$ for $t = 2, 3, \ldots, n$ and consider the variables $w_1, u_2, u_3, \ldots u_n$. The orthogonal prediction errors from these are the same as those for the original series. Moreover, their covariance matrix is fully specified by the nonzero values, which we express as

$$\begin{aligned} \text{Var}\, w_1 &= \left(1 + \tfrac{(\phi-\theta)^2}{(1-\phi^2)}\right)\sigma_a^2 \\ \text{Var}\, u_t &= (1+\theta^2)\sigma_a^2 \\ \text{Cov}(w_1, u_2) &= \text{Cov}(u_t, u_{t+1}) = -\theta \sigma_a^2 \end{aligned} \tag{4.33}$$

The first orthogonal error is just $a_{1,0} = w_1$ so that $h_1^2 = 1 + (\phi - \theta)^2/(1 - \phi^2)$. The statistical interpretation of Choleski factorization is that each orthogonal error is created by subtracting from u_t its regression on the previous error. This is very similar to the regeneration of residuals for the MA(1) model but with a changing coefficient. To reinforce the similarity with (4.13), we write this as

$$a_{t,t-1} = u_t + \theta_t a_{t-1,t-2} \tag{4.34}$$

where the regression coefficient is actually $-\theta_t$ and is given, using $\text{Cov}(u_t, a_{t-1,t-2}) = \text{Cov}(u_t, u_{t-1}) = -\theta \sigma_a^2$ and $\text{Var}(a_{t-1,t-2}^2) = h_{t-1}^2 \sigma_a^2$, as

$$-\theta_t = \frac{-\theta}{h_{t-1}^2}. \tag{4.35}$$

Further, the new error variance ratio is given by

$$h_t^2 = (1 + \theta^2) - \theta_t^2 h_{t-1}^2. \tag{4.36}$$

The likelihood is thus readily calculated.

The other widely used method uses a state-space representation of the model and the Kalman filter. The results are exactly the same as for the Choleski factorization, but the principles and computations do differ.

The formulation of the state transition and observation equations in the state-space representations differs somewhat from that illustrated in Chapter 2 for the simple structural model. The state-space representation of an ARMA model is a way of rewriting it that still has only one white noise contribution, the innovations a_t, rather than separate white-noise contributions to the state transition and observation equations.

The standard approach is to use for *state variables* at time t quantities that are sufficient to form all future predictions of the series. For an AR(1) model this is just the value w_t; for an AR(2) model w_t and w_{t-1} would suffice and these are known at time t. A general formulation for the ARMA(p, q) model is given in Section 4.7. We consider here the ARMA(1,1) model for which only one state variable x_t is needed at time t. This is the prediction of w_{t+1} made at time t:

$$x_t = \phi w_t - \theta a_t. \tag{4.37}$$

This is *not* known completely at time t from the finite record $w_1, w_2, \ldots, w_t$. The Kalman filter provides a means of calculating the best estimate $\hat{x}_{t,t}$ of x_t given this finite record. The orthogonal residual at time $t+1$ is then given by $a_{t+1,t} = w_{t+1} - \hat{x}_{t,t}$.

Because the state is now designed to predict the *next* observation rather than the current one, the observation equation is different from that used in the illustration at the end of Chapter 2. The procedure for estimation of the states is slightly different, but the basic principles are the same and we illustrate them for the ARMA(1,1) model.

4.3. LIKELIHOODS CALCULATED USING ORTHOGONAL ERRORS

With the state variable defined in (4.37) the observation equation for w_t is then

$$w_t = x_{t-1} + a_t. \tag{4.38}$$

The state transition equation follows by substituting w_t in (4.37):

$$x_t = \phi x_{t-1} + (\phi - \theta)a_t. \tag{4.39}$$

We begin the cycle of updating by assuming that we have an estimate $\hat{x}_{t-1,t-1}$ of x_{t-1}. From this we obtain a prediction of the next state:

$$\hat{x}_{t,t-1} = \phi \hat{x}_{t-1,t-1} \tag{4.40}$$

and a prediction of the next observation:

$$\hat{w}_{t,t-1} = \hat{x}_{t-1,t-1}. \tag{4.41}$$

Let the error in $\hat{x}_{t-1,t-1}$ and its variance be

$$\begin{aligned} e_{t-1,t-1} &= x_{t-1} - \hat{x}_{t-1,t-1} \\ \operatorname{Var} e_{t-1,t-1} &= p_{t-1,t-1}\sigma_a^2. \end{aligned} \tag{4.42}$$

The respective errors in predictions (4.40) and (4.41) and their variances are then, for the state error

$$\begin{aligned} e_{t,t-1} &= x_t - \hat{x}_{t,t-1} = \phi e_{t-1,t-1} + (\phi - \theta)a_t \\ \operatorname{Var} e_{t,t-1} &= p_{t,t-1}\sigma_a^2 = [\phi^2 p_{t-1,t-1} + (\phi - \theta)^2]\sigma_a^2 \end{aligned} \tag{4.43}$$

and for the orthogonal residual

$$\begin{aligned} a_{t,t-1} &= w_t - \hat{w}_{t,t-1} = e_{t-1,t-1} + a_t \\ \operatorname{Var} a_{t,t-1} &= h_t^2 \sigma_a^2 = (p_{t-1,t-1} + 1)\sigma_a^2. \end{aligned} \tag{4.44}$$

The regression of the state error on the orthogonal error again gives the correction to the state estimate as

$$\hat{x}_{t,t} = \hat{x}_{t,t-1} + K_t a_{t,t-1}. \tag{4.45}$$

The regression coefficient K_t is again just the ratio of the covariance between the state error (4.43) and the orthogonal error (4.44) to the variance of the latter:

$$K_t = \frac{\phi p_{t-1,t-1} + \phi - \theta}{h_t^2}. \tag{4.46}$$

The corrected state error $e_{t,t} = e_{t,t-1} - K_t a_{t,t-1}$ has its variance reduced to

$$p_{t,t}\sigma_a^2 = \left(p_{t,t-1} - \frac{K_t^2}{h_t^2}\right)\sigma_a^2. \tag{4.47}$$

This completes the cycle of updating, which provides both the orthogonal residual $a_{t,t-1}$ and its variance factor h_t. The cycle is started by setting $\hat{x}_{0,0} = 0$ and $p_{0,0} = \text{Var } x_0/\sigma_a^2 = (\phi - \theta)^2/(1 - \phi^2)$.

The state-space approach may appear a little more complicated than the Choleski factorization, but that is due in part to the transformation of the series w_t to the MA(1) process u_t before applying Choleski factorization. This same transformation could be used before applying the state-space approach, and the resulting calculations can be seen by setting $\phi = 0$ in the state-space and estimation equations. These are almost identical to the Choleski method for this simple model. In general the state-space approach as presented here has the advantage of being easily adapted to handle missing values of the series w_t. There are also improvements to the Kalman filter that exploit the special nature of the ARMA equations. Both the Choleski factorization and the state-space methods are widely used and methods for the general ARMA model are given by Ansley (1979) and Mélard (1984).

4.4. PROPERTIES OF ESTIMATES AND PROBLEMS IN ESTIMATION

Consider first the estimation of the coefficient ϕ in the stationary AR(1) model

$$w_t = \phi w_{t-1} + a_t \tag{4.48}$$

by simple lagged regression of $w_2, w_3, \ldots, w_n$ on $w_1, w_2, \ldots, w_{n-1}$. The results given by this regression are generally valid; the estimates and the standard errors provided by the ordinary least-squares procedure provide reliable and efficient inference for ϕ. The properties in the case of the general AR(p) model are presented by Anderson (1971). These properties apply in theory for large samples but are reasonable for most applications except when the value of ϕ is close to unity. A problem would be indicated if the usual 95% confidence interval for ϕ included unity. It is worth looking more closely at the estimate and its properties, in order to understand how the situation differs from standard regression. The estimate is

$$\hat{\phi} = \frac{\sum_{t=2}^{n} w_t w_{t-1}}{\sum_{t=2}^{n} w_{t-1}^2}. \tag{4.49}$$

Substituting for $w_t = \phi w_{t-1} + a_t$ gives

$$\hat{\phi} = \phi + \frac{\sum_{t=2}^{n} a_t w_{t-1}}{\sum_{t=2}^{n} w_{t-1}^2}. \tag{4.50}$$

4.4. PROPERTIES OF ESTIMATES AND PROBLEMS IN ESTIMATION 99

If this were standard linear regression, we would treat the values w_{t-1} of the regression (4.49) as fixed quantities, that is, condition, on them, so that the ratio in (4.50) would be a linear combination of the normally distributed model errors $a_2, a_3, \ldots a_n$. Its mean and variance would then be directly evaluated, giving

$$\hat{\phi} \sim \text{normal}\left(\phi, \frac{\sigma_a^2}{\sum_{t=1}^{n} w_{t-1}^2}\right). \tag{4.51}$$

This argument cannot be applied in the context of time series regression, because fixing the values of w_{t-1} would also fix the values of a_t. The sampling properties of the ratio in (4.50) are therefore usually derived by first considering the numerator. Its mean and variance can be readily verified to be respectively 0 and $(n-1)\sigma_a^2\sigma_w^2$ and the central limit theorem extended to establish its large sample normality. The denominator in large samples may be replaced by $(n-1)\sigma_w^2$ with a small relative error. Using the fact that $\sigma_a^2 = (1-\phi^2)\sigma_w^2$ then gives the large sample property:

$$\hat{\phi} \sim \text{normal}\left(\phi, \frac{1-\phi^2}{n-1}\right). \tag{4.52}$$

For most practical purposes the standard linear regression result (4.51) is close enough to (4.52) for it to be generally used.

An important exception arise if an AR(1) model is estimated when in reality the series is not stationary but follows a random walk (i.e., $\phi = 1$). Then the preceding large sample formulas fail. Inference can no longer be made as if the lagged regression had the properties of simple linear regression. In particular, the distribution of the estimate is no longer normal and distributional results developed by Dickey and Fuller (1979) must be used.

The estimation of the parameter θ in the MA(1) model

$$w_t = a_t - \theta a_{t-1} \tag{4.53}$$

is a nonlinear regression problem. From the foregoing derivation of the likelihood, the sum of squares to be minimized is obtained by the recursive regeneration of

$$a_t = w_t + \theta a_{t-1} \quad \text{for } t = 2, 3, \ldots, n. \tag{4.54}$$

We assume for simplicity that a_1 is set to some fixed value. The derivatives a_t^θ of the "residuals" a_t with respect to the parameter θ may also be recursively generated by differentiating (4.54) to give

$$a_t^\theta = a_{t-1} + \theta a_{t-1}^\theta. \tag{4.55}$$

The obvious fact that this derivative depends also on the value of θ demonstrates that the residuals are not linear functions of θ. Note that we may write

$$a_t^\theta = b_{t-1} \tag{4.56}$$

where, in a parallel manner to (4.54),

$$b_t = a_t + \theta b_{t-1}. \tag{4.57}$$

Taking an initial parameter estimate to be θ_0 with corresponding residuals $a_{t,0}$ and derivatives $b_{t-1,0}$, we can produce a local linear approximation

$$a_t \simeq a_{t,0} + (\theta - \theta_0) b_{t-1,0} \tag{4.58}$$

which we write so as to appear like a linear regression for estimating the parameter correction $\delta\theta = \theta - \theta_0$:

$$a_{t,0} = -\delta\theta b_{t-1,0} + a_t \tag{4.59}$$

giving

$$\hat{\delta\theta} = -\frac{\sum_{t=2}^n a_{t,0} b_{t-1,0}}{\sum_{t=2}^n b_{t-1,0}^2}. \tag{4.60}$$

The old parameter is then corrected by this estimate to give the new parameter $\theta_1 = \theta_0 + \hat{\delta\theta}$ and the process repeated to convergence. It is possible for a value of $\hat{\theta}$ to be generated outside the range $-1 < \theta < 1$ in which case only a fraction of the parameter correction is applied. This method appears to be quite reliable even when extended to MA(q) models with high order q.

We gain insight into the properties of the parameter estimate obtained in this way by adding $\theta_0 b_{t-1,0}$ to both sides of (4.59) and using (4.57) to obtain the regression:

$$b_{t,0} = (\theta_0 - \delta\theta) b_{t-1,0} + a_t. \tag{4.61}$$

This is now an autoregressive equation with parameter $\theta_0 - \delta\theta = 2\theta_0 - \theta$. Given *any* value of θ_0 sufficiently close to the true value θ for the linear approximation (4.58) to be good, this tells us that the sampling properties of $2\theta_0 - \hat{\theta}$ are the same as those of an autoregression with the same parameter. In particular, considering θ_0 to be the true (although unknown) parameter value θ leads to the parallel large sample result to (4.51):

$$\hat{\theta} \sim \text{normal}\left(\theta, \frac{1-\theta^2}{n-1}\right). \tag{4.62}$$

A similar approach may be applied in the case of ARMA models. For the ARMA(1,1) model the parameter corrections $-\delta\theta$ and $\delta\phi$ are determined by a regression similar to (4.59), of $a_{t,0}$ on $b_{t-1,0}$, which is generated again by (4.57), and on $c_{t-1,0}$, which is similarly generated by $c_t = w_t + \theta c_{t-1}$. Provided convergence to the global minimum of the sum of squares occurs, the standard errors generated by this regression may be used reliably in large samples as the standard errors of the parameter estimates. For ARMA models, however, such convergence may not take place if the initial parameter values are not close to the global minimum.

One of many possible methods for obtaining preliminary parameter estimates, proposed by Durbin (1960) and developed by Hannan and Rissanen (1982), uses two steps of linear regression. First, a relatively high-order AR model is fitted to the series using simple lagged regression. For the example above of the plastic extrusion width measurements, a lag of 10 is used. The order should be about that at which the sample pacf dies out, so that the residuals $\hat{a}_t$ from this linear regression are reasonably good estimates of the series innovations. Use too low an order of autoregression, and the residuals are not uncorrelated; use too high an order, and they suffer from estimation error. An automatic order selection criterion could be used.

Next, the regression of w_t on $w_{t-1}, w_{t-2}, \ldots, w_{t-p}$ and $-\hat{a}_{t-1}, -\hat{a}_{t-2}, \ldots, -\hat{a}_{t-q}$ is fitted to obtain estimates of the coefficients $\phi_1, \phi_2, \ldots, \phi_p$ and $\theta_1, \theta_2, \ldots, \theta_q$. For the ARMA(1,1) model for the width measurements, this method gave estimates of $\hat{\phi} = 0.9271$ (SE 0.0697) and $\hat{\theta} = 0.6921$ (SE 0.0889) with correlation between the estimates of 0.782. When these were used as starting values in the linearized regressions for the least squares (maximum likelihood) estimates described above, convergence took just two cycles, to the values $\hat{\phi} = 0.9436$ (SE 0.0242) and $\hat{\theta} = 0.6763$ (SE 0.0535) with correlation 0.673. Note that the standard errors and the correlation were reduced for the maximum likelihood estimates.

This is a very useful method, but the reduction of efficiency of the preliminary estimates is greatest when the need for good preliminary estimates is greatest, so that less direct methods based on visual inspection of the sample acf and spectrum can still be valuable. The problems arise when fitting an ARMA model with one or more "signal" components with substantial added noise, and particularly when the signals are cyclical, for example, business cycles. In this case the autoregressive and moving-average parts of the model can have near-canceling factors with roots close to the boundary of stationarity and invertibility. The example used in this chapter is nearly, but not quite, in this category. It is then important to have ways of checking the fit of models fitted by the iterative non-linear least-squares procedure.

4.5. CHECKING THE FITTED MODEL

An estimated model needs to be checked to discern whether it provides a good fit to the data. The estimated model may not fit the data for one of two reasons: because it was not well chosen and *cannot* provide a good fit to the data or because it was poorly estimated, even though it is capable of a good fit to the data.

We consider four aspects of model checking. The first is to check the basic assumption that the residuals, which are estimates of the innovations, show no evidence of autocorrelation. Due consideration must be given to the fact that their estimation will have some effect on their statistical properties even if the estimated model does fit the data well.

This check requires simply that we look at the residuals and their sample statistical properties, just as we inspected the original series. Figure 4.1 shows the plots for the residuals from the ARMA(1,1) model fitted to our example series. These are

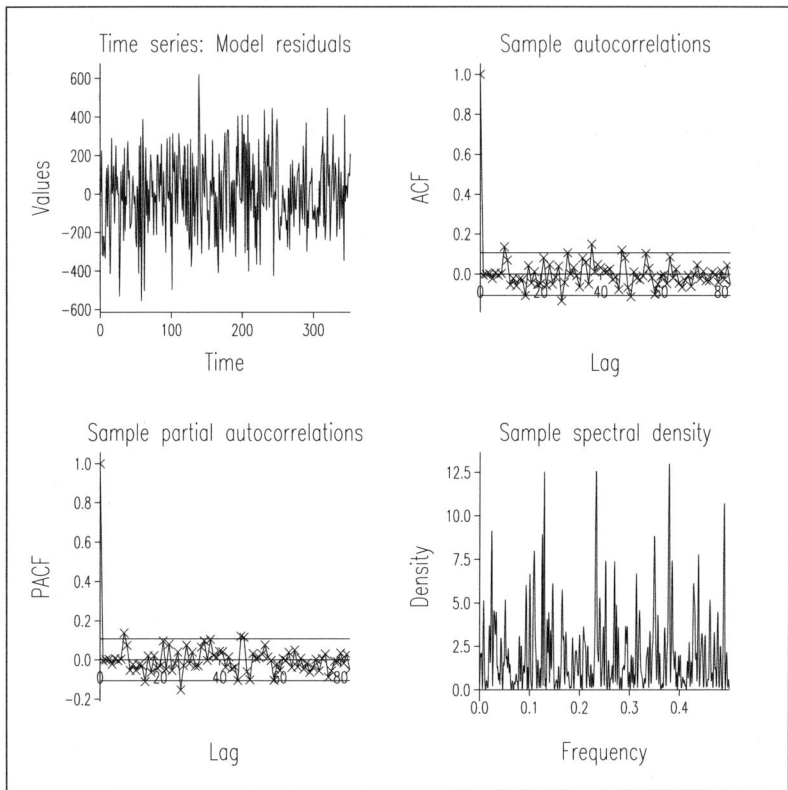

FIGURE 4.1 Analysis of residuals from an ARMA(1,1) model fitted to the series plastic extrusion width measurements.

generally consistent with the residuals being white noise. In particular, the sample autocorrelations r_k lie mostly within the plotted limits about zero. A formal test of whether the series is white noise uses the statistic

$$X = n \sum_{k=1}^{K} r_k^2 \qquad (4.63)$$

where n is the series length. This is based on the large sample properties of $r_k \sim$ normal$[0, (1/n)]$. Under the assumption that the model fits the data the large sample distribution of X is chi-squared on $K - p - q$ degrees of freedom, the reduction in the degrees of freedom allowing for the small extent to which estimating the model "overwhitens" the residuals. A modification to this statistic to improve its properties in smaller samples is presented by Ljung and Box (1978). A choice must be made, regarding the number K of autocorrelations included in the statistic. Evidence for lack of fit generally comes from patterns of larger values of low lag correlations, and choosing K too large could dilute this evidence. When *very* long series are modeled even very slight deficiencies in the fitted model can be revealed by this statistic because

4.5. CHECKING THE FITTED MODEL

a good fit requires the sample autocorrelations to be very small. A judgment must then be made as to whether the lack of fit justifies effort further to improve the model.

For our example, using $K = 40$, the value of X is 50.48, which is exceeded with probability .0847 by a chi-squared variate with 38 degrees of freedom. For comparison, when an AR(2) model is fitted to the same series, the value of X is 73.11 and the probability is .0005. The ARMA(1,1) model therefore appears acceptable and the AR(2) model most definitely not.

The second aspect of model checking is comparison of the properties of the fitted model with those features of the series and its sample statistical properties that had been considered important at the stage of model identification. We focus on the sample spectrum rather than the autocorrelations. Figure 4.2 shows the sample spectrum of our example data with the spectra of two fitted models overlaid. The line which passes through the sample spectra, rising with it sharply at low frequencies, is that of the ARMA(1,1) model, which captures well the low-frequency peak. The other line is the spectrum of the AR(2) model, which fails to fit this peak and also compromises the fit to the higher frequencies. The residual spectrum is in fact the ratio of the sample spectrum of the series to the fitted model spectrum. Inadequacies tend to be more evident when comparing the series and model spectra than when simply inspecting the residual spectrum.

A third aspect of model checking is validation by forecasting out-of-sample values of the series. Forecast construction is described in Section 5.4 (of Chapter 5). A proportion of data at the end of the series is withheld and various forecasts of this data produced using the model fitted to the earlier part. We look for consistency between the forecast limits and the data. This in itself does not demonstrate a good fit, but lack of consistency is clear evidence of model inadequacy—assuming, of course, that the series does have the stationary properties that permit forecasting of future values.

Finally it is suggested that a model that fits well by the foregoing criteria should also be tested by fitting an extended model; one usually in which the autoregressive or moving average order is increased. Formal tests can then be used to check that no significant improvement in fit can be achieved by such an extension.

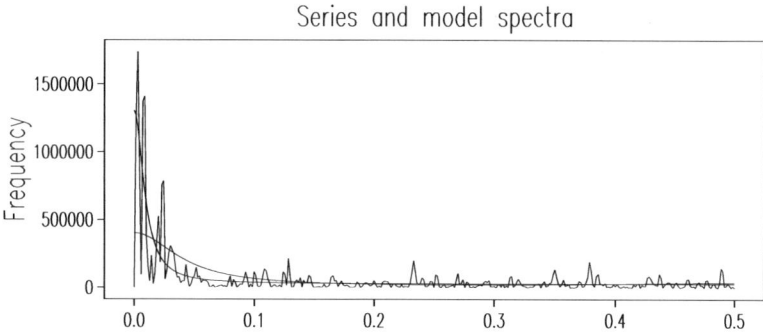

FIGURE 4.2 Sample spectrum of the series of plastic extrusion width measurements with superimposed model spectra.

4.6. ESTIMATION BY FITTING TO THE SAMPLE SPECTRUM

In the previous section we showed the model spectrum set against the sample spectrum of the series. It is possible to obtain estimates of the model parameters by fitting the model spectrum to the sample spectrum. These estimates are, in large samples, equivalent to the maximum likelihood estimates. The values of the sample spectrum $S^*(f)$ at the grid of frequencies $f_j = j/n$ can be considered as equivalent to the time series data for this purpose. We stated in Chapter 2 the distributional property on which estimation may based, that these these values are independent with

$$S^*(f_j) \sim \text{exponential}[S(f_j)]. \tag{4.64}$$

An expression for the model spectrum $S(f)$ in terms of the ARMA model parameters is required. The ARMA(1,1) example in Figure 4.2 illustrates this. Using the backward shift operator the model is written

$$w_t = \frac{(1 - \phi B)}{(1 - \theta B)} a_t = \psi(B) a_t. \tag{4.65}$$

The spectrum may now be obtained from the expression given in (2.46), rewritten in the form

$$S(f) = \psi(B)\psi(B^{-1}) 2\sigma_a^2 \tag{4.66}$$

with the value of B set to $\exp(i 2\pi f)$. Then

$$S(f) = \frac{(1 - \theta B)(1 - \theta B^{-1})}{(1 - \phi B)(1 - \phi B^{-1})} 2\sigma_a^2 = \frac{1 + \theta^2 - \theta(B + B^{-1})}{1 + \phi^2 - \phi(B + B^{-1})} 2\sigma_a^2. \tag{4.67}$$

This can be expressed using $B + B^{-1} = 2\cos(2\pi f)$ in either of the two equivalent forms:

$$S(f) = \frac{M + N \cos 2\pi f}{1 + P \cos 2\pi f} = \frac{Q}{1 + P \cos 2\pi f} + R. \tag{4.68}$$

In the case that $Q > 0$ and $R > 0$ the second of these forms corresponds to the interpretation of the series as the sum of an AR(1) signal and an independent white noise component.

Estimation is then a matter of fitting the curve $S(f)$ to the data $S^*(f)$. Suppose that the value of ϕ, and hence P, is specified; then (4.68) is a linear spectrum model with coefficients Q and R for components $S_1(f) = 1/(1 + P \cos 2\pi f)$ and $S_2(f) = 1$. Because the distribution of the data $S^*(f)$ about the mean $S(f)$ is exponential, the methods of generalized linear modeling (McCullagh and Nelder 1983) can be used to obtain the maximum likelihood fit. This is iterative. Supposing that the best estimate of the fitted function at any stage is $\hat{S}(f)$, a regression of the scaled response $S^*(f)/\hat{S}(f)$ on the scaled regressors $S_1(f)/\hat{S}(f)$ and $S_2(f)/\hat{S}(f)$ is used to determine new values of the coefficients $\hat{Q}$ and $\hat{R}$ and the new $\hat{S}(f)$. Carrying out this procedure to convergence gives a fit identical to that shown in Figure 4.2 for the ARMA(1,1) model, if the MLE estimate of ϕ is used.

The class of ARIMA models having spectra that can be expressed as linear combinations of known components includes many frequently used models. For these models estimation using the spectrum is particularly useful. There is no difficulty, in principle, in extending the abovementioned method to fit all the parameters M, N, and P in the AR(1) model spectrum as expressed in the first part of (4.68). That expression illustrates the general point, that ARIMA model spectra are ratios of polynomials in $\cos(2\pi f)$, a fact that explains the flexibility of the ARIMA model for fitting a wide range of shapes of spectra. It also gives insight into the problem of estimation when the orders of both autoregressive and moving-average parts are large. Iterative methods are required. If the initial parameter estimates are poor so that the initial spectrum poorly matches the shape of the sample spectrum, the estimation procedure will find it difficult to "latch on" to the best parameters. Using the spectrum for model estimation does have the advantage of all curve-fitting methods, that the goodness of fit can be appreciated visually.

4.7. ESTIMATION OF STRUCTURAL MODELS BY THE KALMAN FILTER

In the final section of Chapter 2 the Kalman filter was used to estimate the states for two examples of structural models, and its derivation was given for a simple model. We now present the Kalman filter equations for state estimation and prediction of the general state-space model. These are required for application to more general structural models such as the seasonal example in the same section. A very important point is that the prediction errors (innovations) from the Kalman filter can again be used to form the likelihood for estimation of free model parameters, such as the four variances in that example, using (4.30). The section concludes by explaining some of the important relationships between ARIMA and state-space models.

In the general state-space model the state and observation equations are assumed to be linear with independent Gaussian disturbance or errors. We can extend the models given in the last section of Chapter 2 to allow time-varying coefficients. Following a slightly different convention for time indexing, we write the state equation as

$$x_{t+1} = A_t x_t + \alpha_t \qquad (4.69)$$

where A_t are known, square, *transition* matrices and α_t is an independent sequence of zero mean multivariate normal variables:

$$\alpha_t \sim \text{MN}(0, W_t). \qquad (4.70)$$

The covariance matrix W_t may be singular. This happens when α_t is specified as a combination of a smaller number of disturbance terms.

The observation equation we write as

$$z_t = H_t x_t + \epsilon_t \qquad (4.71)$$

where H_t are known *observation* matrices and ϵ_t is also an independent sequence of variables:

$$\epsilon_t \sim \text{MN}(0, V_t). \qquad (4.72)$$

The sequences ϵ_t and α_t are generally assumed to be completely independent of each other.

We start by assuming that we know the conditional distribution of x_t given the collection, Z_t say, of all current and all past observations $z_1, z_2, \ldots, z_t$. This is specified by a conditional mean and variance in the form

$$x_t \sim \text{MN}(m_t, P_t). \qquad (4.73)$$

The Kalman filter is the steps by which we *update* this estimate to the next timepoint. The first step is that of *prediction*. From the state equation (4.69) the conditional distribution of x_{t+1} given Z_t may then be obtained as

$$x_{t+1} \sim \text{MN}(r_{t+1}, Q_{t+1}) \qquad (4.74)$$

where

$$\begin{aligned} \text{E}(x_{t+1} \mid Z_t) &= r_{t+1} = A_t m_t \\ \text{Var}(x_{t+1} \mid Z_t) &= Q_{t+1} = A_t P_t A_t^T + W_t \end{aligned} \qquad (4.75)$$

using the "sandwich rule" of multivariate normal distributions. Similarly, from the observation equation (4.71) at the *next* timepoint $t+1$, we obtain the conditional mean s_{t+1} and variance G_{t+1} of z_{t+1} given Z_t. We also need the covariance matrix C_{t+1} between z_{t+1} and x_{t+1} in their joint conditional distribution:

$$\begin{aligned} \text{E}(z_{t+1} \mid Z_t) &= s_{t+1} = H_{t+1} r_{t+1} \\ \text{Var}(z_{t+1} \mid Z_t) &= G_{t+1} = H_{t+1} Q_{t+1} H_{t+1}^T + V_{t+1} \\ \text{Cov}(z_{t+1}, x_{t+1} \mid Z_t) &= C_{t+1} = H_{t+1} Q_{t+1} \end{aligned} \qquad (4.76)$$

so that the joint distribution of z_{t+1} and x_{t+1} given Z_t is

$$\begin{pmatrix} z_{t+1} \\ x_{t+1} \end{pmatrix} \sim \text{MN} \left[\begin{pmatrix} s_{t+1} \\ r_{t+1} \end{pmatrix}, \begin{pmatrix} G_{t+1} & C_{t+1} \\ C_{t+1}^T & Q_{t+1} \end{pmatrix} \right]. \qquad (4.77)$$

The prediction step is completed. The correction step derives, from this joint distribution, the conditional distribution of x_{t+1} given z_{t+1} and Z_t, that is, given Z_{t+1}, thereby

4.7. ESTIMATION OF STRUCTURAL MODELS BY THE KALMAN FILTER

incorporating into the state estimate the information in the new observation as well as the past. The standard formula for the regression of x_{t+1} on z_{t+1} is used:

$$x_{t+1} \sim MN(m_{t+1}, P_{t+1}) \qquad (4.78)$$

where

$$\begin{aligned} E(x_{t+1} \mid Z_{t+1}) &= m_{t+1} = r_{t+1} + C_{t+1}^T G_{t+1}^{-1}(z_{t+1} - s_{t+1}) = r_{t+1} + K_{t+1}a_{t+1} \\ Var(x_{t+1} \mid Z_{t+1}) &= P_{t+1} = Q_{t+1} - C_{t+1}^T G_{t+1}^{-1} C_{t+1} = Q_{t+1} - K_{t+1} C_{t+1} \end{aligned} \qquad (4.79)$$

and K_{t+1} is known as the *Kalman gain*, which is applied to the prediction error a_{t+1} of z_{t+1}. This completes the recursive cycle of the filter. Other versions of the filter have been developed. In particular, the *square-root filter* is widely used. Rather than updating the variance matrix P_t of the state estimate, it updates its triangular Choleski factor. The advantage is that it avoids the possibility of negative variances that can otherwise occur through rounding errors when some states may be very accurately, or possibly perfectly, estimated from the observations.

The smoothed estimates of the states x_t are their expectations given the whole of the observed series, specifically, Z_n, not just the present and past values Z_t. There are various forms of the smoothing equations by which these may be constructed. The ones presented here do not require the observed series, only the filtered and predicted state estimates m_t and r_{t+1} and their variances P_t and Q_{t+1}, which are saved after applying the Kalman filter. The following equations are applied in *reverse* time order to obtain the mean μ_t and variance Σ_t of x_t conditional on Z_n, the whole set of observations:

$$\begin{aligned} J_t &= P_t A_{t+1}^T Q_{t+1}^{-1} \\ \mu_t &= m_t + J_t(\mu_{t+1} - r_{t+1}) \\ \Sigma_t &= P_t + J_t(\Sigma_{t+1} - Q_{t+1})J_t^T. \end{aligned} \qquad (4.80)$$

Smoothed estimates are valuable in applications such as trend extraction.

There are several other considerations when applying the Kalman filter. Firstly, the state-space model should be miminal. That means that no state-space model with a lower state dimension can provide an exactly equivalent stochastic description of the observed series. Conditions for this are set out in many texts such as Harvey (1990). The initial state distribution also requires specification of its mean and variance and a common solution is to specify a very disperse distribution with a large initial variance for the states.

In many time series applications the coefficients of the state-space model are assumed to be unchanging, rather than time-varying. In that case the Kalman gain matrix K_t typically converges to a fixed value K as time increases. This happens even if the state process is not stationary but is just time-invariant. It was pointed out for the simple example in Chapter 2. It is accompanied by convergence of the state variance matrices Q_t and P_t to values Q and P, although, of course, the estimated state vector does not converge.

Note then that, on convergence, it is possible to express the filter equations as

$$r_{t+1} = Ar_t + AKa_t \\ z_t = Hr_t + a_t \tag{4.81}$$

which have an appearance very similar to that of the original state transition and observation equations. These equations provide an *innovations representation* in which z_t has *exactly* the same stochastic properties as those implied by the original state-space model. It may be used as an equivalent state-space model for the purposes of forecasting. The main distinction is that the states r_t in this representation are *known* constructs of the Kalman filter. They are obtained by using the second equation to obtain the innovation from the new observation as $a_t = z_t - Hr_t$, then the first equation to *update* r_{t+1}. Similar equations arise for models of multivariate time series, and they have a central role in the systems approach to time series modeling presented in Chapters 15 and 16.

For univariate time series a direct equivalence between constant coefficient state-space models and ARIMA models can be derived from the innovation representation (4.81). The state-space model will always have an ARIMA representation in which the maximum of the generalized autoregressive and moving average orders is the state dimension. Similarly, an ARIMA model always has an innovations representation (4.81) where the state dimension is the maximum ARIMA order. An example was given in Section 4.3 for the application of the Kalman filter to ARMA(1,1) model estimation. For the general ARMA(p,q) model with $q \leq p$, the representation is given by taking the state at time t to be the vector of forecasts $\hat{z}_t = (\hat{z}_{t,t-1}, \hat{z}_{t+1,t-1}, \ldots \hat{z}_{t+p-1,t-1})'$ made at time $t-1$ of the values $z_t, z_{t+1}, \ldots z_{t+p-1}$. The observation equation is then simply

$$z_t = \hat{z}_{t,t-1} + a_t = \begin{pmatrix} 1 & 0 & \cdots & 0 \end{pmatrix} \begin{pmatrix} \hat{z}_{t,t-1} \\ \hat{z}_{t+1,t-1} \\ \vdots \\ \hat{z}_{t+p-1,t-1} \end{pmatrix} + a_t. \tag{4.82}$$

The state transition equation is constructed from the updating rules for ARIMA model forecasts which are presented in the next chapter:

$$\begin{pmatrix} \hat{z}_{t+1,t} \\ \hat{z}_{t+2,t} \\ \vdots \\ \hat{z}_{t+p,t} \end{pmatrix} = \begin{pmatrix} 0 & 1 & \cdots & 0 \\ 0 & \ddots & \ddots & 0 \\ 0 & \cdots & 0 & 1 \\ \phi_p & \phi_{p-1} & \cdots & \phi_1 \end{pmatrix} \begin{pmatrix} \hat{z}_{t,t-1} \\ \hat{z}_{t+1,t-1} \\ \vdots \\ \hat{z}_{t+p-1,t-1} \end{pmatrix} + \begin{pmatrix} \psi_1 \\ \psi_2 \\ \vdots \\ \psi_p \end{pmatrix} a_t. \tag{4.83}$$

4.7. ESTIMATION OF STRUCTURAL MODELS BY THE KALMAN FILTER   109

The moving average parameters enter indirectly through the coefficients ψ_k of the model representation $z_t = \psi(B)a_t$.

From the innovations representation (4.81) it is possible to derive the forecasts of z_t for a general state-space model as linear combinations of the state vector at time t:

$$\hat{z}_{t+k,t-1} = HA^k r_t = d_k r_t \tag{4.84}$$

so that the vector of forecast values $\hat{z}_t = (\hat{z}_{t,t-1}, \hat{z}_{t+1,t-1}, \ldots, \hat{z}_{t+p-1,t-1})'$, where p is here the state dimension, may be written

$$\hat{z}_t = D r_t \tag{4.85}$$

in which the rows of D are the vectors d_k. If this relationship is inverted and the substitution $r_t = D^{-1}\hat{z}_t$ is made, the innovations representation (4.81) is transformed to have the appearance of (4.83). The autoregressive coefficients and the coefficients ψ_k in the ARIMA representation of the state-space model are then obtained directly by inspection. The moving average coefficients may be derived from these.

For example, taking the seasonal state-space model of Chapter 2 with

$$A = \begin{pmatrix} 1 & 1 & 0 & 0 & \cdots & 0 & 0 \\ 0 & 1 & 0 & 0 & \cdots & 0 & 0 \\ 0 & 0 & -1 & -1 & \cdots & -1 & -1 \\ 0 & 0 & 1 & 0 & \cdots & 0 & 0 \\ 0 & 0 & 0 & 1 & \cdots & 0 & 0 \\ \vdots & \vdots & \vdots & \vdots & \ddots & \vdots & \vdots \\ 0 & 0 & 0 & 0 & \cdots & 1 & 0 \end{pmatrix} \tag{4.86}$$

and

$$H = \begin{pmatrix} 1 & 0 & 1 & 0 & \cdots & 0 & 0 \end{pmatrix} \tag{4.87}$$

gives

$$D = \begin{pmatrix} 1 & 0 & 1 & 0 & \cdots & 0 & 0 \\ 1 & 1 & -1 & -1 & \cdots & -1 & -1 \\ 1 & 2 & 0 & 0 & \cdots & 0 & 1 \\ 1 & 3 & 0 & 0 & \cdots & 1 & 0 \\ 1 & 4 & 0 & 1 & \cdots & 0 & 0 \\ \vdots & \vdots & \vdots & \vdots & \vdots & \vdots & \vdots \\ 1 & 12 & 1 & 0 & \cdots & 0 & 0 \end{pmatrix}. \tag{4.88}$$

The first row of D is H, Each following row is obtained by postmultiplying the one above by A.

Substituting $r_t = D^{-1}\hat{z}_t$ in (4.81) gives $\hat{z}_t = DAD^{-1}\hat{z}_{t-1} + DAKa_t$. The matrix DAD^{-1} has the form of the transition matrix in (4.83), where the last row is

$$H = \begin{pmatrix} -1 & 1 & 0 & 0 & \cdots & 0 & 1 \end{pmatrix}. \tag{4.89}$$

This shows that the (generalized) autoregressive operator in the ARIMA representation is

$$1 - B - B^{12} + B^{13} = \nabla \nabla_{12}. \tag{4.90}$$

The elements of DAK would give the values of $\psi_1, \psi_2, \ldots, \psi_{13}$ in that representation. The transition and observation matrices in a state-space model therefore determine the autoregressive part of its ARIMA representation. Its moving-average part depends on the Kalman gain matrix K.

REFERENCES

Anderson, T. W. (1971). *The Statistical Analysis of Time Series*. Wiley, New York.

Ansley, C. F. (1979). An algorithm for the exact likelihood of a mixed autoregressive-moving average process. *Biometrika* **66**, 59–65.

Box, G. E. P. and Jenkins, G. M. (1976). *Time Series Analysis, Forecasting and Control*, revised ed. Holden-Day, Oakland, CA.

Dickey, D. A. and Fuller, W. A. (1979). Distribution of the estimates for autoregressive time series with a unit root. *J. Am. Stat. Assoc.* **74**, 427–431.

Durbin, J. (1960). The fitting of time series models. *Rev. Inte. Stat. Inst.* **28**, 233–244.

Hannan, E. J. and Rissanen, J. (1982). Recursive estimation of mixed autoregressive-moving average order. *Biometrika* **69**, 81–94.

Harvey, A. C. (1990). *Forecasting, Structural Time Series Models and the Kalman Filter.* Cambridge Univ. Press, Cambridge.

Ljung, G. M. and Box, G. E. P. (1978). On a measure of lack of fit in time series models. *Biometrika* **66**, 67–72

McCullagh, P. and Nelder, J. A. (1983). *Generalised Linear Models*. Chapman and Hall, London.

Mélard, G. (1984). A fast algorithm for the exact likelihood of autoregressive-moving average models. *Appl. Stat.* **33**, 104–114.

Payne, R. W. (1993). *Genstat 5 Release 3 Reference Manual*. Clarendon Press, Oxford.

CHAPTER 5

Prediction and Model Selection

Daniel Peña
Universidad Carlos III de Madrid

5.1. INTRODUCTION

This chapter deals with prediction of univariate time series models. We assume that we have observed a sample $(z_1, \ldots, z_T)$ of a time series and we want to find a method to generate predictions of future values of the time series given the observed values. A key question in forecasting, as well as in estimation, is to measure the uncertainty of the predictions. Without this information, the forecasts are not useful because we do not have an idea of their precision. A usual way to express the uncertainty in our forecast is to compute a prediction confidence interval, that is, an interval that will contain the future value we are forecasting with high probability, as .95 or .99. Finally, forecasting is a dynamic task: when new data are available, forecasts need to be updated taking into account the new information.

A linear time series model provides a straightforward way to compute these three components. For nonlinear time series the specification of the forecasting system can be a difficult task, and in this chapter we concentrate on linear time series. The chapter is organized as follows. First the properties of minimum mean-squared error forecasts are analyzed. Second, we discuss the generation of predictions for ARIMA model and the interpretation of the structure of the forecasts. It is shown that the forecast function can be decomposed into components associated with the trend, the seasonality and the transitory component. Third, we show how to build prediction confidence intervals and how to update the forecast when new data are available. Fourth, a general rule is introduced for combining forecasts from several sources. Finally, we present two criteria for automatic model selection that are based on the model forecasting performance: the Akaike AIC and Schwarz's BIC criterion.

A Course in Time Series Analysis, Edited by Daniel Peña, George C. Tiao, and Ruey S. Tsay.
ISBN 0-471-36164-X. © 2001 John Wiley & Sons, Inc.

5.2. PROPERTIES OF MINIMUM MEAN-SQUARE ERROR PREDICTION

5.2.1. Prediction by the conditional expectation

We assume first that we have observed the values $Z_T = (z_1, \ldots, z_T)'$ of a zero mean stationary time series and we want to forecast a future value, z_{T+k}. In order to compare alternative forecasting procedures we need to introduce a criterion of optimally. Obviously we want to have forecast errors as small as possible, and a useful way to establish this condition is by choosing the forecast as the value that minimizes the mean of the squares forecast errors. Forecasts obtained by this criterion are called *minimum mean-square error forecasts* (MMSEF), and they can be computed as follows. Let $g_T(k)$ be the forecast we want to generate, where T is the forecast origin and k the forecast horizon. This forecast must minimize

$$MSE(z_{T+k}, g) = E[z_{T+k} - g_T(k)]^2, \qquad (5.1)$$

where the expected value is taken over the joint distribution of z_{T+k} and Z_T. Using the well-known property of conditional expectations, $E(y) = E_x E_{y/x}(y)$, we can take first the expected value with respect to the distributions of z_{T+k}/Z_T and afterward with respect to the distribution of Z_T. In the first step we consider the sequence Z_T as fixed, and we obtain

$$MSE(z_{t+k} \mid Z_T) = E\left[z_{T+k}^2 \mid Z_T\right] + g_T(k)^2 - 2g_T(k)E[z_{T+k} \mid Z_T]$$

and taking the derivative of this equation with respect to g we obtain

$$g_T(k) = E[z_{T+k} \mid Z_T] = \hat{z}_T(k). \qquad (5.2)$$

This result indicates that, conditioning to the observed sample, the MMSEF is obained by computing the conditional expectation of the random variable we want to forecast given the available information. It is easy to see that this result holds for any sequence Z_T and is, therefore, general. To show this, consider any other predictor, $\hat{z}_T^*(k)$, and let us call, as before, *MSE* to the mean-square forecast error of $\hat{z}_T(k)$ as defined by (5.2), and $MSE^* = E[z_{T+k} - \hat{z}_T^*(k)]^2$ to the mean-square forecast error of $\hat{z}_T^*(k)$. Adding and substracting $\hat{z}_T(k)$ in this last expression we have that

$$MSE^* = MSE + E[\hat{z}_T^*(k) - \hat{z}_T(k)]^2$$

because the double product is zero, as it can be easily seen by taking first the expectation with respect to the distributions of z_{T+k}/Z_T. Then, it is clear that $MSE^* \geq MSE$, and they will be equal if $\hat{z}_T^*(k) = E[z_{T+k} \mid Z_T]$.

5.2.2. Linear predictions

Conditional expectations can be, in same cases, difficult to compute. However, if we restrict our search to forecasting functions that are linear functions of the observations, we can easily obtain the best linear predictor minimizing the MSE. The general equation for a linear predictor is

$$\hat{z}_T(k) = b_{k0}z_T + \cdots + b_{k(T-1)}z_1 = \mathbf{b}_k' Z_T$$

and calling MSEL to the mean square error of a linear forecast

$$MSEL(z_{T+k} \mid Z_T) = E[z_{T+k} - \mathbf{b}_k' Z_T]^2. \tag{5.3}$$

Minimizing this expression with respect to $\mathbf{b}_k$, we have

$$E[(z_{T+k} - \mathbf{b}_k' Z_T) Z_T] = 0$$

which implies that the best linear forecast must be such that the forecast error $z_{T+k} - \hat{z}_T(k)$ is uncorrelated with (orthogonal to) the set of observed variables. This property suggests the interpretations of linear predictors as projections. Calling $\Gamma_T = E(Z_T Z_T')$ to the covariance matrix of the vector Z_T and $\gamma_k = E(z_{T+k} Z_T) = (\gamma(k), \gamma(k+1), \ldots, \gamma(k+T))'$ to the covariance vector between z_{t+k} and Z_T, the coefficients for the best linear predictor will be given by

$$\mathbf{b}_k = \Gamma_T^{-1} \gamma_k. \tag{5.4}$$

This expression assumes that Γ_T is nonsingular, and a sufficient condition for this is that $\gamma(0) > 0$ and $\gamma(h) \to 0$ when $h \to \infty$.

It is clear that

$$MSE(z_{t+k} \mid Z_T) \leq MSEL(z_{t+k} \mid Z_T)$$

and as for gaussian processes the best predictor is always linear, both concepts coincide in normal models.

Although the previous results have been established for stationary process, they also hold for non stationary processes. Note that the proof of (5.2), which says that the prediction that minimizes the MSE given an information set I_t is the expectation of the variable conditional to I_t, requires only that the conditional expectation as well as the conditional variance be finite. Therefore, the result holds for any random variable z_{t+k} such that the second moments conditional to the information set I_t are finite. For nonstationary process (5.4) will not be valid, because the autocovariance function is not defined in this case, although the orthogonality property of the forecast error does hold.

5.3. THE COMPUTATION OF ARIMA FORECASTS

Suppose that we want to forecast a time series $Z_T = (z_1, \ldots, z_T)$ which follows an ARIMA(p, d, q) model. We will work in the general case and so seasonal models are just particular cases with high-order values for the ARIMA parameters. Let us assume first that the parameters of the model are known. Suppose that we have data until time T and we want to compute the one step ahead forecast for z_{T+1}, defined by

$$z_{T+1} = c + \varphi_1 z_T + \cdots + \varphi_h z_{T-h+1} + a_{T+1} - \theta_1 a_T - \cdots - \theta_q a_{T-q+1} \qquad (5.5)$$

where $h = p + d$ and the operator $\varphi(B) = \phi(B)(1-B)^d$ includes the unit roots. The one step ahead forecast from time T will be the conditional expectation of this random variable given the available information. Calling $\hat{z}_T(1) = E[z_{T+1} \mid Z_T]$ to this conditional expectation, we have

$$\hat{z}_T(1) = c + \varphi_1 z_T + \cdots + \varphi_h z_{T-h+1} - \theta_1 a_T - \cdots - \theta_q a_{T-q+1}, \qquad (5.6)$$

because the expected value for the observed sample data or the errors are themselves, and the only unknown random variable in (5.5) is a_{T+1} that has an expected value equal to zero. Note that as we are assuming that the parameters are known, the errors are also known, because we can compute them recursively from the observations given some starting values. This implies that, comparing (5.5) and (5.6)

$$a_{T+1} = z_{T+1} - \hat{z}_T(1)$$

which means that the perturbations a_t can be interpreted as the one step ahead forecast errors of the model.

Let us consider now the *multiple steps ahead forecast*. As before, the MSE forecast of z_{T+k} given the data until time T will be the expectation of this variable conditional to the observed data. We will assume that this conditional expectation exits, and we will call

$$\hat{z}_T(j) = E[z_{T+j} \mid Z_T] \quad j = 1, 2, \ldots, k$$
$$\hat{a}_T(j) = E[a_{T+j} \mid Z_T] \quad j = 1, 2, \ldots, k$$

where T is the origin and j is the horizon of the forecast. The unobserved random variable that we want to forecast, z_{T+k}, is generated by

$$z_{T+k} = c + \varphi_1 z_{T+k-1} + \cdots + \varphi_h z_{T+k-h} + a_{T+k} - \theta_1 a_{T+k-1} - \cdots - \theta_q a_{T+k-q}$$

and taking conditional expectations given Z_T we obtain

$$\hat{z}_T(k) = c + \varphi_1 \hat{z}_T(k-1) + \cdots + \varphi_h \hat{z}_T(k-h)$$
$$- \theta_1 \hat{a}_T(k-1) - \cdots - \theta_q \hat{a}_T(k-q). \qquad (5.7)$$

5.3. THE COMPUTATION OF ARIMA FORECASTS

This expression has two parts. The first one, which depends on the AR coefficients, will determine the form of the long-run forecast. The second one, which depends on the moving-average coefficients, will disappear for $k > q$. Note that for $j > 0$, $\hat{z}_T(-j) = z_{T-j}$, that is, conditional to the sample the expected values of the data that have already been observed are the observed sample data. In the same way $\hat{a}_T(-j) = a_{T-j}$, but $\hat{a}_T(j) = 0$, because the expectation of a future perturbation is zero. Therefore, the forecasts generated from (5.6) for $k > q$ satisfy the equation

$$(1 - \varphi_1 B - \cdots - \varphi_h B^h)\hat{z}_T(k) - c = 0$$

where now the backshift operator B is operating over k, and the origin of the forecast, T, is fix. This equation is called the *eventual forecast function* and defines the long-run forecast generated from the ARIMA model. We will study this function in the next section, but here we illustrate it in two simple but important cases.

1. Suppose the simplest stationary model, the AR(1) model. The one-step-ahead forecast is

$$\hat{z}_T(1) = c + \phi_1 z_T$$

the two-steps-ahead forecast is

$$\hat{z}_T(2) = c + \phi_1 \hat{z}_T(1) = c(1 + \phi_1) + \phi_1^2 z_T$$

and, in the same way, for any $k > 0$ we have

$$\hat{z}_T(k) = c + \phi_1 \hat{z}_T(k-1) = c\left(1 + \phi_1 + \cdots + \phi_1^{k-1}\right) + \phi_1^k z_T.$$

Note that, for large k, as $|\phi_1| < 1$, the term $\phi_1^k z_T$ will go to zero, and the sum of the rest of the terms will go to $c/(1 - \phi_1)$, the mean of the process. We will see in the next section that this result is general, that is, for large k the long-run forecast for any stationary ARMA(p,q) model is the mean of the process, $\mu = c/(1 - \phi_1 - \cdots - \phi_p)$. Note that if $c = 0$, the long-run forecast will go to zero, again the process mean.

2. Now consider a nonstationary process as the random walk. Then the one-step-ahead forecast is

$$\hat{z}_T(1) = c + z_T$$

for the next period

$$\hat{z}_T(2) = c + \hat{z}_T(1) = 2c + z_T$$

and, for any $k > 0$

$$\hat{z}_T(k) = c + \hat{z}_T(k-1) = kc + z_T.$$

We see that all the forecasts are following a straight line with slope c. Also, if $c = 0$, all forecasts are constant and equal to the last observed value. Note the key importance of the constant in the forecast function for nonstationary models. When $d = 1$, the long-run forecast will be a straight line but the constant determines the slope of this line. Also note that the long-run forecasts depend on the last observed points, whereas in the stationary case it is always equal to the mean. In the next section we will show that this property applies to all nonstationary series.

5.4. INTERPRETING THE FORECASTS FROM ARIMA MODELS

5.4.1. Nonseasonal models

Let us start first with the case of nonseasonal models. We have seen in the previous section that the eventual forecast function of a nonseasonal ARIMA model verifies for $k > q$

$$\phi(B)(\nabla^d \hat{z}_T(k) - \mu) = 0 \tag{5.8}$$

where we have introduced the mean of the stationary series, μ instead of the constant $c = \phi(1)\mu$. Remember that B in this equation operates over the horizon k, and the origin T is fixed. As the polynomials $\phi(B)$ and ∇^d do not have roots in common, it was proved by Espasa and Peña (1995) that the general solution of (5.8) can be written, for $k > \max(0, q - d - p)$ as

$$\hat{z}_T(k) = P_T(k) + t_T(k) \tag{5.9}$$

where the first term, called the *permanent component*, is the solution of

$$\nabla^d P_T(k) = \mu, \tag{5.10}$$

and the second term, called the *transitory component*, is the solution of

$$\phi(B) t_T(k) = 0. \tag{5.11}$$

It is straightforward to verify that (5.9) is a solution of (5.8). Let us now analyze the form of each component.

The permanent component

The permanent component is the solution of (5.10) and it will be given by

$$P_T(k) = \beta_0^{(T)} + \beta_1^{(T)} k + \cdots + \beta_d k^d \tag{5.12}$$

where $\beta_d = \mu/d!$ is determined by the mean of the stationary process, whereas the rest of the parameters in this function, $\beta_i^{(T)}$, depend on the initial values and change with the

5.4. INTERPRETING THE FORECASTS FROM ARIMA MODELS

forecast origin. It is straightforward to check that (5.12) verifies the condition (5.10). When $d=0$, the permanent component is $P_T(k) = \mu$, and so it will be a constant equal to μ for all horizons. When $d=1$, the permanent component is $P_T(k) = \beta_0^{(T)} + \mu k$, and we have a deterministic linear trend with slope μ; if $\mu = 0$, then the solution is $P_T(k) = \beta_0^{(T)}$ and the permanent component is just a constant. When $d=2$, the solution is $P_T(k) = \beta_0^{(T)} + \beta_1^{(T)} k + \mu k^2/2$ and the model has a quadratic trend with the leading term determined by the mean of the stationary process μ; if $\mu = 0$, the equation reduces to a linear trend $P_T(k) = \beta_0^{(T)} + \beta_1^{(T)} k$, but now the slope will not be constant and will depend on the origin of the forecast.

It is interesting to compare the straight-line forecasts generated by a $I(1)$ model with a constant term μ and by the model $I(2)$ without constant. Let β_1 be the slope in the straight line generated by the forecasts from the $I(1)$ model. This slope is the mean of the stationary series $w_t = (1-B)z_t = \nabla z_t$ and so it is estimated by

$$\hat{\beta}_1 = \frac{1}{n-1} \sum_{t=2}^{t=n} \nabla z_t = \frac{z_n - z_1}{n-1}$$

which means that the increase in the forecast for one period to the next is the average of the observed growths, ∇z_t, in the sample. Let us compare this forecast with the one generated by a model with two differences and a MA(1) part, the often used ARIMA(0,2,1) model, $\nabla^2 z_T = (1-\theta B) a_T$, that also generates forecasts following a straight line. Then

$$\hat{z}_T(1) = 2z_T - z_{T-1} - \theta a_T = z_T + \hat{\beta}_2$$

where we have called

$$\hat{\beta}_2 = z_T - z_{T-1} - \theta a_T. \quad (5.13)$$

It is easy to see that for any $k > 0$ the forecasts are

$$\hat{z}_T(k) = z_T + k\hat{\beta}_2$$

and therefore they will follow a straight line with slope $\hat{\beta}_2$. Let us analyze how this slope incorporates the sample information. As $a_T = (1-\theta B)^{-1} \nabla^2 z_T$ we obtained from (5.13) that the slope is estimated by

$$\hat{\beta}_2 = \nabla z_T - \theta(1-\theta B)^{-1}(\nabla z_T - \nabla z_{T-1})$$

which can be written as

$$\hat{\beta}_2 = (1-\theta) \sum_{i=0}^{i=T-1} \theta^i \nabla z_{T-i}.$$

This expression shows that the slope is a weighted mean of the observed growths with weights decreasing with the lag. In general, it can be shown that $I(2)$ models compute the slope, β_2, as a weighted average of the observed growth values but given more weight to the last observed growths and less to the most remote ones.

We conclude that although both models, the $I(1)$ model with constant and the $I(2)$ without, generate forecasts by making some weigthed average of past growth, they do so in a different way. The $I(1)$ model with constant makes a simple average; that is, past growths are as relevant as the latest growths to forecast the next growth. The $I(2)$ model makes a weigthed average with weights that decrease exponentially with time, so that past growths have smaller weights than do the latest growths. Note also that forecasts from the $I(2)$ model without constant are adaptive, because they are always a weighted function of all observed growths, whereas those from model $I(1)$ with a constant are not, because the forecast growth is the sample mean, and unless we reestimate this parameter when new observations are available, it will be constant for different forecasts. This is an important advantage of models without constant; they are more adaptive than models that include a constant. A practical implication of this analysis is that, when in doubt, it is better to differentiate in order to have a model without a constant to make the model more robust and flexible. Sánchez and Peña (2000) have presented a rigorous proof of the advantages of overdifferencing in limiting cases.

The two integrated models compared above incorporate the time information in the sample in an intuitively sensible way. This is an important difference with respect to nondynamic models, as linear regression. Suppose, for instance, that we had followed the naive approach of fitting a straight line by least squares to the data. Then we will have again a model that generates forecasts following a straight line as

$$\hat{z}_T(k) = a + b_R(T+k)$$

where b_R is the estimated slope by least squares. In order to compare this model with the previous ones, suppose, to simplify the analysis, that we have five observations, $T = 5$, and let us write $t = (-2, -1, 0, 1, 2)$ so that $\bar{t} = 0$, and the sample points are $(z_{-2}, z_{-1}, z_0, z_1, z_2)$. Then, the slope of the regression model is computed by least squares as

$$b_R = \frac{-2z_{-2} - z_{-1} + z_1 + 2z_2}{10}$$

which can be expressed as

$$b_R = .2(z_{-1} - z_{-2}) + .3(z_0 - z_{-1}) + .3(z_1 - z_0) + .2(z_2 - z_1)$$

that is, the slope is a weigthed average of the observed growths but giving minimum weight to the last observed values. In fact, it can be shown (Peña 1995) that the slope

5.4. INTERPRETING THE FORECASTS FROM ARIMA MODELS

in a regression line can be written as

$$b_R = \sum_{t=2}^{t=T} w_t \nabla z_t$$

where $\sum w_t = 1$ and the weights, w_t, are symmetric and take the minimum value at the beginning and the end of the sample and the maximum value in the middle. This leads to the not very convincing result that if you use a regression model to forecast next year sales using a sample of, for instance, 10 years, you are giving the minimum weight to last year sales and the maximum to the increase 5 years ago. On the other hand, an $I(2)$ model always gives more weight to the last growths showing the advantage of forecasting with time series models over naive deterministic regression ones.

In summary, the long-run forecast from an ARIMA model is the mean if the series is stationary and a polynomial for nonstationary models. In this last case, the leading term of the polynomial is a constant if $\mu \neq 0$, whereas it depends on the forecast origin (and so it is adaptive) if $\mu = 0$.

Note that if $\theta \to 1$, this model will be very close to the previous random walk with drift.

The transitory component
The transitory component is the solution of (5.11) and so it will be given by

$$t_T(k) = \sum_{i=1}^{P} A_i G_i^k$$

where G_i^{-1} are the roots of the autoregressive polynomial and A_i are coefficients depending on the forecast origin. As $|G_i| \leq 1$, this term will disappear for large horizons, which justifies the name of transitory term. Note that if $p > 1$, two complex roots will determine a damped sine wave. Therefore, in the general case, the transitory part will be a combination of exponential terms and sine waves.

For instance, consider the model $(1-\phi B)\nabla z_t = a_t$. Then $G_1 = \phi$ and the forecasts must have the form

$$\hat{z}_T(k) = c_T + A_1 \phi^k$$

where c_T, the constant that appear as the solution of $\nabla P_T(k) = 0$, and A_1, the constant in the transitory equation, must be determined from the initial conditions. The two equations required to determine the two unknowns can be obtained by computing the two first forecasts (one and two steps ahead). Then

$$\hat{z}_T(1) = c_T + A_1 \phi = z_T + \phi(z_T - z_{T-1})$$

and

$$\hat{z}_T(2) = c_T + A_1\phi^2 = z_T + \phi(z_T - z_{T-1}) + \phi^2(z_T - z_{T-1})$$

and the solution of these two equations is

$$c_T = z_T + \frac{\phi(z_T - z_{T-1})}{1 - \phi}$$

and

$$A_1 = -\frac{\phi(z_T - z_{T-1})}{1 - \phi}.$$

These results indicate that the forecasts for $T+1, \ldots, T+k$ are slowly approaching the long-run forecast c_T. Note that as $A_1\phi^k$ goes to zero, the adjustment made by the transitory component on the permanent forecast decreases geometrically. If ϕ is small, the long-run forecast is close z_T, and the adjustment of the transitory components plays a small role. However, if ϕ is close to one, the long-run forecast can be much larger than the last observed value, z_T, and the transitory part is very important to define the way in which the forecast move from z_T to the final forecast value c_T.

5.4.2. Seasonal models

For seasonal processes the forecast will satisfy the equation

$$\Phi(B^s)\phi(B)\left(\nabla_s^D \nabla^d \hat{z}_t(k) - \mu\right) = 0.$$

This equation can also be decomposed into a term associated to the nonstationary part and another linked to the stationary part. However, here it is interesting to decompose further these two terms into the part due to the regular operators and the one due to the seasonal ones. This is especially useful for the permanent component, which now can be split into a term linked to the trend and another linked to the seasonal structure.

Let us assume the usual case of $D=1$. Then, the seasonal difference can be written as

$$(1 - B^s) = (1 + B + B^2 + \cdots + B^{s-1})(1 - B)$$

and calling

$$S_s(B) = 1 + B + \cdots + B^{s-1}$$

5.4. INTERPRETING THE FORECASTS FROM ARIMA MODELS

to the term without the real root equal to one, the forecast equation can be written as

$$\Phi(B^s)\phi(B)(S_s(B)\nabla^{d+1}\hat{z}_T(k) - \mu) = 0, \qquad (5.14)$$

which has the property that all the operators involved do not share roots in common. The solution of this equation for $k > \max(0, q + sQ - d - s - p - sP)$ is given by

$$\hat{z}_T(k) = T_T(k) + E_T(k) + t_T(k) \qquad (5.15)$$

where now the permanent component has been split into two terms; the first one is the trend component, and it is the solution of

$$\nabla^{d+1} T_T(k) = \frac{\mu}{s} \qquad (5.16)$$

and the second is the seasonal component, that is, the solution of

$$S_s(B) E_T(k) = 0. \qquad (5.17)$$

Finally, the transitory component is now the solution of

$$\Phi(B^s)\phi(B) t_T(k) = 0 \qquad (5.18)$$

and will die out for large horizon.

It is straightforward to check that (5.15), as defined by (5.16) to (5.22), is the solution of (5.14). The trend component has the same form as for non seasonal data and is given by (5.12). Note, however, that for seasonal models the order is $d+1$ and also $\beta_{d+1} = \mu/s(d+1)!$.

Seasonal component
The seasonal component will be given by

$$\sum_{j=1}^{s} E_T(j) = \sum_{s+1}^{2s} E_T(j) = 0$$

and the solution of this equation is a function of period s and values summing zero each s lags. The coefficients of this function are called seasonal coefficients, and they will be changing over time because they depend on the forecast origin. The long-run forecast will be determined by the permanent component, and it will have the structure

$$\hat{z}_T(k) = T_T(k) + E_T(k)$$

where the coefficients of both the polynomial trend component and the seasonal component are changing over time, depending on the forecast origin.

The airline model. Let us analyze the forecast structure of one of the models most often used with economic and business monthly data. The model is

$$\nabla\nabla_{12}z_t = (1 - \theta B)(1 - \Theta B^{12})a_t$$

and was used by Box and Jenkins (1976) to fit the time series of airline international traffic presented as an example in this chapter. Since then the IMA(0, 1, 1) × (0, 1, 1)$_s$ is called the airline model. The equation of the forecast generated by this model is

$$\hat{z}_t(k) = \hat{z}_t(k-1) + \hat{z}_t(k-12) - \hat{z}_t(k-13) - \theta\hat{a}_t(k-1)$$
$$- \Theta\hat{a}_t(k-12) + \theta\Theta\hat{a}_t(k-13)$$

and according to the previous analysis, we know that this equation can be written for $k > 0$ as

$$\hat{z}_t(k) = \beta_0^{(t)} + \beta_1^{(t)}k + S_k^{(t)}$$

that is, a linear trend plus a seasonal component with coefficients that are changing over time. In order to determine the parameters, we need to know the initial conditions. As we can write, for $j = 1, \ldots, 13$

$$\hat{z}_t(j) = \hat{\beta}_0^{(t)} + j\hat{\beta}_1^{(t)} + S_j^{(t)}$$

with $S_j^{(t)} = S_{j+12}^{(t)}$, we obtain that the slope is given by

$$\beta_1^{(t)} = \frac{\hat{z}_t(13) - \hat{z}_t(1)}{12} \tag{5.19}$$

and calling

$$\bar{z}_t = \frac{1}{12}\sum_1^{12}\hat{z}_t(j) = \hat{\beta}_0^{(t)} + \hat{\beta}_1^{(t)}\left(\frac{1 + \cdots + 12}{12}\right)$$

we have that

$$\hat{\beta}_0^{(t)} = \bar{z}_t - \frac{13}{2}\hat{\beta}_1^{(t)} \tag{5.20}$$

The seasonal coefficient are

$$S_j^{(t)} = \hat{z}_t(j) - \hat{\beta}_0^{(t)} - \hat{\beta}_1^{(t)}j \tag{5.21}$$

and will be given by the deviations of the forecast from the trend component.

5.5. PREDICTION CONFIDENCE INTERVALS

5.5.1. Known parameter values

The uncertainty of the forecast when the parameters are known is easy to compute from the MA(∞) representation of the process. Let us write

$$z_t = \psi(B)a_t$$

where the ψ_i parameters are obtained by using the relationship

$$\phi(B)(1-B)^d \psi(B) = \theta(B).$$

Then, we can write

$$z_{T+k} = \sum_0^\infty \psi_i a_{T+k-i} \qquad (\psi_0 = 1)$$

and taking expected values conditional to the observed data, we have that

$$\hat{z}_T(k) = \sum_0^\infty \psi_{k+j} a_{T-j}. \tag{5.22}$$

The forecast error is

$$e_T(k) = z_{T+k} - \hat{z}_T(k) = a_{T+k} + \psi_1 a_{T+k-1} + \cdots + \psi_{k-1} a_{T+1}$$

with variance

$$\text{Var}(e_T(k)) = \sigma^2 \left(1 + \psi_1^2 + \cdots + \psi_{k-1}^2\right)$$

Note that this equation indicates that the uncertainty of the long-run forecasts is different for stationary and nonstationary models. For a stationary model $\psi_k \to 0$ when $k \to \infty$, so that the series converge. For instance for an AR(1) model $\psi_k = \phi^k$ and $\text{Var}(e_T(k)) = \sigma^2/(1 - \phi^2)$. As we have seen, the long-run forecast goes to the mean, and the uncertainty of this forecast is finite. Note that although this uncertainty can be much larger than σ^2, the uncertainty of the one-step-ahead forecast, it remains bounded. However, when ϕ goes to one, that is, when we are close to the nonstationary case, the variance of the forecast grows without bounds. This means that we cannot make useful long-run forecasts for nonstationary models because the uncertainty will go to infinite.

If the distribution of the forecast error is known, we can compute confidence intervals for the forecast or prediction confidence intervals. For instance, assuming normal errors, the 95% confidence interval for the random variable z_{T+k} is

$$\hat{z}_T(k) \pm 1,96\hat{\sigma}\left(1 + \psi_1^2 + \cdots + \psi_{k-1}^2\right)^{1/2}.$$

Sometimes we are interested in forecasting a vector of future values $(z_{T+1}, \ldots, z_{T+k})$. Then we have to take into account that these forecasts are going to be correlated, that is when we observe z_{T+1} the forecasts for the rest of the observations have to be updated. We will see in Section 5.6 how this is done. We can anticipate here that the updating will depend on the covariances between the forecasts. From (5.22) we see that for $h > 0$

$$cov(\hat{z}_T(i), \hat{z}_T(i+h)) = E(e_T(i), e_T(i+h)) = \sigma^2 \sum_{s=0}^{i-1} \psi_{h+s}\psi_s,$$

and in particular if $h = 1$, $cov(\hat{z}_T(i), \hat{z}_T(i+1)) = \psi_1 \sigma^2$.

5.5.2. Unknown parameter values

In real applications the parameters are unknown. However, it can be shown that the uncertainty introduced in the forecast for this additional source of uncertainty is small for moderate sample size, and can be ignored in practice. We will illustrate the problem for $k = 1$ in the zero mean AR(1) case [more general analysis can be found in Box and Jenkins (1976)]. Then the forecast is

$$\hat{z}_T(1) = \hat{\phi} z_T$$

and the true forecast error, $e_T(1) = a_T$, is related to the observed forecast errror, $e_T^*(1) = z_{T+1} - \hat{\phi} z_T$, by

$$e_T(1) = e_T^*(1) + (\hat{\phi} - \phi)z_T.$$

Assuming to simplify that z_T is fixed, and using that $\text{Var}(\hat{\phi}) = \sigma^2 / \sum z_{t-1}^2$ we have that

$$\text{Var}(e_T^*(1)) = \sigma^2 \left(1 + z_T^2 / ns_z^2\right)$$

where $ns_z^2 = \sum z_{t-1}^2$. This equation indicates that the forecast error has two components. The first one, σ^2, is the uncertainty due to the random behavior of the observation we want to forecast. This uncertainty will be present even if we knew the parameters of the model that generates the observations. The second component measures the parameter uncertainty because the parameters are estimated from the sample. Note that this second term is of order $1/n$, and it can be safely ignored for medium or large sample size. A similar result can be proved in the general case.

5.6. FORECAST UPDATING

5.6.1. Computing updated forecasts

Let us show how forecasts are adapted when new observations become available. By (5.22)

$$\hat{z}_T(k) = \psi_k a_T + \psi_{k+1} a_{T-1} + \cdots$$
$$\hat{z}_{T+1}(k-1) = \psi_{k-1} a_{T+1} + \psi_k a_T + \cdots$$

which leads to

$$\hat{z}_{T+1}(k-1) - \hat{z}_T(k) = \psi_{k-1} a_{T+1}$$

where

$$a_{T+1} = z_{T+1} - \hat{z}_T(1)$$

and so the forecasts are adapted by

$$\hat{z}_{T+1}(k-1) = \hat{z}_T(k) + \psi_{k-1} a_{T+1} \tag{5.23}$$

Note that the forecasts are updated by adding some part of the observed last forecast error to the previous forecast, and the coefficients for forecast updating are the $\{\psi_i\}$ weights. This equation has a straightforward interpretation. Given the data until time T, the two random variables z_{T+1} and z_{T+k} follow jointly a normal distribution with expected values $\hat{z}_T(1)$ and $\hat{z}_T(k)$, variances σ^2 and $\sigma^2(1 + \psi_1^2 + \cdots + \psi_{k-1}^2)$, and covariance $\sigma^2 \psi_{k-1}$. Then the best estimate of z_{T+k} given z_{T+1} will be the conditional expectation, which is given by the standard regression equation (5.23). This analysis provides immediately with the updating equation for any possible situation: we compute the linear regression (conditional expectation) of the new forecast given the observed values.

5.6.2. Testing model stability

A test for model stability was developed by Box and Tiao (1976). If the model is correct and we call $\hat{a}_{T+j}$ to the one-step-ahead forecast errors computed from the estimated parameter values, we have that the statistic

$$Q = \frac{\sum_{j=1}^{h} \hat{a}_{T+j}^2}{\sigma^2}$$

will be distributed as a χ^2 distribution with h degrees of freedom. As σ^2 will be estimated by the sample residual variance $\hat{\sigma}^2$, the statistic

$$Q* = \frac{\sum_{j=1}^{h} \hat{a}_{T+j}^2 / h}{\hat{\sigma}^2} \tag{5.24}$$

will be distributed as an F distribution with h and $n - p - q$ degrees of freedom, where n is the sample size and $p + q$ the number of estimated parameters. If $Q*$ is large, we can conclude that the model is not adequate.

An example

We will make forecast using the airline data from Box and Jenkins (1976). The data include 12 years of monthly data of the log of the number of passengers in thousand international flights between January 1949 and December 1960. We will start assuming that we have just the first 6 years of data, which are plotted in Figure 5.1. The following model is estimated for this sample

$$\nabla \nabla_{12} z_t = (1 - .40B)(1 - .67B^{12}) a_t$$

and the residual standard deviation is .0185.

Figure 5.2 plots the data and the forecasts for 1955, 1956, and 1957 generated from this model. It can be seen that the forecast of every year follows the same structure: a linear trend and a set of seasonal factors. For instance, the forecasts for the first 2 years (1955 and 1956) are presented in Table 5.1.

We can compute the trend and the seasonal factors estimates easily as follows. The forecast for January 1955 from December 1954 is $\hat{z}_{72}(1) = 2.3671$, with a

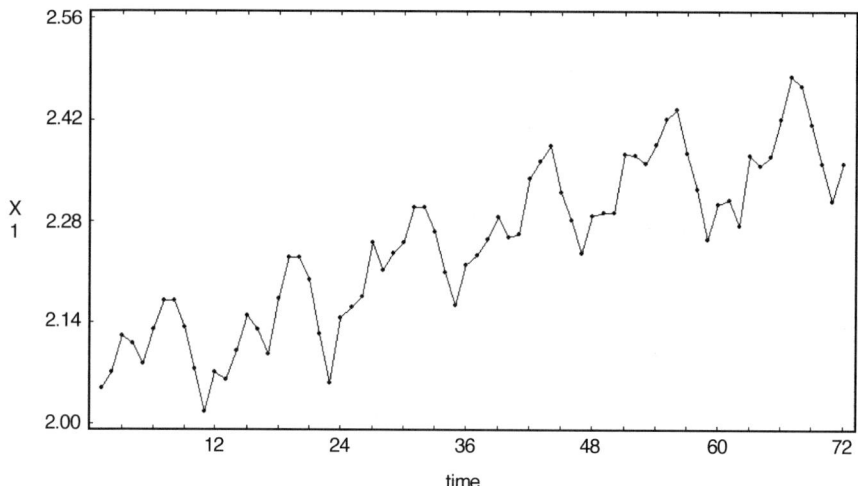

FIGURE 5.1 First 6 years of data (1949–1954) for the airline data.

5.6. FORECAST UPDATING

TABLE 5.1. Forecasts for 1955 and 1956 from Sample Data for 1949–1954

J	F	M	A	M	J	J	A	S	O	N	D
2.3671	2.3656	2.4378	2.4211	2.4214	2.4689	2.5140	2.5160	2.4648	2.4119	2.3559	2.4110
2.4176	2.4161	2.4883	2.4716	2.4719	2.5194	2.5645	2.5665	2.5153	2.4624	2.4064	2.4615

standard error of .0185, and for January 1956 is $\hat{z}_{72}(13) = 2.4176$, with a standard deviation of .0446. This corresponds to a rate of growth (the data is in logs) of $2.4176 - 2.3671 = .0505$, that is of 5.05%. The same growth is forecasted for all the following years. For instance, $\hat{z}_{72}(25) = 2.4681$, with a standard deviation of .0694. Note that the uncertainty in the forecast increases with the horizon. In the first forecast for January 1955 the standard deviation is 1.85% and it goes up to 6.94% for January 2 years later.

The forecasted montly growth is $.0505/12 = .00042$. The forecasted seasonal factors can be obtained by substracting for each forecast the trend effect (see 5.21) as follows. We first compute $\hat{\beta}_0$ by (5.20) as

$$\hat{\beta}_0 = \frac{2.3671 + 2.3656 + \cdots + 2.3559 + 2.4110}{12} - \frac{13}{2}\left(\frac{.0505}{12}\right) = 2.4022$$

and then the trend for each month. Table 5.2 shows the computations required to get the seasonal factors. For instance, the first trend is $2.4022 + .0004 = 2.4026$ and the difference with the forecast gives the seasonal effect. It can be seen that the lowest month for the seasonal effect is November (9% below the average) and the highest July and August (8% above the average).

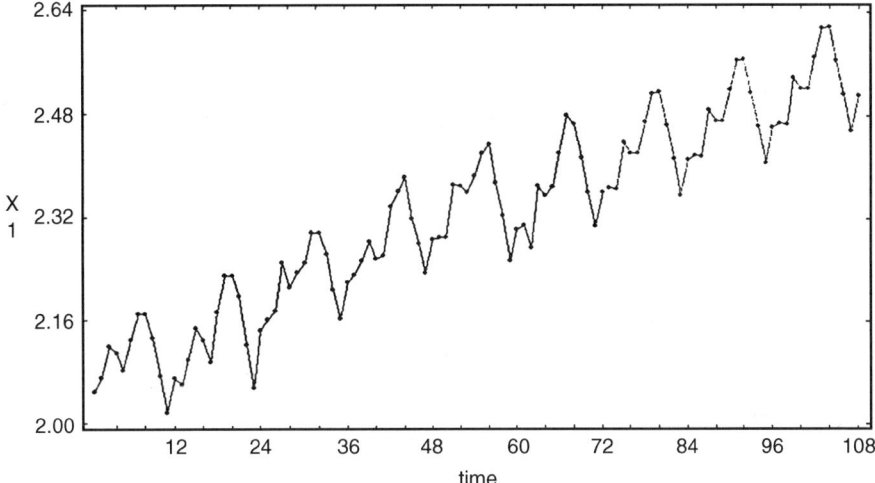

FIGURE 5.2 First 6 years of data (1949–1954) and 3 years of forecasts (1955–1957) for the airline data.

TABLE 5.2. Computation of Seasonal Forecasts Using (5.21)

	J	F	M	A	M	J	J	A	S	O	N	D
Trend	2.406	2.411	2.415	2.419	2.423	2.428	2.432	2.436	2.440	2.444	2.449	2.453
Forecast	2.367	2.366	2.438	2.421	2.421	2.469	2.514	2.516	2.465	2.412	2.356	2.411
Seasonal	−.039	−.045	.023	.002	−.002	.041	.082	.080	.025	−.032	−.093	−.042

We can check for model changes when more data are available. For instance, suppose that once we have observed one year more of data we compute the forecast errors and the Q statistics presented in Section 5.6.2. In this case $Q(12) = 21.186$ and $Q(12) = 1.765$. Also with 2 years of data $Q(24) = 61.916$ and $Q(24) = 2.58$. These values indicate that the model seem to be adequate.

Computing the rate of growth and seasonal factors from different origins can be useful to monitor slow changes of the components of the model over time. Suppose for instance that we forecast having 4 more years of data. Then the fitted model to the sample of 120 data point (1949–1958) is

$$\nabla \nabla_{12} z_t = (1 - .34B)(1 - .54B^{12})a_t$$

Figure 5.3 plots the assumed observed time series with 10 years of data and the two forecasted years. Table 5.3 shows the first 13 forecasts generated with origin $T = 120$ and the computation of the trend and seasonal factors from this new origin.

It is clear from the plot that the rate of growth has slowed down in the last period. The yearly forecasted growth is $(2.572 − 2.542) = .03$, which corresponds to a monthly growth of .0025. Table 5.3 shows the computation of the trend for each month and the

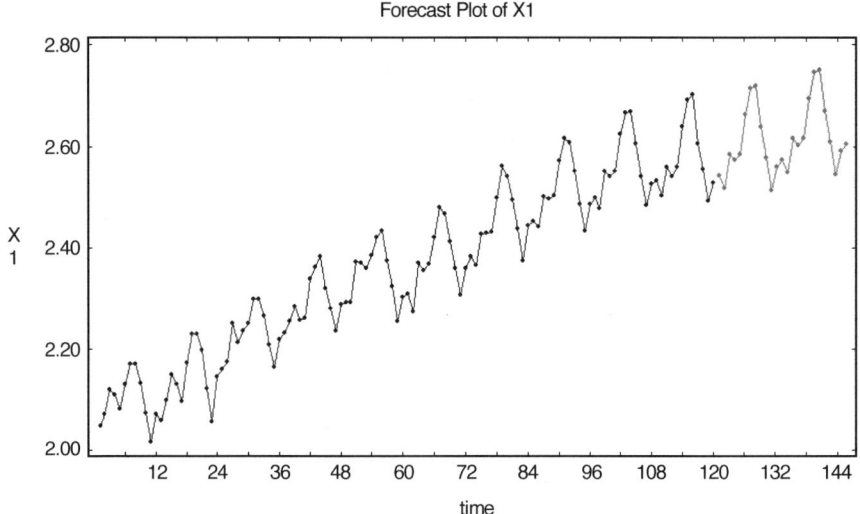

FIGURE 5.3 Ten years of data plus two forecasted years (1959, 1960).

5.7. THE COMBINATION OF FORECASTS

TABLE 5.3. Seasonal Factors for the Last 2 Years

	J	F	M	A	M	J	J	A	S	O	N	D	J
F	2.542	2.521	2.583	2.571	2.583	2.656	2.7052	2.709	2.633	2.576	2.516	2.560	2.572
T	2.583	2.585	2.588	2.590	2.593	2.595	2.597	2.600	2.602	2.605	2.607	2.610	2.613
S	−.041	−.065	−.004	−.019	−.009	.061	.108	.109	.031	−.029	−.091	−.050	−.041

seasonal factor. Comparing Tables 5.2 and 5.3 we see that the seasonality has changed a little during the period. In fact, July and August are becoming more important and are accounting for a larger proportion of passengers.

5.7. THE COMBINATION OF FORECASTS

In many forecasting problems in addition to the forecasts generated by the ARIMA time series model, we have the possibility to consider additional information. An important problem is how to combine different sources of infomation. These problem was studied first by Newbold and Granger (1974), who provide some very useful rules for combining forecasts. The problem of combining information from different sources has been in the last years subject to a large amount of research (see Draper et al. 1992).

We assume here that the additional information available may come from different sources. It may correspond to forecasts generated by a model with different level of aggregation; for instance, we have forecasts for monthly data but also a forecast for the total of the year made from a model with yearly data. In other cases, the forecasts come from a model that includes several explanatory variables. Also we may have subjective information we want to take into account to adjust the ARIMA forecasts together with a measure of the precision of this information with respect to the time series forecast. These examples can be put in the following setup—we have two or more independent forecasts of a given random variable, Z, and we want to combine them in order to have a better forecast.

A *general rule* to carry out this combination is as follows. Given n independent forecast $\hat{Z}_i$ of an unknown vector random variable Z of dimension p such that $E(\hat{Z}_i) = E(Z)$ with covariance matrices V_i, that may be singular, the best unbiased forecast (minimizing the trace of the error covariance matrix) is given by

$$\hat{Z}_T = \sum_{i=1}^{n} \left(\sum_{j=1}^{n} V_j^- \right)^{-1} V_i^- \hat{Z}_i \qquad (5.25)$$

where V_i^- is the Moore–Penrose generalized inverse of V_i, and where we have assumed that V_j^- is nonsingular. The covariance matrix of $\hat{Z}_T$ is then easily seen to be

$$V_T^{-1} \sum_{i=1}^{n} V_i^- . \qquad (5.26)$$

The proof of this rule can be found in Peña (1997). For instance, suppose that we have generated a vector of forecasts $\hat{Z}_h$ from an ARIMA model for the future observations of a time series $Z_h = (z_{T+1}, \ldots, z_{T+h})$ and we know the variance covariance matrix of the forecasts (see Section 5.5), which we will call V_h. Suppose that we know something about some linear combination of these forecasts, that is, we have a set of m restrictions on variable Z as

$$Y = CZ \tag{5.27}$$

but these restrictions have some uncertainty because Y is a random variable that we assume comes from a multivariate normal distribution with mean $CE(Z)$ and covariance matrix Σ_y. For instance, suppose that we have a model for monthly data and other for yearly data. Then the restriction is that the sum of the monthly forecasts should be equal to the forecast for the year. That is, $m = 1$, $C = (1, 1, \ldots, 1)$, and $V_y = \sigma_y^2$ is the variance of the yearly forecast.

Equation (5.27) implies a new forecast for Z. Any vector of the form

$$\hat{Z}_R = AC'(CAC')^{-1}Y$$

where we assume that the square matrix (CAC') has full rank, provides a forecast from the restriction equation. As we may choose any of them we take the corresponding to $A = I$. The covariance matrix of this forecast vector is singular and is given by

$$V_R = C'(CC')^{-1}V_y(CC')^{-1}C.$$

In order to apply rule (5.25) to obtain the MMSEF, we need a generalized inverse of V_R. It can be checked that V_R^-, the Moore–Penrose generalized inverse of V_R, is given by

$$V_R^- = C'V_y^{-1}C$$

and so the MMSEF will be

$$\hat{Z}_T = \left(C'V_y^{-1}C + V_h\right)^{-1} C'V_y^{-1}C\hat{Z}_R + \left(C'V_y^{-1}C + V_h\right)^{-1} V_h^{-1}\hat{Z}_h$$

or, in more simple way we can write this as

$$\hat{Z}_T = \hat{Z}_h + \left(C'V_y^{-1}C + V_h\right)^{-1} C'V_y^{-1}(Y - C\hat{Z}_h)$$

which indicates how to adjust the difference between Y and $C\hat{Z}_h$ in order to improve the forecast.

A particular interesting case of this rule is when $V_y = 0$, that is, when the restriction is exact. For instance, we are forecasting monthly expenses that are subject to a budget constraint, so that we know that the sum of the monthly forecast must be equal to some given constant. These types of problems have been studied by Guerrero and Peña (2000). It is interesting to note that the general rule (5.25) provides a direct solution for many forecasting time series problems. For instance, the Kalman filter introduced in Chapter 4 can be easily derived by applying this rule (see Peña 1997).

5.8. MODEL SELECTION CRITERIA

Several criteria have been proposed for selecting time series models since the seminal work of Akaike (1969, 1974). Among them are the Bayesian information criteria BIC of Schwarz (1978) and Akaike (1979), the penalty methods of Hannan and Quinn (1979), the predictive least-squares criterion of Rissanen (1986), extended by Lai and Lee (1997), and the modified AIC of Hurvich and Tsai (1989) and Cavanaugh and Shumway (1997). Surveys on the performance of these criteria for ARMA order selection can be found in Bhansali (1993) and Postcher and Srinivasan (1994).

In this section we present the FPE and AIC criteria introduced by Akaike and the Schwarz (1978) proposal, which seems to be overall the one to be recommended for time series model selection (Koreisha and Yashimoto 1991).

5.8.1. The FPE and AIC criteria

Suppose that we want to select the order of an AR(p) model in such a way that the out-of-sample one-step-ahead prediction mean-squared error is minimized. This MSE is given by

$$MSE(z_{T+1}) = E[z_{T+1} - \hat{\phi}' Z_p]^2$$

where now $Z_p = (z_T, \ldots, z_{T-p})$ and the expectation is taken with respect to the joint distribution of the variables $(z_{T+1}, \hat{\phi}', Z_p)$. The forecast error can be decomposed as

$$e_{T+1} = z_{T+1} - \phi' Z_p + (\phi - \hat{\phi})' Z_p,$$

and so

$$MSE(z_{T+1}) = \sigma^2 + E[(\phi - \hat{\phi})' Z_p Z_p' (\phi - \hat{\phi})] \quad (5.28)$$

which decompose the forecast error as the sum of the variable uncertainty and the parameter uncertainty. We can assume, to simplify, that the variables (z_{T+1}, Z_p) are independent of $\hat{\phi}$. This is equivalent to assuming that the parameters are computed from a sample $Z_0 = (z_1, \ldots, z_{T-p-1})$ that does not include the values that enter into the forecasting equation. This is a reasonable assumption for large sample size. First, Z_p and $\hat{\phi}$ will have then a small correlation because the estimated parameter is computed by using all the sample and Z_p will be a small fraction of the total sample. For instance, if $n = 500$ and the model is $AR(1)$, we do not expect that knowing that the parameter vector has been understimated is informative for guessing the last observed value. Second, the correlation between z_{T+1} and $\hat{\phi}$ will also be small because, although z_T will depend obviously on Z_p, given this last value, it may have an insignificant relationship with the other data used to compute the estimate.

Using this assumption, we can compute the expectation $E[(\phi - \hat{\phi})' Z_p Z_p' (\phi - \hat{\phi})]$ with respect to (Z_0, Z_p) by first taking the expectation with respect to the distribution of Z_p, which leads to $E(Z_p Z_p') = \Gamma_p$, and afterward by taking the expectation with respect to Z_0. The first step produces

$$MSE(z_{T+1}/Z_0) = \sigma^2 + E[(\phi - \hat{\phi})' \Gamma_p (\phi - \hat{\phi})].$$

Now, in order to compute the expectation with respect to Z_0, we first note that this function depends on the sample only through $\hat{\phi}$, and so we can compute the expectation with respect to this variable. As $\sqrt{n}(\phi - \hat{\phi})$ has a normal asymptotic distribution with zero mean and covariance $\sigma^2 \Gamma_p^{-1}$, the quadratic form $(\phi - \hat{\phi})' \Gamma_p (\phi - \hat{\phi}) n / \sigma^2$ is asymptotically a χ_p^2 distribution, and the expectation (5.28) can be approximated by

$$MSE(z_{T+1}) = \sigma^2 \left(1 + \frac{p}{n}\right).$$

An unbiased estimate of σ^2 is $n\hat{\sigma}^2/(n-p)$, where $\hat{\sigma}^2$ is the MLE estimate. Inserting this estimate in the last equation, we have an estimation of the out-of-sample forecast error. If we want to minimize this value, it implies that the order p of the AR model should be chosen by minimizing

$$FPE = \frac{\hat{\sigma}^2(n+p)}{n-p}.$$

This criterion, *final prediction error* (FPE), combines fitting, as given by $\hat{\sigma}^2$, with parsimony, due to the penalty introduced by the term $(n+p)/(n-p)$ for increasing the order p. An equivalent form for this criterion is

$$\log FPE = \log \hat{\sigma}^2 + \log n \left(1 + \frac{p}{n}\right) - \log n \left(1 - \frac{p}{n}\right),$$

and using that $\log(1+x) \approx x$ for x small this expression can be approximated for large n by $\log \hat{\sigma}^2 + 2p/n$. Multiplying the equation for n, we obtain the AIC criterion

$$AIC = n \log \hat{\sigma}^2 + 2p.$$

This Akaike information criterion can be derived using entropy considerations (Akaike 1974). The AIC chooses the model that gives the best approximation to the true model asymptotically with the Kullback–Leibler measure of distance. It is obtained by substituting the out-of-sample mean-squared criterion by $-2(\log$ maximized likelihood), which is a more general measure of model fitting. The estimation of this quantity leads to the AIC criterion that has the general form

$$AIC = -2(\log \text{ maximized likelihood}) + 2(\text{number of parameters}) \quad (5.29)$$

and for ARMA models this reduces, dropping constants, to

$$AIC = n \log \hat{\sigma}^2 + 2(p+q) \quad (5.30)$$

where $(p+q)$ is the number of parameters in the model.

The problem with AIC is that tends to overestimate the number of parameters, even asymptotically. This problem was noticed by a number of authors, including Schwarz (1978). More recently, Hurvich and Tsai (1989) have shown that in small samples or when the number of fitted parameters is a moderate to large fraction of the sample size, it can drastically overfit the data. To solve this problem, Akaike (1979) proposed a modification of this criterion that he called BIC and that is equivalent, for large samples, to the Schwarz criterion presented in the next section. On the other hand, Shibata (1980) proved that if the predictor is selected by AIC, it is asymptotically efficient, in the sense of minimizing the one-step-ahead mean squared prediction error.

5.8.2. The Schwarz criterion

Schwarz (1978) presented a Bayesian way to estimate the dimension of a model. He assumed that we have a set of models M_j with prior probabilities $p(M_j)$ with parameters θ_j, and we want to select the model which maximizes the posterior probability of the model given the data, $p(M_j/T)$. This probability is computed by

$$p(M_j/Y) = cp(M_j) \int p(Y/\theta_j) p(\theta_j/M_j) d\theta_j$$

where $p(Y/\theta_j)$ is the likelihood of the data and $p(\theta_j/M_j)$ the prior for the parameters.

Making an asymptotic approximation to this integral, he showed that the model to be chosen is the one that minimizes

$$BIC = -2(\log \text{ maximized likelihood}) + (\log n)(\text{number of parameters}). \quad (5.31)$$

In this criterion the penalty for introducing new parameters is greater that AIC, so that BIC tends to select simpler models than those chosen by AIC. The difference between both criteria can be very large if n is large. It can be proved that the BIC criterion has asymptotic consistency under general conditions

5.9. CONCLUSIONS

Forecasting methods are conditional to the given model and do not take into account the model uncertainty. This point has been stressed by several authors. A way to overcome this problem is by using a linear combination of models with weights given by the relative probability of each model. In this way the forecasts will be generated by

$$\hat{y}_T(k) = \sum_{i=1}^{h} w_i \hat{y}_{iT}(k)$$

where w_i are the weights and $\hat{y}_{iT}(k)$ the forecast generated by model ith. This approach has been mainly used from a Bayesian perspective under the name of Bayesian model averaging (BMA). Several strategies to implement this approach from the Bayesian point of view can be found in Madigan and Raftery (1994).

Finally, some authors have proposed to adapt the estimation criterion to the objective of the forecast. This means that if we want to generate forecasts for lead time $j = 1, \ldots, J$ we could estimate J values for the parameters of the model minimizing the j steps ahead forecast ($j = 1, \ldots, J$). This approach was advocated by Findley (1983) and is discussed in Tiao and Xu (1993), Tiao and Tsay (1994), Hurvich and Tsai (1997), and Bhansali (1998), among others.

REFERENCES

Akaike, H. (1969). Fitting autorregresive models for prediction. *Ann. Inst. Stat. Math* **21**, 343–347.

Akaike, H. (1974). A new look at the statistical model identification. *IEEE Trans. Aut. Contr.* AC **19**, 203–217.

Akaike, H. (1979). A Bayesian extension of the minimum AIC procedure of autoregressive model fitting. *Biometrika* **66**(2), 237–242.

Bhansali, R. J. (1993). Order selection for linear time series models: A review. In T. Subba Rao (ed.). *Developments in Time Series Analysis*, pp. 50–66. Chapman and Hall, London.

Bhansali, R. J. (1998). Parameter estimation and model selection for multistep prediction in time series: A review. In S. Ghosh (ed.). *Asymtotics Nonparametrics and Time Series*. Marcel Dekker, New York.

Box, G. E. P. and Jenkins, G. M. (1976). *Time Series Analysis, Forecasting and Control*. Holden-Day, San Francisco.

Box, G. E. P. and Tiao, C. G. (1976). Comparison of forecast and actuality. *Appl. Stat.*, **25**, 195–200.

Cavanaugh, J. E. and Shumway, R. H. (1997). A Bootstrap variant of AIC for state space model selection. *Statistica Sinica* **7**, 473–496.

Draper, D. et al. (1992), *Combining Information. Statistical Issues and Opportunities for Research*. National Academy Press, Washington DC.

Espasa, A. and Peña, D. (1995). The decomposition of the forecast function in seasonal ARIMA models. *J. Forecasting* **14**, 565–583.

Findley, D. F. (1983). On the use of multiple models for multi-period forecasting. *Proc. Bus. Econ. Stat, ASA.*, 528–531.

# REFERENCES

Guerrrero, V. and Peña, D. (2000). Linear combination of restrictions and forecasts in time series. *J. Forecasting*, **19**, 103–122.

Hannan E. J. and Quinn, B. J. (1979). The determination of the order of an autorregresssion. *J. Roy. Stat. Soc. B* **41**, 190–95.

Hurvich, C. M. and Tsai, C. L. (1989). Regression and time series model selection in small samples. *Biometrika* **76**, 297–307.

Hurvich, C. M. and Tsai, C. L. (1997). The selection of multiple linear predictor for short time series. *Statistica Sinica* **7**, 395–406.

Koreisha, S. and Yashimoto, G. (1991). A comparison among identifications procedures for autorregressive moving average models. *Int. Stat. Rev.* **59**(1), 37–58.

Lai, T. L. and Lee, C. P. (1997). Information and prediction criteria for model selection in stochastic regression and ARMA models, *Statistica Sinica* **7**, 285–309.

Madigan, D. and Raftery, A. (1994). Model selection and accounting for model uncertainty in graphical models using Occam's window. *J. Am. Stat. Assoc.* **89**, 1535–1546.

Newbold, P. and Granger, C. W. J. (1974). Experience with forecasting univariate time series and the combination of forecasts. *J. Roy. Stat. Soc. A* 131–146.

Peña, D. (1997). Combining information in statistical models. *Am. Stat.* **51**(4), 326–332.

Peña, D. (1995). Forecasting growth with time series models. *J. Forecasting* **14**, 97–105. 1995.

Postcher, B. M. and Srinivasan, S. (1994). A comparison of order determination procedures for ARMA models. *Statistica Sinica* **4**, 29–50.

Rissanen, J. (1986). Stochastic complexity and modelling. *Ann. of Stat.* **4**, 461–464.

Sánchez, I. y and Peña, D. (2000). Properties of predictors in overdifferenced nearly nonstationary autoregression. *J. Time Series Anal.*

Schwarz, G. (1978). Estimating the dimension of a model. *Ann. Stat.* **6**(2), 461–464.

Shibata, R. (1980). Asymptotically efficient selection of the order of the model for estimating parameters of a linear process. *Ann. Stat.* **8**, 147–164.

Tiao, C. G. and Xu, D. (1993). Robustness of MLE for multi-step predictions: The exponential smoothing case. *Biometrika* **80**, 623–641.

Tiao, C. G. and Tsay, R. S. (1994). Some advances in non-linear and adaptive modelling in time-series. *J. Forecasting* **13**, 109–131

West, M. and Harrison, J. (1989). *Bayesian Forecasting and Dynamic Models*, Springer-Verlag, New York.

CHAPTER 6

Outliers, Influential Observations, and Missing Data

Daniel Peña
Universidad Carlos III de Madrid

6.1. INTRODUCTION

Time series data, as all types of statistical data, are often subject to outliers or discordant observations. Their study has been approached from two points of view. The first is the diagnostic approach, in which diagnostic methods are applied to the residuals of the estimated model to identify possible outliers that are tested afterward. Once the outliers are identified, a model that incorporates them is proposed, and the outlier effects and the parameters of the model are estimated jointly. In this way we obtain both the effect of the outliers, which can be in some cases the main objective of the analysis, and a robust parameter estimate. The second is the robust approach, in which the estimation method is modified so that the estimates are not contaminated by the presence of outliers. Once we have a robust parameter estimate, the outliers can be easily identified and tested. Both methodologies complement each other, and ideas from one approach can be used to improve the other. In this chapter we will concentrate on the diagnostic approach, which seems to be the most widely used in applications.

Outliers in time series data can be represented within the framework of ARIMA models or state-space models. As there is a well-known relationship between both representations (see Chapter 4), the results from one model can be transferred to the other, and in this chapter we will use the ARIMA representation. Thus, we assume that the outliers happen on a time series, y_t, which can be modelled by

$$\phi(B)\nabla^d y_t = \theta(B)a_t \tag{6.1}$$

A Course in Time Series Analysis, Edited by Daniel Peña, George C. Tiao, and Ruey S. Tsay.
ISBN 0-471-36164-X. © 2001 John Wiley & Sons, Inc.

6.1. INTRODUCTION

where B is the backshift operator such that $By_t = y_{t-1}$, $\phi(B) = 1 - \phi_1 B - \cdots - \phi_p B^p$, and $\theta(B) = 1 - \theta_1 B - \cdots - \theta_q B^q$, are polynomials in B of degrees p and q, respectively, with roots outside the unit circle, $\nabla = 1 - B$ is the difference operator, $\nabla^d y_t$ is a stationary series, and a_t is a white-noise sequence of iid $N(0, \sigma_a^2)$ variables. The model can also be written in the AR(∞) form as

$$\pi(B)y_t = a_t \tag{6.2}$$

where $\pi(B) = \nabla^d \phi(B)/\theta(B) = 1 - \pi_1 B - \pi_2 B^2 - \cdots$ or in the MA(∞) form as

$$y_t = \psi(B)a_t \tag{6.3}$$

where $\psi(B) = \theta(B)/\phi(B)\nabla^d$.

Fox (1972) defined the additive and innovative outliers in time series and proposed the use of maximum likelihood ratio tests for detecting them. Guttman and Tiao (1978), Miller (1980), Chang (1982), and Chan (1995) studied the effect of outliers in the autocorrelation of the series. Chang and Tiao (1983) and Chang et al. (1988) extended the results of Fox (1972) to ARIMA models and proposed a likelihood ratio test and an iterative procedure for detecting outliers and estimating the model parameters. Score test were proposed by Abraham and Yatawara (1988). Tsay (1988) generalized the Chang–Tiao–Chen procedure to include the detection of level shifts and temporary changes. Random level shifts were studied by Chen and Tiao (1990). Balke (1993) proposed a modification to Tsay's procedure for solving the confusion between level shift and innovative outliers using an additional search of outliers with a white-noise model. Chen and Liu (1993) presented an outlier detection and parameter estimation procedure for ARIMA models that seems to be widely used.

Abraham and Chuang (1989) considered deletion statistics based on influence measures in regression for outlier identification. Peña (1986, 1987, 1990) proposed a missing-value approach to the study of influence in time series, presented statistics to measure the influence of different types of outliers, and discussed the link between outliers and missing data. Bruce and Martin (1989) studied the identification of outlier patches in ARIMA models using ideas of influential observations in time series. Lefrançois (1991) proposed a deleted one influence measure for the autocorrelation function. Abraham and Chuang (1993) applied the EM algorithm to the estimation of outliers. Ljung (1982, 1989, 1993) studied the likelihood function of ARMA models with missing data and its relation to outlier analysis.

This chapter is organized as follows. In Section 6.2 different types of outliers are reviewed. In Section 6.3 some procedures for outlier identification and robust estimation are discussed. Section 6.4 is dedicated to a review of influential observations in ARIMA models. Section 6.5 discusses the problem of multiple outliers. The relationship between outliers, influential observations and missing-value estimation is discussed in Section 6.6. Section 6.7 includes some brief comments about forecasting with outliers.

6.2. TYPES OF OUTLIERS IN TIME SERIES

6.2.1. Additive outliers

An additive outlier corresponds to an external error or exogenous change of the observed value of the time series at a particular timepoint; that is, instead of observing the series y_t, we observe a new series, z_t, which is related to the original one by

$$z_t = \begin{cases} y_t & t \neq T \\ y_t + \omega_A & t = T. \end{cases} \quad (6.4)$$

An additive outlier can be interpreted in general terms as a measurement error at time T, $1 \leq T \leq n$, or as an impulse effect due to exogenous causes. For instance, when the original series describes the output from some system, an additive outlier corresponds to a particular unexpected event that happens at time T, such as a strike, an accident or a breakdown, and which modifies the output of the system at this point of time, without further effects on the future values of the time series.

An alternative representation to (6.4) of the relationship between the original series, y_t, and the observed series, z_t, is given by the model

$$z_t = \omega_A I_t^{(T)} + \psi(B) a_t \quad (6.5)$$

where $I_t^{(T)}$ is a dummy variable which is zero at all lags except at time $t = T$ in which $I_t = 1$. An equivalent way to write (6.5) is

$$\pi(B)\left(z_t - \omega_A I_t^{(T)}\right) = a_t.$$

For instance Figure 6.1 shows an AR(1) simulated time series with parameter .8 and the same time series with an additive outlier at time $t = 20$ of size equal to seven times the standard deviation of the series. The figure also shows the autocorrelation function of the original and contaminated series, and it can be seen that this function is seriously affected by a single AO.

An additive outlier can have very serious effects on the properties of the observed time series. It will affect (1) the estimated residuals and also (2) the estimates of the parameter values. In order to show the first effect, let us assume that the parameters of the model are known and let us write the ARIMA model followed by y_t in the AR(∞) representation (6.2). If an additive outlier happens at $t = T$, the residuals of the observed time series for $t \geq T$ will be computed in the usual way by

$$e_{T+j} = z_{T+j} - \pi_1 z_{T+j-1} - \cdots - \pi_p z_{T+j-p} - \cdots.$$

Before the AO occurred, these residuals will have been identical to the residuals from the original process, computed by

$$a_{T+j} = y_{T+j} - \pi_1 y_{T+j-1} - \cdots - \pi_p y_{T+j-p} - \cdots.$$

6.2. TYPES OF OUTLIERS IN TIME SERIES

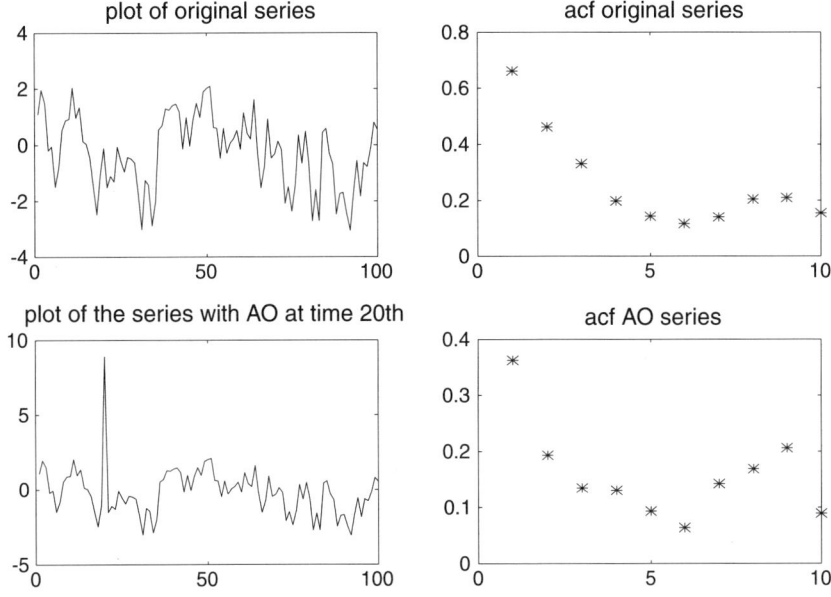

FIGURE 6.1 A simulated AO on a AR(1) series.

However, after time T, they will be different from the white-noise perturbations, a_t, and the relationship between them will be

$$e_{T+j} = a_{T+j} - \pi_j \omega_A \quad j \geq 0 \tag{6.6}$$

where $\pi_0 = -1$. This means that the p posterior residuals will be affected. For instance, a 10% proportion of outliers will contaminate $10p\%$ of the residuals. If the sample size is 100, and the model is AR(5), the presence of 10% of additive outliers will contaminate half of the residuals.

Additive outliers can also have a strong effect on the estimation of the model parameters. For instance, consider a simple AR(1) model, and let us assume for simplicity that it has zero mean so that the only parameter to be estimated is the ϕ parameter. Then the least-squares estimate is given by

$$\hat{\phi}_0 = \frac{\sum y_t y_{t-1}}{\sum y_t^2}. \tag{6.7}$$

Suppose now that an additive outlier occurs at time T. Then instead of observing y_t we observe z_t given by (6.4), and the parameter estimate is obtained by

$$\hat{\phi} = \frac{\sum z_t z_{t-1}}{\sum z_t^2}. \tag{6.8}$$

Replacing in (6.8) y_t by z_t as given in (6.4), the parameter estimate can be written as

$$\hat{\phi} = \frac{\hat{\phi}_0 + n^{-1}\tilde{\omega}_A(\tilde{y}_{T-1} + \tilde{y}_{T+1})}{1 + 2\tilde{\omega}_A\tilde{y}_T + n^{-1}\tilde{\omega}_A^2}, \tag{6.9}$$

where $\tilde{\omega}_A = \omega_A/s_y$ and $\tilde{y}_t = y_t/s_y$ are respectively the size of the additive outlier and the observed values standardized by the sample variance of the true process $s_y^2 = \sum y_t^2/n$. It is clear that for any fixed sample size

$$\text{if} \quad \omega_A \to \infty, \Rightarrow \hat{\phi} \to 0.$$

Thus, a large additive outlier will bias the estimated parameter toward zero. This effect is shown in Figure 6.1, in which the autocorrelation coefficients of the contaminated series are smaller that the one in the original series.

It can be proved in general that a large additive outlier will push all the autocorrelation coefficients toward zero. Note also that the effect on the parameter estimate of a finite additive outlier depends on the previous and posterior values of the series. Finally, as one would expect, the effect of the outlier decreases for large sample size.

For instance, Figure 6.2 shows the scatterplots to compute the first three autocorrelation coefficients for the series of Figure 6.1. It can be seen that the outlier generates two outliers in the scatterplot. The first outlier is not influential, but the second one is a high leverage outlier with strong effect on the computation of the correlation coefficient between the two variables.

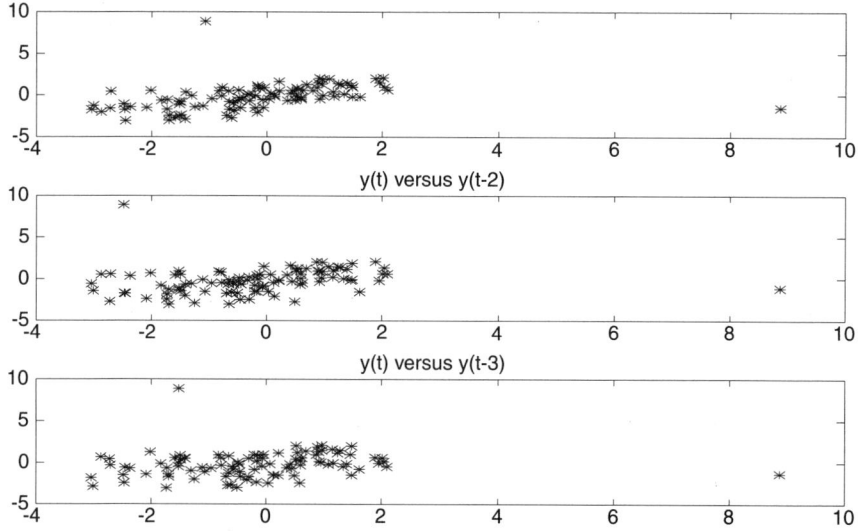

FIGURE 6.2 Scatterplots for the first three lags for the contaminated AO series.

6.2.2. Innovative outliers

The second type of outlier introduced by Fox (1972) is called *innovative* or an *innovational outlier* (IO). It can be generated by some internal change or endogenous effect on the noise of the process. The model for an IO is built by adding an impulse effect to the noise of the original process, as follows

$$z_t = \psi(B)\left(\omega_I I_t^{(T)} + a_t\right) \tag{6.10}$$

where z_t is the observed time series, ω_I is the outlier size, and $I_t^{(T)}$ is a dummy variable that is zero at all lags except at time $t = T$ in which $I_t = 1$. This model is called the *innovation outlier* (IO) model. It can also be written as

$$\pi(B)z_t = \omega_I I_t^{(T)} + a_t. \tag{6.11}$$

From this definition we obtain that the original process, y_t, is related to the observed process, z_t, for

$$z_t = \begin{cases} y_t & t < T \\ y_t + \omega_I \psi_j & t = T + j, \quad j > 0. \end{cases} \tag{6.12}$$

where the coefficients ψ_j come from the MA(∞) representation of the ARIMA process. This result shows that the effect on the observed time series of an innovative outlier depends on the ARIMA model.

Note that the noise of an ARIMA model represents the joint effect of all the nonsystematic changes on the variables that are causing the time series, y_t. From this point of view an IO can be interpreted as an outlier effect on the time series which are causing y_t.

Figure 6.3 shows an example of an IO on a simulated AR(1) with parameter .8. The figure also shows the autocorrelation function of the original and contaminated process, and it can be seen that the effect of the IO is smaller than in the case of the AO. In general terms, innovative outliers have less damaging effects on the time series than do additive outliers. (Compare Figures 6.1 and 6.3.)

Let us analyze the effect of an IO on the residuals and on the estimation. Assuming first that the parameters of the model are known, considering the case of an AR(p) and using the same notation than in the case of additive outliers, the relationship between the observed contaminated residuals, e_t, and the true residuals, a_t, is

$$e_T = a_T - \omega_I$$

but for any $j > 0$

$$e_{T+j} = a_{T+j}$$

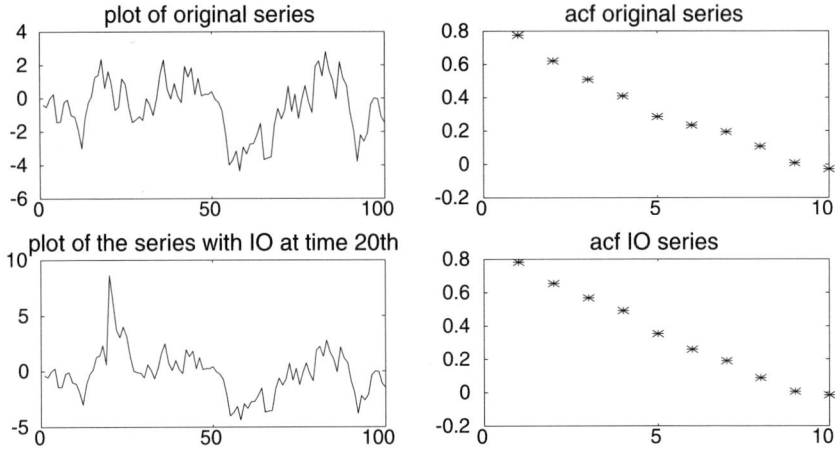

FIGURE 6.3 An IO on a simulated AR(1) series.

and the rest of the residuals will not be affected. Thus, only the residual at time T would be affected.

As shown in Figure 6.3, innovative outliers are expected to produce a small effect on the autocorrelation and hence on the parameter estimates. To see this, suppose the AR(1) model. Then, using the same notation than in (6.9), replacing z_t by $y_t + \omega_I \phi^j$ in

$$\hat{\phi} = \frac{\sum z_t z_{t-1}}{\sum z_t^2}$$

and after some straightforward algebra, it is easy to show that

$$\hat{\phi} = \frac{\hat{\phi}_0 + n^{-1} S_1 + n^{-1} \tilde{\omega}_I^2 (\phi/(1-\phi^2))}{1 + n^{-1} 2 S_2 + n^{-1} \tilde{\omega}_I^2/(1-\phi^2)}$$

where $\hat{\phi}_0 = (\sum y_t^2)^{-1} \sum y_t y_{t-1}$, $\tilde{\omega}_I = \omega_I / \Delta y$ $S_1 = \tilde{\omega}_I \sum_{j=0} (\tilde{y}_{T-1+j} + \tilde{y}_{T+1+j}) \phi^j$ and $S_2 = \tilde{\omega}_I \sum_{j=0} \tilde{y}_{T+j} \phi^j$. Then, for fixed sample size

$$\text{if} \quad \omega_I \to \infty, \Rightarrow \hat{\phi} \to \hat{\phi}_0,$$

and we obtain a consistent estimate of the parameter value. Therefore, for large sample size the effect of IO on the parameter values can be neglected. To illustrate this point, Figure 6.4 shows the scatter plots of the values of the IO contaminated AR(1) time series and the three first lags.

It can be seen that now we have several large leverage points, but these are located in the directions indicated by the relationship, and so the distortion of the correlation coefficient is small.

6.2. TYPES OF OUTLIERS IN TIME SERIES

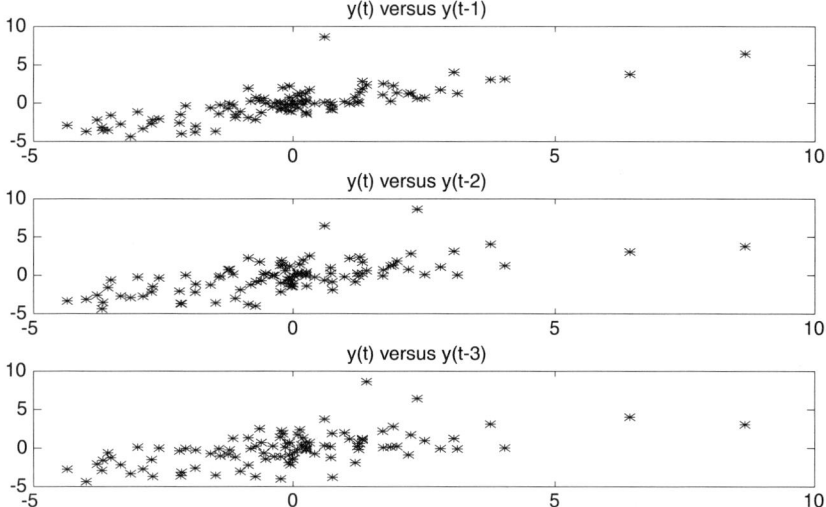

FIGURE 6.4 Scatterplots for the first three autocorrelation coefficients for a series with an IO.

Note that additive and innovational outliers are equivalent for a white-noise time series. For a random walk the effect of an innovative outlier is a level shift, which will be studied in the next section.

An alternative way to introduce innovative outliers is to assume that the noise of the process follows a normal contaminated distribution (see, e.g., Abraham and Box 1979). In this way the distribution of a_t is

$$(1 - \alpha)N\left(0, \sigma_a^2\right) + \alpha N\left(0, k\sigma_a^2\right)$$

where k is greater than 1. This model assumes again an outlier on the distribution of the noise of the process and the analysis is equivalent to the one presented in this section.

6.2.3. Level shifts

A *level shift* (LS) corresponds to a modification of the local mean or level of the process starting from a specific point and continuing until the end of the time period observed. For a stationary process, a level shift implies a change in the process mean after some point, and therefore the process is transformed into a nonstationary one. When a level shift appears, the observed series is related to the original series by

$$z_t = \begin{cases} y_t & t < T \\ y_t + \omega_L & t \geq T \end{cases} \qquad (6.13)$$

and so the level shift can be seen as a sequence of additive outliers of the same size starting at some point of time and lasting until the end of the observed time period.

The model for this type of outlier is

$$z_t = \omega_L S_t^{(T)} + \psi(B)a_t$$

where $S_t^{(T)}$ is a step function that takes the value 0 before T and 1 by $t \geq T$. This function is related to the impulse function used in the previous sections by

$$S_t^{(T)} = 1/(1-B) I_t^{(T)}$$

because if we apply a difference to the step function, we obtain the impulse function. This model can also be written as

$$\pi(B)\left(z_t - \omega_L S_t^{(T)}\right) = a_t. \tag{6.14}$$

Sometimes a LS can appear as the effect of an IO on a nonstationary time series. Consider the effect of an innovative outlier on a random-walk process

$$(1-B)z_t = \omega I_t^{(T)} + a_t, \tag{6.15}$$

and as the inverse of the difference operator is the sum operator this model can be written as

$$z_t = \omega S_t^{(T)} + \sum_{j=0}^{t} a_j$$

which implies a LS on the time series.

The effect of a LS on the residuals and on the parameter estimates can be strong. Assuming the parameters known, the observed residuals, $e_t = \pi(B)z_t$ are related to the true residuals or perturbations, $a_t = \pi(B)(z_t - \omega_L S_t^{(T)})$ by

$$e_t = a_t + \pi(B)\omega_L S_t^{(T)} = a_t + l(B)\omega_L I_t^{(T)}$$

where $l_j = 1 - \pi_1 - \cdots - \pi_j$ are the coefficients of $l(B) = (1 + l_1 B + l_2 B^2 + \cdots) = \pi(B)/(1-B)$. Then we have

$$e_t = \begin{cases} a_t & t < T \\ a_t + \omega_L l_j & t = T + j. \end{cases} \tag{6.16}$$

This means that all residuals after the LS can be affected. If the original model is a stationary AR(p), we have

$$e_T = a_T + \omega_L$$
$$e_{T+j} = a_{T+j} + \omega_L(1 - \phi_1 - \cdots - \phi_j) \quad j = 1, \ldots, p$$
$$e_{T+j} = a_{T+j} + \omega_L(1 - \phi_1 - \cdots - \phi_p) \quad j \geq p+1$$

6.2. TYPES OF OUTLIERS IN TIME SERIES

and after time $T + p + 1$ all the residuals are affected by the same amount. For a nonstationary process, so that $\pi(1) = 0$, the residuals from T to $T + h$, where h is the order of the AR approximation of the stationary part, are affected, but the residuals after $T + h$ are not. For instance, for a random walk only one residual will be affected. Therefore, the effect of a level shift depends on (1) the model and is expected to be larger for stationary than for nonstationary processes (2) the distance between the period in which the LS appears, T, and the last observation. For instance, if the LS happens on the last observation, the LS over the time series z_t is equivalent to an additive outlier, and so only the last residual will be affected.

The effect of a LS on the autocorrelation coefficients can be studied as before. Assuming again an AR(1) model and calling $r_z(1) = (\sum z_t^2)^{-1} \sum z_t z_{t-1}$ to the first autocorrelation coefficient for the observed series z_t and $r_y(1)$ the corresponding to y_t, using that now $z_t = y_t + \omega_L$, it is easy to show that

$$r_z(1) = \frac{r_y(1) + n^{-1}S_3 + n^{-1}\tilde{\omega}_L^2(n-T)}{1 + n^{-1}2S_4 + n^{-1}\tilde{\omega}_L^2(n-T+1)}$$

where $S_3 = \tilde{\omega}_L(\tilde{y}_{T-1} + \tilde{y}_T + 2\tilde{y}_{T+1} + \cdots + 2\tilde{y}_n)$ and $S_4 = \tilde{\omega}_L \sum_{j=0} \tilde{y}_{T+j}$. Then if $n - T$ is not too small

$$\text{if} \quad \tilde{\omega}_L \to \infty, \Rightarrow r_z(1) \to 1.$$

That is, the effect of a level shift is to push the first autocorrelation coefficient to one. It is easy to show that this effect will also appears at all lags if $n - T$ is large and, therefore, when a stationary series suffers a level shift the series will seem to have a unit root. An intuitive explanation of this effect can be obtained from Figure 6.5, in which the scatterplots of the values of the series and some lags are presented. It can seen that the effect of the LS is to increase the autocorrelation coefficients.

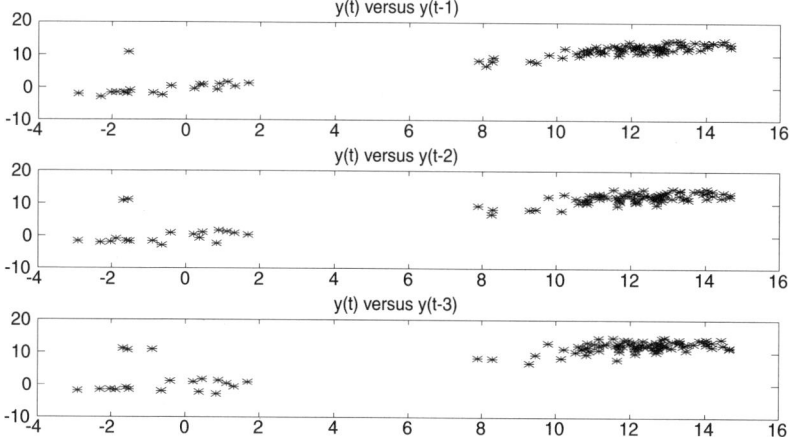

FIGURE 6.5 Scatterplot for the first three autocorrelation coefficients for a series with a LS.

6.2.4. Outliers and intervention analysis

The types of outliers studied can be considered as particular cases of interventions or deterministics effects in a time series. *Intervention analysis* is a procedure introduced by Box and Tiao (1975) to model the effect of a dynamic change on a time series at a known point of time. For instance, these authors studied the series of pollution level in downtown Los Angeles and investigated whether a known intervention, the diversion of traffic due to the opening of the Golden Gate freeway, has had an effect on the time series. Calling z_t to the observed time series the intervention analysis model is

$$z_t = \omega V(B) I_t^{(T)} + \psi(B) a_t \qquad (6.17)$$

in which ω is a constant and $V(B) = (1 + v_1 B + v_2 B^2 + \cdots)$ is the *transfer function* of the intervention at time T, which, in general, will be represented by a polynomial operator in the backshift operator B obtained as a rational ratio of two finite operators $V(B) = a(B)/b(B)$. Box and Tiao (1975) analyzed different structures for the dynamic intervention effect given by $V(B)$ and studied the fitting of the model by maximum likelihood.

As indicated by Chang et al. (1988), the additive outlier is a particular case of equation (6.17), in which $V_i(B) = 1$. In this case the intervention does not have any dynamics, and the result is called an AO of size ω. The innovational outlier appears if $V_i(B) = 1/\pi(B)$; that is, when the dynamic of the intervention is the same as the dynamic of the original process, we have an IO at time T. The intervention analysis approach suggests other types of outliers that might be considered. For instance, we may assume that the effect of a LS decreases with time and after some time it disappears. A way to model this is with a *transitory change*, introduced by Tsay (1988), as a LS that dies out in an exponential way. This effect is defined as an intervention effect in which $V(B) = (1 - \delta B)^{-1}$:

$$z_t = \frac{\omega_{TC}}{1 - \delta B} I_t^{(T)} + \frac{1}{\pi(B)} a_t. \qquad (6.18)$$

Note that if $\delta = 1$, the transitory change is identical to the LS, whereas if $\delta = 0$, it is an additive outlier. Other interesting type of outlier is the *ramp shift outlier* (Chen and Tiao 1990). In the same way as integrating (summing) an additive outlier to obtain a LS, we may integrate the level shift to obtain a ramp shift. The model for this outlier will be

$$z_t = \omega_L R_t^{(T)} + \psi(B) a_t$$

where $R_t^{(T)}$ is a ramp effect, that is, $R_t^{(T)} = 0$ for $t < T$ and $R_t^{(T)} = j$ for $t = T + j$, $j = 1, 2, \ldots$. Then the outlier will produce a change in the slope of the series after T. This model can also be written as a LS over the first difference

$$(1 - B) z_t = \omega_L S_t^{(T)} + \psi(B) a_t.$$

In the same way we can define a *temporary ramp* shift by

$$z_t = \frac{\omega_{TC}}{1 - \delta B} S_t^{(T)} + \frac{1}{\pi(B)} a_t$$

which indicates a transitory change in the slope of the process. These outliers can be useful in models $I(1)$ with constant or in models $I(2)$, as both of them show a linear trend behavior.

We may also consider the possibility of several outliers at the same period. This possibility has been present since the seminal paper by Fox (1972), and in the case of additive and innovational outliers was studied by Muirhead (1986), Chang et al. (1988), and Abraham and Yatawara (1988), among others. For stationary models, adding innovative and additive outliers will lead to a more general form of intervention effect. For nonstationary models, the innovative outlier always has a LS component, and it will lead to the consideration of level shifts and additive outliers at the same time. However we can also obtain a bad parametrization of the data that may lead to a lack of identification problem. For instance, an additive outlier plus a LS at T are equivalent to an additive outlier at T and a LS at $T + 1$. Kaiser (1995) has studied *mixed outliers* of the form

$$z_t = \psi_1(B)\left(\omega_I I_t^{(T)} + \psi_2(B) a_t\right) \tag{6.19}$$

where the model $\psi(B)$ is factorized into two terms, $\psi_1(B)$ and $\psi_2(B)$; the first one defines the transfer function of the intervention model. In this way the outlier is not a complete innovative outlier neither an additive one. These types of outliers can be useful to define effects on different components of the ARIMA model.

6.3. PROCEDURES FOR OUTLIER IDENTIFICATION AND ESTIMATION

In order to eliminate the effect of an outlier in a given time series it is necessary to (1) detect the time at which the outlier happens; (2) identify the type of outlier, and (3) remove its effect by estimating a model in which the outlier is incorporated. The problem is complicated because we know neither the location nor the outlier type. An ingenious procedure proposed by Chang and Tiao (1983) and Chang et al. (1988) is as follows. At each sample point, we analyze what, if any, will be the most likely type of outlier. This is done by studying at each time the likelihood ratio for the types of outliers considered. We can compute the p value in the test for each type of outlier and take as more likely outlier at each point the one that leads to a smaller p-value. This will produce a sequence of most likely outliers at each timepoint, for instance, (IO, AO, IO, ..., LS, AO) and a time series of p values. Then we can choose as the candidate outlier timepoint the one that has associated the smallest p value, and as outlier type the corresponding outlier effect. Once the location and the type of outlier are defined, the effect of the outlier can be removed by fitting the appropriate intervention analysis model. We study in the following sections the implementation of this idea.

6.3.1. Estimation of outlier effects

Suppose first that the parameters of the ARIMA model are known, and let us analyze how to estimate the outlier size at a given time T when the type of outlier AO, IO, LS, or TC is known. The model from the observed series z_t is given by (6.17), where $V(B)$ is 1, $1/\pi(B)$, $1/(1-B)$, or $1/(1-\delta B)$, depending on the type of outlier. Let $e_t = \pi(B)z_t$ be the residuals from the observed series given the true values for the model parameters. We can write the model as

$$e_t = \omega_i x_t + a_t \qquad (6.20)$$

where for additive outliers $\omega_i = \omega_A$ and $x_t = \pi(B)I_t^{(T)}$, for innovational outliers $\omega_i = \omega_I$ and $x_t = I_t^{(T)}$, for level shift $\omega_i = \omega_L$ and $x_t = \pi(B)/(1-B)$, and for transitory changes $\omega_i = \omega_{TC}$ and $x_t = \pi(B)/(1-\delta B)$. As the model parameters are assumed to be known, ω_i can be estimated by least squares, leading to $\hat{\omega}_i = \sum e_t x_t / \sum x_t^2$, with variance $\sigma_a^2 (\sum x_t^2)^{-1}$.

In the AO case the estimation leads to

$$\hat{\omega}_A = \rho_A^2 \pi(F) e_T \qquad (6.21)$$

where F is the forward operator defined by $Fz_t = z_{t+1}$, and $\rho_A^2 = (1 + \pi_1^2 + \cdots + \pi_{n-T}^2)^{-1}$. Note that this result is consistent with the property that all the residuals after T are affected by the outlier and therefore all of them have information about it. For instance, for an AR(1) the estimate is

$$\hat{\omega}_A = \frac{e_T - \phi e_{T+1}}{1 + \phi^2}. \qquad (6.22)$$

Note that from (6.6) e_T is an unbiased estimate for ω_A with variance σ_a^2. However, again from (6.6), $-e_{T+1}/\phi$ is also an unbiased estimate for ω_A with variance σ^2/ϕ^2. The least-squares estimate combines this two sources of information and as they are independent (because they depend on the errors a_T and a_{T+1} that are independent) they should be weighted by their relative precisions (the inverse of their variances). In this way we will have to write

$$\hat{\omega}_A = \frac{\frac{1}{\sigma_a^2} e_T + \frac{\phi^2}{\sigma_a^2}(-e_{T+1}/\phi)}{\frac{1}{\sigma_a^2} + \frac{\phi^2}{\sigma_a^2}}$$

and we go back to (6.22). In the general case, as $e_{T+j} = -\pi_j \omega_A + a_{T+j}$, where the a_t are iid, we may build for each residual after T an estimate of the parameter by $\hat{\omega}_A^{(T+j)} = -e_{T+j}/\pi_j$. This set of estimates is $(e_T, -e_{T+1}/\pi_1, \ldots, -e_{T+j}/\pi_j)$, and all of them are unbiased and independent, with variance σ_a^2/π_j^2. Therefore the estimate (6.21) is just a linear combination of the available estimators with weights proportional to their relative precision. The variance is $\text{Var}(\hat{\omega}_A) = \rho_A^2 \sigma_a^2$.

For innovational outliers the only residual that has information about the outlier size is the one at T, and so, as one would expect, the estimate is

$$\hat{\omega}_I = e_T$$

and the variance is $\text{Var}(\hat{\omega}_I) = \sigma_a^2$.

In the LS case, all the residuals after the shift have information, and we would expect that the estimate combines all of them in some optimal way. The estimate is

$$\hat{\omega}_L = \rho_L^2 l(F) e_T$$

where $l(B) = \pi(B)/(1-B)$ and $\rho_L^2 = (1 + l_1^2 + l_2^2 + \cdots + l_{n-T}^2)^{-1}$, where l_i are the coefficients of $l(B)$. Note that, by the definition of level shift, we can obtain an unbiased estimate of its size at each time after T by e_{T+j}/l_j with variance σ_a^2/l_j^2. Then, as in the previous cases, the estimate is a linear combination of the unbiased estimates that can be obtained from the residuals, with weights that are equal to their relative precision. It can also be shown (Cheng and Tiao 1990) that this statistic measures the difference of levels before and after time T. The variance of the estimate is $V(\hat{\omega}_L) = \rho_L^2 \sigma_a^2$.

Finally, in the TC case

$$\hat{\omega}_{TC} = \rho_T^2 \beta(F) e_T$$

where $\rho_T^2 = (1 + \beta_1^2 + \beta_2^2 + \cdots + \beta_{n-T}^2)^{-1}$, in which β_i are the coefficients of $\beta(B) = \pi(B)/(1-\delta B)$ and $V(\hat{\omega}_{TC}) = \rho_T^2 \sigma_a^2$. This estimate has an interpretation similar to that in the previous ones.

6.3.2. Testing for outliers

In order to test whether one outlier of known type has occurred at time T, the standard test is

$$H_0: \quad \omega_j = 0$$
$$H_j: \quad \omega_j \neq 0$$

where $j = I, A, L, TC$. When the parameters are known we can use the e_t given by (6.20) and the test is equivalent to testing the slope in a simple regression model. As it is well known, the likelihood ratio criterion leads to comparing the estimated parameter to its standard error, and so the test statistic is

$$H_0 \text{ vs } H_j: \quad \lambda_{j,t} = \frac{\hat{\omega}_{j,t}}{\rho_{j,t} \sigma_a} \qquad (6.23)$$

where $\rho_{I,T} = 1$, and the distribution of $\lambda_{j,t}$ is student t. When the location of the outlier is unknown, Fox (1972) suggested to use the likelihood ratio test $\eta_t = \max_{1 \leq t \leq n}\{\lambda_{j,t}\}$. However, the sample distribution of η_t is complicated due to the correlation between the $\lambda_{j,t}$. The percentiles of the distribution can be found by simulation, and they were obtained by Chang et al. (1988). Ljung (1993) have suggested some approximations to this distribution.

When the parameters are unknown, the likelihood ratio test requires the estimation of them under both hypothesis. The estimation under the alternative hypothesis must be carried out at every T, which makes the testing process computationally very lengthy. Chang and Tiao (1983) proposed an iterative procedure in which the parameters are only estimated at each iteration under H_0. They showed by Monte Carlo that if the parameters are substituted by their consistent estimates under H_0, the $\lambda_{j,T}$ ($j = A, I, L, TC$) statistics are asymptotically distributed as $N(0, 1)$. Abraham and Yatawara (1988) obtained similar results using the score test. This procedure was generalized by Tsay (1988) to include LS and TC and runs as follows.

- In the first stage a model is fitted by maximum likelihood (ML) to the time series assuming that there is no outliers. Using the estimated parameters, the residuals e_t, its variance estimate (which may be a robust estimate, as the median of the absolute value of the residuals), and the likelihood ratios using the initial estimates as values for the parameters are computed at each timepoint.
- For all $t = 1, \ldots, n$ the statistic

$$\eta_t = \max\{|\hat{\lambda}_{i,t}|, i = A, I, L, TC|\}.$$

is computed. If $\max \eta_t = |\hat{\lambda}_{A,T}| \geq C$, where C is a predetermined constant, we assume an AO in $t = T$; if $\max \eta_t = |\hat{\lambda}_{I,T}| \geq C$, we assume an IO in $t = T$; and so on. The value of C is usually 3.5 or 4. When an outlier is detected, the residuals are corrected taking into account the type of outlier. For instance, if an IO is identified, a new residual is defined at this point by $\check{e}_t = \hat{e}_T - \hat{\omega}_I$, if an AO is detected, the residual is defined as $\check{e}_t = \hat{e}_t - \hat{\omega}_A \hat{\pi}(B) I_t^{(T)}$, for $t \geq T$, and if an LS (or TC) is detected, we have $\check{e}_t = \hat{e}_t - \hat{\omega}_{LS} g(B) I_t^{(T)}$, where $g(B) = l(B)$ for LS and $\beta(B)$ for TC.
- Using these new residuals, a new variance estimate $\check{\sigma}_e^2$ is obtained, and the likelihood ratios $\hat{\lambda}_{i,t}$ are again computed by using the new residuals $\check{e}_t$ and their variance $\check{\sigma}_e^2$. The identification of outliers and the computation of new residuals is repeated until no further points appear as outliers.
- In the second stage the sizes of the identified outliers and the parameters are jointly estimated. Let us assume that k possible outliers have been identified in positions $T_1, T_2, \ldots, T_k$. Then the following model is estimated:

$$z_t = \sum_{j=1}^{k} \omega_j V_{T_j}(B) I_t^{(T_j)} + \psi(B) a_t \qquad (6.24)$$

6.3. PROCEDURES FOR OUTLIER IDENTIFICATION AND ESTIMATION

where $V_{T_j}(B) = 1$ for AO, $V_{T_j}(B) = \psi(B)$ for IO, and so on. Using these new parameters estimates the process is repeated until no new outliers are found. The authors showed in a simulation study that this procedure seems to work very well when we have isolated outliers. Kabaila (1994) has shown that the optimal invariant detector of a single additive outlier in an unknown position performs closely to this likelihood ratio procedure.

An example

We will aply this procedure to the airline data considered in the previous chapter. The model fitted for the whole period from January 1949 to December 1960 (144 observations) is

$$\nabla\nabla_{12}z_t = (1 - .402B)(1 - .56B^{12})a_t$$

and the residuals of the fitted model are displayed in Figure 6.6.

The application of the outlier detection procedure using the SCA software leads to the following results:

Time	Estimate	T Value	Type
29	.041	4.07	AO
54	−.042	−3.50	LS
62	−.035	−3.43	AO
135	−.045	−3.85	AO

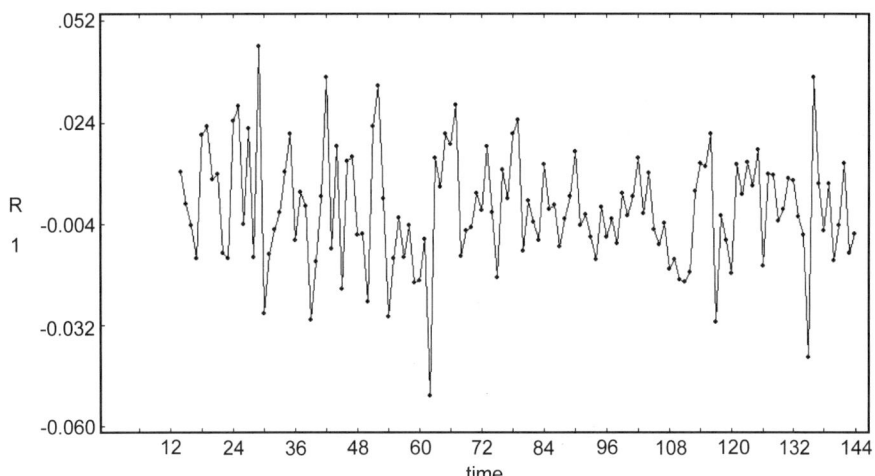

FIGURE 6.6 Residuals from the airline model.

6.4. INFLUENTIAL OBSERVATIONS

6.4.1. Influence on time series

It is well known that in a regression model we may have points that are not identified as outliers by the usual likelihood ratio tests for mean shift or variance shift and that have a strong influence on the parameter values of the model and hence on the forecast. These points have been called *influential observations* (Cook 1977, Cook and Weisberg 1982). The detection of influential observations is important to understand the sensitivity of the parameter values to a small fraction of the data.

Influential observations are found in static models in two ways. In the global approach each observation is deleted from the computations and a measure of the change that this deletion produces in some properties of the model (as the parameters or the forecast) is computed. In the local approach, (Cook 1986), the observation is assumed to have more variance (less weight) than the others. Several authors have proposed measuring the influence on an observation by the same idea as in regression, that is, deleting it from the computation of the correlations or the parameter estimates. For instance, Lefrançois (1991) proposed a measure of influence of an observation on the autocorrelation coefficient by computing the difference between the k-autocorrelation coefficient with and without observation T, that is, the observation is made equal to its unconditional mean so that the terms $(y_T - \bar{y})$ are zero. A similar approach is proposed by Abraham and Chuang (1989), who used as measure of influence and outlier detection for an AR(p) model the Draper and John (1981) statistic for regression. Again this measure is equivalent to substituting a value in the time series for its unconditional mean.

Peña (1986) and Brillinger (1987) proposed studying the global influence of an observation by making it a missing value. In this way the observation is substituted by its conditional expectation given the rest of the data instead of its unconditional mean. The difference can be quite strong if the correlation is high. Peña (1987) suggested a measure of influence on the parameter values of an ARIMA model as follows. First write the ARIMA model in the AR(h) approximation, that is, the ARIMA model is written as in (6.2):

$$y_t = \sum_{i=1}^{h} \pi_i y_{t-i} + a_t.$$

Then measure of the influence of observation T on the parameter values is

$$P_\pi(T) = \frac{(\hat{\pi} - \hat{\pi}_{(T)})' \hat{\Sigma}_\pi^{-1} (\hat{\pi} - \hat{\pi}_{(T)})}{h \hat{\sigma}_a^2}, \tag{6.25}$$

6.4. INFLUENTIAL OBSERVATIONS

where $\hat{\pi}$ is the maximum likelihood estimate of the parameter vector π assuming no outliers, $\hat{\pi}_{(T)}$ is the ML estimation assuming that observation at $t = T$ is missing, $\sum_\pi \sigma_a^2$ is the variance–covariance matrix of the $\hat{\pi}$ estimated vector, and h is the number of parameters. Note that this measure is general and can be used for stationary as well as for nonstationary processes. It is straightforward to show that calling $\hat{Z} = \mathbf{X}_z \hat{\pi}$ and $\hat{Z}_{(T)} = \mathbf{X}_z \hat{\pi}_{(T)}$ to the estimated vectors of forecasts computed from the two parameter vectors (with all the data and assuming that the observation in $t = T$ is missing) and $\mathbf{X}_z$ to the matrix

$$\mathbf{X}_z = \begin{pmatrix} z_h & z_{h-1} & \cdots & z_1 \\ z_{h+1} & z_h & \cdots & z_2 \\ \cdot & \cdot & \cdots & \cdot \\ \cdot & \cdot & \cdots & \cdot \\ \cdot & \cdot & \cdots & \cdot \\ z_{t-1} & z_{t-2} & \cdots & z_{t-h} \end{pmatrix},$$

since $(\hat{Z} - \hat{Z}_{(T)})'(\hat{Z} - \hat{Z}_{(T)}) = (\hat{\pi} - \hat{\pi}_{(T)})'(\mathbf{X}_z'\mathbf{X}_z)(\hat{\pi} - \hat{\pi}_{(T)})$, the influence measure (6.25) can be written as

$$P_{\hat{Z}}(T) = \frac{(\hat{Z} - \hat{Z}_{(T)})'(\hat{Z} - \hat{Z}_{(T)})}{h\hat{\sigma}_a^2} \tag{6.26}$$

and the advantage of this expression is that it can be computed by the ARIMA representation of the model and the AR approximation is not required. In this last expression $h = p + q$, the number of parameters used to compute the vector of forecasts.

Peña (1987) also proposed to measure the change on the variance by

$$D_v(T) = \frac{\hat{\sigma}^2 - \hat{\sigma}_{(T)}^2}{\hat{\sigma}_{(T)}^2} \tag{6.27}$$

where $\hat{\sigma}_{(T)}^2$ corresponds to a model in which observation T is assumed to be missing. He stressed that this last measure is a measure of outlyness of the point, and so its use is equivalent to an outlier test. This statistic has also been used by Bruce and Martin (1989) and by Ledolter (1990), who derive it by studying local influence measures in time series, and by Ljung (1993).

6.4.2. Influential observations and outliers

A simple way to measure the influence of an additive outlier at a given point over the parameters of a time series is to compute the Mahalanobis distance between the vector of ML estimated parameters assuming no outliers in the series and the same vector estimated by assuming an additive outlier at this point. In this way we obtain the effect that an additive outlier can have at each point of the sample. The same idea can be applied for other types of outliers.

Peña (1990) showed that the ML estimation of the parameters when an additive outlier is assumed at time T is carried out by considering the observation at time T as a missing value, which leads to substitute it by its conditional mean given the rest of the data. However, the ML estimation when an innovational outlier is assumed at time T is carried out by deleting the observation from the computations, which leads to substitute it by its marginal mean as in the static case. Since additive outliers have more influence on the parameters than innovative outliers, this author proposed as a standard measure of influence the statistic (6.25), where $\hat{\pi}_{(T)}$ is estimated by assuming that the observation is a missing data.

In (6.26) the value of the series at time T is not completely disregarded because it appears in the forecast vector $\hat{Z}_{(T)}$. An alternative measure of influence that avoids this can be built by looking directly to the change in the forecast vectors with and without the additive outlier. This measure is

$$D_{\hat{Z}}(T) = \frac{(\hat{Z} - \hat{Z}_T^{(INT)})'(\hat{Z} - \hat{Z}_T^{(INT)})}{h\hat{\sigma}_a^2} \qquad (6.28)$$

where h is the order of the ARIMA model or number of parameters, $\hat{\sigma}_a^2$ is the estimation of the white-noise variance, $\hat{Z}$ is the vector of forecasts assuming no outliers, and $\hat{Z}_T^{(INT)}$ is the vector of forecasts computed by assuming an additive outlier at $t = T$. This vector of forecasts is provided by the intervention model

$$\pi(B)\bigl(z_t - \omega_A I_t^{(T)}\bigr) = a_t \qquad (6.29)$$

where $\pi(B)$, ω_A and $I_t^{(T)}$ are as defined previously.

Note that in linear regression models a measure of influence based on the change in the parameter estimates is always equivalent to one based on the change in forecast, whereas in time series this equivalence is lost, because (6.25) and (6.28) are different. In fact, it is proved in Sánchez and Peña (1997) that they are related by

$$D_{\hat{Z}}(T) \simeq P_{\hat{Z}}(T) + \frac{\hat{\lambda}_{A,T}^2}{h} \qquad (6.30)$$

so that $D_{\hat{Z}}(T)$ can be interpreted as the effect of the change of the parameters plus the effect of the additive outlier. Monte carlo studies have shown that (6.28), which includes the outlier size, seems to be more effective in detecting outliers that have a strong influence on the model than is (6.25).

6.5. MULTIPLE OUTLIERS

6.5.1. Masking effects

The method presented in the previous section works very well when the series has a single outlier or a few isolated outliers. However, sometimes the series is subject to

6.5. MULTIPLE OUTLIERS

patches of outliers that may produce masking. The generalization of the outlier model (6.17) for k outliers is

$$z_t = \sum_{i=1}^{k} \omega_i V_i(B) I_t^{(T_i)} + y_t \qquad (6.31)$$

Assuming first, to simplify the presentation, that the parameters are known, and calling $e_t = \pi(B) z_t$, model (6.31) can be written as

$$e_t = x_t' \beta + a_t \qquad (6.32)$$

where $\beta' = (\omega_1, \ldots, \omega_k)$ and $x_t' = (x_{1t}, \ldots, x_{kt})$, with $x_{it} = \pi(B) V_i(B) I_t^{(T_i)}$.

Outlier identification methods based on estimating the effects of the outliers one by one use model (6.17) instead of model (6.31). These procedures are expected to work well when the matrix $(\sum_{t=1}^{n} x_t x_t')^{-1}$ is roughly diagonal, but may lead to serious biases when the series have patches of additive outliers and level shifts. Note that for an innovational outlier $x_{it} = I_t^{(T_i)}$, and therefore the estimation of its effect is typically uncorrelated with other effects. However, for additive outliers $x_{it} = \pi(B) I_t^{(T_i)}$, and the correlation between the effects of consecutive additive outliers can be very high. This is expected to happen when we have patches of outliers, an empirical fact found by Bruce and Martin (1989).

For instance, suppose that we have two consecutive additive outliers of magnitudes ω_1 and ω_2 at times T and $T+1$. Then $k=2$ and the expected value for the estimator of ω_1 using model (6.17), and assuming that it is the only outlier, which will be called $\hat{\omega}_1^{(s)}$, is given by

$$E(\hat{\omega}_1^{(s)}) = \omega_1 + \omega_2 \frac{\sum_{i=0}^{n-T-1} \pi_i \pi_{i+1}}{\sum_{i=0}^{n-T} \pi_i^2},$$

where $\pi_0 = -1$. As an example, if $\omega_1 = \omega_2 = \omega$ and the process is a random walk, the estimation assuming a single outlier at $t = T$ will be half of the true outlier value, and the variance will be $\sigma^2/2$, and so the expected value for the likelihood ratio will be $\omega/(\sqrt{2}\sigma)$. On the other hand, in the correct model the expected value for the estimation is ω and has a variance of $2\sigma^2/3$ leading to a expected likelihood ratio of $\sqrt{3}\omega/(\sqrt{2}\sigma)$. We see in this simple case, where the parameters are assumed to be known, that the presence of the second outlier is producing a masking effect that can lead to wrong outlier identification.

When the parameters are unknown, the problem is still more serious because a sequence of additive outliers can produce important biases on the parameter estimates. For instance, suppose that we have a sequence of k outliers of sizes $\omega_1, \ldots, \omega_k$, at times $T, T+1, \ldots, T+k-1$. It is shown in the appendix that calling $r_z(h)$ to the computed autocorrelation coefficients and $r_y(h)$ to the true autocorrelation coefficients without outliers effects, we have

$$r_z(h) = \frac{r_y(h) + n^{-1}S_1 + n^{-1}\left[\sum_{i=1}^{k-h}\omega_i\omega_{i+h} - \frac{k^2}{n}\bar{\omega}^2\right]}{1 + n^{-1}\left[\sum \tilde{\omega}_i^2 - \frac{k^2}{n}\bar{\omega}^2\right] + 2n^{-1}S_2}, \qquad (6.33)$$

where $\tilde{y}_t = (y_t - \bar{y})/s_y$, $\tilde{\omega}_i = \omega/s_y$, $\bar{\tilde{\omega}} = \sum \tilde{\omega}_i/k$, $ns_y^2 = \sum y_t^2 - n\bar{y}^2$, $S_1 = \sum \tilde{\omega}_i (\tilde{y}_{T-1+i-h} + \tilde{y}_{T-1+h+i})$, and $S_2 = \sum \tilde{\omega}_i \tilde{y}_{T-1+i}$. We may consider the two following cases:

1. All outliers have the same sign and similar size. Then, if $\omega \to \infty$, $r_z(h) \to (k-h)/k$ for $k > h$, and $r_z(h) \to 0$ for $k \leq h$. This implies that if k is small (1 or 2) the series will seem to be white noise, whereas if k is large, it will seem a nonstationary process.
2. Outliers have large random values. For instance, suppose they are drawn from a distribution of zero mean and very large variance. Then $r_z(h) \to 0$ and the series will seem to be white noise.

In the general case of patches of outliers of arbitrary size, the effect depends on the relative size of the patch and the sizes and length of the other patches and can be obtain from (6.33). In any case, if the outliers were correctly identified, the problem of the possible bias in their size can be overcome by a step of joint estimation of all the outlier candidates. However, the bias may be so strong that some outliers are masked and remain unidentified.

6.5.2. Procedures for multiple outlier identification

The first procedure for outlier patches detection is due to Bruce and Martin (1989). These authors generalized the method by Peña (1987) for measuring influential observations in time series to consecutive groups of influential data and proposed to identify outliers by using (6.27). They show that for AR(p) models with an isolated outlier the influence measure (6.25) may indicate as influential p values before the outlier and p values after and called this effect the smearing effect. They proposed a procedure based on computing (6.27) for groups of $k = 1, 2, 3\ldots$ but it is difficult to apply and does not seem to work well in general.

Chen and Liu (1993) extended the procedure of Chang et al. (1988) and Tsay (1988) to avoid masking in time series. They recognize that outliers may produce initial biases in the parameter estimates that affect all the procedure and that some outlier will not be identified because of a masking effect. These authors proposed a modification of previous procedures is as follows.

Stage 1. Initial parameter estimates are computed, and outliers are detected one by one by using the likelihood ratio criterion until no further outliers are found.

6.5. MULTIPLE OUTLIERS 157

Stage 2. The size of the identified outliers are estimated jointly. Then a backward elimination procedure is applied; that is, the minimum likelihood ratio for the detected outliers is computed, and if this value is smaller than the critical value C used to detect outliers, the point is eliminated of the set of outlier candidates. After any point is eliminated, the remaining outliers and the parameter values are again estimated by maximum likelihood, and the backward deletion is again applied.

Stage 3. Stages 1 and 2 are repeated until no further outliers are found.

This method has three main problems. The first one is the confusion between level shift and innovative outliers (in favor of the latter) when a level shift is present in a time series. This situation was pointed out by Balke (1993) with respect to Tsay (1988) procedure, but it can be applied as well to the procedure by Chen and Liu (1993). Balke indicated that (1) the presence of level shifts causes serious problems in the initial specification and estimation of the model; and (2) the use of the maximum likelihood ratio to distinguish between level shifts and innovation outlier does not work, because the expected value of the likelihood ratio for an innovative outlier exceeds the expected value of the likelihood ratio for level shifts. Thus, level shifts will be often wrongly identified as innovative outliers.

The second is the biased estimation of the initial parameter values. This problem arises because the initial estimation of the parameters is made under the hypotheses of no outliers in the data, which may lead to begin the search for outliers by using a very biased set of parameters and, as a consequence of this, the procedure may fail. This situation is spatially important, as shown by Balke (1993), when the series has a level shift, but it can also appear for patches of outliers of similar size that, as shown before, produce a similar effect to a level shift.

The third problem is masking. It appears mainly when there is a sequence of consecutive additive outliers, because the usual procedures based on the identification of outliers one by one may fail in the identification of some of the members of the group.

We analyze with more detail these problems and some possible solutions in the next sections. The proposed solutions have been implemented in a procedure for multiple outlier identification and robust estimation proposed by Sánchez and Peña (1997).

Confusion between innovative outliers and level shifts under H_0

Suppose that we have a LS on a AR(1) stationary series. Then the estimated parameter will be close to one and the series will seem to follow a random walk. As in a random walk an IO is identical to a LS, there is a clear possibility of confusion between both. In the general case, the larger AR root will go to one, and the problem will be similar. The problem is complicated because when choosing between an IO and a LS by the likelihood ratio test the distribution of these statistics under the null hypothesis makes it very easy to identify a LS as an IO. This problem was identified by Balke (1993), who proposed to add an additional search for outliers using a white-noise model.

However, this idea has two problems. First the white noise model does not allow to distinguish additive from innovative outliers, and second, it may lead to spurious level shifts that will be not easy to detect.

Let us show that the distribution of both statistics is different under the null. The statistics for testing for an IO is

$$\hat{\lambda}_{I,T} = \frac{\hat{e}_T}{\hat{\sigma}_a}$$

whereas for an LS is

$$\hat{\lambda}_{L,T} = \frac{\hat{e}_T - \sum_{i=1}^{n-T} \hat{l}_i \hat{e}_{T+i}}{\hat{\sigma}_a \left(1 + \sum_{i=1}^{n-T} \hat{l}_i^2\right)^{1/2}}$$

where $\hat{l}_1 = -1 + \hat{\pi}_1$, $\hat{l}_2 = -1 + \hat{\pi}_1 + \hat{\pi}_2, \ldots, \hat{l}_{n-T} = -1 + \sum_{i=1}^{n-T} \hat{\pi}_i$. Then the relationship between $\hat{\lambda}_{I,T}$ and $\hat{\lambda}_{L,T}$ is:

$$\hat{\lambda}_{L,T} = \frac{\hat{\lambda}_{I,T} + \frac{\sum_{i=1}^{n-T} \left(\hat{e}_{T+i}\left(1-\sum_{j=1}^{i} \hat{\pi}_j\right)\right)}{\hat{\sigma}_a}}{\left(1 + \sum_{i=1}^{n-T} \left(-1 + \sum_{j=1}^{i} \hat{\pi}_j\right)^2\right)^{1/2}}. \tag{6.34}$$

For AR(p) models, the closer to unity each of the elements of the sequence ϕ_1, $\phi_1 + \phi_2, , \sum_{i=1}^{p} \phi_i$ are, the nearer the LS critical values will be to the IO critical values. For an invertible ARMA model, under H_0 (no outliers), when $t = T$ is not close to the end of the series, for large n the second term will go to zero and the likelihood ratio for level shifts, $\hat{\lambda}_{L,T}$, is expected to be smaller than the likelihood ratio for innovational outliers, $\hat{\lambda}_{I,T}$.

This result suggests that for invertible ARMA models the statistics for level shift and innovative outliers should not be compared together, because the critical values under the null hypothesis can be quite different. In order to check this result in finite samples Sánchez and Peña (1997) carried out a simulation study of the distribution of these two statistics. These authors obtained the critical values for the likelihood ratio statistics confirming that the detection method based on $\eta_t = \max_t\{|\hat{\lambda}_{i,t}|\}$ seems to be inadequate, because the sampling behavior of the maximum value of the statistic for the LS is different to the corresponding ones for IO and AO. The confusion between LS and IO can be avoided by, on the one hand, comparing IO versus AO, and, on the other, dealing with LS alone.

Improving the initial parameter estimates
The possible bias in the initial search for outliers can cause the procedure to fail. It would be convenient to start the search with parameter estimates that are not strongly affected by some data points that may be outliers. A way to detect the observations which have the strongest effects on the parameter estimates is to use the measure of influence defined in the previous section. We can assume then that the influential

6.5. MULTIPLE OUTLIERS

observations are missing (which is equivalent to assume additive outliers at these points) and compute the parameters under these hypothesis. Sánchez and Peña (1997) checked that this idea works well with isolated outliers but when the time series has a LS or a sequence of consecutive additive outliers of a similar size, which produces a behavior similar to a LS, the influence measure (6.28) detects as influential observations a low percentage of the observations affected by the LS or included in that sequence. This is the masking effect. Then, if we delete observations according only to $D_{\hat{Z}}(T)$ several outliers will be undetected, and will biased the initial parameter estimates. To avoid this situation these authors define an influence measure for LS or sequences of outliers which can be used jointly with (6.28) to carry out the initial cleaning of the sample data.

The following measure is proposed to check the effects of patches of observations

$$DL(T) = \frac{(\hat{Z} - \hat{Z}_T^{(ILS)})'(\hat{Z} - \hat{Z}_T^{(ILS)})}{h\hat{\sigma}_a^2} \tag{6.35}$$

where $\hat{Z}$ is the vector of forecasts assuming no outliers effects ever the series and $\hat{Z}_T^{(ILS)}$ is the vector of forecasts assuming a level shift at time T that is estimated by the intervention model.

As in the case of additive outliers, we could have measure the influence of LS by analyzing the change in the parameter estimates. Calling $\hat{\pi}$ to the MLE of π supposing no outliers, and $\hat{\pi}_L$ to the MLE considering that there exists a LS in $t = T$ we could use the Mahalanobis distance between these estimates as in (6.25) to build a measure of the change in the parameters. Sánchez and Peña (1997) showed that

$$DL(T) \simeq PL(T) + \frac{\hat{\lambda}_{L,T}^2}{h} \tag{6.36}$$

and, as in the additive outlier case, $DL(T)$ can be interpreted as the effect of the change of the parameters plus the effect of a level shift.

In summary, we can improved the initial estimation of the parameters by using the influence measures (6.28) and (6.35) to correct the series of all the points, which seems to have a strong effect on the parameter estimates, and then compute the initial estimated parameters from this corrected series.

Outlier patches

As indicated before, patches of outliers are very difficult to detect when searching for outliers one by one. As an example of this consider three consecutives AO at times T, $T + 1$, and $T + 2$ of the same size ω in the process $(1 - \phi B)z_t = a_t$. It is straightforward to show using (6.6) and (6.22) that when we know the true parameter value, the estimation of the size of the outlier at these three position will be $\omega(1 - \phi(1 - \phi))/(1 + \phi^2)$, $\omega(1 - \phi)^2/(1 + \phi^2)$, and $\omega(1 - \phi(1 - \phi))/(1 + \phi^2)$. For instance, if $\phi = .7$, these estimates will be $.53\omega$, $.06\omega$, and $.53\omega$. It is clear that we may hope to identify the beginning and the end of the sequence, but it is very unliky that we can identify the outlier in the middle. The problem is more serious if the parameter is estimated from the data. Then φ will be close to one and the estimated values will be ω/2, 0, and ω/2.

Note that a set of k consecutive outliers of size ω from T to $T + k - 1$, or an outlier patch of size k, is equivalent to two level shifts. The first at time T and size ω and the second at time $T + k - 1$ of size $-\omega$. Therefore outlier patches can be eliminated approximately by cleaning the series of level shifts. More flexibility is obtained by allowing the possibility of more than one outlier at any given period. Then the possibility of different sizes in an outlier patch is taken into account by incorporating additive outliers at some points of the level shift period. Justel et al. (1998) have studied alternative procedures for patches of outlier detection in the Bayesian framework.

6.6. MISSING-VALUE ESTIMATION

The study of additive outliers and influential observations in time series is closely related to missing-value analysis because an outlier at T implies that the true value at this point is not observed. Thus we have a missing-value problem. For that reason we will review briefly some of the most important results in this area.

6.6.1. Optimal interpolation and inverse autocorrelation function

Suppose first that we have a stationary time series with a missing observation at time T. Then the estimation of the missing value is called the *interpolation problem* and it is solved by computing the expectation of the unobserved random variable given the rest of the data. Grenander and Rosenblatt (1957) showed that this expectation is given by

$$E(z_T / Z_{(T)}) = -\sum_{i=1}^{\infty} \delta_i (z_{T+i} + z_{T-i}) \qquad (6.37)$$

where δ_i are the inverse autocorrelation coefficients, and $Z_{(T)}$ includes all the data but the missing value. Brubacher and Tunniclife-Wilson (1976) obtained also this result by a least-squares approach. A simple way to define the inverse autocorrelation function is as follows (Peña and Maravall, 1991). Define the dual process of an invertible ARIMA model (6.1) as the ARMA process

$$\theta(B) z_t = \phi(B) \nabla^d a_t \qquad (6.38)$$

that is, the dual process is built by interchanging the role of the AR and MA operators. Then the autocorrelation function of the dual process (6.38) is the inverse autocorrelation function (IAF) of the original process (6.1). For instance, the AR(1) process $(1 - \phi B) z_t = a_t$ will have an inverse autocorrelation function (IAF) that is the autocorrelation function of the MA(1) process $z_t = (1 - \phi B) a_t$. Therefore the

6.6. MISSING-VALUE ESTIMATION

IAF has the first autocorrelation coefficient equal to $-\phi/(1 + \phi^2)$ and all the other values are equal to zero. Using (6.37), the *optimal interpolator* for a zero mean AR(1) process is

$$E(z_T/Z_{(T)}) = \frac{\phi}{1 + \phi^2}(z_{T+1} + z_{T-1}). \tag{6.39}$$

Note that the dual process of an invertible one is stationary and therefore the inverse autocorrelation function always exits. Peña and Maravall (1991) showed that the result (6.37) can be used for stationary as well as nonstationary processes. The optimal interpolator has a simple interpretation. Let us write the time series in the general AR(∞) representation

$$z_t = \sum_{i=1}^{\infty} \pi_i z_{t-i} + a_t. \tag{6.40}$$

Then, if the value z_T is missing, we can obtain an unbiased estimate of it by using

$$\hat{z}_T^{(0)} = \sum_{i=1}^{\infty} \pi_i z_{t-i} \tag{6.41}$$

and this estimate, which is built from the previous observations to the missing value will have variance σ_a^2. However, we have more information z_T. This information is contained on all observations after the missing value. That is, we can write for all j such that $\pi_j \neq 0$

$$z_T = \pi_j^{-1}\left(z_{T+j} - \sum_{\substack{i=1 \\ i \neq j}}^{\infty} \pi_i z_{T+t-i}\right) - \frac{a_{T+j}}{\pi_j} \tag{6.42}$$

and, therefore, we can obtain additional "backward" unbiased estimates of z_T from this equation by

$$\hat{z}_T^{(j)} = \pi_j^{-1}\left(z_{T+j} - \sum_{i \neq j} \pi_i z_{T+t-i}\right) \tag{6.43}$$

with variance σ_a^2/π_j^2. As all these estimates are conditionally unbiased and independent given the observed data, the best linear unbiased estimate of the missing value z_T will be

$$\hat{z}_T = \sum_{j=0}^{\infty} \frac{\pi_j^2}{\sum \pi_j^2} \hat{z}_T^{(j)} \tag{6.44}$$

where $\pi_0 = -1$. It is easy to show that this estimate is equivalent to the well known expression for the missing value estimation in a Gaussian stationary time series given by (6.37). An advantage of formulation (6.44) is that it provides a clear understanding of how to proceed when the missing value is near the extremes of the series so that the two-sided symmetric filter (6.37) has to be truncated. Then, we have to combine forward estimates with the $n - T$ backward estimates that are available, and the exact formula for the finite sample interpolator is

$$\hat{z}_{T,F} = \sum_{j=0}^{n-T} \frac{\pi_j^2}{\sum_0^n \pi_j^2} \hat{z}_T^{(j)} \qquad (6.45)$$

The previous results are easily generalized to groups of missing observations. We will illustrate the computation of the optimal interpolator for more than one missing data by a simple example: suppose we have an AR(1) process in which the values z_T and z_{T+1} are missing. Then, we can compute the optimal interpolators as follows. For z_T we have the forward estimate:

$$\hat{z}_T^{(0)} = \phi z_{T-1} \qquad (6.46)$$

with error variance σ_a^2, and now, as $z_{T+2} = \phi^2 z_T + \phi a_{T+1} + a_{T+2}$, we can compute the backward estimate

$$\hat{z}_T^{(2)} = \phi^{-2} z_{T+2} \qquad (6.47)$$

with variance $\sigma_a^2(1 + \phi^2)/\phi^4$. Therefore, the best linear unbiased estimate will be

$$\hat{z}_T = \frac{\phi(1 + \phi^2)}{1 + \phi^2 + \phi^4} z_{T-1} + \frac{\phi^2}{1 + \phi^2 + \phi^4} z_{T+2} \qquad (6.48)$$

which agrees with the general formula obtained by a different approach in Peña and Maravall (1991). The estimate of $\hat{z}_{T+1}$ will be similar to (6.48) but with the roles of z_{T-1} and z_{T+2} reversed.

6.6.2. Estimation of missing values

The previous analysis suggests the following procedure for computing missing values in time series: (1) performed a first interpolation of the missing values, identify the ARIMA model and estimate its parameters by ML in the completed series and (2) obtain the inverse autocorrelation coefficients, that are straightforward given the model, and compute the optimal interpolators of the missing values by (6.37). This procedure can be iterated, that is once the series have been completed by the optimal interpolators we can compute another set of parameter estimates, which will lead to new missing-value interpolation and so on. The iteration is important when the number of missing values is large, because the first parameter estimation based on some crude interpolation may lead to biased parameter estimates.

If our objective is to estimate the model parameters, we can apply maximum likelihood directly to the observed data. In order to undestand the procedure, we will illustrate it in the simple case of an AR(1). The likelihood function computed by the

6.6. MISSING-VALUE ESTIMATION

prediction error decomposition assuming that z_T is missing is easy to obtain using that

$$f(z_1, \ldots z_{T-1} z_{T+1} \ldots z_n) = f(z_1) f(z_2/z_1) \cdots f(z_{T+1}/z_{T-1}) f(z_{t+2}/z_{T+1}) \cdots f(z_n/z_{n-1}).$$

Assuming normality and that the process has zero mean and conditioning on the first observation, the conditional loglikelihood function can be easily obtained by noting that for all $t \geq 2$ but $t \neq T$ we have that $f(z_t/z_{t-1})$ is $N(\phi z_{t-1}, \sigma^2)$, and the distribution $f(z_{T+1}/z_{T-1})$ is also normal with parameters $N(\phi^2 z_{T-1}, \sigma^2(1 + \phi^2))$. Thus the conditional loglikelihood function to be maximized is

$$l(\phi, \sigma^2/z_1) = -\frac{(n-2)}{2} \ln \sigma^2 - \frac{1}{2} \ln(1 + \phi^2) - \sum_{t \in A} \frac{(z_t - \phi z_{t-1})^2}{2\sigma^2}$$

$$- \frac{(z_{T+1} - \phi^2 z_{T-1})^2}{2\sigma^2(1 + \phi^2)} \tag{6.49}$$

where the set A is $\{1, 2, \ldots, T-1, T+2, \ldots, n\}$. Let us compare this function with one obtained for a series without missing values but that has an additive outlier at time T. Then the model can be written as

$$z_t = \phi z_{t-1} + \omega I_t^{(T)} - \phi \omega I_{t-1}^{(T)} + a_t$$

and the conditional loglikelihood function is

$$l(\omega, \phi, \sigma^2/z_1) = -\frac{(n-1)}{2} \ln \sigma^2 - \sum_{t \in A} \frac{(z_t - \phi z_{t-1})^2}{2\sigma^2} - \frac{(z_T - \phi z_{T-1} - \omega)^2}{2\sigma^2}$$

$$- \frac{(z_{T+1} - \phi(z_T - \omega))^2}{2\sigma^2}. \tag{6.50}$$

The estimation in (6.50) can be carried out in two steps. In the first one, conditioning on ϕ and obtain the estimate for ω given ϕ. Differentiating (6.50) with respect to ω and setting the result equal to zero we obtain

$$(z_T - \phi z_{T-1} - \omega) = (z_{T+1} - \phi(z_T - \omega))\phi$$

which leads to

$$\hat{\omega} = z_T - \frac{\phi}{(1 + \phi^2)}(z_{T+1} + z_{T-1}). \tag{6.51}$$

Note that this estimate can be interpreted as the difference between the observed value and its optimal interpolator by using the AR(1), as given by (6.39). Inserting this equation in (6.50), we obtain

$$l(\phi, \sigma^2/z_1) = -\frac{(n-1)}{2}\ln\sigma^2 - \sum_{t\in A}\frac{(z_t - \phi z_{t-1})^2}{2\sigma^2} - \frac{(z_{T+1} - \phi^2 z_{T-1})^2}{2\sigma^2(1+\phi^2)} \quad (6.52)$$

and it is clear that (6.50) and (6.52) will be equivalent for large n. Note that this function does not depend on z_T.

We have shown in this simple example that for moderate sample size the likelihood function for estimating the parameters of a series with missing values will be similar to the likelihood function of a model that fills the hole in the series with any arbitrary value and then assumes an additive outlier at this point. This relationship was first found by Peña (1987). This relationship suggests a simple procedure to estimate missing values in time series. Fill the holes with arbitrary values (e.g., equal to zero) and assume that this complete time series has AOs at the missing value positions. The parameters estimates obtained form the intervention analysis model are the parameter estimates for the series with missing values, and the estimation of the missing values are the AO sizes estimates with opposite sign. Note that we can generalize (6.51) as

$$\hat{\omega} = z_T - \hat{z}_{T/R}$$

where $\hat{z}_{T/R} = E(z_T/z_1, \ldots, z_{T-1}, z_{T+1}, \ldots, z_n)$ is the optimal interpolator. Therefore, if $z_T = 0$, we have that $\hat{\omega} = -\hat{z}_{T/R}$. Gomez et al. (1999) have shown that when the number of missing values is small this additive outlier approach can be faster and as accurate as the direct computation of the likelihood with missing values by the Kalman filter. Nieto and Martinez (1996) have presented an alternative recursive method based on restricted forecast to compute the missing values. Related work can be found in Shumway and Stoffer (1982), Harvey and Pierce (1984), Kohn and Ansley (1986), Battaglia and Bhansali (1987), Abraham and Chuang (1993), Ljung (1993), Gómez and Maravall (1994), and Peña and Tiao (1991).

6.7. FORECASTING WITH OUTLIERS

It has been often stressed that the prediction intervals computed from ARIMA models are too short, that is, forecasts are out of the bounds more often than it will be expected (see, e.g., Draper 1995). It is clear that the usual measures of forecast uncertainty take into account two sources of variability. The first is the probabilistic uncertainty due to the presence of noise in the model, which implies some uncertainty even if the model were known exactly. The second is the sample uncertainty, which is due to the fact that parameters are unknown and must be estimated. However, no attention is given to model uncertainty, which implies that the structure of the model is unknown and it has been either assumed or selected form the data by some criterion. This last

6.7. FORECASTING WITH OUTLIERS

source of uncertainty is, in many cases, the most important of the three source of uncertainty.

For instance, real-time series data quite often have outliers and this source of uncertainty can be taken into account by forecasting from a model that allows for this possibility. Suppose that we have data $(1, \ldots, t-1)$ and we assume that the future value of the time series, z_t, may be affected by an outlier of unknown type. This implies that the model for this observation is

$$z_t = \sum_{i=1}^{4} \omega_{it} V_i(B) u_{it} + \psi(B) a_t,$$

where the terms $V_i(B)$ correspond to the four outliers types defined before [see (6.17)] and the sizes are given by random variables ω_{it} with some specified distribution. The variables u_{it} are Bernoulli variables independent of the ω_{it} indicating the probability for each type of oulier, so that $P(u_{it} = 1) = \alpha_i$ and $P(u_{it} = 0) = 1 - \alpha_i$, and $\sum \alpha_i = \alpha$ gives the probability of any type of outlier affecting the series. We assume for simplicity (although this assumption can be dropped without much trouble) that at each time only a type of outlier can happen and so $P(u_{it} = 1, u_{jt} = 1) = 0$ for any pair $ij, i \neq j$. The sequence of variables $\{u_{it}, t = n, \ldots\}$ will have a covariance matrix M_i, and, as particular cases, we can consider (1) independence, so that this matrix is diagonal, or (2) some kind of dependency as a Markov chain. The specification of the probabilities and the distribution for the outliers sizes can be done by looking at the past history of the series.

To illustrate the behavior of a model of this type, let us consider the simplest case in which we allow for the possibility of additive outliers with probability $\alpha = P(u_t = 1)$ with a size ω coming from some distribution with mean $\eta\sigma$ and variance $\lambda\sigma^2$, where σ^2 is the variance of the noise. Suppose that we have data until time n. Then for $t > n$

$$z_t = \omega_t u_t + y_t$$

where $y_t = \psi(B) a_t$ represent the model without outliers. The one-step-ahead forecast will be

$$\hat{z}_t(1) = \eta\alpha\sigma + \hat{y}_t(1).$$

The mean square forecast error (MSFE) will be

$$\text{MSFE}(1) = E(z_{t+1} - \hat{z}_t(1))^2 = \sigma^2(1 + \lambda\alpha + \alpha(1-\alpha)\eta^2)$$

where we have used the independence of ω_{t+1} and u_{t+1} and that $E[\omega_{t+1}^2] = \sigma^2(\lambda+\eta^2)$ and $E[u_{t+1}^2] = \alpha$. For instance, suppose that $\alpha = .1$, $\eta = 5$, and $\lambda = 2$. Then the variance of the one-step-ahead forecast will increase more than 3 times, due to the small uncertainty of the presence of an additive outlier. We also see that the main source of increase in uncertainty comes from the mean and no for the variance of the distribution of the outlier size.

This analysis is related to Hamilton (1989), who used Markov chains to model changes in regime in economic time series, and to McCulloch and Tsay (1994), who use a Bayesian LS model to improve prediction in AR models.

6.8. OTHER APPROACHES

A comparison of several robust estimation procedures can be found in Schick and Mitter (1994). A Bayesian approach to deal with outliers can be seen in Albert and Chib (1993), McCulloch and Tsay (1994), and Justel et al. (1998). Multivariate outliers have been study by Tsay et al. (2000).

6.9. APPENDIX

Suppose that we have the sequence of outliers $\omega_1, \ldots, \omega_k$ at time $T, \ldots, T+k-1$. Then

$$r_z(h) = \frac{\sum z_t z_{t-h} - n\bar{z}^2}{\sum z_t^2 - n\bar{z}^2} = \frac{A}{B}.$$

The components of the numerator are

$$\sum z_t z_{t-h} = \sum y_t y_{t-h} + \sum_{i=1}^{k} \omega_i (y_{T-1+i-h} + y_{T-1+h+i}) + \sum_{i=1}^{k-h} \omega_i \omega_{i+h}$$

and

$$n\bar{z}^2 = n\bar{y}^2 + \frac{k^2}{n}\bar{\omega}^2 + 2\bar{y}k\bar{\omega}$$

leading to

$$A = \sum y_t y_{t-h} - n\bar{y}^2 + \sum \omega_i (y_{T-1+i-h} - \bar{y} + y_{T-1+h+i} - \bar{y})$$
$$+ \sum_{i=1}^{k-h} \omega_i \omega_{i+h} - \frac{k^2}{n}\bar{\omega}^2.$$

In the denominator, we have

$$\sum z_t^2 = \sum y_t^2 + \sum \omega_i^2 + 2\sum \omega_i y_{T-1+i}$$

and

$$B = \sum y_t^2 - n\bar{y}^2 + \sum \omega_i^2 - \frac{k^2}{n}\bar{\omega}^2 + 2\sum \omega_i(y_{T-1+i} - \bar{y}).$$

Then, calling $\tilde{y}_t = (y_t - \bar{y})/s_y$, $\tilde{\omega} = \omega/s_y$, $\tilde{\bar{\omega}} = \sum \tilde{\omega}_i$, $ns_y^2 = \sum y_t^2 - n\bar{y}^2$, $S_1 = \sum \tilde{\omega}_i (\tilde{y}_{T-1+i-h} + \tilde{y}_{T-1+h+i})$, $S_2 = \sum \tilde{\omega}_i \tilde{y}_{T-1+i}$ we have

$$r_z(h) = \frac{r_y(h) + n^{-1}S_1 + n^{-1}\left[\sum_{i=1}^{k-h} \tilde{\omega}_i \tilde{\omega}_{i+h} - \frac{k^2}{n}\tilde{\bar{\omega}}^2\right]}{1 + n^{-1}\left[\sum \tilde{\omega}_i^2 - \frac{k^2}{n}\tilde{\bar{\omega}}^2\right] + 2n^{-1}S_2}.$$

Suppose now that all the outliers have the same size, $\tilde{\omega}_i = \omega$, then

$$r_z(h) = \frac{r_y(h) + n^{-1}\omega \sum S_1 + n^{-1}\left[(k-h)\omega^2 - \frac{k^2}{n}\omega^2\right]}{1 + n^{-1}\left[k\omega^2 - \frac{k^2}{n}\omega^2\right] + 2n^{-1}\omega \sum S_2}.$$

If $\omega \to \infty$

$$\lim_{\omega \to \infty} r_z(h) = \frac{k - h - \frac{k^2}{n}}{k - \frac{k^2}{n}} \to \frac{k-h}{k}, \quad k > h.$$

On the other hand, for $k \leq h$

$$\lim_{\omega \to \infty} r_z(h) = \frac{-n^{-1}\frac{k^2}{n}\omega^2}{n^{-1}\left[k\omega^2 - \frac{k^2}{n}\omega^2\right]} = -\frac{\frac{k^2}{n}}{k - \frac{k^2}{n}} \to 0.$$

When the outliers are different, calling $\sum \tilde{\omega}_i^2 - k\bar{\omega}^2 = ks_\omega^2$, and $k\text{Cov}(h) = \sum \tilde{\omega}_i \tilde{\omega}_{i+h} - k\tilde{\bar{\omega}}^2$, we have that

$$r_z(h) = \frac{r_y(h) + n^{-1}S_1 + n^{-1}k\text{Cov}(h) + n^{-1}k\tilde{\bar{\omega}}^2\left(\frac{n-k}{n}\right)}{1 + n^{-1}ks_\omega^2 + n^{-1}k\tilde{\bar{\omega}}^2\left(\frac{n-k}{n}\right) + 2n^{-1}S_2}.$$

ACKNOWLEDGMENT

The author is very grateful to Maria Angeles Carnero, Esther Ruiz, and Ruey Tsay for many useful comments on a previous draft of this chapter.

REFERENCES

Abraham, B. (1980). Missing observations in time series. *Commun. Stat. (Theory Meth.)* **16**, 1643–1653.

Abraham, B. and Box, G. E. P. (1979). Bayesian analysis of some outlier problems in time series. *Biometrika* **66**, 229–236.

Abraham, B. and Chuang, A. (1989). Outlier detection and time series modelling. *Technometrics* **31**, 241–248.

Abraham, B. and Chuang, A. (1993). Expectation-maximization algorithms and the estimation of time series models in the presence of outliers. *J. Time Series Anal.* **14**(3), 221–234.

Abraham, B. and Yatawara, N. (1988). A score test for detection of time series outliers. *J. Time Series Anal.* **9**, 109–119.

Albert, J. H. and Chib, S. (1993). Bayes inference via Gibbs sampling of autoregressive time series subject to Markov mean and variance shifts. *J. Business Econ. Stat.* **11**, 1–15.

Balke, N. S. (1993). Detecting level shitfs in time series. *J. Business Econ. Stat.* **11**(1), 81–92.

Battaglia, F. and Bhansali, R. J. (1987). Estimation of the interpolation error variance and an index of linear determinism. *Biometrika* **74**, 771–779.

Box, G. E. P. and Jenkins, G. M. (1976). *Time Series Analysis, Forecasting and Control.* Holden-Day, San Francisco.

Box, G. E. P. and Tiao, G. C. (1975). Intervention analysis with applications to economic and environmental problems. *J. Am. Stat. Assoc.* **70**, 70–79.

Brillinger, D. (1987). Discussion of influence functional for time series by Martin and Yohai. *Ann. Stat.* 819–822.

Brubacher, S. R. and Tunnicliffe-Wilson, G. (1976). Interpolating time series with application to the estimation of holiday effects on electricity demand. *J. Roy. Stat. Soc. C* **25**, 107–116.

Bruce, A. G. and Martin, D. (1989). Leave-k-out diagnostics for time series (with discussion). *J. Roy. Stat. Soc. B* **51**(3), 363–424.

Chan, W. (1995). Understanding the effect of time series outliers on sample autocorrelations. *Test* **4**(1), 179–186.

Chang, I. (1982). *Outliers in time series.* Unpublished Ph.D. thesis. Univ. Wisconsin, Dept. Madison.

Chang, I. and Tiao, G. C. (1983). *Estimation of Time Series Parameters in the Presence of Outliers.* Technical Report 8, Statistics Research Center, Univ. Chicago.

Chang, I., Tiao, G. C., and Chen, C. (1988). Estimation of time series parameters in the presence of outliers. *Technometrics* **30**(2), 193–204.

Chen, C. and Liu, L. (1993). Joint estimation of model parameters and outlier effects in time series. *J. Am. Stat. Assoc.* **88**(421), 284–297.

Chen, C. and Tiao, C. G. (1990). Random level-shift time series models, ARIMA approximations, and level-shift detection. *J. Business Econ. Stat.* **8**(1), 83–97.

Chuang, A. (1987). *Outliers in Time Series.* Unpublished manuscript. Univ. Waterloo, Dept. Statistics and Actuarial Science.

Cook, D. R. (1977). Detection of influential observations in linear regression. *Technometrics* **19**, 15–18.

Cook, D. R. (1986). Assessement of local influence. *J. Roy. Stat. Soc. B* **48**, 133–169.

REFERENCES

Cook, R. D. and Weisberg, S. (1982). *Residuals and Influence in Regression*. Chapman and Hall, London.

Draper, D. (1995). Assessment and propagation of model uncertainty. *J. Roy. Stat. Soc. B* **57**, 45–70.

Draper, N. R. and John, J. A. (1981). Influential observations and outliers in regression. *Technometrics* **23**, 21–26.

Fox, A. J. (1972). Outliers in time series. *J. Roy. Stat. Soc. B* **34**, 350–363.

Gómez, V. and Maravall, A. (1994). Estimation, prediction and interpolation for nonstationary series with the Kalman filter. *J. Am. Stat. Assoc.* **89**, 611–624.

Gómez, V., Maravall, A., and Peña, D. (1999). Missing observation in ARIMA models: Skipping approach versus additive outlier approach. *J. Econ.* **88**, 341–363.

Grenander, V. and Rosenblatt, M. (1957). *Statistical Analysis of Stationary Time series*, Wiley, New York.

Guttman, I. and Tiao, G. C. (1978). Effect of correlation on the estimation of a mean in the presence of spurious observations. *Can. J. Stat.* **6**, 229–247.

Hamilton, J. (1989). A new approach to the economic analysis of nonstationary economic time series and the business cycle. *Econometrica* **57**, 357–384.

Harvey, A. C. and Pieree, R. G. (1984). Estimation procedures for structural time series models. *J. Am. Stat. Assoc.* **79**, 125–131.

Justel, A., Peña, D., and Tsay, R. S. (1998). *Detection of Outlier Patches in Autoregressive Time Series*. Technical Report, Statistics and Econometric series, Univ. Carlos III de Madrid.

Kabaila, P. (1994). The detection of a single additive outlier of unknown position. *J. Time Series Anal.* **15**(5), 507–522.

Kaiser, R. (1995). *Observationes Atipicas en Series Temporales. El Tipo Mixto*. Ph.D. thesis. European Institute, Florence.

Kohn, R. and Ansley, C. F. (1986). Estimation, prediction and interpolation for ARIMA models with missing data. *J. Am. Stat. Assoc.* **81**, 751–761.

Ledolter, J. (1989). The effect of additive outliers on the forecasts from ARIMA models. *Int. J. Forecasting* **5**, 231–240.

Ledolter, J. (1990). Outlier diagnostics in time series models. *J. Time Series Anal.* **11**, 317–324.

Lefrançois, B. (1991). Detecting over-influential observations in time series. *Biometrika* **78**(1), 91–99.

Ljung, G. (1982). The likelihood function for a stationary Gaussian autoregressive moving average process with missing observations. *Biometrika* **69**, 265–268.

Ljung, G. M. (1989). Outliers and missing observations in time series. *ASA Proc. Business Econ. Sect.* 397–401.

Ljung, G. (1993). On outlier detection in time series. *J. Roy. Stat. Soc. B* **55**(2), 559–567.

McCulloch, R. E. and Tsay R. S. (1994). Bayesian analysis of autoregressive time series via the Gibbs sampler. *J. Time Series Anal.* **15**, 235–250.

Miller, R. B. (1980). Comment on "Robust estimation of autoregressive models." In D. R. Brillinger and G. C. Tiao (eds.), *Directions in Time Series*. Institute of Mathematical Statistics, Hayward, CA.

Muirhead, C. R. (1986). Distinguishing outlier types in time series. *J. Roy. Stat. Soc. B* **48**(1), 39–47.

Nieto, F. H. and Martinez, J. (1996). A recursive approach for estimating missing observations in a univariate time series. *Commun. Stat. (Theory Meth.)* **25**(9), 2101–2116.

Peña, D. (1986). Discussion of local influence by D. Cook. *J. Roy. Stat. Soc. B* **48**(2), 164–165.

Peña, D. (1987). Measuring the importance of outliers in ARIMA models. In Puri (ed.), *New Perspectives in Theoretical and Applied Statistics,* pp. 109–118. Wiley, New York.

Peña, D. (1990). Influential observations in time series. *J. Business Econ. Stat.* **8**(2), 241.

Peña, D. (1991). Measuring influence in dynamic regression models. *Technometrics* **33**(1), 93–101.

Peña, D. and Maravall, A. (1991). Interpolation, outliers and inverse autocorrelations. *Commun. Stat. (Theory Meth.)* **20**, 3175–3186.

Peña, D. and Tiao, G. C. (1991). A note on Likelihood estimation of missing values in Time series. *Am. Stat.* **45**, 212–213.

Sánchez, M. J. and Peña, D. (1997). *The Identification of Multiple Outliers in ARIMA Models*. Technical Report, Dept. Estadística y Econometría, Univ. Carlos III de Madrid.

Schick, I. C. and Mitter, S. K. (1994). Robust recursive estimation in the presence of heavy-tailed observation noise. *Ann. Stat.* **22**(2), 1045–1080.

Shumway, R. H. and Stoffer, D. S. (1982). An approach to time series smoothing and forecasting using the EM algorithm. *J. Time Series Anal.* **13**, 353–375.

Tsay, R. S. (1986). Time series models specification in the presence of outliers. *J. Am. Stat. Assoc.* **81**, 132–141.

Tsay, R. S. (1988). Outliers, level shifts, and variance changes in time series. *J. Forecasting* **7**, 1–20.

Tsay, R. S., Peña, D., and Pankratz, A. (2000). Outliers in Multivariate Time Series. *Biometrika*, (in press).

CHAPTER 7

Automatic Modeling Methods for Univariate Series

Víctor Gómez
Ministerio de Hacienda

Agustín Maravall
Banco de España

In this chapter, a unified approach to automatic modeling for univariate series is presented. First, ARIMA models and the classical methods for fitting these models to a given time series are briefly reviewed. Second, some automatic methods for model identification are described and an algorithm for automatic model identification is proposed. Third, outliers are incorporated into the model and an algorithm for automatic outlier detection and correction is proposed. Fourth, combining the proposed algorithms for automatic model identification and automatic outlier detection and correction, an algorithm is proposed for automatic model identification in the presence of outliers. Finally, the previous algorithm is extended to cope with missing observations, trading day and Easter effects, and intervention and regression effects.

7.1. CLASSICAL MODEL IDENTIFICATION METHODS

The modeling procedure for ARIMA models proposed by Box and Jenkins (1976) presented in Chapter 3 was by no means a process that could be fully automated with the help of computers. In particular, model identification was rather an art that required an expert in time series analysis to carry it out. In spite of the advances that have taken place since the late 1970s, it continues to be the most difficult part of the

A Course in Time Series Analysis, Edited by Daniel Peña, George C. Tiao, and Ruey S. Tsay.
ISBN 0-471-36164-X. © 2001 John Wiley & Sons, Inc.

model building process. Besides, we have to be aware of the fact that the majority of time series encountered in practice usually have outliers, which makes the modeling procedure even more difficult.

Later we will briefly review the most important methods for automatic detection and correction of outliers that are currently in use. As far as ARIMA model identification is concerned, the presence of outliers can make it very difficult due to the important biases induced by the outliers in the parameter estimates and in the sample autocorrelation and partial autocorrelation functions. For this reason, any good strategy for ARIMA model identification has to account for the presence of outliers.

There are many reasons why one should try to automate as much as possible the ARIMA model identification stage, but they can be basically reduced to two. The first one is that one should eliminate as much as possible all mundane and mechanical chores, which can be performed by the computer, thus increasing the analyst's productivity. Users who are accomplished analysts may invest more of their precious time on troublesome data sets that they have to model. On the contrary, those who are not experts in time series models, can use a powerful methodology that they couldn't even dream of using before. The second reason has to do with the objectivity of the identification stage, since it is desirable that this stage not be subject to heuristic methods and ad hoc procedures that vary with each time series expert. For example, if a National Statistical Office has to produce some statistical data that require the modeling of some time series data sets and an expert is involved in the production process who uses subjective techniques, it may be criticized for publishing data that are neither objective nor reproducible.

7.1.1. Subjectivity of the classical methods

The Box–Jenkins method for model identification relies heavily on the inspection of plots of data over time and the inspection of the graphs of the sample autocorrelation and partial autocorrelation functions. These last tools can be effective to identify pure autoregressive or pure moving-average models, but no so effective with mixed ARMA models. Besides, the determination of the stationary transformation, that is, the numbers d and D in the general seasonal multiplicative ARIMA model

$$\phi(B)\Phi(B^s)\nabla^d\nabla_s^D z_t = C + \theta(B)\Theta(B^s)a_t \tag{7.1}$$

where C is a constant, s is the number of seasons, $\nabla = 1 - B$ is a regular difference, $\nabla_s = 1 - B^s$ is a seasonal difference, and B is the backshift operator, $Bz_t = z_{t-1}$, can be very difficult.

With the exception of very few cases in which the data show a very distinctive pattern it is usually rather difficult to identify a model for the series at hand. For example, given a sample of finite length, it may be extremely difficult to distinguish between the nonstationary model $(1 - .7B)\nabla z_t = a_t$ and the process $(1 - 1.704B + .706B^2)z_t = (1 - .715B)(1 - .989B)z_t = a_t$, which has very similar coefficients but for which the autoregressive polynomial has all its zeros outside the units circle.

Therefore, it is most probable that several time series experts who use the Box–Jenkins method, when confronted with the same data, will specify different models. This makes the whole process of classical model identification dependent on the person who applies the techniques. More experienced individuals are more likely to select an adequate model for the series. This subjectivity is inherent in the classical model identification methods.

7.1.2. The difficulties with mixed ARMA models

Assume that in (7.1) the degrees of the autoregressive polynomials $\phi(B)$ and $\Phi(B)$ are p and P and those of the moving average polynomials $\theta(B)$ and $\Theta(B)$ are q and Q. It was mentioned in the last section that the sample autocorrelation and partial autocorrelation functions can effectively identify pure moving-average ($p + P = 0$) and pure autoregressive ($q + Q = 0$) models. On the other hand, when both the degrees of the autoregressive polynomial ($p + P$) and the moving-average polynomial ($q + Q$) are not 0, the previous functions are much more difficult to interpret. In this case, other model identification methods, different from the classical methods, are called for.

The difficulty of identifying mixed ARMA models is further increased when seasonality is also present in the time series at hand. Several major advances have been made since the late 1970s to identify ARIMA models for nonseasonal time series. Among these, we can mention the extended autocorrelation function and the smallest canonical correlation methods developed by Tsay and Tiao (1984, 1985). These methods are very informative in the identification of ARIMA models for nonseasonal time series, but they are less successful when they are directly applied to seasonal time series. It is to be noted that these methods can also be used with nonstationary series.

Since the early 1970s, some penalty function methods have been proposed for ARMA model identification. These methods can be used with seasonal time series and their popularity is constantly increasing. The reason for this is that they are automatic and can be effective and computationally cheap. However, although some results have been extended to nonstationary series, these methods are in principle only applicable to stationary series.

7.2. AUTOMATIC MODEL IDENTIFICATION METHODS

In this section, we deal with automatic model identification methods, in contrast to the classical model identification methods considered in the last section. First, in order to obtain the degrees d and D of the stationary transformation in (7.1), we can use unit root tests. Then, several methods can be applied to identify an ARMA model for the stationary (differenced) series. We review in this section the penalty function and the pattern identification methods. Both of these methods are automatic and can be regarded as objective. However, there is always some degree of subjectivity also in these methods, such as when selecting the highest orders, p, P, q, and Q, to be considered for model (7.1). For this reason, we prefer to use the term *automatic* rather than *objective* when we refer to them. A good reference for ARMA model identification is the book by Choi (1992).

7.2.1. Unit root testing

In the Box–Jenkins methodology, the decision concerning the need for differencing is based on the characteristics of the plot of the data and of its sample autocorrelation function. For example, failure of this last function to die out sufficiently quickly indicates that differencing is required.

There has been a growing interest in more formal inference procedures concerning the appropriateness of differencing operators in the model. Since all the roots of the differencing operators $\nabla = 1 - B$ and $\nabla_s = 1 - B^s$ lie on the unit circle, testing for differencing is usually referred to as unit root testing.

It is interesting to note that, as Dickey and Pantula (1987) point out, the results obtained by several authors suggest that overdifferencing is not a problem as far as forecasting is concerned. However, there appear to be uses for unit root tests in investigating some economic hypothesis. The practical implication of this is that when one is interested in the routine treatment of many series for forecasting purposes, one should not care very much about whether some of the series are overdifferenced. It is our practical experience that much the same thing happens with regard to model based seasonal adjustment. We can say that overdifferencing is compensated by moving average parameters that go to unity.

We will not review here the vast amount of existing literature concerning unit root testing. The reader can consult, for example, Reinsel (1997) and the references therein. We will content ourselves with making a few remarks on existing procedures.

The two "classical" unit root tests of Dickey–Fuller and Phillips–Perron tend to exhibit rather poor behavior in the presence of certain types of serial correlation. See the Monte Carlo analysis by Schwert (1989).

When there is no seasonality in the series at hand and only regular differences, that is, differences of the form ∇^d, are considered, it seems that the sequential testing procedure suggested by Dickey and Pantula (1987) is the best strategy to follow. According to these authors, only tests that compare a null hypothesis of k unit roots with an alternative of $k - 1$ unit roots are considered. In the sequential procedure, one should start with the largest k under consideration and work down, that is, decrease k by one each time the null hypothesis is rejected.

The situation is different for seasonal time series. In this case, further research is needed and no general agreement exists on how to proceed as far as unit root testing is concerned. It seems that a generalization of the Dickey–Pantula (1987) approach to the seasonal case would be an interesting topic to investigate.

7.2.2. Penalty function methods

In the identification stage, once the differencing orders d and D in (7.1) have been obtained for the nonstationary series $\{z_t\}$, the problem remains of finding an ARMA model for the differenced series $w_t = \nabla^d \nabla_s^D z_t$.

7.2. AUTOMATIC MODEL IDENTIFICATION METHODS

Since the early 1970s, some procedures to determine the orders k and i of an ARMA(k, i) model have been proposed that minimize a function of the form

$$P(k, i) = \ln \hat{\sigma}_{k,i}^2 + (k + i)\frac{C(n)}{n}, \quad k \leq K, \quad i \leq I \quad (7.2)$$

where $\hat{\sigma}_{k,i}^2$ is the maximum likelihood estimate of the variance of the white noise variance, $C(n)$ is some function of the number of observations n of the series, and K and I are upper bounds for the orders, usually imposed a priori. Because $\hat{\sigma}_{k,i}^2$ decreases as the orders increase, it cannot be a good criterion to select the orders by minimizing it. This is the reason why the penalty term $(k + i)C(n)/n$ is included.

If $C(n)$ in (7.2) is replaced with 2, we obtain the famous AIC criterion, which stands for *Akaike's information criterion*. Other possible choices are $C(n) = \ln n$, which corresponds to the BIC (*Bayesian information criterion*), and $C(n) = 2\ln(\ln n)$, which gives the HQ criterion (*Hannan and Quinn*). The BIC criterion imposes a greater penalty term than does AIC.

One criterion for selection of AR(p) models is the FPE (*final prediction error*) criterion, which is given by FPE$(p) = \{1 + (p/n)\}\hat{\sigma}_p^2$.

The BIC criterion estimates the orders of an ARMA model consistently, whereas the AIC does not. However, this is not a reason to prefer BIC instead of AIC because consistency is based on the assumption that there is a "true" ARMA model for the series and this is a doubtful proposition. Models are artificial constructs and probably there is no such a thing as a true model.

The FPE, AIC, and BIC criteria have been described in more detail in Chapter 5.

It is our practical experience and also the experience of some other authors, like, for example, Lütkepohl (1985), that the BIC criterion works better in practice than AIC, in terms of selecting more often the original model when working with simulated series and selecting models with a better fit when working with real series.

Although the penalty function methods are in principle computationally expensive, because they need maximum likelihood estimates for all possible ARMA models, there are methods, like the Hannan–Rissanen method described later in this chapter, which use cheaper estimates based on linear regression techniques only. Also, in the case of multiplicative seasonal ARMA models, it will be seen that it is possible to further reduce the computational burden by proceeding sequentially. That is, by iterating between selections of the regular and of the seasonal parts.

The penalty function methods can also be used to identify vector ARMA models (Reinsel 1997). The penalty functions to use with multivariate data are direct generalizations of the ones for the univariate case. This is a great advantage, not shared by many of the other identification methods.

7.2.3. Pattern identification methods

Since the early 1980s, some methods have been applied for determining the orders of an ARMA process that use the extended Yule–Walker equations. For the

ARMA(p, q) process

$$z_t + \phi_1 z_{t-1} + \cdots + \phi_p z_{t-p} = C + a_t + \theta_1 a_{t-1} + \cdots + \theta_q a_{t-q}$$

these last equations are given by

$$\gamma_j = -\phi_1 \gamma_{j-1} - \cdots - \phi_p \gamma_{j-p}, \quad j = q+1, q+2, \ldots$$

where γ_j, $j = 0, 1, \ldots$, is the autocovariance function of the process. These methods are often called *pattern identification methods* (Choi 1992). It is to be noted that, contrary to Choi's remark about penalty function methods being computationally exorbitant and pattern identification methods being computationally cheap, it will be shown later in this chapter that the proposed sequential application of the Hannan–Rissanen method, which is based on the BIC criterion, for stationary seasonal models is computationally cheap and can be very effective.

The pattern identification methods are so called because they are based on certain functions that give rise to two-way arrays with distinctive patterns. For each ARMA(p, q) model, the corresponding two-way array shows a unique pattern. Using the sample analog of this two-way array, an ARMA model is identified by looking for a theoretical pattern that is closely resembled by the sample one. Among the many pattern identification methods that have been proposed in the literature, we can mention the R and S array method by Gray et al. (1978), the Corner method by Beguin et al. (1980), the extended sample autocorrelation method by Tsay and Tiao (1984), and the smallest canonical correlation method by Tsay and Tiao (1985). These last two methods can be effective with nonseasonal time series and can also be used with nonstationary series. However, the R and S array method and the Corner method, which can be used only with stationary series, do not seem to be very useful even for data with no seasonality. The Corner method has been applied to identification of transfer function models by Liu and Hanssens (1982).

7.2.4. Uniqueness of the solution and the purpose of modeling

In the identification stage of model building, it is often the case that there are several models for which the fit is acceptable. For example, if the BIC criterion is used, there may be a very small difference between the BIC of an AR(2) model and the BIC of an ARMA(1, 1) model. In this case, we can probably use any of these two competitive models to model the data.

When some competitive models exist, one should try to select the more parsimonious one, that is, the one with less parameters. On occasion, it may be useful to select models that are also balanced. This means that the degree of the autoregressive part, included the nonstationary transformation, equals the degree of the moving average part. Balanced models are useful when one is going to perform model-based seasonal adjustment.

7.3. TOOLS FOR AUTOMATIC MODEL IDENTIFICATION

In summary, models should be considered as artificial constructs, which are useful for certain purposes, but are only a crude approximation to reality. In this respect, the criteria used to select models, especially when some competing models exist, depend on the applications. Some criteria may be good for forecasting, but not so good for signal extraction, for example. One should always have in mind that usually there is not a unique solution to the identification problem.

7.3. TOOLS FOR AUTOMATIC MODEL IDENTIFICATION

In this section, some practical procedures will be described for automatic model identification. The emphasis is on the word "practical," so that the methods presented will aim at simplicity, efficiency and speed when applied to real data.

We will start with a test that we propose for the log-level specification. The test is based on the maximum likelihood principle applied to a series that is supposed to follow the model $(0, 1, 1)(0, 1, 1)_s$. This is the airline model of Box and Jenkins (1976). The reason why we select that model in this and other tests later in this chapter is that it encompasses many other models and is a model very often found in practice.

We will then review the two-stage method proposed by Gómez (1998) to estimate unit roots. After that, the Hannan–Rissanen method, hereafter referred to as the HR method, will be reviewed. This method is used to identify an ARMA model for the stationary (differenced) series. It is based on the BIC criterion and is computationally cheap. Some improvements to the HR method, proposed by Gómez (1998), will be described.

7.3.1. Test for the log-level specification

The test for the log-level specification is based on the maximum likelihood estimation of the parameter λ in the Box–Cox transformation. We fit an airline model with mean to the data, first in logs ($\lambda = 0$) and then without logs ($\lambda = 1$). Let $w = (w_1, \ldots, w_n)'$ be the differenced series and let T be a transformation of the data, which can be any of the Box–Cox transformation. Assuming for simplicity that $T(w)$ is normally distributed with mean 0 and $\text{Var}(T(w)) = \sigma^2 \Sigma$, the logarithm of the density function $f(w)$ of w is

$$\ln(f(w)) = k - \frac{1}{2}\left\{ n\ln(\sigma^2) + \ln|\Sigma| + T(w)'\Sigma^{-1}T(w)/\sigma^2 + \ln\left(\frac{1}{J(T)}\right)^2 \right\},$$

where k is a constant and $J(T)$ is the Jacobian of the transformation. Considering the parameter λ in the T transformation fixed, the previous density function is maximized first with respect to the other model parameters. It is easy to see that σ^2 can be concentrated out of this function by replacing it with the maximum likelihood estimator

$\hat{\sigma}^2 = T(w)'\Sigma^{-1}T(w)/n$. The concentrated function is

$$l(w) = -\frac{1}{2}\left\{n\ln(T(w)'\Sigma^{-1}T(w)) + \ln|\Sigma| + \ln\left(\frac{1}{J(T)}\right)^2\right\} + \cdots,$$

where the dots indicate terms that do not depend on the model parameters. After having maximized with respect to all model parameters different from λ, we maximize with respect to λ. Let $\Sigma = LL'$, where L is a lower triangular matrix, be the Cholesky decomposition of Σ. Then, the expression $T(w)'\Sigma^{-1}T(w) \times |\Sigma|^{1/n} = |L|^{1/n}T(w)'\Sigma^{-1}T(w)|L|^{1/n}$ is a nonlinear sum of squares that we denote by $S(w, T)$. The maximum likelihood principle leads to the minimization of the quantity $S(w, T)(1/J(T))^{2/n}$. It is easy to see that $(1/J(T))^{1/n}$ is the geometric mean in the case of the logarithmic transformation, and unity in the case of no transformation. Therefore, the test compares the sum of squares of the model without logs with the sum of squares multiplied by the square of the geometric mean in the case of the model in logs. Logs are taken in case this last function is the minimum.

7.3.2. Regression techniques for estimating unit roots

Let the observed series $\{z_t\}$ follow the ARIMA(p, d, q) model

$$\phi(B)(\delta(B)z_t - \mu) = \theta(B)a_t \tag{7.3}$$

where $\phi(B) = 1 + \phi_1 B + \cdots + \phi_p B^p$, $\delta(B) = 1 + \delta_1 B + \cdots + \delta_d B^d$ and $\theta(B) = 1 + \theta_1 B + \cdots + \theta_q B^q$ are polynomials in the backshift operator B of degrees p, d and q, $\{a_t\}$ is a iid. $N(0, \sigma^2)$ sequence of random variables and μ is the mean of the differenced process. The roots of $\delta(B)$ are assumed to lie on and those of $\phi(B)$ outside the unit circle, so that the process $w_t = \delta(B)z_t$ follows a stationary ARMA(p, q) process. As mentioned earlier, most economic series follow so-called multiplicative seasonal models, where

$$\begin{aligned}\delta(B) &= \nabla^d \nabla_s^D, \\ \phi(B) &= \phi_r(B)\phi_S(B^s) \\ \theta(B) &= \theta_r(B)\theta_S(B^s)\end{aligned} \tag{7.4}$$

s is the number of observations per year, $\nabla^d = (1-B)^d$, and $\nabla_s^D = (1-B^s)^D$. In practice, for economic time series, the inequalities $0 \le d \le 2$ and $0 \le D \le 1$ hold. For simplicity, we will use in the rest of the section the notation (7.3), even for multiplicative seasonal models. We will make specific reference to these models when necessary.

In the following, a procedure to obtain the differencing orders is reviewed which is based on the estimation of unit roots. This last procedure is the first step of an

7.3. TOOLS FOR AUTOMATIC MODEL IDENTIFICATION

automatic model identification method which has been proposed by Gómez (1998) and is implemented in programs TRAMO and SEATS, see Gómez and Maravall (1997). The estimation of the unit roots is done by first estimating autoregressive models of the form

$$(1 + \phi_1 B + \phi_2 B^2)(1 + \Phi B^s)(z_t - \mu) = a_t, \tag{7.5}$$

where $\{z_t\}$ is the observed series, s is the number of observations per year, μ in the mean of the process, and $\{a_t\}$ is a sequence of iid. $N(0, \sigma^2)$ random variables. Then, the series is differenced using the differencing orders given by the unit roots obtained after estimating (7.5) and an ARMA(1, 1) × (1, 1)$_s$ model with mean, that is, a model of the form

$$(1 + \phi B)(1 + \Phi B^s)(w_t - \mu) = (1 + \theta B)(1 + \Theta B^s) a_t, \tag{7.6}$$

is fitted to the differenced series $\{w_t\}$. If any new unit roots appear after estimating (7.6), the differencing orders are properly increased and a new model (7.6) is fitted. The process is continued until no more unit roots are found. Then, the residuals of the last estimated model are used to decide whether to specify a mean for the model or not. The choice of models (7.5) and (7.6) will be justified later.

Suppose that the series $\{z_t\}$ follows model (7.3), where it is assumed $\mu = 0$ to simplify matters. Then, by Theorems 3.2 and 4.1 of Tiao and Tsay (1983), the ordinary least squares (OLS) estimators obtained from an AR(k) regression, where $k \geq d$, asymptotically verify

$$\hat{\Phi}_k(B) \doteq \hat{\delta}(B) \hat{\phi}_m(B)$$

where $\doteq$ denotes asymptotic equivalence in probability, $m = k - d$ and $\hat{\Phi}_k(B)$, $\hat{\delta}(B)$ and $\hat{\phi}_m(B)$ are, respectively, the polynomials estimated by OLS in the autoregressions

$$\Phi_k(B) z_t = a_t$$
$$\delta(B) z_t = a_t$$
$$\phi_m(B) w_t = a_t$$

where $w_t = \delta(B) z_t$ is a stationary process that follows the ARMA(p, q) model $\phi(B) w_t = \theta(B) a_t$ and the subindex in $\Phi_k(B)$ and $\phi_m(B)$ denotes the polynomial degree. In addition, the equality $\hat{\delta}(B) = \delta(B) + O_p(n^{-1})$ holds, where n is the series length.

The practical implication of this result is that if we perform an autoregression of order greater than or equal to the (unknown) degree of the polynomial $\delta(B)$, we obtain a consistent estimate of $\delta(B)$ as a component of $\hat{\Phi}_k(B)$. If we specify a model of the form AR(2) × (1)$_s$ for $\Phi_k(B)$, we cover the cases $\delta(B) = 1$, $\delta(B) = \nabla$, $\delta(B) = \nabla_s$, $\delta(B) = \nabla \nabla_s$ and $\delta(B) = \nabla^2 \nabla_s$, which are the ones of most applied interest.

In the case of non-seasonal models, where $\delta(B) = \nabla^d$ and $0 \leq d \leq 2$ is assumed, if we specify an AR(2) model, all important cases are covered.

Based on the previous considerations, the algorithm used to identify the differencing polynomial is as follows:

1. Specify a model of the form AR(2) × (1)$_s$ with mean, given by equation (7.5) if the process is multiplicative seasonal, or an AR(2) model with mean, also given by (7.5) but without the second factor, if the process is regular. This autoregressive process is estimated using the HR method, which will be described later, unless the user decides to use unconditional least squares. If the roots estimated with the HR method lie outside the unit circle, the autoregression is estimated again using unconditional least squares. A root is considered to be a unit root if its modulus is greater than a specified value, which by default is .97. Go to step 2.

2. In addition to the differencing degrees identified in step 1 as a result of the estimated unit roots, a model of the form ARMA(1, 1) × (1, 1)$_s$ with mean for seasonal series, or a model ARMA(1, 1) with mean for non-seasonal series, is specified. Letting w_t be the series that results from differencing z_t with the differencing polynomial obtained after the estimation of the initial autoregression, the equations for these models are given by (7.6) in the seasonal case, and by (7.6) without the factors involving B^s in the regular case. The model is estimated using the HR method or exact maximum likelihood, depending on the option selected by the user, and if any of the estimated autoregressive parameters is close to 1, the degree of differencing is increased accordingly. A parameter is considered to be close to 1 if its modulus is greater than a specified value, which by default is .88. To avoid cancellation of terms in the model, the absolute value of the difference between each autoregressive parameter and its corresponding moving average parameter should be greater than .15. For multiplicative seasonal models, it is not possible to pass from 0 differencing to $\nabla\nabla_s$ directly. If this happens, the roots of the autoregressive polynomial obtained in step 1 are considered again, the one with greatest modulus is selected, and the series is differenced accordingly. If the series has been differenced in this step, repeat this step (2). Otherwise, go to step 3.

3. Using the residuals of the last estimated model, it is decided whether to specify a mean for the model of the series or depending on the significance of the estimated residual mean. Stop.

The ARMA(1, 1) × (1, 1)$_s$ model used in step 2 is very flexible and constitutes a generalization of the airline model of Box and Jenkins (1976). For stationary series, it approximates well many of the ARMA models encountered in practice. When it is used with nonstationary series, it can detect autoregressive unit roots that have not been detected by the autoregressive model used step 1. Imagine, for example, a model of the form $(1 - B)z_t - \mu = (1 - .8B)a_t$, where the autoregressive and the moving average part almost cancel out. In this case, an ARMA(1, 1) model would probably estimate the unit root better than an AR(2) model.

7.3. TOOLS FOR AUTOMATIC MODEL IDENTIFICATION

Now consider the case of a regression model with ARIMA errors. The question naturally arises as to whether the previous analysis is still valid and if, in consequence, the procedure just described is also applicable in this case. By the results of Tsay (1984, pp. 119–120), it is possible, under very general conditions, to work with the original series in order to identify the differencing polynomial.

7.3.3. The Hannan–Rissanen method

After having obtained the stationary transformation, the next step in the model building process is the identification of an ARMA(p, q) model for the differenced series, possibly corrected for outliers and other regression effects. We will start by assuming that there are neither outliers nor other regression effects and we will extend the results later in this chapter to the general case.

In the following, the HR method and a procedure to identify ARMA(p, q) models based on it are reviewed. This last procedure is the second step of an automatic model identification method proposed by Gómez (1998) and is implemented in programs TRAMO and SEATS; see Gómez and Maravall (1997). The HR method is a penalty function method based on the BIC criterion, where the estimates of ARMA model parameters are computed by means of linear regressions. Therefore, these estimates are computationally cheap, although it can be shown that the estimators have similar properties to those obtained by maximum likelihood (see Hannan and Rissanen 1982).

Let $z = (z_1, \ldots, z_n)'$ the observed series, which follows model (7.3), where we assume $\mu = 0$ for simplicity. After $\delta(B)$ has been identified, we can compute the differenced series $w_t = \delta(B)z_t$, $t = d+1, \ldots, n$, which follows the ARMA(p, q) model

$$\phi(B)w_t = \theta(B)a_t \qquad (7.7)$$

where $\phi(B), \theta(B)$ and $\{a_t\}$ are like in (7.3). If the model is multiplicative seasonal, the decomposition (7.4) holds. In order to avoid notational problems, let the differenced series be $w = (w_1, \ldots, w_{n-d})'$. If the orders of the fitted model (7.7) are (p, q), the BIC statistic is

$$\text{BIC}_{p,q} = \log(\hat{\sigma}^2_{p,q}) + (p+q)\log(n-d)/(n-d), \qquad (7.8)$$

where $\hat{\sigma}^2_{p,q}$ is the maximum likelihood estimator of σ^2. The BIC criterion estimates the orders (p, q) by selecting $(\hat{p}, \hat{q})$, which minimizes (7.8).

The method just described to select the orders, which is based on the traditional BIC criterion, is computationally expensive because one has to perform a nonlinear optimization for each (p, q) to compute $\hat{\sigma}^2_{p,q}$. For this reason, Hannan and Rissanen (1982) propose to perform the estimation using linear regression techniques in three steps, although the third step is used to compute estimators of the ARMA model selected by the BIC criterion with properties similar to those of maximum likelihood estimators. Therefore, only the first two steps are used to select the orders (p, q).

Computation of $BIC_{p,q}$

In the first step of the HR method, which takes place only if there is a moving-average part $(q > 0)$, estimates $\hat{a}_t$ of the innovations a_t in (7.7) are obtained by fitting a long autoregressive model to the series. That is, given a big positive integer N, the $\hat{a}_t$ are computed using

$$\hat{a}_t = \sum_{j=0}^{N} \hat{\phi}_N(j) w_{t-j}, \quad \hat{\phi}_N(0) = 1, \quad t \geq 1$$

where $w_t = 0$ if $t \leq 0$ and the $\hat{\phi}_N(j)$ are computed using the Durbin–Levinson algorithm. This last algorithm consists of first estimating the sample autocovariances

$$c_t = \frac{1}{n-d} \sum_{s=1}^{n-d-t} w_s w_{s+t}$$

and then recursively computing the $\hat{\phi}_N(j)$ using the equations

$$\hat{\phi}_N(N) = -\sum_{j=0}^{N-1} \frac{\hat{\phi}_{N-1}(j) c_{N-j}}{\hat{\sigma}^2_{N-1}}, \quad \hat{\phi}_N(j) = \hat{\phi}_{N-1}(j) + \hat{\phi}_N(N) \hat{\phi}_{N-1}(N-j),$$

$$\hat{\sigma}^2_N = \{1 - \hat{\phi}^2_N(N)\} \hat{\sigma}^2_{N-1}, \quad \hat{\phi}_1(1) = \frac{c_1}{c_0}, \quad \hat{\sigma}^2_0 = c_0.$$

In the procedure proposed by Gómez (1998), the value of N is selected to be $N = \max\{[\log^2(n-d)], 2\max\{p, q\}\}$, where (p, q) are the orders of the ARMA model for which the BIC is being computed and $[\log^2(n-d)]$ is the integer part of $\log^2(n-d)$. This choice is based on the fact that Hannan and Rissanen (1982, p. 88) assume that n is greater than $\log(n-d)$, but not greater than $\log^b(n-d)$, for some $b < \infty$.

In the second step of the HR method, given the orders (p, q), first the parameters of model (7.7) are estimated by minimizing

$$S(p, q) = \sum_{t=m}^{n-d} \left\{ \sum_{j=0}^{p} \phi_j w_{t-j} - \sum_{j=1}^{q} \theta_j \hat{a}_{t-j} \right\}^2 \tag{7.9}$$

where $m = \max\{p+1, q+1\}$ and $\phi_0 = 1$. Then, the estimator $\hat{\sigma}^2_{p,q}$ is computed by the formula $\hat{\sigma}^2_{p,q} = S(p, q)/(n-d)$ and the $BIC_{p,q}$ statistic is computed using (7.8). The use of an efficient numerical method, like, for example, the application of the QR algorithm based on Housholder transformations, to minimize (7.9) is important to avoid singularity problems when both p and q are overspecified.

7.3. TOOLS FOR AUTOMATIC MODEL IDENTIFICATION

In the procedure proposed by Gómez (1998), the following modifications are made. If there is no moving average part ($q = 0$), the estimation of the parameters of the ARMA model finishes here. Note that, in this case, the estimates obtained for the autoregressive part coincide with the ones obtained by OLS.

If there is a moving average part ($q > 0$), the estimators $\tilde{\phi}_j$ and $\tilde{\theta}_j$, obtained by minimizing (7.9), are consistent but have a bias and, therefore, they are not asymptotically efficient. In order to obtain bias–corrected, consistent and asymptotically efficient estimators, see Zhao–Guo (1985), first form

$$\tilde{a}_t = -\sum_{j=1}^{q} \tilde{\theta}_j \tilde{a}_{t-j} + \sum_{j=0}^{p} \tilde{\phi}_j w_{t-j}, \quad t \geq 1,$$

where $\tilde{a}_t = 0$ and $w_t = 0$ if $t \leq 0$. Then put

$$\eta_t = -\sum_{j=1}^{p} \tilde{\phi}_j \eta_{t-j} + \tilde{a}_t, \quad \xi_t = -\sum_{j=1}^{q} \tilde{\theta}_j \xi_{t-j} + \tilde{a}_t, \quad t \geq 1,$$

where $\eta_t = 0$ and $\xi_t = 0$ if $t \leq 0$. Finally, regress $\tilde{a}_t$ on $-\eta_{t-j}$, $j = 1, \ldots, p$, and ξ_{t-j}, $j = 1, \ldots, q$. The estimated regression coefficients are added to the estimators $\tilde{\phi}_j$ and $\tilde{\theta}_j$ to obtain the desired estimators $\hat{\phi}_j$ and $\hat{\theta}_j$.

When there are a moving average ($q > 0$) and an autoregressive ($p > 0$) part, Gómez (1998) proposes to obtain better estimates of the moving average part by repeating the previous procedure with the series filtered with the autoregressive filter. That is, the series is first filtered with the autoregressive filter $\hat{\phi}(B)$ estimated in the two previous steps to obtain the series $x_t = \hat{\phi}(B)w_t$. Then, the series x_t, which asymptotically follows the model $x_t = \theta(B)a_t$ and, therefore, does not have an autoregressive part, is subject to the two previous steps.

Once the parameter estimates of model (7.7) have been obtained for some orders (p, q), the estimator $\hat{\sigma}^2_{p,q}$ is needed to compute the $\text{BIC}_{p,q}$ statistic. In the procedure proposed by Gómez (1998), the residuals r_t, $t = 1, \ldots, n = \max\{p, q\}$ of the series w_t are first computed using a fast Kalman filter routine based on the algorithm of Morf et al. (1974). Then, the rest of the residuals r_t, $t = n + 1, \ldots, n - d$ are recursively obtained using the difference equation (7.7). Finally, the estimator $\hat{\sigma}^2_{p,q}$ is computed by the formula

$$\hat{\sigma}^2_{p,q} = \frac{1}{n-d} \sum_{t=1}^{n-d} r_t^2$$

and the $\text{BIC}_{p,q}$ statistic is computed using (7.8).

Optimization of $BIC_{p,q}$

After having described the algorithm to compute $\text{BIC}_{p,q}$ for each (p, q), we now review the algorithms used by the HR method and the procedure proposed by Gómez

(1998) to obtain the optimal model of the form (7.7). In the HR method, the model is selected as that ARMA($\tilde{p}, \tilde{q}$) model for which BIC$_{\tilde{p},\tilde{q}}$ is minimum among all ARMA(p, q) models satisfying $p \leq P$ and $q \leq Q$, where P and Q are fixed upper bounds. These authors recommend to search first among models with $p = q$ and refine the search later.

To describe the procedure proposed by Gómez (1998), suppose the general case, where the series follows a multiplicative seasonal model given by (7.4). In practice, it is assumed that the orders of the ARMA(p_r, q_r) × (p_s, q_s)$_s$ model followed by the series verify $0 \leq p_r, q_r \leq 3$ and $0 \leq p_s, q_s \leq 2$, and the BIC statistic should be computed for all these combinations. Since the resulting number of combinations is high, the search is performed sequentially. The algorithm is

1. First specify an ARMA(3, 0) model for the regular part. Then, compute the BIC statistic for models where the seasonal part verifies $0 \leq p_s, q_s \leq m_s$, and select the minimum. The number m_s is selected by the user; the default value is 1.
2. Fix the seasonal part to that selected in step 1, compute the BIC statistic for models where the regular part verifies $0 \leq p_r, q_r \leq m_r$, and select the minimum. The number m_r is selected by the user; the default value is 3.
3. Fix the regular part to that selected in step 2, compute the BIC statistic for models where the seasonal part verifies $0 \leq p_s, q_s \leq m_s$, and select the minimum. The number m_s is that of step 1.

The a justification for the previous algorithm is as follows. In step 1, the regular part is assumed to be an ARMA(3, 0) model. This is usually a good approximation to many regular models found in practice, so that step 1 amounts to first filtering the series with the approximate regular model and then finding a seasonal model for the filtered series. This seasonal model will probably be a good approximation to the seasonal part. In step 2, we filter the series with the seasonal model found in step 1 and find an appropriate regular model. In step 3, we filter the series with the regular model found in step 2 and look for an appropriate seasonal model. Clearly we could iterate this procedure further, but usually the three steps are enough to find a satisfactory model.

The previous algorithm allows for a substantial reduction in computing time and, however, the results obtained with it are very satisfactory. Once the previous algorithm has finished, and in order to avoid the tendency of BIC to overparametrize, especially in the seasonal part, the smallest five BIC are first ordered in ascending order. Then, the first one is compared to the other four and if the difference in absolute value is less than a certain number and the biggest of the two BIC corresponds to a more parsimonious seasonal part, this last one is selected. Among all the BIC that satisfy this condition, the one that corresponds to the more parsimonious part is selected, provided that the seasonal part exists ($p_s > 0$ or $q_s > 0$). The procedure also favors balanced models (models where the degrees of the autoregressive and the moving average parts coincide).

In the previous algorithm, if the parameters estimated for an ARMA model are such that the roots of the autoregressive or the moving-average polynomials lie within the unit circle, this fact is considered as an indication of model inadequacy and the model is rejected.

The tentative model ARMA(3, 0) specified in step 1 of the previous algorithm seems to be robust and the sequential search of the algorithm has given very satisfactory results in all performed tests of the proposed procedure, with real and simulated series.

If there is a mean or other regression effects in model (7.7), the procedure proposed by Gómez (1998) obtains first OLS estimators of the regression parameters. Then, these effects are subtracted from the differenced series before computing the parameter estimates of model (7.7) and also before computing the residuals r_t needed in the computation of $\hat{\sigma}^2_{p,q}$ and the BIC statistic.

7.3.4. Liu's filtering method

The SCA software package has incorporated a module for automatic ARIMA model identification, called "SCA-Expert." This module uses a procedure based on the filtering method proposed by Liu (1989) and certain heuristic rules. Briefly, this method consists of the following:

1. Examine first the sample autocorrelation functions (SACF) of z_t, $(1 - B)z_t$, $(1 - B^s)z_t$ and $(1 - B)(1 - B^s)z_t$ to assert the differencing orders and to see whether seasonality is present. After that, examine the SACF of the properly differenced series. If an obvious seasonal ARIMA model can be specified from the SACF, stop. Otherwise, go to the following step. Denote by w_t the differenced series.

2. If an obvious tentative model cannot be deduced from the SACF of w_t, estimate an intermediate model of the type ARMA(1, 1) × (1, 1)$_s$. If no one of the autoregressive parameters is close to 1, generate the series R_t and S_t, which are the result of filtering w_t with the ARMA(1, 1)$_s$ and ARMA(1, 1) models that make up the intermediate model.

    If any of the autoregressive parameters is close to 1, then difference properly. After differencing, a new intermediate model of the same type is estimated and new R_t and S_t series are generated.

3. Use the SACF and sample partial autocorrelation functions, as well as the extended sample autocorrelation function, of R_t to identify an ARMA model adequate for the R_t series.

4. In order to identify a model for S_t, the SACF of S_t can be used. If a model is not clear for S_t, examine also the estimated parameters for the seasonal part in the intermediate model and use them to specify a model for S_t.

One problem with the previous algorithm is that the computerized specification of either the differencing orders or a seasonal or regular model from the SACF or the sample partial autocorrelation function does not seem at all clear. On the other hand, the idea of filtering the series with an approximate regular or seasonal model to find the other part of the model is a good one and usually gives satisfactory results in practice.

7.4. AUTOMATIC MODELING METHODS IN THE PRESENCE OF OUTLIERS

Many time series encountered in practice have outlying observations. These may be due to errors in the data, strikes, changes in regulations, etc. The presence of outliers can make extremely difficult the process of model identification. For this reason, any automatic model identification method has to incorporate some kind of outlier treatment.

In the rest of this chapter, we will make use of the notation and definitions introduced in Chapter 6 in connection with outliers. In this section, after examining some algorithms for outlier treatment, we first review the method proposed by Gómez (1998) for automatic outlier detection and correction. It is pointed out that in the previous algorithms an exact filter should be used, instead of the inverse of the model, which is the filter usually applied in practice. Second, some estimation and filtering techniques are reviewed which are used to speed up the algorithms of the previous methods. Third, some reasons are given for the need to robustify automatic modeling methods. Finally, an algorithm is proposed for automatic model identification in the presence of outliers.

7.4.1. Algorithms for automatic outlier detection and correction

We will start by considering that there is only one outlier. After having described how the effect of the outlier can be estimated and adjusted for, the case of multiple outliers will be considered. Finally, the algorithm proposed by Gómez (1998) will be reviewed. The emphasis here will be on exact filtering, as opposed to the usual practice of filtering with the inverse of the model followed by the series.

Estimation and adjustment for the effect of an outlier

Suppose that the parameters in model (7.3) are known, the observed series is $z^* = (z_1^*, \ldots, z_n^*)'$, the outlier free series is $z = (z_1, \ldots, z_n)'$ and put $Y = (v(B)I_1^T, \ldots, v(B)I_n^T)'$, where $v(B) = \theta(B)/(\delta(B)\phi(B))$, for an IO, $v(B) = 1$ for an AO, $v(B) = 1/(1 - cB)$ for a TC (usually $c = .7$), and $v(B) = 1/(1 - B)$ for an LS. Then, the model is

$$z_t^* = \omega v(B) I_t^T + z_t \qquad (7.10)$$

which is a regression model with ARIMA errors and can be rewritten in more compact notation as

$$z^* = Yw + z. \qquad (7.11)$$

7.4. AUTOMATIC MODELING METHODS IN THE PRESENCE OF OUTLIERS

To simplify the exposition, we will assume that z in (7.11) follows an ARMA model or, what amounts to the same thing, $\delta(B) = 1$ in (7.3). If this is not the case, we would work with the series obtained by differencing z^*, z and Y in (7.11). Let $\text{Var}(z) = \sigma^2 \Omega$ and $\Omega = LL'$, with L lower triangular, the Cholesky decomposition of Ω. Premultiplying (7.11) by L^{-1}, the following ordinary least squares model is obtained:

$$L^{-1}z^* = L^{-1}Yw + L^{-1}z. \tag{7.12}$$

Letting $r = L^{-1}z$, the equality $\text{Var}(r) = \sigma^2 I_n$ holds and vector r is the residual vector of the series (not observed). If we let the estimated residuals be $r^* = L^{-1}z^*$ and write $X = L^{-1}Y$, we can write (7.12) as

$$r^* = Xw + r. \tag{7.13}$$

If Y is 0 in (7.11), the model would be $z^* = z$ and if we applied the Kalman filter to this model, we would obtain $L^{-1}z^*$. This result, which a standard result of control theory, allows us to see the Kalman filter as an algorithm that, applied to any vector v instead of z^*, yields $L^{-1}v$. Therefore, if we apply the Kalman filter to the vector of observations z^* and to the vector Y, we can move from (7.11) to (7.12) or, what amounts to the same thing, from (7.11) to (7.13).

We can estimate ω by OLS in (7.13) to obtain

$$\tilde{\omega} = (X'X)^{-1}X'r^* \tag{7.14}$$

where the estimator variance is $\text{Var}(\tilde{\omega}) = (X'X)^{-1}\sigma^2$. To test the null hypothesis that there is no outlier at $t = T$, we can use the statistic

$$\tau = \frac{(X'X)^{1/2}\tilde{\omega}}{\sigma}, \tag{7.15}$$

which is distributed $N(0, 1)$ under the null.

In practice, the parameters of model (7.3) will not be known and they will have to be estimated. Under these circumstances, the usual procedure consists of estimating first the parameters of model (7.3) by exact maximum likelihood, as if there were no outliers, and then using instead of (7.14) and (7.15), their sample counterparts

$$\hat{\omega} = (\hat{X}'\hat{X})^{-1}\hat{X}'\hat{r}^*, \qquad \hat{\tau} = \frac{(\hat{X}'\hat{X})^{1/2}\hat{\omega}}{\hat{\sigma}}$$

which are obtained by replacing in (7.14) and (7.15) the unknown parameters with their estimates. It can be shown that $\hat{\tau}$ is asymptotically equivalent to τ (see Chang et al. 1988, p. 196). Each matrix X and, therefore, $X'X$, depends on the type of the outlier.

Up to now, we have assumed that r^* and X were computed by means of an "exact" filter, which was the Kalman filter. This is the correct thing to do, since the number of observations in a time series is always finite and we cannot apply the semiinfinite filter, given by the inverse of the series model $\pi(B) = 1 + \pi_1 B + \pi_2 B^2 + \cdots = \phi(B)\delta(B)/\theta(B)$, to (7.10) to obtain

$$\pi(B) z_t^* = \omega \left[\pi(B) v(B) I_t^T \right] + a_t, \quad t = 1, \ldots, n,$$

instead of (7.12). In practice, the usual procedure consists of truncating the filter $\pi(B)$ and disregarding some observations at the beginning of the series (see Chen and Liu 1993, p. 285). In the procedure proposed by Gómez (1998), the residuals are filtered with an exact filter to obtain r^* and the filter $\pi(B)$ is used to filter the vector Y in (7.11). Note that using the Kalman filter to filter the vector Y for each possible combination of outliers would be computationally burdensome.

The case of multiple outliers
When multiple outliers are present, we should use instead of (7.10) the model

$$z_t^* = z_t + \sum_{i=1}^{k} \omega_i v_i(B) I_t^{t_i}. \tag{7.16}$$

As shown by Chen and Liu (1993), the estimators of the ω_i obtained simultaneously using (7.16), can be very different from the ones obtained by an iterative process using the results of the previous section—that is, by obtaining first $\hat{\omega}_1$, then $\hat{\omega}_2$, and so on. For this reason, it is important that every algorithm for outlier detection perform at some point multiple regressions to detect spurious outliers and correct the bias produced in the estimators sequentially obtained.

In order to estimate the parameters in the multiple regressions, when the parameters of the ARIMA model (7.3) are assumed to be known, the algorithm proposed by Gómez (1998) uses first the Kalman filter like when we moved from (7.11) to (7.12). Then, the estimators of the ω_i and the corresponding statistics are computed using (7.14) and (7.15). This is done in an efficient manner, using the QR algorithm and Housholder transformations. A more detailed description will be given at the end of this section.

Estimation of the standard deviation σ of the residuals
When outliers are present in the series, the usual sample estimator can overestimate σ. For this reason, it is advisable to use a robust estimator. In the procedure proposed by Gómez (1998) the estimator used is the MAD estimator, defined by

$$\hat{\sigma} = 1.483 \times \text{median}\{|r_t^* - \tilde{r}^*|\},$$

where $\tilde{r}^*$ is the median of the estimated residuals $r^* = L^{-1} z^*$. The parameters of the model were assumed to be known in the previous formula. If they were unknown, they would be replaced with their estimates, as usual.

7.4. AUTOMATIC MODELING METHODS IN THE PRESENCE OF OUTLIERS

For outlier treatment, the procedure proposed by Gómez (1998) assumes that the orders (p, d, q) of model (7.3) are known and it proceeds iteratively. In the first stage, outliers are detected one by one and the model parameters are modified after each outlier has been detected. When no more outliers have been detected, the procedure goes to the second stage, where a multiple regression is performed. The outliers with the lowest t-value is discarded and the procedure goes back to the first stage to iterate.

The procedure used to incorporate or reject outliers is similar to the stepwise regression procedure for selecting the "best" regression equation. This results in a more robust procedure than that of Chen and Liu (1993), which uses "backward elimination" and may therefore detect too many outliers in the first stage of the procedure.

Up to now, we have supposed that there were no regression effects, but it is easy to incorporate these effects into the procedure. Let the series follow the regression model with ARIMA errors

$$z_t = y_t'\beta + v_t, \quad t = 1, \ldots, n, \tag{7.17}$$

where $\beta = (\beta_1, \ldots, \beta_k)'$ is the vector containing the regression parameters, which may include the mean as the first component, $\{z_t\}$ is the observed series, $\{y_t\}$ are the vectors containing the regression variables and $\{v_t\}$ follows the ARIMA model (7.3) with $\mu = 0$. Then, the algorithm proposed by Gómez (1998) for automatic detection and correction of outliers, described in detail, is as follows:

Initialization. If there are any regression variables in the model, including the mean, the regression coefficients are estimated by OLS and the series is corrected for their effects. Go to stage I.

Stage I: detection and estimation of outliers one by one.

I.1. The ARIMA parameters are estimated, using the HR method, and the series is corrected for all regression effects present at the time, including the outliers so far detected. If desired by the user, exact maximum likelihood can be used for estimation, instead of the HR method.

I.2. Considering the estimates of the ARIMA parameters obtained in I.1 as fixed, the regression coefficients are estimated by GLS and their t statistics are computed. To this end, the fast algorithm of Morf et al. (1974) is used, followed by the QR algorithm. New estimated residuals are obtained.

I.3. With the estimated residuals obtained in I.2, the robust MAD estimator of the standard deviation of the residuals is computed.

I.4. If $w = (w_{d+1}, \ldots, w_n)'$, where d is the degree of the differencing operator, denotes the differenced series, the statistics $\hat{\tau}_{IO}^t$, $\hat{\tau}_{AO}^t$, $\hat{\tau}_{LS}^t$ and $\hat{\tau}_{TC}^t$ are computed for $t = d+1, \ldots, n$. To this end, the residuals computed in I.2 and the MAD obtained in I.3 are used. Let, for each $t = d+1, \ldots, n$,

$\lambda_t = \max\{|\hat{\tau}^t_{IO}|, |\hat{\tau}^t_{AO}|, |\hat{\tau}^t_{TC}|, |\hat{\tau}^t_{LS}|\}$. If $\lambda = \max_t \lambda_t = |\hat{\tau}^T_{tp}| > C$, where C is a preselected critical value, then there is a possible outlier of type tp at T. The subindex tp can be IO, AO, TC, or LS. If no outlier has been found the first time the algorithm passes through this point, then stop. The series is free from outlier effects. If no outlier has been found, but it is not the first time that the algorithm passes through this point, then go to II. If, on the contrary, an outlier has been found, then correct the series for all regression effects, using the estimates obtained in I.2 and the last outlier coefficient estimate obtained while computing λ, and go back to I.1 to iterate.

Stage II: multiple regression. Using the estimates of the multiple regression and their t statistics obtained the last time the algorithm passed through I.2, check whether there are any outliers with a t statistic $< C$, where C is the same critical value than in I.4. If there aren't any, stop. If, on the contrary, there are some, then remove the one with the lowest absolute t value and go back to I.2 to iterate.

7.4.2. Estimation and filtering techniques to speed up the algorithms

To estimate the regression parameters in model (7.16), when the autoregressive and moving-average parameters of the ARIMA model are assumed to be known, the procedure proposed by Gómez (1998) uses the following algorithm. Let the observed series $z = (z_1, \ldots, z_n)'$ follow the regression model with ARIMA errors

$$z = Y\beta + u, \qquad (7.18)$$

where $\beta = (\beta_1, \ldots, \beta_k)'$ is the vector containing the regression parameters, which may include the mean as the first component, Y is an $n \times k$ matrix of full column rank and u follows the ARIMA model (7.3) with $\mu = 0$, which is supposed to be known. After differencing z, the columns of Y and u in (7.18), it is obtained that

$$w = X\beta + v \qquad (7.19)$$

where $w = (w_{d+1}, \ldots, w_n)'$, X is an $(n-d) \times k$ matrix, the components of $v = (v_{d+1}, \ldots v_n)'$ follow the ARMA model $\phi(B)v_t = \theta(B)a_t$ and it is assumed that the degree of the differencing polynomial $\delta(B)$ is d.

If $\text{Var}(v) = \sigma^2 \Omega$ and $\Omega = LL'$, with L lower triangular, is the Cholesky decomposition of Ω, then, premultiplying (7.19) by L^{-1}, it is obtained that

$$L^{-1}w = L^{-1}X\beta + L^{-1}v \qquad (7.20)$$

which is an OLS model. As described in Section 7.4.1, the Kalman filter can be applied to w and the columns of the X matrix to move from (7.19) to (7.20). The Kalman

7.4. AUTOMATIC MODELING METHODS IN THE PRESENCE OF OUTLIERS

filter algorithm used is the fast algorithm of Morf et al. (1974). β can now efficiently estimated in (7.20) by means of the QR algorithm. This last algorithm produces an orthogonal matrix Q such that $Q'L^{-1}X = (R', 0)'$, where R is a nonsingular upper triangular matrix. Partitioning $Q' = (Q_1, Q_2)'$ conforming to $(R', 0)'$, one can move from (7.20) to

$$Q_1'L^{-1}w = R\beta + Q_1'L^{-1}v$$
$$Q_2'L^{-1}w = + Q_2'L^{-1}v,$$

from which $\hat{\beta} = R^{-1}Q_1'L^{-1}w$ and $\hat{\sigma}^2 = w'(L^{-1})'Q_2Q_2'L^{-1}w/(n-d-k)$ are easily obtained. The Q matrix is obtained by means of Housholder transformations.

7.4.3. The need to robustify automatic modeling methods

The presence of outliers in the series can affect tremendously all automatic model identification procedures, starting with the specification of unit roots and ending with the identification of an ARMA model for the differenced series. For this reason, there is a need to robustify automatic modeling methods. This can be achieved by the following scheme. Specify first a robust model for the series. This model could be the airline model, since, as mentioned earlier, it encompasses many models and is a model very often found in practice. Then, use this model to detect and correct the series for outlier effects. The critical value at this stage should not be low because we want to correct the series for the effects of the biggest outliers, which are the outliers that can distort most the automatic model identification procedure. With the series corrected for the outlier effects detected with the airline model, apply the automatic model identification procedure. With the model identified by this last procedure, specify a lower critical value and detect and correct the series for outliers. This cycle can be repeated several times until a satisfactory model is found. Usually, two iterations are enough.

7.4.4. An algorithm for automatic model identification in the presence of outliers

Taking into account the procedure proposed by Tsay (1986) and the previous considerations on how it could be improved, we propose an algorithmical procedure [implemented in programs TRAMO and SEATS; see Gómez and Maravall (1997)], which, briefly described, is the following:

1. *Preliminary tests*. If desired by the user, the procedure can test for the log-level specification, trading day, and Easter effects. These last two tests are performed using the default model (airline model).
2. *Initialization*. If the user wants the series to be corrected for outliers, accept the model specified by the user (the default model is the airline model) and go to

step 3. Otherwise, go to step 1. The critical value C for outlier detection can be either entered by the user or specified by the procedure. In this last case, the value of C is selected depending on the length of the series.

3. *Step 1*. If the user has specified the differencing orders and whether there should be a mean in the model, go to step 2. Otherwise, the series is first corrected for all regression effects, if any. Then, using the corrected series, the differencing orders for the ARIMA model are automatically obtained and, also automatically, it is decided whether to specify a mean for the series or not. Go to step 2.
4. *Step 2*. Perform automatic identification of an ARMA(p, q) model for the differenced series, corrected for all outliers and other regression effects, if any. If the user wants to test for trading day and Easter effects and any of these effects was specified in the preliminary tests, check whether the specified effects are significant for the new model. If the user wants to correct the series for outliers, go to step 3. Otherwise, stop.
5. *Step 3*. Assuming the model known, perform automatic detection and correction of outliers using C as critical value. If a stop condition is not satisfied, perhaps decrease the critical value C and go to step 1.

In the previous algorithm, the procedures for obtaining the differencing orders, automatic model identification and automatic detection and correction of outliers are the ones proposed by Gómez (1998), which have been described in previous sections. The test for the log-level specification is the one considered in the previous section. The trading day and Easter effects, as well as tests for their presence in the model, will be described in detail in the next section.

7.5. AN AUTOMATIC PROCEDURE FOR THE GENERAL REGRESSION–ARIMA MODEL IN THE PRESENCE OF OUTLIERS, SPECIAL EFFECTS, AND, POSSIBLY, MISSING OBSERVATIONS

In this section, the algorithm for automatic model identification in the presence of outliers of last section is extended to the case in which there are missing observations. The algorithm was seen to handle any kind of regression effect. Special effects, such as trading day and Easter effects, are considered in detail, as well as intervention and other regression effects. Tests for the presence of trading day and Easter effects are given.

7.5.1. Missing observations

The procedure proposed in the last section for automatic model identification in the presence of outliers can be extended easily to the case of missing observations. Missing observations are treated as additive outliers. This implies that we can work with a complete series, because the missing values are first assigned tentative values. Then, after the model has been estimated, the difference between the tentative value

7.5. AN AUTOMATIC PROCEDURE

and the estimated regression coefficient is the interpolated value. See Gómez et al. (1998) for details.

Since we work with a complete series (there are no holes in it), we can use the same algorithms described previously for automatic model identification and for automatic detection and correction of outliers. The tentative values assigned to the missing observations are the semisum of the two adjacent values.

7.5.2. Trading day and Easter effects

Traditionally, six variables have been used to model the trading day effect; These are: $(N\ Mondays) - (N\ Sundays), \ldots, (N\ Saturdays) - (N\ Sundays)$ (where $N =$ number of).

The motivation for using these variables is that it is desirable that the sum of the effects of each day of the week cancel out. Mathematically, this can be expressed by the requirement that the trading day coefficients β_j, $j = 1, \ldots 7$, verify $\sum_{j=1}^{7} \beta_j = 0$, which implies $\beta_7 = -\sum_{j=1}^{6} \beta_j$.

Sometimes, a variable, called the *length-of-month* variable, is also included. This variable is defined as $m_t - \bar{m}$, where m_t is the length of the month (in days) and $\bar{m} = 30.4375$ is the average month length.

Another variable that can be used is the leap-year variable. This variable is equal to 0 for all months different from February. In February, it takes the value $-.25$ if February has 28 days, and $.75$ if February has 29 days (which is a leap year).

There is the possibility of considering a more parsimonious modeling of the trading day effect by using one variable instead of six. In this case, the days of the week are first divided into two categories: working days and non-working days. Then, the variable is defined as $(N\ (M, T, W, Th, F)) - (N\ (Sat, Sun) \times 5/2)$.

Again, the motivation is that it is desirable that the trading day coefficients β_j, $j = 1, \ldots, 7$ verify $\sum_{j=1}^{7} \beta_j = 0$. Since $\beta_1 = \beta_2 = \cdots = \beta_5$ and $\beta_6 = \beta_7$, we have $5\beta_1 = -2\beta_6$.

The Easter variable models a constant change in the level of daily activity during the d days before Easter. The value of d is usually supplied by the user.

The variable has zeros for all months different from March and April. The value assigned to March is equal to $p_M - m_M$, where p_M is the proportion of the d days that fall on that month and m_M is the mean value of the proportions of the d days that fall on March over a long period of time. The value assigned to April is $p_A - m_A$, where p_A and m_A are defined analogously. Usually, a value of $m_M = m_A = 1/2$ is a good approximation.

Since $p_A - m_A = 1 - p_M - (1 - m_M) = -(p_M - m_M)$, the sum of the effects of both months, March and April, cancel out, a desirable feature.

Since trading day and Easter effects are modeled by means of regression variables, a possible test for these effects is the following. If no model has been identified, specify an airline model with mean. Otherwise, use the identified model. Then, using the differenced series w, apply first the Kalman filter to move from model (7.19) to model (7.20), where β is the vector of regression parameters, that includes the trading day and/or Easter parameters. Since model (7.20) is an OLS model, we can use an ordinary F test to test if all trading day parameters are zero or not. A student t test can be used to test if the Easter parameter is zero.

7.5.3. Intervention and regression effects

Intervention variables are regression variables that are used to model certain abnormal effects, like strikes, major changes in economic policy, natural disasters (Box and Tiao 1975).

Examples of intervention variables are

- Impulses
- Level shifts
- Temporary changes
- Ramps

These variables usually consist of sequences of ones and zeros. Other regression effects, like economic variables thought to be related to the observed series, can also be incorporated.

7.6. EXAMPLES

The automatic model identification procedure proposed by Gómez (1998) and described earlier in this chapter was applied to 35 series that follow models covering a very broad spectrum. The TRAMO program, which, as mentioned earlier, implements the automatic model identification and automatic outlier detection procedures proposed by Gómez (1998), was applied with the parameters "$IDIF = 3$, $INIC = 3$" specified in the input file. This means that "the program will search first for regular differences up to order 2 and for seasonal differences up to order 1. Then, it will continue with the identification of an ARMA model for the differenced series, searching for regular polynomials up to order 3 and for seasonal polynomials up to order 1". The test for the log-level specification was not applied, so that the parameter "LAM" was set to 1 (no logs) whenever necessary. The default value of "LAM" is 0 (logs). Also, the parameter "MQ," which is the seasonal period, was set to the appropriate value whenever the seasonal period was different from 12, the default value.

The results are reported in Section 7.7. Of the 35 series, 13 are series which have appeared in published articles and for which an ARIMA model has been identified by some expert in time series analysis. The rest are simulated series. For the simulated series, the identified models coincide with the models from which the series were generated. For the real series, TRAMO identifies either the same model as the one identified by the time series expert or an also acceptable, sometimes better, model.

7.6. EXAMPLES

In order to illustrate the use of the algorithm proposed by Gómez (1998) for automatic detection and correction of outliers, we consider the example of the ozone (O_3) mean levels in Los Angeles city during the period of January 1955 to December 1972. This series was analyzed by Box and Tiao (1975) as an example of a series for which intervention analysis could be applied.

Box and Tiao (1975) identified three intervention variables and a multiplicative moving-average model for the series differenced with seasonal difference. More specifically, the model is

$$z(t) = \omega_1 \, INT1(t) + \frac{\omega_2}{1 - B^{12}} INT2S(t)$$

$$+ \frac{\omega_3}{1 - B^{12}} INT2W(t) + \frac{(1 + \theta_1 B)(1 + \theta_{12} B^{12})}{1 - B^{12}} a(t)$$

where $INT1$ is 1 in January 1960 and the following months and 0 otherwise, $INT2S$ is 1 in the summer months, starting in June 1966, and 0 otherwise, and $INT2W$ is 1 in the winter months, starting in 1966, and 0 otherwise.

The series, together with its intervention variables, was subject to the procedure proposed by Gómez (1998) for automatic detection and correction of outliers. The TRAMO program was applied with the parameters "$IATIP = 1$, $IMVX = 1$, $VA = 3$" specified in the input file. This specification means the following: (1) the automatic outlier detection procedure will search for outliers of the three types, LS, AO, and TC; (2) exact maximum likelihood will be used to estimate the parameters of ARIMA models during the outlier detection stage; and (3) the critical level 3, will be used for the identification of outliers.

The results are displayed in Table 7.1. Four outliers have been identified. Two outliers of type AO, at $t = 11$ and $t = 21$, and two outliers of type TC, at $t = 39$ and $t = 43$.

To illustrate the algorithm of Section 7.4.4, we consider the example of the monthly variety stores sales considered by Hillmer et al. (1983). For the logged series, these authors identified the ARIMA model

$$\nabla \nabla_{12} z(t) = \frac{1 + \theta_{12} B^{12}}{1 + \phi_1 B + \phi_2 B^2} a(t). \tag{7.21}$$

TABLE 7.1. Outliers Identified for the Ozone Series

Outlier	Estimate	t Value	Type
$t = 11$	3.2773	4.82	AO
$t = 39$	−1.9287	−3.45	TC
$t = 21$	2.4878	3.71	AO
$t = 43$	−1.8824	−3.36	TC

TABLE 7.2. Outliers Identified for the Variety Store Sales Series

Outlier	Estimate	t Value	Type
$t = 45$	.096	5.23	TC
$t = 96$	.084	−4.38	AO
$t = 112$	−.176	−10.18	LS

The TRAMO program was run with the parameters "$LAM = -1$, $ITRAD = -1$, $IEAST = -1$, $IDIF = 3$, $INIC = 3$, $IATIP = 1$," specified in the input file. The first three parameters tell the program to perform the test for the log-level specification, trading day and Easter effects, respectively. The last parameter is used to specify automatic outlier detection. When this parameter is used in conjunction with $IDIF = 3$ and $INIC = 3$, the program will apply the algorithm of Section 7.4.4.

To implement the algorithm of Section 7.4.4, after performing the tests for the log-level specification, trading day and Easter effect, the TRAMO program can go through up to three rounds. In the first round, it uses the default model and default critical value C, or the model and critical value entered by the user, and detects and corrects the series for outliers. As mentioned earlier, the default model is the airline model of Box and Jenkins (1976). The default critical value C depends on the series length. In the second round, using the outlier corrected series, the program automatically identifies a model and, with that model, it performs a second automatic detection and correction of outliers. Usually, these two rounds are sufficient to identify a model with a good fit. If this is not the case, the program iterates. After the third round, it the fit is still not acceptable, the program specifies a general model. This general model is an ARMA(3, 1)(0, 1)$_s$ for the differenced series, where the differencing orders are the same of the last round. At some point of the procedure, the identified model is compared to the airline model and the model with the best fit is selected. This is done because the airline model is a robust model and departures from this model can be unstable.

Using the TRAMO program in the manner just described, the following results were obtained for the variety stores sales series. The test for the log–level specification specified the logarithmic transformation for the data. Neither trading day nor Easter effect were detected. In the first round, using the default model (the airline model) and a critical value $C = 3.5$, the program detected outliers at $t = 45$, of type TC, at $t = 96$, of type AO and at $t = 112$, of type LS. After correcting the series for the outlier effects, the program identified first the differencing polynomial $\delta(B) = \nabla \nabla_{12}$, without specifying a mean for the differenced process. Then, the program identified model (7.21). With this model, the program detected the same outliers than before, as can be seen in Table 7.2.

7.7. TABULAR SUMMARY

Automatic model identification is summarized in Table 7.3.

7.7. TABULAR SUMMARY

TABLE 7.3. Summary of the Automatic Model Identification for 35 Real or Simulated Series

Series	Simulated Models or Manually Identified Models	Model Obtained by TRAMO
1 Maddala (1972): Grunfeld's inversion series ($n=20$)	$(1+\phi_1 B + \phi_2 B^2)z(t) = C + a(t)$	Same as left
2 Hillmer et al. (1983): clothing sales ($n=153$)	$\nabla\nabla_{12}z(t) = (1+\theta_1 B + \theta_2 B^2) \times (1+\theta_{12}B^{12})a(t)$	Same as left
3 Hillmer et al. (1983): hardware sales ($n=155$)	$\nabla\nabla_{12}z(t) = (1+\theta_1 B) \times (1+\theta_{12}B^{12})a(t)$	Same as left
4 Hillmer et al. (1983): variety stores sales ($n=153$)	$(1+\phi_1 B + \phi_2 B^2)\nabla\nabla_{12}z(t) = (1+\theta_1 B)a(t)$	Same as left
5 Box and Tiao (1975): ozone series ($n=216$)	$\nabla_{12}z(t) = C + (1+\theta_1 B) \times (1+\theta_{12}B^{12})a(t)$	Same as left
6^a Box and Tiao (1975): CPI series ($n=234$)	$\nabla z(t) = C + (1+\theta_1 B)a(t)$	$\nabla\nabla_{12}z(t) = (1+\theta_1 B) \times (1+\theta_{12}B^{12})a(t)$
7^b Chatfield and Prothero (1973): monthly sales series ($n=77$)	$(1+\phi_1 B)\nabla\nabla_{12}z(t) = (1+\theta_{12}B)a(t)$	$\nabla_{12}z(t) = C + (1+\theta_1 B + \theta_2 B^2)a(t)$
8^c Hamilton and Watts (1978): weekday coffee data ($n=178$)	$(1+\phi_1 B)\nabla z(t) = (1+\theta_5 B^5)a(t)$	$\nabla z(t) = C + (1+\theta_1 B) \times (1+\theta_5 B^5)a(t)$
9^d Box and Jenkins (1976): series A ($n=197$)	$\nabla z(t) = (1+\theta_1 B)a(t)$, or $(1+\phi_1 B)z(t) = C + (1+\theta_1 B^1)a(t)$	$\nabla z(t) = (1+\theta_1 B)a(t)$
10^d Box and Jenkins (1976): series C ($n=226$)	$(1+\phi_1 B)\nabla z(t) = a(t)$, or $\nabla^2 z(t) = (1+\theta_1 B^1 + \theta_2 B^2)a(t)$	$(1+\phi_1 B)\nabla z(t) = a(t)$
11^e Box and Jenkins (1976): series E ($n=100$)	$(1+\phi_1 B + \phi_2 B^2)z(t) = C + a(t)$, or $(1+\phi_1 B + \phi_2 B^2 + \phi_3 B^3)z(t) = C + (1+\theta_1 B^1)a(t)$	$(1+\phi_1 B + \phi_2 B^2)z(t) = C + (1+\theta_1)a(t)$
12 Box and Jenkins (1976): series F ($n=70$)	$(1+\phi_1 B)z(t) = C + a(t)$	Same as left

(Continues)

TABLE 7.3. (*Continued*)

Series	Simulated Models or Manually Identified Models	Model Obtained by TRAMO
13 Box and Jenkins (1976): series G ($n = 144$)	$\nabla\nabla_{12}z(t) = (1 + \theta_1 B) \times (1 + \theta_{12} B^{12})a(t)$	Same as left
14 Ljung and Box (1979): simulated series ($n = 75$)	$z(t) = (1 + \theta_1 B)a(t)$	Same as left
15 Tsay and Tiao (1984): simulated series with AR complex unit roots ($n = 100$)	$(1 + \phi_1 B + \phi_2 B^2)\nabla^2 z(t) = (1 + \theta_1 B)a(t)$	Same as left
16 Box and Tiao: simulated series R1 ($n = 150$)	$z(t) = C + (1 + \theta_1 + \theta_2 B^2)a(t)$	Same as left
17 Box and Tiao: simulated series R2 ($n = 162$)	$(1 + \phi_1 + \phi_2 B^2)z(t) = C + a(t)$	Same as left
18 Box and Tiao: simulated series R3 ($n = 147$)	$\nabla z(t) = C + (1 + \theta_1 B)a(t)$	Same as left
19[f] Box and Tiao[g]: simulated series R4 ($n = 161$)	$\nabla z(t) = (1 + \theta_1 B + \theta_2 B^6)a(t)$	$(1 + \phi_6 B^6)\nabla z(t) = (1 + \theta_1 B)a(t)$
20 Box and Tiao[g]: simulated series R5 ($n = 155$)	$\nabla^2 z(t) = (1 + \theta_1 B + \theta_2 B^2)a(t)$	Same as left
21 Box and Tiao[g]: simulated series R6 ($n = 178$)	$(1 + \phi_1 B + \phi_2 B^2)\nabla z(t) = a(t)$	Same as left
22 Box and Tiao[g]: simulated series R7 ($n = 149$)	$(1 + \phi_1 B)z(t) = C + a(t)$	Same as left
23 Box and Tiao[g]: simulated series R8 ($n = 148$)	$(1 + \phi_1 B)\nabla z(t) = C + a(t)$	Same as left
24 Box and Tiao[g]: simulated series R9 ($n = 151$)	$\nabla z(t) = (1 + \theta_1 B)a(t)$	Same as left
25 Box and Tiao[g]: simulated series R10 ($n = 146$)	$\nabla z(t) = C + (1 + \theta_1 B + \theta_2 B^2)a(t)$	Same as left

(*Continues*)

TABLE 7.3. (*Continued*)

Series	Simulated Models or Manually Identified Models	Model Obtained by TRAMO
26 Box and Tiao[g]: simulated series S1 ($n = 150$)	$\nabla\nabla_{12}z(t) = (1 + \theta_1 B)$ $\times (1 + \theta_2 B^{12})a(t)$	Same as left
27 Box and Tiao[g]: simulated series S2 ($n = 162$)	$(1 + \phi_1 B)\nabla_{12}z(t)$ $= (1 + \theta_1 B^{12})a(t)$	Same as left
28 Box and Tiao[g]: simulated series S3 ($n = 147$)	$\nabla z(t) = (1 + \theta_1 B^{12})a(t)$	Same as left
29 Box and Tiao[g]: simulated series S4 ($n = 161$)	$(1 + \phi_1 B^{12})z(t)$ $= C + (1 + \theta_1 B)a(t)$	Same as left
30 Box and Tiao[g]: simulated series S5 ($n = 155$)	$(1 + \phi_1 B)\nabla_{12}z(t) = C + a(t)$	Same as left
31 Box and Tiao[g]: simulated series S6 ($n = 178$)	$\nabla\nabla_4 z(t) = (1 + \theta_1 B)a(t)$	Same as left
32 Box and Tiao[g]: simulated series S7 ($n = 149$)	$\nabla^2(1 + \phi_1 B^6)z(t) = a(t)$	Same as left
33 Box and Tiao[g]: simulated series S8 ($n = 148$)	$(1 + \phi_1 B + \phi_2 B^2)z(t)$ $= C + (1 + \theta_1 B^6)a(t)$	Same as left
34 Box and Tiao[g]: simulated series S9 ($n = 151$)	$\nabla_{12}z(t) = C + (1 + \theta_1 B)a(t)$	Same as left
35[f] Box and Tiao[g]: simulated series S10 ($n = 146$)	$\nabla\nabla_{12}z(t)$ $= (1 + \theta_1 B + \theta_2 B^{12})a(t)$	$\nabla\nabla_{12}z(t) = (1 + \theta_1 B)$ $\times (1 + \theta_2 B^{12})a(t)$

[a] The model obtained by TRAMO is also acceptable. The seasonality is rather stable ($\theta_{12} = -.92223$). Using the SEATS program, it can be verified that the seasonality is also small and may be neglected.
[b] The model obtained by TRAMO is better. The model used in the original article is overdifferenced.
[c] The model obtained by TRAMO is better, although the original model is also acceptable.
[d] In the original book, two alternative models were considered. TRAMO obtains the best one.
[e] In the original book, two alternative models were considered. TRAMO obtains a model better than any of them.
[f] TRAMO uses, in its automatic option, multiplicative models because their simplicity and that they have less problems with nonstationarity and noninvertibility. However, by selecting a nonautomatic option, the analyst may use nonmultiplicative models if he prefers to do so.
[g] All of these series are documented in Scientific Computing Associates Corp. (1984).

REFERENCES

Beguin, J. M., Gourieroux, C., and Monfort, A. (1980). Identification of a mixed autoregressive moving average process: The corner method. In O. D. Anderson (ed.), *Time Series*, North-Holland, Amsterdam, pp. 423–436.

Box, G. E. P. and Cox, D. R. (1964). An analysis of transformations. *J. Roy. Stat. Soc. B* **26**, 211–243.

Box, G. E. P. and Jenkins, G. M. (1976). *Time Series Analysis, Forecasting and Control*. Holden-Day, San Francisco.

Box, G. E. P. and Tiao, G. C. (1975). Intervention analysis with applications to economic and environmental problems. *J. Am. Stat. Assoc.* **70**, 70–79.

Brockwell, P. and Davis, R. (1992). *Time Series: Theory and Methods, 2nd ed.* Springer-Verlag, Berlin.

Chang, I. and Tiao, G. C. (1983). *Estimation of Time Series Parameters in the Presence of Outliers*. Technical Report 8, Univ. Chicago, Statistics Research Center.

Chang, I., Tiao, G. C., and Chen, C. (1988). Estimation of time series parameters in the presence of outliers. *Technometrics* **30**, 193–204.

Chatfield, C. (1979). Inverse autocorrelations. *J. Roy. Stat. Soc. A* **142**, 363–377.

Chatfield, C. and Prothero, D. L. (1973). Box-Jenkins seasonal forecasting: Problems in a case-study. *J. Roy. Stat. Soc. (A)* **136**, 295–336.

Chen, C. and Liu, L. (1993). Joint estimation of model parameters and outlier effects in time series. *J. Am. Stat. Assoc.* **88**, 284–297.

Choi, B. (1992). *ARMA Model Identification*. Springer-Verlag, New York.

Cleveland, W. S. (1972). The inverse autocorrelations of a time series and their applications. *Technometrics* **14**, 277–298.

Dickey, D. A. and Pantula, S. G. (1987). Determining the order of differencing in autoregressive processes. *J. Business Econ. Stat.* **5**, 455–461.

Fox, A. J. (1972). Outliers in time series. *J. Roy. Stat. Soc. B* **34**, 350–363.

Gómez, V. (1998). *Automatic Model Identification in the Presence of Missing Observations and Outliers*. Working Paper D-98009, Ministerio de Economia y Hacienda, Dirección General de Anilisis y Programación Presupuestaria, Madrid.

Gómez, V. and Maravall, A. (1997). *Programs TRAMO and SEATS, Instructions for the User* (Beta Version: June 1997). Working Paper 97001, Dirección General de Anilisis y P. P., Ministerio de Economia y Hacienda, Madrid.

Gómez, V., Maravall, A., and Peña, D. (1998). Missing observations in ARIMA models: Skipping strategy versus additive outlier approach. *J. Econ.* **88**, 341–363.

Gray, H. L., Kelley, G. D., and McIntire, D. D. (1978). A new approach to ARMA modeling. *Communi. Stat.* **B7**, 1–77.

Hamilton, D. C. and Watts, D. G. (1978). Interpreting partial autocorrelation function of seasonal time series models. *Biometrika* **65**, 135–140.

Hannan, E. J. and Rissanen, J. (1991). Recursive estimation of mixed autoregressive-moving average order. *Biometrika* **69**, 81–94.

Hillmer, S. C., Bell, W. R., and Tiao, G. C. (1983). Modeling considerations in the seasonal adjustment of economics time series. In A. Zellmer (ed.), *Applied Time Series Analysis of Economic Data*, Washington, D. C.: U.S. Departmental of Commerce, Bureau of the Census, 74–100.

REFERENCES

Ljung, G. M. and Box, G. E. P. (1979). The likelihood function of stationary autoregressive-moving average models. *Biometrika* **66**, 265–270.

Liu, L. M. (1989). Identification of seasonal ARIMA models using a filtering method. *Commun. Stat.* **18**, 2279–2288.

Liu, L. M. and Hanssens, D. M. (1982). Identification of multiple-input transfer function models. *Commun. Stat.* **A11**, 297–314.

Lütkepohl, H. (1985). Comparison of criteria for estimating the order of a vector autoregressive process. *J. Time Series Anal.* **6**, 35–52.

Maddala, G. S. (1977). *Econometrics*. McGraw-Hill, New York, pp. 280–281.

Molard, G. (1984). A fast algorithm for the exact likelihood of autoregressive-moving average models. *Appl. Stat.* **35**, 104–114.

Morf, M., Sidhu, G. S., and Kailath, T. (1974). Some new algorithms for recursive estimation on constant, linear, discrete-time systems. *IEEE Trans. aut. contr.* **AC-19**, 315–323.

Pearlman, J. G. (1980). An algorithm for the exact likelihood of a high-order autoregressive-moving average process. *Biometrika* **67**, 232–233.

Reinsel, G. C. (1997). *Elements of Multivariate Time Series Analysis*, 2nd ed. Springer-Verlag, New York.

Schwert, G. W. (1989). Tests for unit roots: A Monte Carlo investigation. *J. Business Econ. Stat.* **7**, 147–159.

Scientific Computing Associates Corp. (1984). Collection of time series for research and training. Working Paper No. 109, Chicago.

Tiao, G. C. and Tsay, R. S. (1983). Consistency properties of least squares estimates of autoregressive parameters in ARMA models. *Ann. Stat.* **11**, 856–871.

Tsay, R. S. (1984). Regression models with times series errors. *J. Am. Stat. Assoc.* **79**, 118–124.

Tsay, R. S. (1986). Time series model specification in the presence of outliers. *J. Am. Stat. Assoc.* **81**, 132–141.

Tsay, R. S. and Tiao, G. C. (1984). Consistent estimates of autoregressive parameters and extended sample autocorrelation functions for stationary and nonstationary ARMA models. *J. Am. Stat. Assoc.* **79**, 84–96.

Tsay, R. S. and Tiao, G. C. (1985). Use of canonical analysis in time series model identification. *Biometrika* **72**, 299–316.

Zhao-Guo, C. (1985). The asymptotic efficiency of a linear procedure of estimation for ARMA models. *J. Time Series Anal.* **6**, 53–62.

CHAPTER 8

Seasonal Adjustment and Signal Extraction in Economic Time Series

Víctor Gómez
Ministerio de Hacienda

Agustín Maravall
Banco de España

8.1. INTRODUCTION

Seasonal adjustment has a long and well-documented tradition; see, for example, Nerlove et al. (1979), Zellner (1978), Moore et al. (1981), Den Butter and Fase (1991), and Hylleberg (1992). In essence, it consists in the removal of the seasonal variation from a time series. Since neither the seasonally adjusted (SA) series nor the seasonal component are directly observed, both can be seen as "unobserved components" (UC) of the series, and seasonal adjustment becomes a problem of UC estimation. Because the SA series is supposed to provide a cleaner signal of the underlying evolution of the variable, seasonal adjustment can also be viewed as a signal extraction problem in a "signal plus noise" decomposition of the series, where the noise is the seasonal component.

The widespread use of seasonal adjustment reflects powerful reasons. The most basic one is simply the need to understand better our present situation and to adjust our forecasts. As an example, in Cervantes (1605), Sancho Panza, overwhelmed by the disasters that befall on them, asks (the senior) Don Quijote whether their misfortunes occur randomly or at periodic, forecastable, intervals. Of course, seasonal adjustment is also performed because of more sophisticated purposes. For example, in the preamble of the Federal Reserve Act of 1913, the U.S. Congress sets as one of the main objectives of the Federal Reserve to accommodate seasonal variations

A Course in Time Series Analysis, Edited by Daniel Peña, George C. Tiao, and Ruey S. Tsay.
ISBN 0-471-36164-X. © 2001 John Wiley & Sons, Inc.

8.1. INTRODUCTION

in credit so as to maintain interest rates stable (Federal Reserve Board 1915). The fact is that seasonal adjustment of economic series has become a nearly universal practice and millions of series are routinely adjusted. Moreover, economic analysis and research make heavy use of SA series, in the belief that they help interpretation and simplify modelling.

This chapter is not an attempt to summarize some of the last research developments, still at an early testing stage, but to present what we see as the state of the art concerning seasonal adjustment methods that satisfy two general constraints: (1) that the method be of general availability and (2), that they can be, at present, reliably and efficiently used in large-scale applications by data-producing agencies. An implication of these two general requirements is that they restrict us to a world of univariate analysis. Multivariate extensions are few, still of limited capacity, and at an experimental stage [an interesting example is contained in the program STAMP; see Koopman et al. (1996)].

It is a fact that the methods used to estimate UC in applied research often have little to do with the methods used by official data producing agencies, and this is a source of problems. The method presented in this chapter provides a relatively powerful tool that can be of interest in both cases. But, first, a word of caution may be appropriate.

The idea of living in a SA world is somewhat dangerous. It would, of course, cure Seasonal Auto Depression afflictions. But for a family of colibris whose brain size varies seasonally (enlarging for the winter, so as to be able to remember the places where food was stored,) seasonal adjustment of the brain size would prove disastrous. Within the economic field, the economics of seasonality (and some implications for seasonal adjustment) has attracted some attention; see, for example, Ghysels (1993a), Maravall (1983), Plosser (1978), Canova (1992), and Miron (1986). We shall not pursue this issue further, except to stress an important conclusion that will also emerge from our discussion, namely, that, as was the case with the brain of colibris, data used in econometric models should not be, as a rule, seasonally adjusted. [Further arguments that favour this conclusion can be found, e.g., in Wallis (1974), Osborn (1988), Ghysels and Perron (1993), Maravall (1995), and Findley et al. (1998).]

There are several seasonal adjustment methods that satisfy the two general requirements mentioned above [see, e.g., Fisher (1995) and Balchin (1995)]. We shall not survey them, but center on a particular class whose origins can be found in Nerlove et al. (1979), Cleveland and Tiao (1976), Engle (1978), Harrison and Stevens (1976), Box et al. (1978), Piccolo and Vitale (1981), Burman (1980), Hillmer and Tiao (1982), Harvey and Todd (1983), and Gersh and Kitagawa (1983), to quote some important contributions. This class of methods is based on parametric models for the series and components, and computes the latter as the minimum mean-squared error (MMSE) estimators given the observations (this is the "signal extraction" procedure). The models used are linear stochastic processes, often parametrized in the ARIMA-type format (Box and Jenkins 1976). The methods that fall into this class will be called *model-based signal extraction* (MBSE) methods.

A linear stochastic process is understood to mean a linear filter of gaussian innovations. Therefore, we shall not deal with nonlinear extensions, such as the ones in Harvey et al. (1992), Kitagawa (1987), Nelson (1996), and Sheppard (1994), among others. Since what we have in mind is monthly (or lower-frequency) data, nonlinearity is seldom a serious problem and, as seen in Fiorentini and Maravall (1996), proper outlier correction seems powerful enough to linearize most of those series. Moreover, one of the convenient features of the MBSE approach is that it permits to solve, in an internally coherent way, additional problems that might be relevant for the correct extraction of the signal. Examples are outlier correction, interpolation of missing values, trading day and Easter effect correction, incorporation of regression or intervention variable effects, and, of course, forecasting; see, for example, Hillmer et al. (1983) and Harvey (1989).

8.2. SOME REMARKS ON THE EVOLUTION OF SEASONAL ADJUSTMENT METHODS

8.2.1. Evolution of the methodologic approach

The crucial problem underlying the evolution of seasonal adjustment methods is the lack of a precise answer to the question regarding what is seasonality. The absence of a well-defined and generally accepted definition has fostered proliferation of procedures, and made it difficult to find common grounds for comparison. We shall briefly review some basic features of some approaches that provide the evolutionary line of the MBSE approach. In so doing, we leave aside important methods such as, for example, the Bayesian BAYSEA procedure developed by Akaike and Ishiguro (1980), or the nonparametric SABL and STL procedures of the Bell Laboratories (Cleveland et al. 1978, 1990). Description and/or discussion of various of these methods can be found in Zellner (1978, 1983), Den Butter and Fase (1991), Hylleberg (1992), Ghysels (1993b), and Eurostat (1998a).

It will prove helpful to establish first some simple definitions. One is that of a *deterministic* model, which is meant to denote a model that can be forecast without error if the parameters are known. The second is the concept of white noise, which will denote a zero mean, finite variance, normally identically independently distributed (niid) variable. Finally, a moving-average (MA) filter applied to the observations will mean a linear combination of the latter.

The simplest way to model the seasonal component is as a deterministic function with seasonal dummy variables, as in (for monthly data) $s_t = \sum_{i=1}^{12} \beta_i d_{it}$, where $d_{it} = 1$ for month i and 0 otherwise, and the β coefficients satisfy $\beta_1 + \cdots + \beta_{12} = 0$. An equivalent formulation uses cosine functions with the seasonal harmonics as frequencies. What characterizes these deterministic components is that

$$s_t + s_{t-1} + \cdots + s_{t-11} = 0 \qquad (8.1)$$

that is, their sum over 12 consecutive months is zero. The SA series may be further decomposed into a deterministic function of time (the trend) and a noise or irregular component. The trend (p_t) may be some polynomial in time, in its simplest form $p_t = a + bt$, which would imply

$$p_t - p_{t-1} = b, \qquad (8.2)$$

8.2. SOME REMARKS ON THE EVOLUTION OF SEASONAL ADJUSTMENT METHODS

or

$$(p_t - p_{t-1}) - (p_{t-1} - p_{t-2}) = 0. \tag{8.3}$$

The performance of these deterministic models proved unsatisfactory. The estimators of the β parameters were typically unstable and did not seem to converge as observations increased. Residual seasonality could often be detected, and the out-of-sample forecasting performance of the overall model was poor. Although some extensions of the deterministic regression model have been developed (see, e.g., Stephenson and Farr 1972, Nourney 1986, Statistisches Bundesamt 1997) attention moved in a different direction. Fixed deterministic components seemed to be inadequate because components "move" in time (an obvious example of a moving seasonal component is the weather, precisely one of the major causes of seasonality). Attention shifted to MA filters, which seemed capable of capturing some of the moving features of the components. MA filters could be rationalized in several ways: (1) as "local" approximations to deterministic functions of time (see, e.g., Kendall 1976); (2) since the moving features can be seen as the result of randomness, a natural way to think about the components is in the frequency domain. Obviously, the spectrum [by this term we also refer to the pseudospectrum when unit autoregressive roots are present; see Harvey (1989)] of a seasonal component would basically consist of peaks for the seasonal frequencies. The trend component, in turn, would be a peak around the zero frequency and, in general, a peak in the spectrum of the series for a cyclical frequency would indicate the presence of a periodic cyclical component. It follows that one could design "bandpass" filters in the frequency domain that would only capture the variation of the series within a specific frequency band. MA filters are also obtained as the time domain representation of bandpass filters (see, e.g., Oppenheim and Schaffer 1989). Since proper timing of events, and in particular of turning points requires that the complete filter induces a zero-phase effect in the adjusted series, and this, in turn, implies symmetric and centered filters, for now, we shall restrict our attention to this type of filters; and (3) symmetric MA filters are also derived from optimizing some criterion that attempts to balance a tradeoff between fitting and smoothness (see, e.g., Gourieroux and Monfort 1990).

As we shall see later, the three rationalizations of MA filters are closely linked, and the design of the filter requires, in all cases, a priori decisions. For example, what function should be used as local approximation? Which width should be selected for the frequency band? Which should be the penalty function? Once these a priori decisions have been taken, a so-called ad hoc MA filter can be derived. The filter will have a fixed structure, independent of the structure of the series to which it is being applied.

In the field of seasonal adjustment, the most important filter designed has been unquestionably the one in the program X11 (Shiskin et al. 1967). Program X11 basically consists of a linear filter, to which some additional features (e.g., possible trimming of presumed outliers) and options (mostly the selection of a few alternatives concerning the length of the filters) have been added. Program X11 has generated a family of programs [X11 ARIMA and X12 ARIMA; see Dagum (1980), Bureau of

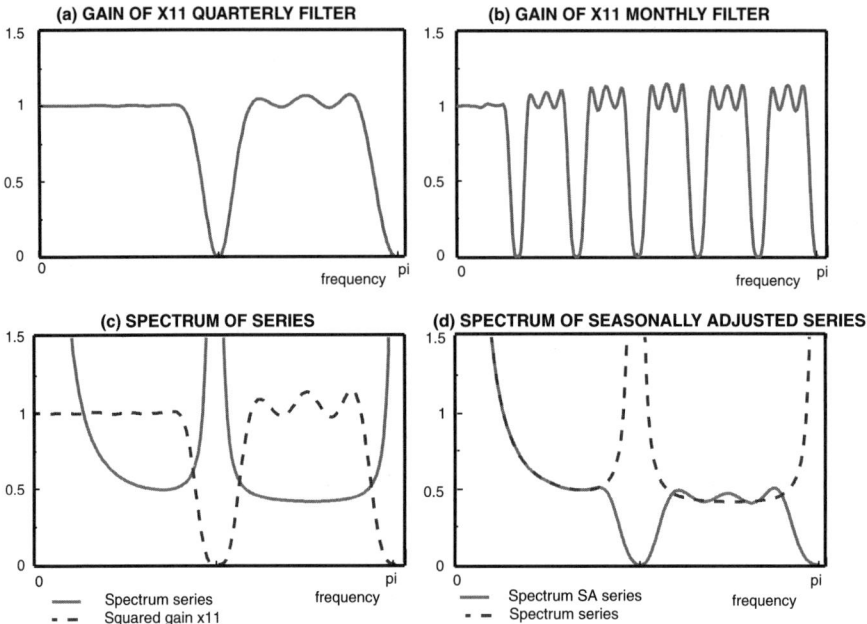

FIGURE 8.1 Gain of X11 filter and selected series spectra.

the Census (1997), and Findley et al. (1998)], where the basic seasonal adjustment filter still is the linear filter in X11; we shall refer to the default value of this filter as the X11 filter. Figures 8.1a and 8.1b display the gain of the X11 filter, specifically, the way X11 filters the frequencies of the series spectrum, for a quarterly and monthly series, respectively. When the gain is 1, the frequency is fully transmitted; when the gain is 0, the frequency is ignored. If applied to a series with the spectrum of Figure 8.1c, the filter removes the variation around the seasonal frequencies, and provides a SA series with the spectrum of Figure 8.1d.

The empirical fact that many economic series have a similar dynamic structure and that this structure is broadly adequate for the X11 filter, evidences the ingenuity of the X11 designers and explains the success of the X11 program. But as the number of series treated increased and experience accumulated, the limitations of the filter became more apparent. The main limitation, in essence, is the rigidity implied by its fixed character. For some series, spurious results will be obtained, in particular those associated with under and overadjustment.

For a series containing a highly stochastic seasonal component, as evidenced by the width of the seasonal peaks in the series spectrum of Figure 8.2a, the width of the dips in the squared gain of the X11 filter seem too narrow. Application of the filter to the series yields a SA series with the spectrum of Figure 8.2b. The underadjustment causes the awkward peaks for frequencies that are in the neighborhood of the seasonal ones. On the other hand, for a series containing a close-to-deterministic seasonal

8.2. SOME REMARKS ON THE EVOLUTION OF SEASONAL ADJUSTMENT METHODS 207

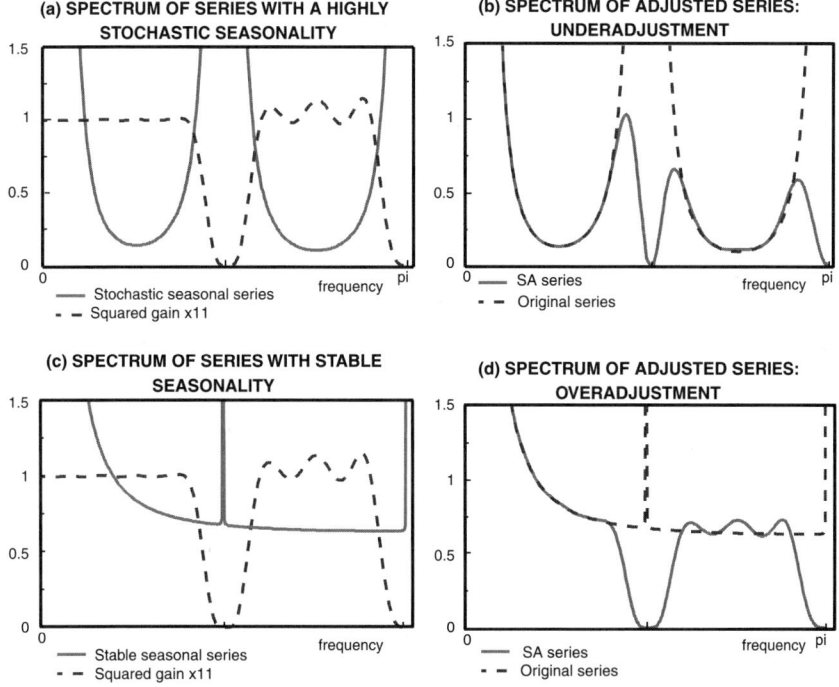

FIGURE 8.2 Under and over adjustment produced by X11 filters.

component, as evidenced by the narrow peaks in the spectrum of Figure 8.2c, the width of the dips in the filter gain are too wide and, as seen in Figure 8.2d, X11 removes variance that is not a associated with the seasonal peaks of the series. In this case the result is overadjustment.

Clearly, the filter to seasonally adjust white noise should simply be 1, since there is no seasonality. Alternatively, the filter to seasonally adjust a purely seasonal series (perhaps a seasonal component produced by X11) should simply be zero. The conclusion that the filter should depend on the structure of the series seems obvious. The MBSE approach solves this problem by tailoring the filter according to the model fit to the series.

8.2.2. The situation at present

Although, as mentioned before, we do not review seasonal adjustment methods, it is of interest to make a brief reference to the main ones currently used by data-producing agencies. First, there are some isolated uses of several methods at specific institutions, which are in the process of being replaced sometime in the near future. Examples are the program GLAS, based on a spectral polishing of the series and used at the Bank of England (see Balchin 1995), program SABL and program DAINTIES, the latter based on one-sided moving regressions (see Eurostat 1998a), both still used at some sections of Eurostat, and program BAYSEA used at the Bank of Japan. Statistics Germany uses the moving-regression type method Berlin BV4 (see Statistiches Bundesamt 1997), although this is not the case for the Bank of Germany, which, as the vast majority of agencies and institutions, uses a member of the X11 family of programs. In many cases the standard X11 is used; in many other cases, the Statistics Canada modification X11 ARIMA is used; in some cases (Organization of National Statistics and Bank of Germany, e.g.) X11 with some added modifications is used. The U.S. Bureau of the Census has just made available a new member of the family, X12 ARIMA, which presumably will replace in many cases the older members of the family. Besides incorporating additional tools for diagnosis in both cases, X11 ARIMA improved on X11 by incorporating ARIMA forecasts and backcasts, so as to obtain better estimates at both ends of the series. X12 ARIMA has added a preadjustment program (REGARIMA), which deals with outliers and special effects (such as trading day) by means of a regression-ARIMA-type model. Further, the number of ad hoc filters available is larger, and the selection of the appropriate filter should depend on the particular series being adjusted. We mentioned before that, over the 40 years of the X11 empire, awareness of its limitations had inevitably increased. The extensions of X11 are attempts at solving some of the main limitations. It is worth noticing that the basic tools employed are ARIMA-model-based tools, often reflecting the need to adjust the filter to the structure of the particular series.

During the 1990s work was done on developing methods based on the (so-called) model-based approach. This work followed the basic methodology of Burman (1980), Hillmer and Tiao (1982), and Harvey and Todd (1983). Two directions emerged: one that begins by directly specifying the model for the components, which has been termed the structural time series (STS) approach (see Engle 1978, Harvey 1989); the other approach, termed the *ARIMA-model-based* (AMB) method, starts by identifying a model for the observed series, and derives from that the appropriate models for the components (see Box et al. 1978, Bell and Hillmer 1984). The models are linear stochastic processes, often parametrized in the ARIMA format. Fruit of that work are, within the STS approach, program STAMP, and, within the AMB approach, the pair of programs TRAMO-SEATS [TRAMO is a preadjustment program; part of SEATS emerged from an original program of Burman; see Gómez and Maravall (1996)]. Their use has spread beyond academic and research applications to the production of official statistics. On a small scale, STAMP is used at some agencies (examples are the ONS and the Statistical Institute of Cantabria, Spain). TRAMO and SEATS have been used routinely on large data sets at Eurostat since 1994, and their use extends at present to various european countries. The simultaneity of the appearance of X12 ARIMA and of the first large-scale experiences with an AMB method has fostered a renewed interest in the topic of seasonal adjustment (interest now extends to trend-cycle estimation and to preadjustment of the series). This interest has been further reinforced by the effort of European countries to harmonize the production of data. The outcome of this interest has concentrated mostly on comparisons of X12 ARIMA with TRAMO-SEATS, and considerable amount of information on this work can be found at the internet site *http://europa.eu.int/en/comm/eurostat/research/noris4/*. Two recent task forces created for the purpose of reaching a recommendation concerning seasonal adjustment methods presented their reports at two conferences, in June 1998, in Rome, organized by the Italian Statistical Institute, and in October 1998 in Bucharest, organized by Eurostat and the International Statistical Institute. The first task force ("Seasonal Adjustment Research Appraisal" committee) was formed by representatives from different fields of professional activity and institutions; the second one was a Eurostat task force on seasonal adjustment policy. Both committees recommended the use of the AMB (TRAMO-SEATS) method; see Eurostat (1998b) and SARA (1998). Although the issue of selecting a seasonal adjustment method is far from being universally settled, a trend seems discernable. The model-based method has come of age and this may eventually lead to the replacement of the X11 paradigm.

8.3. THE NEED FOR PREADJUSTMENT

The model used, as already mentioned, is that of a linear stochastic process. Before this assumption can be made, some modifications to the series often are needed, that is, the series needs preadjustment. Some of these modifications are

- Interpolation of missing values.
- Outlier correction.
- Removal of special effects, such as trading day and easter effects. The first refers to the difference in the number of weekdays per month; the second, to the location of the easter period in different years.
- Correction for special events known a-priori. These effects will be referred to as "intervention variable" effects (Box and Tiao 1975).
- Correction for the effect of other variables (examples can be national and regional festivities, or some indicator whose effect one wishes to remove).

Those types of effects (including missing values), traditionally neglected or dealt with by some empirical procedure, can all be expressed as regression variables. In the MBSE approach, a convenient tool is the regression ARIMA model

$$y_t = W_t \beta + x_t \qquad (8.4)$$

where y_t is the observed series, W_t is the matrix with rows the regression variables, β is a vector of coefficients, and x_t follows a possibly nonstationary (NS) ARIMA model. [For the case of missing observations, an equivalent procedure is to leave them out of the likelihood, and estimate them with a fixed-point smoother; see Gómez et al. (1999)]. The series $x_t = y_t - W_t \beta$ is the "linearized" series, in the sense that it can be assumed to be generated by a linear process.

For the general case of possible missing observations and possibly NS x_t series, estimation of model (8.4) has been discussed in previous chapters (see also Gómez and Maravall 1994). For this type of preadjustment to be operational in large-scale use, it requires an automatic model identification and outlier correction procedure. At present, these requirements can be met in a straightforward manner. (Chapter 7). Two programs that perform preadjustment based on models of the type (8.4) are the programs REGARIMA and TRAMO.

In presenting the MBSE method we shall assume that the ARIMA model is known and that the observed series is a linear series.

8.4. MODEL SPECIFICATION

We consider the additive decomposition (perhaps for the log of the series)

$$x_t = s_t + n_t \tag{8.5}$$

where s_t denotes the SA series (the "signal") and n_t the seasonal component (the "noise"). Often, the SA series is expressed as

$$s_t = p_t + u_t$$

where p_t is denoted the trend (or trend cycle) and u_t is the irregular component. This last component is supposed to absorb highly erratic variation, often simply white noise. In so far as the main purpose of removing seasonality is to obtain a better signal of the underlying evolution of the series, and since the addition of white noise will hardly improve the signal, for the rest of the paper, we assume that the irregular component u_t is white noise. Proceeding in this way, the trend is defined as the residual after removal of the seasonal and the white-noise components. It follows that an AR(2) factor associated, for example, with a 2-year cycle would be part of the trend, as would be an AR factor with a relatively small modulus. These factors, which cause short-term and transitory movements in p_t can be separated from the trend, as in $p_t = m_t + c_t$, where c_t represents a stationary transitory component and m_t, the smoother trend. What should enter c_t and how smooth the trend should be depends on the analyst horizon. Until Section 8.11, our perspective will be a short-term use, and hence we consider short-term trends, also called *trend-cycle components*. We shall refer to them simply as *trends*; their aim is to provide a smoothed SA series; the smoothing removes the noise and perhaps some, relatively small, autocorrelation.

In the MBSE approach, the components are modeled as parametric linear stochastic processes, chosen so as to capture the spectral peaks associated with each component. Denote by B the backward operator (such that $B^j x_t = x_{t-j}$,) and let $\nabla = 1 - B$ and $S = 1 + B + \cdots + B^{\tau-1}$ denote the differencing and the annual aggregation operators, respectively ($\tau =$ number of periods per year). The parametric model expressions can be rationalized as follows.

A stochastic trend can be seen as the equilibrium relationships (8.2) or (8.3), that characterize a deterministic trend, perturbated every period by some random disturbance with zero mean and moderate variance. Thus (8.2) may become the

8.4. MODEL SPECIFICATION

random walk-plus-drift trend model

$$\nabla p_t = \mu + a_{pt}, \quad a_{pt} \sim niid(0, \sigma_p^2)$$

while (8.3) could become the model $\nabla^2 p_t = a_{pt}$ or, more generally, the IMA (8.1) model $\nabla^2 p_t = (1 + \theta_p B)a_{pt}$, all of them well-known models for the trend (see, e.g., Stock and Watson 1988, Gersch and Kitagawa 1983, Harvey and Todd 1983). More generally, one can think of models for the trend of the type

$$\phi_p(B)\nabla^d p_t = \theta_p(B)a_{pt} \tag{8.6}$$

where $\phi_p(B)$ and $\theta_p(B)$ are low-order polynomials, with all roots of $\theta_p(B)$ real, positive, and stable, and $d = 1, 2$, or, very occasionally, 3 (see Maravall 1993).

Concerning the seasonal component, n_t, condition (8.1), satisfied by a deterministic seasonal component, can be restated as $Sn_t = 0$. Perturbating every period this equilibrium with zero-mean random shocks of moderate variance, a stochastic component is obtained, with model

$$Sn_t = w_t \tag{8.7}$$

where w_t is a stationary process, often a finite MA. Examples can be found in Harvey and Todd (1983), Burridge and Wallis (1984), Gersch and Kitagawa (1983), Aoki (1990), and Kohn and Ansley (1987). More generally, one can think of models of the type

$$\phi_n(B)Sn_t = \theta_n(B)a_{nt}, \quad a_{nt} \sim niid(0, \sigma_n^2) \tag{8.8}$$

where the roots of $\phi_n(B)$ are associated with seasonal frequencies (see Maravall 1989).

The irregular component is assumed to be white noise. When a separate stationary transitory component is included, we shall simply assume an ARMA expression

$$\phi_c(B)c_t = \theta_c(B)a_{ct}, \quad a_{ct} \sim niid(0, \sigma_c^2).$$

On some relatively rare occasions, the polynomial $\phi_c(B)$ has roots associated with a fixed-period cyclical component [examples are found in Crafts et al. (1989) and Jenkins (1979)]. In economics, however, the term *cycle* is often used to denote the seasonally adjusted and detrended series (see, e.g., Stock and Watson 1988). What is relevant to our purpose is that, while the very concepts of trend and seasonality imply a persistence or a regularity associated with nonstationarity, the transitory and irregular components are associated with stationary behavior.

In general, if k components are present, the model will consist of the set of equations

$$x_t = x_{1t} + \cdots + x_{kt} \tag{8.9}$$

$$\phi_i(B)x_{it} = \theta_i(B)a_{it}, \quad i = 1, \ldots, k \tag{8.10}$$

where $\phi_i(B)$ and $\theta_i(B)$ are finite polynomials in B of orders p_i and q_i, respectively, with no root in common and with all roots on or outside the unit circle, and the variable a_{it} is a $(0, \sigma_i^2)$ white noise. The following assumptions are made:

Assumption A: The variables a_{it} and a_{jt}, $i \neq j$, are uncorrelated for all values of (t,t').
Assumption B: The ϕ_i polynomials are prime.
Assumption C: The θ_i polynomials do not share unit roots in common.

Assumption A is based on the a priori belief that what causes, for example, seasonal fluctuations (weather, holidays) has little to do with what causes the evolution of the trend (productivity, technology). Of course, the assumption may be questioned on some applications [as an example, Ghysels (1994) finds possible correlation between seasonality and cycle for U.S. GNP]. Assumption B seems sensible given that different components are associated with different spectral peaks [violation of the assumption, besides, would produce estimators with unbounded MSE; see Pierce (1979)]. Finally, assumption C guarantees invertibility of the model for x_t. This last assumption could be relaxed, but it is rather innocuous and simplifies considerably notation.

Since aggregation of ARIMA models yields ARIMA models, the series x_t will also follow an ARIMA model, say

$$\phi(B)x_t = \theta(B)a_t \tag{8.11}$$

where a_t is white noise with variance σ_a^2 and $\phi(B)$—but not $\theta(B)$—may contain unit roots. From (8.9)–(8.11), it is straightforward to show that the AR polynomial in the model for x_t satisfies

$$\phi(B) = \phi_1(B)\phi_2(B) \cdots \phi_k(B) \tag{8.12}$$

and the MA one can be obtained from the relationship

$$\theta(B)a_t = \sum_{i=1}^{k} \phi_{ni}(B)\theta_i(B)a_{it} \tag{8.13}$$

where $\phi_{ni}(B)$ is the product of all $\phi_j(B)$, $j = 1, \ldots, k$, not including $\phi_i(B)$. [Thus, for example, $\phi_{n1}(B) = \phi_2(B) \cdots \phi_k(B)$.]

The model consisting of equations (8.9) and (8.10), together with assumptions A, B, and C will be referred to as an *unobserved component ARIMA* (UCARIMA) model. It will prove convenient to express the UCARIMA model also in a more compact way, as the signal-plus-noise model (8.5), where s_t is the signal of interest and n_t groups all other components.

The specification of the UCARIMA model has followed two main directions. As mentioned earlier, the STS approach starts by directly specifying the models for the components, thus avoiding identification problems; as a counterpart, it assumes a particular structure for the time series at hand. (Identification of a component is typically ensured by restricting the order of its MA polynomial, q_i, to be smaller than that of its AR polynomial, p_i.)

To avoid possible misspecification problems, the AMB approach starts by identifying the ARIMA model for the observed x_t, and derives the components from the structure of that model. For the "trend + seasonal + irregular" components case, the AMB approach, in essence, does the following. Given the ARIMA model for the observed data (8.11), factorization of the AR polynomial yields the AR polynomials for the component models, which are of the type (8.6) and (8.8). Most often, the model for the seasonal component is given by (8.5) with w_t an MA process of order $(\tau - 1)$, which is exactly the structure a seasonal component should have according to Roberts and Harrison (1984). If the spectra of all components are nonnegative the decomposition is called *admissible*. For a given observed ARIMA model (8.11), in general there is not a unique UCARIMA representation that can generate it. The AR polynomials can be obtained from the factorization of $\phi(B)$, but the θ_i polynomials and the innovation variances (σ_i^2) are not identified. The AMB approach solves this underidentification problem by, first, assuming $q_i \leq p_i$. Then it can be seen that the different (admissible) decompositions differ in the way white noise is allocated among the components (Hillmer and Tiao 1982, Bell and Hillmer 1984). By adding all additive white noise to the irregular component, a unique decomposition is achieved. This decomposition is termed canonical and, in it, all components except the irregular have a spectral minimum of zero, and are thus noninvertible. Hillmer and Tiao (1982) show that the canonical decomposition maximizes the variance of the irregular and minimizes the variance of the other component innovations, providing thus components as stable as possible given the model for the series.

Although the specifications vary, the models in the STS and the AMB approaches are both UCARIMA-type models and are closely related (Maravall 1985). Table 8.1 contains some examples of model specification for monthly series. Whereas in the STS approach the models for the components are parsimonious and the ARIMA model for the observed model is not, the inverse is true for the AMB approach. For the rest of the paper only the UCARIMA structure is of relevance; the additional assumptions made to identify the particular models used for the components play no role.

8.5. ESTIMATION OF THE COMPONENTS

Using the two-component representation of the UCARIMA model, let s_t be the signal of interest and n_t the rest of the series ("the noise"). The model is given by equation (8.5), the models

$$\phi_s(B)s_t = \theta_s(B)a_{st} \tag{8.14}$$

$$\phi_n(B)n_t = \theta_n(B)a_{nt} \tag{8.15}$$

TABLE 8.1. Some Examples of Model Specification (Monthly Series)

	A[a]	B[b]	C[c]
Trend component	$\nabla^2 p_t = (1 + \alpha B) a_{pt}$	$\nabla^2 p_t = (1 + \alpha B) \times (1 + B) a_{pt}$	$\nabla^2 p_t = (1 + .26B + .30B^2 - .32B^3) a_{pt}$
Seasonal component	$S(B) s_t = a_{st}$	$S(B) s_t = \theta_s(B) a_{st}$ $\theta_s(B)$ of order 11	$S(B) s_t = (1 + .26 B^{12}) a_{st}$
Irregular component	White noise	White noise	White noise
Overall series	$\nabla \nabla_{12} x_t = \theta(B) a_t$ $\theta(B)$ of order 13, 3 parameters	$\nabla \nabla_{12} x_t = (1 + \theta_1 B)(1 + \theta_{12} B^{12}) a_t$ $\theta(B)$ of order 13, 2 parameters	$\nabla \nabla_{12} x_t = \theta(B) a_t$ $\theta(B)$ of order 14, 0 parameters

[a]Example A: basic structural model (Harvey and Todd 1984); ARIMA specification.
[b]Example B: ARIMA-model-based decomposition of airline model (Hillmer and Tiao 1982).
[c]Example C: ARIMA-model-based interpretation of XII (Cleveland 1972).

8.5. ESTIMATION OF THE COMPONENTS

where a_{st} and a_{nt} are white noises with variances σ_s^2 and σ_n^2, plus assumptions A, B, and C of the previous section.

The model for the observed series is given by (8.11) and the aggregation relationships (8.12) and (8.13) become

$$\phi(B) = \phi_s(B)\phi_n(B)$$
$$\theta(B)a_t = \phi_n(B)\theta_s(B)a_{st} + \phi_s(B)\theta_n(B)a_{nt}.$$

Our purpose is, given X_T, a particular realization of the time series x_t, to obtain the estimator $\hat{s}_t$ such that $E[(s_t - \hat{s}_t)^2 \mid X_T]$ is minimized, that is, the MMSE estimator of s_t. Under the joint normality assumption, $\hat{s}_t$ is also equal to the conditional expectation $E(s_t \mid X_T)$, and hence, a linear function of the elements in X_T.

The model-based signal extraction (MBSE) procedure consists of estimating the signal by its MMSE estimator with the UCARIMA framework described above. Since nonnormality should have been dealt with at the preadjustment level, in this chapter we shall stick to the normality assumption. (When the series is not normal, the estimators remain the best linear projections.)

8.5.1. Stationary case

Rewrite the models in their MA expression as $s_t = \psi_s(B)\,a_{st}$, $n_t = \psi_n(B)\,a_{nt}$, and $x_t = \psi(B)\,a_t$, where $\psi_s(B) = \theta_s(B)/\phi_s(B)$, and similarly for $\psi_n(B)$ and $\psi(B)$.

Projection on a complete realization $x = [x_{-\infty} \cdots x_t \cdots x_\infty]$
Let j_t denote one of the components (hence $j = s, n, p, u$). For the rest of the chapter the ratio of variances σ_i^2/σ_a^2 will be denoted k_i (i.e., $k_s = \sigma_s^2/\sigma_a^2$). Denote by F the "forward" operator, $F = B^{-1}$, such that $F^j x_t = x_{t+j}$. As shown in Whittle (1963), $\hat{s}_t$ is obtained with the symmetric filter

$$\hat{s}_t = \left[k_s \frac{\psi_s(B)\psi_s(F)}{\psi(B)\psi(F)}\right] x_t$$

$$= v(B,F)x_t = \left[v_0 + \sum_{j=1}^{\infty} v_j(B^j + F^j)\right] x_t \qquad (8.16)$$

The filter $v(B, F)$ is the so-called Wiener–Kolmogorov (WK) filter.

Let ACGF(z) denote the autocovariance-generating function of the variable z, and $g_z(\omega)$ its associated spectrum (ω is measured in radians and defined in the interval $-\pi < \omega \leq \pi$). The filter can be expressed as

$$v(B,F) = \frac{\text{ACGF}(s_t)}{\text{ACGF}(x_t)}$$

or, in the frequency domain, as

$$\tilde{v}(\omega) = \frac{g_s(\omega)}{g_x(\omega)}.$$

The function $\tilde{v}(\omega)$ is also referred to as the *gain* of the filter. Thus, for the spectrum of the estimator of the signal

$$g_{\hat{s}}(\omega) = \left[\frac{g_s(\omega)}{g_x(\omega)}\right]^2 g_x(\omega) \qquad (8.17)$$

so that the squared gain of the filter determines how the variance of the series contributes to the variance of the signal for the different frequencies. Note that since $g_x(\omega) = g_s(\omega) + g_n(\omega)$, the gain can also be expressed as

$$\tilde{v}(\omega) = \left(1 + \frac{1}{r(\omega)}\right)^{-1}$$

where $r(\omega) = g_s(\omega)/g_n(\omega)$ is the signal-to-noise ratio. When for some frequency the signal dominates the noise, $\tilde{v}$ approaches 1; when the noise dominates the signal, $\tilde{v}$ approaches zero.

For the two-component model we consider, the WK filter can be expressed, after simplification, as

$$v(B, F) = k_s \frac{\theta_s(B)\phi_n(B)\theta_s(F)\phi_n(F)}{\theta(B)\theta(F)}. \qquad (8.18)$$

Notice that invertibility of the model for x_t guarantees convergence of the filter (in B and in F) irrespective of the ϕ polynomials.

Projection on a finite realization $X_T = [x_1, x_2, \ldots, x_T]$

Having already $\hat{s}_t$, the projection of $\hat{s}_t$ onto X, we can now project $\hat{s}_t$ onto the subset $[x_1, \ldots, x_T]$. One way to do it (Cleveland and Tiao 1976) is by extending X_T with backcasts and forecasts, and then applying the WK filter to the "extended series." The Burman–Wilson algorithm (Burman 1980) allows for the full projection to be efficiently computed with just a few forecasts and backcasts. Proceeding in that way yields $\hat{s}_{t|T} = E(s_t | X_T)$.

An alternative way of computing $\hat{s}_{t|T}$ is by means of the Kalman filter (KF); see, Harvey (1993) or Anderson and Moore (1979). First, the model is put into a state-space representation (many are available,) consisting of an observation equation, say, $x_t = H'z_t$, and a transition equation of the type $z_{t+1} = Fz_t + Gv_t$, where the vectors z_t and v_t, and the matrices H, F, G have been appropriately defined. Then the KF is run with starting conditions derived from the marginal distribution of the variables in the model. Finally a smoother is applied (fixed point or fixed interval smoother)

8.5. ESTIMATION OF THE COMPONENTS

to obtain $E(s_t | X_T)$. For stationary series, proofs of the equivalence between the WK filter and the KF can be found in Kailath (1976) and in Burridge and Wallis (1988).

8.5.2. Nonstationary series

Concepts such as a trend or seasonality inherently imply a time-varying mean associated with NS series. For example, the sum of the seasonal component over 12 consecutive months should not be far from zero. The model-based expression of this condition is given by an expression of the type (8.7), which implies the presence of the S operator in the AR part of the model for the seasonal. The type of nonstationary we consider is the one associated with unit roots (UR) in AR polynomials, such as the ones implied for example by a $\nabla \nabla_{12} (= \nabla^2 S)$ differencing. These roots will capture the NS behavior of trends and of seasonal components.

Bell (1984) shows that under standard assumptions for computing ARIMA forecasts for NS series (Brockwell and Davis 1987), the WK filter given by (8.18) still provides the optimal (MMSE) estimator of the signal s_t for the ∞ realization X in the NS case. For a finite realization X_T, since X_T is a subset of X, it follows that $E(s_t | X_T) = E[E(s_t | X) | X_T)] = E(\hat{s}_t | X_T)$, and the MMSE estimator of s_t for the finite realization can be obtained by projecting $\hat{s}_t$ onto X_T. This is equivalent to replacing the unknowns x_t in X by their forecasts or backcasts (given the observations in X_T). Further, the projection onto the finite series X_T can still be obtained following the Burman–Wilson algorithm (Gómez 1999).

The frequency domain representation of the filter remains also valid, with $g(\omega)$ denoting the pseudospectrum. Despite the ∞ peaks of $g_s(\omega)$ corresponding to UR in $\phi_s(B)$, $\tilde{\upsilon}(\omega)$ is everywhere well defined. In fact, (8.18) shows that $\upsilon(B, F)$ is the ACGF of the stationary (finite variance) ARMA model.

$$\theta(B) y_t = [\theta_s(B) \phi_n(B)] b_t$$

where $b_t \sim$ white noise $(0, k_s)$. The gain $\tilde{\upsilon}(\omega)$ is thus the spectrum of this model.

Extension of the Kalman filter approach to signal extraction in NS series poses a problem with starting conditions, since nonstationarity prevents the use of the marginal distributions. Several solutions have been developed; see for example, Harvey (1993), Ansley and Kohn (1985), de Jong (1991), and de Jong and Chu-Chun Lin (1994). Very broadly, starting conditions are modeled as a random vector (α) with an unknown distribution. A modified KF and a modified smoother is applied to the first observations to get rid of the starting conditions, after which the filter collapses to the ordinary KF and smoother. In brief, if $\hat{\alpha}$ denotes an assumed value for the starting conditions, the KF provides the estimator $E(s_t | X_T, \hat{\alpha})$. By assuming α be fixed and letting $\hat{\alpha}$ be a GLS projection of α onto X_T, the conditional expectation stated above becomes simply $E(s_t | X_T)$, that is, the estimator provided by the WK filter. Thus extension to NS components (and hence NS series) is, under both approaches, straightforward. As in the stationary case, if properly applied, the WK filter and the KF yield the same result [for a general proof, see Gómez (1999)]. The

WK approach is enforced, for example, in the programs PROPHET (Burman 1995) and SEATS. The KF approach is enforced, for example, in the program STAMP.

8.6. HISTORICAL OR FINAL ESTIMATOR

The WK filter given by (8.18) is symmetric and centered, convergent in B and in F and, unless the observed model is a pure AR, the filter will extend from $-\infty$ to $+\infty$. Convergence, however, guarantees that it can always be approximated by a finite two-sided filter. Although estimation uses the full filter, its finite approximation is useful for discussion. We assume that the WK filter (8.18) can be approximated by the $(2L+1)$ – term centered and symmetric filter:

$$v(B,F) = v_0 + \sum_{j=1}^{L} v_j (B^j + F^j). \tag{8.19}$$

In practice, for seasonal adjustment, L typically expands between 3 and 5 years; trends usually converge faster. Therefore, when $T > 2L + 1$, final estimators can be assumed for the central observations of the series.

8.6.1. Properties of final estimator

From (8.16), (8.18), and (8.11), it is obtained that

$$\phi_s(B)\hat{s}_t = k_s \theta_s(B) \frac{\theta_s(F)\phi_n(F)}{\theta(F)} a_t \tag{8.20}$$

or, in short

$$\phi_s(B)\hat{s}_t = \theta_s(B)\alpha_s(F) a_t. \tag{8.21}$$

Thus the model generating $\hat{s}_t$ is known. It will prove helpful to write (8.21) in the (symbolic) representation

$$\hat{s}_t = \xi_s(B, F) a_t \tag{8.22}$$

where the weights of $\xi_s(B, F)$ can be obtained from the identity

$$\phi_s(B)\theta(F)\, \xi_s(B, F) = k_s \theta_s(B)\theta_s(F)\, \phi_n(F) \tag{8.23}$$

see Maravall (1994). The filter $\xi_s(B, F)$ is divergent in B, and convergent in F; only the part in F will be of relevance.

8.6.2. Component versus estimator

As pointed out in Nerlove et al. (1979), the shape of the spectrum of the MMSE estimator of a component in UCARIMA models is different from that of the component. This is a consequence of the fact that, whereas the component s_t follows model (8.14), its MMSE estimator $\hat{s}_t$ follows model (8.21). Comparison of the two models shows that, although they share the same polynomials in B and the same stationarity inducing transformation, their ACGFs and spectra will be different. The most noticeable differences are the following. First, expression (8.17) can be rewritten as

$$g_{\hat{s}}(\omega) = \left(\frac{g_s(\omega)}{g_x(\omega)}\right) g_s(\omega). \tag{8.24}$$

Since $g_s(\omega)/g_x(\omega) \leq 1$, the estimator will always underestimate the variance of the component. Relatively more stochastic components will imply smaller underestimation, and hence the estimator displays a bias toward stability.

The second noticeable difference between the component and estimation spectra is the presence of "dips" in the spectrum of the estimator. In the usual case of a seasonal component satisfying (8.7), from (8.15) and (8.21),

$$\alpha_s(F) = k_s \frac{\theta_s(F)S(F)}{\theta(F)}.$$

Thus the unit roots in S will show up as unit MA roots in the model generating $\hat{s}_t$ and will produce spectral zeroes for the associated seasonal frequencies. The frequency domain derivation also explains the appearance of the spectral zeroes in the estimator model. Consider the case where the signal s_t is the SA series and the noise is a NS seasonal component. Let ω_0 denote a seasonal frequency; then $g_s(\omega_0)$ is finite, while $g_n(\omega_0) \to \infty$, and from (8.17)

$$g_{\hat{s}}(\omega_0) = \frac{g_s(\omega_0)^2}{g_s(\omega_0) + g_n(\omega_0)}$$

will be zero. These spectral zeros are the frequency counterpart of the unit MA roots. More generally, the spectral zeros in the spectrum of the estimator of the SA series will be a feature of any method that removes a nonstationary seasonal component.

The difference between the models for the signal and for its estimator has some relevant implications. The first one has to do with the standard practice of building models on seasonally adjusted data. This practice is based on the belief that, by removing seasonality, model dimensions can be reduced. Yet this belief is unjustified. While the model for the SA series s_t is of the type (8.15), the estimator of the SA series, $\hat{s}_t$, has the structure (8.21), more complicated than the one for s_t or x_t. Table 8.2 compares the MA expansions (the stationary transformation) of the three variables x_t, s_t, and $\hat{s}_t$ for the model $\nabla\nabla_{12} x_t = (1-.4B)(1-.6B^{12})a_t$, a relatively common model, and one for which AMB seasonal adjustment yields results similar to those of X11

TABLE 8.2. MA Weights

Lag	Original Series	SA	SA Series
1	−.4	−1.37	−1.33
2	0	.39	.38
12	−.6	0	−.40
13	.24	0	.53
14	0	0	−.15
24	0	0	−.24
25	0	0	.32
26	0	0	−.09

(Cleveland and Tiao 1976). As seen in Table 8.2, which lists only the lags for which there are some nonzero coefficients, relatively high coefficients may appear for large lags [as was actually detected in Ghysels and Perron (1993)]. Hence no reduction in dimension can be expected from using the SA series.

As for the second implication, consider the difference between (8.14) and (8.21), that is, the factor $\alpha_s(F)$. Direct inspection shows that when n_t is NS or s_t noninvertible, $\alpha_s(F)$ will induce unit MA roots, and hence the estimator will be a noninvertible series. In particular, the estimator $\hat{s}_t$ is noninvertible if n_t is NS, and $\hat{n}_t$ is noninvertible if $\hat{s}_t$ is NS. Hence in a standard trend + seasonal + irregular decomposition, with NS trend and NS seasonality, the estimators of the three components, as well as that of the SA series, will be noninvertible. An important consequence of the previous result is that the estimators of the SA series, trend, seasonal and irregular components will not accept, in general, an AR (or VAR) approximation to its Wold representation.

The third implication is that, in the MBSE approach, knowledge of the theoretical model for the optimal estimator offers a natural tool for additional diagnostics. To illustrate the point, we use as example the white-noise $(0, \sigma_u^2)$ irregular component. Proceeding as before, the model for its MMSE estimator $\hat{u}_t$ is found to be the "inverse" model of the ARIMA model for the series (Bell and Hillmer 1984):

$$\theta(F)\hat{u}_t = \phi(F)a'_t, \quad (a'_t = k_u a_t). \tag{8.25}$$

In practice, $\hat{u}_t$ is obtained as the residual, once the other components have been estimated. If, in an application, the irregular is to be used for residual diagnostics, however, its ACF and variance should not be compared to those of the component u_t, but to those of the theoretical estimator, given by model (8.25). Large departures from white noise in the ACF of $\hat{u}_t$ may be acceptable. Significant differences, however, between the theoretical and empirical ACF of $\hat{u}_t$ would indicate misspecification. The structure of the differences, besides, may provide a clue as to the type of misspecification. For example, if the theoretical ACF of the stationary transformation of the trend has $\rho_1 = -.4$, positive autocorrelation for low lags in the empirical ACF would clearly point towards underestimation of the trend (Maravall 1987).

8.6.3. Covariance between estimators

The models for $\hat{s}_t$ and $\hat{n}_t$ can also be used to derive the joint distribution of the estimators. In particular, the cross-covariance generating function (CCGF) for a stationary series is straightforward to obtain from (8.20) and the equivalent expression for $\phi_n(F)\hat{n}_t$. It is seen that CCGF $(\hat{s}_t, \hat{n}_t)$ is the ACF of the ARIMA model

$$\theta(B)y_t = \theta_s(B)\theta_n(B)b_t \qquad (8.26)$$

with b_t white noise, and $\sigma_b^2 = k_s k_n$. Thus the CCGF is symmetric and the lag 0 covariance between the estimators will always be positive, despite the fact that the theoretical components are orthogonal. This positive covariance between the estimators is the time domain explanation of the underestimation of the components covariance mentioned in Section 8.6.2.

Expression (8.26) does not contain the AR polynomials $\phi_s(B)$ and $\phi_n(B)$. If, say, the first polynomial contains one (or more) unit root, we proceed as follows. First, replace this root by one with the same frequency and modulus $m < 1$, and denote by $\hat{s}_t(m)$ the estimator obtained after having replaced the root. By defining

$$CCGF\,(\hat{s}_t, \hat{n}_t) = \lim_{m \to 1} CCGF\,(\hat{s}_t(m), \hat{n}_t),$$

expression (8.26) is once more obtained. In this sense, in the standard case of NS trend and seasonal components, since the two estimators cannot be cointegrated (the unit AR roots are different), they will diverge in time, each one with a NS variance, but their covariance will remain stationary. In practice, thus, the crosscorrelation between the estimates of NS components will typically be small. Finally, as was the case with the autocovariances, comparison between the crosscovariances of the theoretical estimators and of the estimates actually obtained may provide an additional tool for diagnostics.

8.7. ESTIMATORS FOR RECENT PERIODS

The properties of the estimators have been derived for the final (or historical) estimators. For a finite (long enough) realization, they can be assumed to characterize the estimators for the central observations of the series, but for periods close to the beginning or the end, the filter cannot be completed and some preliminary estimator has to be used. Let the observed series be $X_T = [x_1 \cdots x_t \cdots x_T]$. As shown by Cleveland and Tiao (1976), the MMSE signal estimator (given X_T) can be expressed as

$$\hat{s}_{t|T} = \upsilon(B,F)\hat{x}^e_{t|T}$$

where $\upsilon(B,F)$ is the WK filter (8.18), and $\hat{x}^e_{t|T}$ denote the series extended with forecasts and backcasts. Seasonal adjustment of the time series $X_T = [X_1, \ldots, X_T]$ yields the SA series $[\hat{s}_{1|T}, \ldots, \hat{s}_{T|T}]$, where $\hat{s}_{j|T}$ denotes the estimator of s_j

obtained with X_T. Using the finite filter approximation (8.19), assume that $T > 2L + 1$, so that the estimator for the central observations of the series can be considered final, and that the part in B of the filter can be completed when applied to the last observation. (Thus, for the second half of the series we can ignore starting conditions and the estimator of the signal can be seen as the projection onto the semiinfinite realization $[x_\infty, \ldots, x_{T-1}, x_T]$.) This projection will be represented by the operator E_t.

We center on preliminary estimators for recent periods. Let $k = T - t$, $0 \le k < L$; applying E_T to expression (8.19), we obtain

$$\hat{s}_{T-k|T} = v_L x_{T-L-k} + \cdots + v_k x_T + v_{k+1} \hat{x}_{T+1|T} + \cdots + v_L \hat{x}_{T+L-k|T}$$

or, since $\hat{x}_{T+j|T} = \pi_1^{(j)} x_T + \pi_2^{(j)} x_{T-1} + \cdots$, in terms of the observations

$$\hat{s}_{T-k|T} = \left[v_k + \sum_{j=1}^{L-k} v_{k+j} \pi_1^{(j)}\right] x_T + \left[v_{k-1} + \sum_{j=1}^{L-k} v_{k+j-1} \pi_2^{(j)}\right] x_{T-1} + \cdots$$

and hence the preliminary estimator can be expressed as

$$\hat{s}_{T-k|T} = v_k(B, F, k) x_{T-k} = \sum_{j=-L}^{k} v_{jk} x_{T-k+j},$$

where $v_k(B, F, k)$ is finite and asymmetric, of degree L in B and k in F. The coefficients v_{jk} depend on k, as does the length of the filter. It follows that the models that generate the different preliminary estimators ($\hat{s}_{t|t}, \hat{s}_{t|t+1}, \ldots, \hat{s}_{t|t+L-1}$) will all be different, different also from the model for the final estimator, given by (8.21), and from the model for the component, given by (8.14). Bell (1995) shows, for example, that the model for the concurrent estimator is always of the form

$$\phi_s(B) \hat{s}_{t|t} = \lambda(B) a_t$$

where the order of $\lambda(B) = \max(p_s - 1, q_s)$. It is worth noticing that estimators and component share the AR polynomial $\phi_s(B)$, and hence the same stationary transformation. They differ in the MA part.

As a consequence, the SA series available at a certain time, $[\hat{s}_{1|T}, \ldots, \hat{s}_{t|T}, \ldots, \hat{s}_{T|T}]$, are nonhomogenous. The elements at the beginning, at the end, and in the middle of the series are generated by different models; the SA series has a nonlinear structure, with time-varying parameters [for other nonlinearities in SA series, see Ghysels et al. (1996)].

As a simple example, consider the UCARIMA model with $\nabla s_t = a_{st}$ and n_t white noise (a "random walk-plus-noise" model). Trivially the model for x_t is $\nabla x_t = (1 + \theta B) a_t$, $-1 < \theta < 0$, and the parameters θ and σ_a^2 are determined from $(1 + \theta B) a_t = a_{st} + \nabla n_t$. The model for the component s_t is an IMA(1,0); the model

8.8. REVISIONS IN THE ESTIMATOR

for the final estimator is, from (8.20), the ARIMA(1,1,0) model

$$(1 + \theta F)\nabla \hat{s}_t = k_s a_t \tag{8.27}$$

while for the concurrent estimator the model is also an IMA (1,0), with the innovation a constant fraction of a_t.

8.8. REVISIONS IN THE ESTIMATOR

8.8.1. Structure of the revision

Starting with the concurrent estimator, $\hat{s}_{t|t}$, as new observations become available, the estimator of s_t is revised, yielding the sequence $(\hat{s}_{t|t}, \hat{s}_{t|t+1}, \ldots, \hat{s}_{t|t+k}, \ldots)$. As $k \to \infty$ (in practice, $k > L$), $\hat{s}_{t|t+k}$ converges to $\hat{s}_t$, the final or historical estimator. To look at the revision that the concurrent estimator will undergo, write expression (8.22) as

$$\hat{s}_t = \xi_s(B)^- a_t + \xi_s(F)^+ a_{t+1}. \tag{8.28}$$

When x_t is the last observation, the first term in (8.28) contains the effect of the starting conditions and of the present and past innovations in the series. The second term reflects the effect of future innovations. Taking conditional expectations at time t, $\hat{s}_{t|t} = \xi_s(B)^- a_t$ and the revision in the concurrent estimator $(\hat{s}_t - \hat{s}_{t|t})$ is given by

$$r_t = \xi_s(F)^+ a_{t+1}, \tag{8.29}$$

a zero mean stationary process. Hence the distribution of r_t can be derived. Similar derivation applies to other preliminary estimators, $\hat{s}_{t+k|t}$, including forecasts; see Pierce (1980).

For the random walk-plus-noise example, from the identity

$$\frac{1}{\nabla(1+\theta F)} = \frac{1}{1+\theta}\left(\frac{1}{1-B} - \frac{\theta F}{1+\theta F}\right)$$

we can write, considering (8.27),

$$\hat{s}_t = c\left[\frac{1}{1-B} - \frac{\theta F}{1+\theta F}\right] a_t$$

where $c = k_s/(1+\theta)$. Therefore

$$\xi_s(F)^+ = \frac{-c\theta}{1+\theta F} \tag{8.30}$$

and, from (8.29)

$$(1 + \theta F)r_t = a'_t, \quad a'_t = -c\theta a_{t+1}. \tag{8.31}$$

Hence the revision r_t has the ACF of a stationary AR(1) process.

8.8.2. Optimality of the revisions

Revisions in preliminary estimators are implied by the use of a two-sided filter, as in

$$\hat{s}_t = \cdots + v_1 x_{t-1} + v_0 x_t + v_1 x_{t+1} + v_2 x_{t+2} + \cdots. \tag{8.32}$$

Starting with the concurrent estimator, if the observations are $[x_1, \ldots, x_t]$, then

$$\hat{s}_{t|t} = \cdots + v_0 x_t + v_1 \hat{x}_{t+1|t} + v_2 \hat{x}_{t+2|t} + \cdots \tag{8.33}$$

and when the new observation (x_{t+1}) arrives, the revised estimator is

$$\hat{s}_{t|t+1} = \cdots + v_0 x_t + v_1 x_{t+1} + v_2 \hat{x}_{t+2|t+1} + \cdots$$

and so on. Two-sided filters are necessary to avoid phase effects; they are also implied by MMSE ("optimal") estimation of the components. Of course, to revise series is always disturbing and an inconvenience, and revisions can indeed be large [for a case study, see Maravall and Pierce (1983)]. But revisions simply reflect the fact that knowledge of the future will help in understanding the present, a very basic fact of life. (Concurrent estimators are, like "first impressions," usually insufficient for forming an accurate judgment.) Thus revisions are necessary, and to suppress them is to ignore relevant information, to refuse to improve our knowledge, and to distort our timing of events.

From (8.32) and (8.33), the revision $r_t = \hat{s}_t - \hat{s}_{t|t}$ can be expressed as

$$r_t = v_1(x_{t+1} - \hat{x}_{t+1|t}) + v_2(x_{t+2} - \hat{x}_{t+2|t}) + \cdots = \sum_{j=1}^{\infty} v_j e_t(j) \tag{8.34}$$

where $e_t(j)$ is the jth-period-ahead forecast error of the series. This expression shows that the revision depends on the forecast errors and the weights of the WK filter. This justifies the interest in "small" forecast errors (in essence, the rationale behind the X11ARIMA modification of X11), but revisions still depends on the $v'_j s$, which depend, in turn, on the stochastic structure of the series (i.e., on the ARIMA model). For some series, the revisions should be large; for other series, they should be small. Also, for some series the revisions will last long; for others, they will disappear quickly. Thus, for a given series, there is an appropriate amount of revision. The revision should not be larger than that, nor should it be smaller.

In the MBSE approach, the revisions are "optimal" (in terms of both, size and duration) in the following way. They are implied by optimal (MMSE) forecasting, and optimal (MMSE) estimation of the components. Since the former implies minimum forecast errors, revisions will tend to be small. But the vague (and often made) recommendation of "small revisions" should be replaced by that of "optimal revisions," associated with optimal estimation of the components.

8.9. INFERENCE

8.9.1. Optimal forecasts of the components

Similarly to the case of preliminary estimation, the k-periods-ahead forecast is given by

$$\hat{s}_{T+k|T} = \cdots v_k x_T + v_{k-1}\hat{x}_{T+1|T} + \cdots + v_0\hat{x}_{T+k|T} + \cdots + v_L\hat{x}_{T+k+L|T}$$

hence, in practice, one simply needs to further extend the series with some additional ARIMA forecasts. The properties of the forecast error $s_{T+k} - \hat{s}_{T+k|T}$ can be obtained in exactly the same way as the error in the preliminary estimator that we discuss in the next section. Since, on occasion, one may wish to forecast the trend rather than the original series, a convenient feature of the MBSE method is that it provides optimal forecasts of the components, as well as their associated MSE.

8.9.2. Estimation error

An issue of considerable applied concern has been to obtain a measure of the precision of the component estimator, in particular of the SA series (Bach et al. 1976, Moore et al. 1981; Bank of England 1992). This need is especially felt for key variables that are (explicitly or implicitly) being subject to some type of targeting (e.g., a monetary aggregate or a consumer price index). In these cases, intrayear monitoring and policy reaction is based on the SA series (e.g., see Maravall 1988). We consider now the precision of the concurrent, successively revised, and final estimators, and of the forecasts. Bell and Hillmer (1984), Burridge and Wallis (1985), and Hillmer (1985) have shown how to obtain standard errors for the component MMSE estimators in UCARIMA models of the type we consider. Here we sketch how, under the semi-infinite realization assumption, the models for the errors can be obtained and used in inference.

Because of the stochastic nature of s_t, its final estimator $\hat{s}_t$ contains an error, $e_t = s_t - \hat{s}_t$ ($=\hat{n} - n_t$), to be denoted "final estimation error." Although e_t is unobservable, it can be seen (Pierce 1979) as the output of the stationary ARMA model (8.26). Therefore the distribution of e_t is easily obtained. Since the ACGF of e_t is identical to the CCGF of the estimators $\hat{s}_t$ and $\hat{n}_t$, the final estimation error variance is equal to the lag 0 covariance between the estimators.

For the concurrent estimator, the one of most applied relevance, let $r_T = \hat{s}_T - \hat{s}_{T|T}$ denote the "revision error"; we already saw how its distribution can be obtained. The total estimation error, ε_T, is

$$\varepsilon_T = s_T - \hat{s}_{T|T} = (s_T - \hat{s}_T) + (\hat{s}_T - \hat{s}_{T|T}) = e_T + r_T.$$

Since e_T and r_T are orthogonal (Pierce 1980), the model for ε_T is immediately obtained. The derivation of the model for the error in any preliminary estimator or forecast, $\varepsilon_{t|T} = s_t - \hat{s}_{t|T}$ can be done in an identical manner. From this model, the variance and ACF of the error can be obtained.

For a given ARIMA model for the observed series, analytical expressions for the component estimation error as a function of the particular decomposition chosen is found in Maravall and Planas (in press).

8.9.3. Growth rate precision

Short-term analysis of the evolution of economic variables, as well as the setting of targets, is often based on rates of growth, rather than levels. Assume that we wish to obtain the MSE of the error in the concurrent estimator of the rate of growth over the last m months of a SA series. Since more often than not, ARIMA models are appropriate for the log of macroeconomic time series, let $S_t =$ SA series and $s_t = \log(S_t)$. The rate of growth of the SA series over the last m months is given by $R_t = (S_t - S_{t-m})/S_{t-m}$. Using the linear approximation $R_t = s_t - s_{t-m}$, the concurrent estimator of R_t is $\hat{R}_{t|t} = \hat{s}_{t|t} - \hat{s}_{t-m|t}$. To compute the estimation error variance, consider the identity

$$\hat{s}_{t|t} - \hat{s}_{t-m|t-m} = (\hat{s}_{t|t} - \hat{s}_{t-m|t}) + (\hat{s}_{t-m|t} - \hat{s}_{t-m|t-m}) \qquad (8.35)$$

The left-hand side (LHS) is the difference between two concurrent estimators. Let ε_t and ε_{t-m} be the associated estimation errors. We saw how to derive their variance and ACF. As for the RHS, the first term is $\hat{R}_{t|t}$, and the second term is the m-period revision in the concurrent estimator.

Replacing t by $t - m$ in (8.28), letting $\xi(F)^+ = \sum_{i=0}^{\infty} \xi_j F^j$, and applying the operators E_{t-m} and E_t yields, after simplification,

$$\hat{s}_{t-m|t} - \hat{s}_{t-m|t-m} = \sum_{i=1}^{m} \xi_{i-1} a_{t-m+i}$$

and the identity (8.35) can be rewritten as

$$\hat{s}_{t|t} - \hat{s}_{t-m|t-m} = \hat{R}_{t|t} + \sum_{i=1}^{m} \xi_{i-1} a_{t-m+i}. \qquad (8.36)$$

Denote the error of interest by $D_t = \hat{R}_{t|t} - R_t$. Subtracting $s_t - s_{t-m}$ from both sides

8.9. INFERENCE

of (8.36) yields

$$\varepsilon_t - \varepsilon_{t-m} = D_t + \sum_{i=1}^{m} \xi_{i-1} a_{t-m+i}. \tag{8.37}$$

Because D_t is a function of a_{t+j}, $j > 0$, the two terms on the RHS of (8.37) are orthogonal, and hence

$$\sigma^2(D_t) = 2\sigma_\varepsilon^2 \left(1 - \rho_m^\varepsilon\right) - \sum_{i=1}^{m} \xi_{i-1}^2 \sigma_a^2 \tag{8.38}$$

where ρ_m^ε denotes the m-lag autocorrelation of ε_t. Expression (8.38) can be derived from the UCARIMA model; in the AMB approach, simply from the ARIMA model for the series.

As an example, consider the random walk-plus-noise model, and assume that we are interested in $\sigma^2(D_t)$ for $m = 1$, namely, in the variance of the error in the measurement of the signal rate of growth for the last period.

To compute (8.38), we need σ_ε^2, ρ_m^ε, and ξ_0. The models for the uncorrelated r_t and e_t processes are (8.31) and, from (8.26), $(1 + \theta F)e_t = b_t$, with $\sigma_b^2 = k_s k_n$. From these two AR(1) models one trivially obtains σ_e^2, γ_1^e, σ_r^2, γ_1^r, where γ_1^e and γ_1^r are the lag 1 autocovariances of e_t and r_t, respectively. Thus $\sigma_\varepsilon^2 = \sigma_e^2 + \sigma_r^2$, $\rho_1^\varepsilon = (\gamma_i^e + \gamma_1^r)/\sigma_\varepsilon^2$, and ξ_0 is the coefficient in the expansion of (8.30), that is $\xi_0 = -\theta k_s/(1 + \theta)$.

8.9.4. The gain from concurrent adjustment

A point of concern for data-producing agencies is the frequency at which seasonal adjustment should be performed. Since concurrent adjustment is costly and implies changing the data frequently, seasonal adjustment is often performed once a year (or twice a year), and forecasted seasonal factors are used until the next seasonal adjustment is done.

Naturally, the use of forecasted factors increases the MSE of the SA series, and the question of what would be gained in practice moving from a once-a-year adjustment to a concurrent one is important. The MBSE approach provides a simple answer to the question. From (8.28), it is seen that

$$\text{MSE}(\hat{s}_{t+k|t} - \hat{s}_{t|t}) = \sigma_a^2 \sum_{j=0}^{k-1} \xi_{-j}^2$$

where $\xi_0, \ldots, \xi_{-k+1}$ are the first k coefficients in the polynomial $\xi_s(B)^+$. Thus the loss in precision due to the use of forecasted factors can be easily measured.

8.9.5. Innovations in the components (pseudoinnovations)

If the UCARIMA model parameters are known and the semiinfinite realization is considered, then a_t, the 1-period-ahead forecast error of $x_t (= x_t - \hat{x}_{t|t-1})$, is eventually observed. But since s_t and n_t are never observed, neither will be a_{st} and a_{nt}, the innovations in the components of the models (8.14) and (8.15). We refer to them as "pseudoinnovations" [Harvey and Koopman (1992) use the term "cuasi-residuals"). Although unobservable, their MMSE estimators can be obtained. Taking conditional expectations in (8.14) yields

$$\phi_s(B)\hat{s}_t = \theta_s(B)\hat{a}_{st} \qquad (8.39)$$

where $\hat{a}_{st} = \mathrm{E}(a_{st}|X)$. Using (8.20), (8.39) can be expressed as

$$\hat{a}_{st} = k_s \frac{\theta_s(F)\phi_n(F)}{\theta(F)} a_t. \qquad (8.40)$$

Compared to (8.18), expression (8.40) shows that the filter that provides the MMSE estimator of the standarized pseudoinnovation $\hat{a}_{st}/\sigma_s$ is the one-sided WK filter for obtaining $\hat{s}_t$. In other words, ACF $(\hat{a}_{st}/\sigma_s) = v_s(B, F)$; therefore, although a_{st} is white noise, the estimator $\hat{a}_{st}$ can be highly correlated. Care should be taken thus when interpreting the series $\hat{a}_{st}$, for example, when testing for randomness of a_{st} or for detecting outliers. For the semiinfinite realization, applying E_T to (8.40) yields

$$\theta(F)\hat{a}_{st|T} = k_s\theta_s(F)\phi_n(F)\hat{a}_{t|T}$$

where $\hat{a}_{t|T} = a_t$ when $T \geq t$, and 0 otherwise. Therefore, the concurrent estimators are given by $\hat{a}_{st|t} = k_s a_t$, and $\hat{a}_{nt|t} = k_n a_t$, so that both are a fraction of the series innovation. It is worth noticing that the models for the final estimation error, the revision error, the irregular estimator, and the p-innovation estimator, all have $\theta(F)$ as the AR polynomial. As a consequence, as a general rule, large MA roots in the model for the observed series are associated with slowly converging revisions, and highly autocorrelated irregular and p-innovations.

8.10. AN EXAMPLE

We consider, as an example, the quarterly series of the Spanish industrial production index (IPI) for the period 1981/1–1997/1 (in reverse order, i.e., difference, from January 1997 to January 1981); the series is displayed in Figure 8.3a. To specify the UCARIMA model following the AMB approach we start with the ARIMA model for the observed series. A good fit is provided by the model

$$\nabla\nabla_4 x_t = (1 - .11B)(1 - .96B^4)a_t, \qquad (\sigma_a = 2.03).$$

8.10. AN EXAMPLE

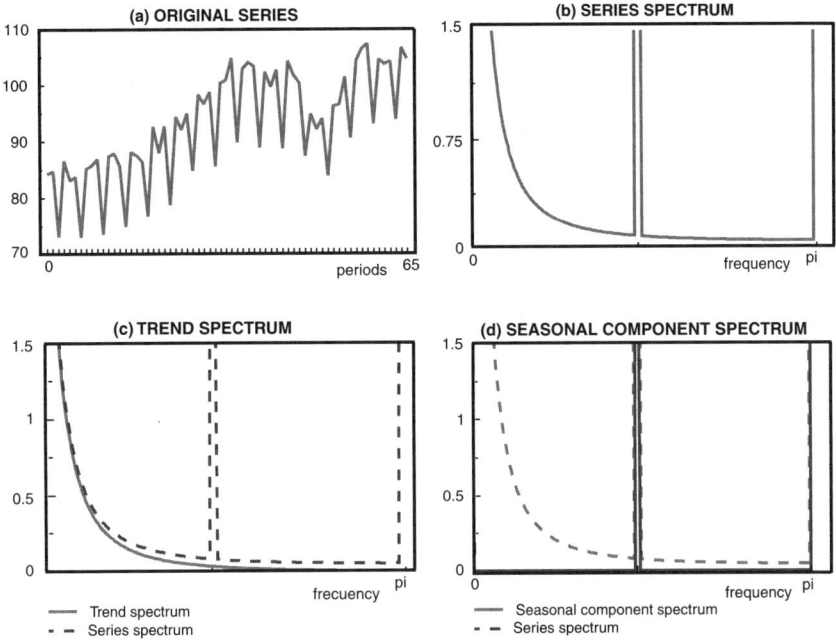

FIGURE 8.3 IPI series and associated spectra.

Direct inspection of the MA parameters indicates the presence of a fairly stochastic trend and a very small or very stable seasonality. In the factorization of the AR polynomial

$$\nabla \nabla_4 = \nabla^2 S \qquad (S = 1 + B + B^2 + B^3),$$

the factors ∇^2 and S imply the presence of a trend and a seasonal component, respectively. Therefore we can decompose the series into

$$x_t = p_t + n_t + u_t$$

where $\phi_p(B) = \nabla^2, \phi_n(B) = S$, u_t is white noise, and $\theta_p(B)$ and $\theta_n(B)$ are polynomials in B of degrees 2 and 11, respectively, which satisfy the identity

$$(1 - .11B)(1 - .96B^4) a_t = \theta_p(B) S a_{pt} + \theta_n(B) \nabla^2 a_{nt} + \nabla \nabla_4 u_t.$$

A simple and efficient procedure to obtain the canonical decomposition (with noninvertible trend and seasonal components) is given in Burman (1980), using a partial fraction expansion of the model in the frequency domain. Easy procedures to compute the ACGF of an ARMA model and to factorize the spectrum of an MA model are given in Box et al. (1978) and Maravall and Mathis (1994), respectively. The UCARIMA model obtained is given by

$$\nabla^2 p_t = (1 + .01B - .99B^2) a_{pt} \qquad (k_p = .19)$$
$$Sn_t = (1 + .50B - .35B^2 - .94B^3) a_{nt}, \qquad (k_n = .0001)$$

and for, the irregular component $k_u = .30$. Notice that $\theta_p(B = -1) = 0$, which implies a spectral zero for the trend at the π (twice-a-year) frequency, while the seasonal component displays a spectral zero for a frequency between the two seasonal ones. Looking at the variance of the component innovations, it is clear that seasonality will be very stable, the trend fairly stochastic, and the irregular relatively important. The SA series, equal to $(p_t + u_t)$, follows the model

$$\nabla^2 s_t = (1 - 1.10B + .11B^2) a_{st} \qquad (k_s = .97), \qquad (8.41)$$

which can be expressed as

$$\nabla^2 s_t = (1 - .11B)(1 - .99B) a_{st}$$

and hence the model is seen to be very close to the random walk-plus-drift process. Further, since k_s is close to 1, seasonal adjustment will not reduce much the stochastic nature of the series. The spectra of the series, the trend, and the seasonal component are displayed in Figures 8.3a–8.3d.

From (8.18), the WK filters to obtain the final estimators of the SA series and seasonal component are given by

$$v_s(B, F) = .97 \frac{||(1 - 1.10B + .11B^2)S||^2}{||(1 - .11B)(1 - .96B^4)||^2}$$

$$v_n(B, F) = .0001 \frac{||(1 + .50B - .35B^2 - .94B^3)\nabla^2||^2}{||(1 - .11B)(1 - .96B^4)||^2}$$

where, if $\varrho(B)$ denotes a polynomial in B, $||\varrho(B)||^2 = \varrho(B)\varrho(F)$. The two filters and the associated squared gains are displayed in Figure 8.4. The narrowness of the dip for the gain function of the SA series and of the peak for the gain of the seasonal component reflects the fact that the seasonality in the series is of a highly stable nature.

Using expression (8.20), the process generating the estimator of the SA series is given by

$$(1 - .11F)(1 - .96F^4)\nabla^2 \hat{s}_t = .97(1 - 1.10B + .11B^2)$$
$$\times (1 - 1.10F + .11F^2)S(F)a_t. \qquad (8.42)$$

The spectra of the SA series (s_t) and its estimator $(\hat{s}_t)$ are shown in Figure 8.5a. It is seen how estimation induces spectral zeros for the seasonal frequencies, and hence noninvertibility of the estimator. The associated spectral dips imply a very small

8.10. AN EXAMPLE

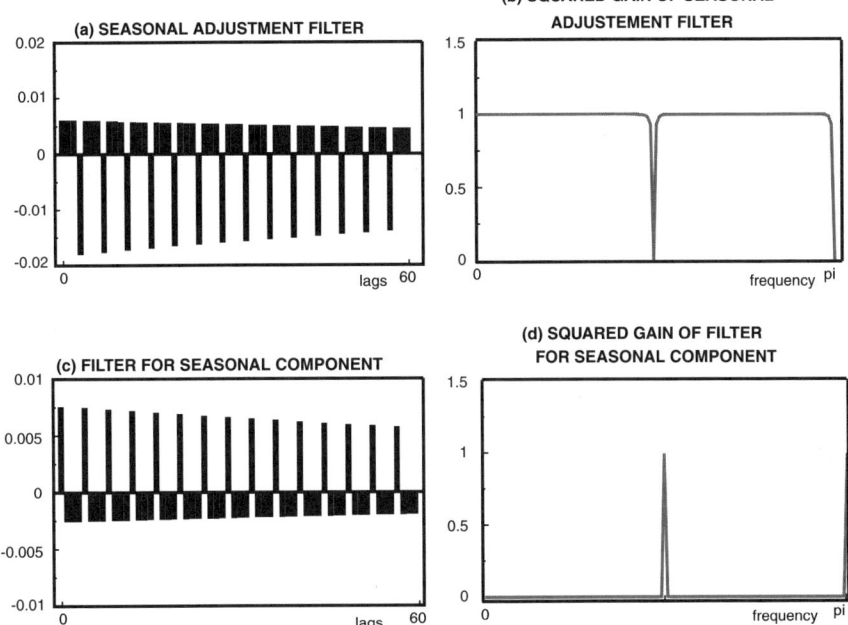

FIGURE 8.4 Selected filters for the IPI series.

underestimation of the variance of the seasonal component. In particular, from (8.41) and (8.42), $\text{Var}(\nabla^2 s_t) = 2.17\,\sigma_a^2$, while $\text{Var}(\nabla^2 \hat{s}_t) = 2.12\,\sigma_a^2$.

The spectrum of the irregular component estimator is shown in Figure 8.5b. Given that the theoretical irregular component is white noise, MMSE is seen to considerably distort its spectrum. The variance underestimation is now more pronounced and Table 8.3 also exhibits the value of the lag 1 and lag 4 autocorrelations (ρ_1 and ρ_4) for the irregular component, its theoretical MMSE estimator, and the estimate actually obtained. (One year has been removed at both end of the series to decrease the distortion due to preliminary estimators.)

MMSE estimation induces a negative and large lag 1 autocorrelation. Comparison of the theoretical estimator of the irregular with the estimate actually obtained (computed as the residual) can be used as a diagnostic tool; the close agreement between estimator and estimate points toward validation of the results.

TABLE 8.3. Irregular Component: Comparison of Second Moments

	Component	MMSE Estimator	Estimate
ρ_1	0	−.44	−.52
ρ_4	0	.02	.07
Variance (in units of σ_a^2)	.30	.16	.15

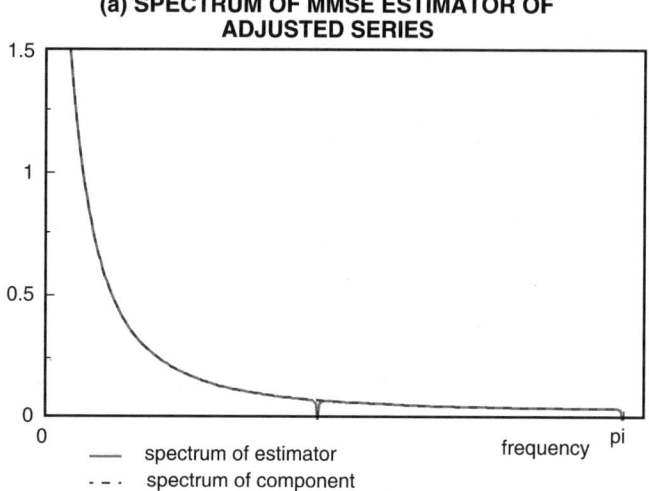
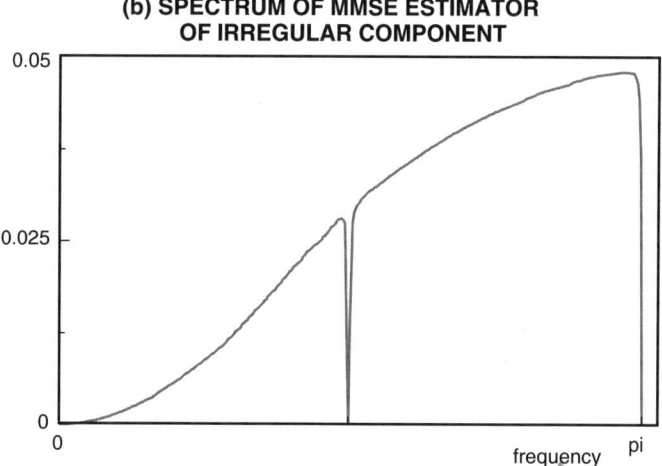

FIGURE 8.5 Select spectra for the IPI series.

Expression (8.26) can be used to derive the covariance between the component estimators. Table 8.4 displays the correlations between the (stationary transformations) of the estimators and of the estimates actually obtained. The correlations are, in all cases, negligible, the estimator and estimate provide, again, similar results.

As for the estimation errors, the variance of model (8.26), particularized for the three components, yields the variance of the final estimation error. To look at the revision errors, the weights of the filter $\xi(B, F)$ can be obtained through (8.23). Table 8.5 presents the estimation error variances of the trend and SA series, for the final and concurrent estimators.

8.10. AN EXAMPLE

TABLE 8.4. Correlation Between Estimators

	Trend and Seasonal Component	Seasonal and Irregular Components	Trend and Irregular Component
Estimator	−.06	.03	−.04
Estimate	−.09	.08	.01

Because of the close-to-deterministic nature of the seasonal component, the estimation error of the SA series, be that the final estimation error or the revision, is very small. The error in estimating the trend is larger because a relatively important irregular component has been removed. Still, the revision is of a moderate size, and the variance of the error in the concurrent estimator is approximately 20% of the innovation variance of the series. Concerning convergence of the revision, as is typically the case, the very small revision in the concurrent estimator of the SA series converges very slowly, while in just one year the trend has practically converged to the final estimator. The slow convergence of the SA series estimator to the final estimator suggests that very little would be gained from moving from a once-a-year adjustment to a concurrent one, and in fact the average decrease in root MMSE would be 1.5%. For this series, infrequent adjustment would imply little loss in precision for the SA series.

Figure 8.6a displays the last 2 years of observations for the series and the next 2 years of forecasts, with the associated 95% confidence intervals. Figures 8.6b and 8.6c exhibit for the trend and seasonal component, the estimates for the last 2 years and the forecasts for the next 2 years, together with the 95% confidence interval. Seasonality is seen to be highly significant and stable, and its forecast is fairly precise. As for the trend, although the forecasts are more precise than those of the original series, they deteriorate fast and would only be useful for short-term horizons.

Finally, analysis of the short-term evolution of the series is based mostly on changes, not on levels. Expressions (8.29) and (8.26) permit us to obtain the ACGF of the revision and final estimation errors, from which, proceeding as in Section 8.9.3, it is straightforward to find that, for example, 90% confidence intervals for the quarterly change implied by the last observation are equal to $(\pm.47)$ when the trend is used, and to $(\pm.19)$ when the SA series is used. Further, if the present rate of annual growth is measured as the rate of change over a one-year period centered at the present month

TABLE 8.5. Estimation Error Variance (in Units of σ_a^2)

	Final Estimation Error	Revision Error	Concurrent Estimation Error	% Reduction in Revision SE[a] after 1 Year of Data
Trend	.13	.08	.21	91
SA series	.01	.01	.02	4

[a] Standard error.

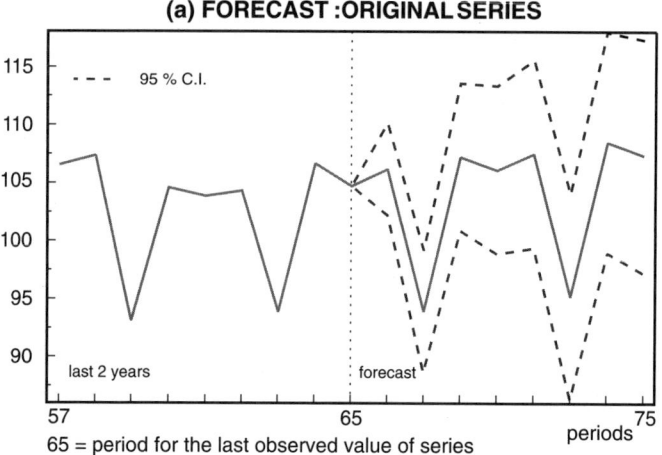
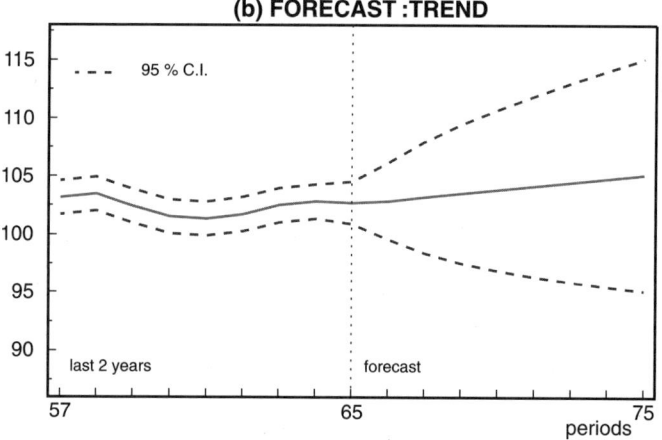
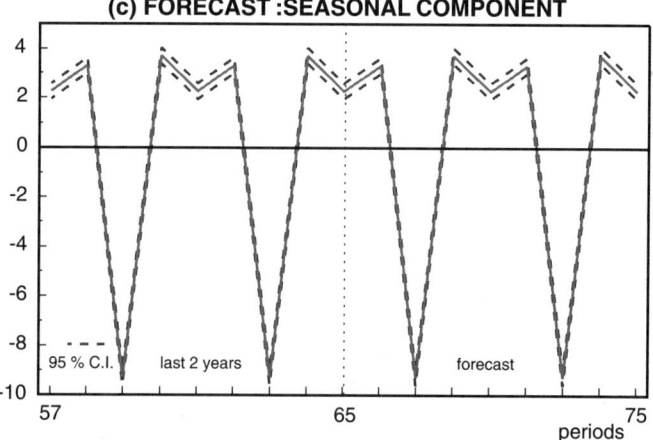

FIGURE 8.6 Selected forecasts for the IPI series.

(a measure that uses two forecasts), the standard error of the annual rate of growth is 3.64 when measured with the original or the SA series, and 3.46 when measured with the trend. For longer spans, thus, the trend signal turns out to be more precise.

8.11. RELATIONSHIP WITH FIXED FILTERS

The MBSE approach we have outlined provides a rich procedure for the derivation of linear filters to estimate signals of interest, and fixed-type filters can often be seen as the result of a particular MBSE application, at least to a reasonable approximation. A well-known case is the approximation to the X11 filter developed by Cleveland and Tiao (1976) and Burridge and Wallis (1984). The model found in these approximations for the aggregate observed series is broadly similar to a class of ARIMA models often found in practice, namely, those of the type $\nabla \nabla_{12} x_t = \theta(B) a_t$, where $\theta(B)$ displays moderately large negative values of ρ_1 and ρ_{12}. The spectral shape of this type of model presents the stylized features of the typical spectra of economic time series, as noticed by Granger (1966). Other model-based interpretations of some ad hoc filters can be found in Tiao (1983), Tiao and Hillmer (1978), King and Rebelo (1993), and Watson (1986).

To illustrate this relationship, we consider a family of fixed filter of the low-pass type (aimed at capturing low-frequency signals, i.e., long-term trends), namely, the Butterworth family of filters, popular in electrical engineering (often in the one-sided expression). For the two-sided filter, the gain is defined by

$$G(\omega) = \frac{1}{1 + \left(\frac{\sin(\omega/2)}{\sin(\omega_c/2)}\right)^{2d}} \quad (0 \leq \omega \leq \pi) \quad (8.43)$$

when based on the sine function (BFS), and by the same expression (8.43), with "sin" replaced by "tan," when based on the tangent function (BFT). The filter depends on two parameters: ω_c, the frequency for which $G(\omega_c) = \frac{1}{2}$, and $d = 1, 2, 3 \ldots$, where larger values of d produce sharper filters.

The time domain expression of (8.43) has been obtained by Gómez (Gómez 2000). Using the identity $2\sin^2(\omega/2) = (1 - e^{-i\omega})(1 - e^{i\omega})$, and replacing $e^{i\omega}$ by B yields

$$g(B, F) = \frac{1}{1 + k[(1 - B)(1 - F)]^d}$$

where $k = [2\sin^2(\omega_c/2)]^{-d}$. It is easily seen that $g(B, F)$ is the WK filter for estimating s_t in the decomposition (8.5), with $\nabla^d s_t = a_{st}$ and n_t white noise ($k = \sigma_n^2/k_s^2$). For the BFT version of the filter, using $\tan^2(\omega/2) = (1 - e^{-i\omega})(1 - e^{i\omega})/(1 + e^{-i\omega})(1 + e^{i\omega})$, the time domain expression becomes

$$g(B,F) = \left\{1 + k\left[\frac{(1-B)(1-F)}{(1+B)(1+F)}\right]^d\right\}^{-1}$$

which is the WK filter to estimate s_t in the decomposition (8.5), with $\nabla^d s_t = (1+B)^d a_{st}$ and n_t white noise. Therefore, both versions of the Butterworth filter accept simple MBSE interpretations. Notice that the signal provided by the BFT will be "canonical" in the sense of displaying a spectral zero for $\omega = \pi$.

When $d = 1$, the BFS yields the "random walk-plus-noise" decomposition. When $d = 2$, from results in King and Rebelo (1993), the BFS yields the popular Hodrick–Prescott (HP) filter (Hodrick and Prescott 1980). Since the HP was derived originally from the minimization of a function that attempts to balance the tradeoff between fitting and smoothness criteria, the example also illustrates the relationship between MA filters derived in this way and the MBSE method. It is worth noting that, although the same filter is obtained with the different approaches, only the MBSE one provides MSE of the estimators as well as forecasts, [for a more complete discussion, see Gómez (1999)].

8.12. SHORT-VERSUS LONG-TERM TRENDS; MEASURING ECONOMIC CYCLES

In the MMBE approach we have followed, the trend can be seen as a smoothed SA series, since it is obtained by removing additive white noise and perhaps some highly transitory effect as described in Section 8.4. As a consequence, the trend will, in general, have power over the range of cyclical frequencies (i.e., the range between the zero and the fundamental seasonal frequency). Trends of this type are also called *trend-cycle components* or *short-term trends*.

From a long-term perspective, short-term trends are of little use since they will not separate the long-term growth from cyclical oscillations. In economics, the study of cycles is an important field, and a simple and standard way to estimate cycles has been by using some low-pass fixed filter, often the HP filter, to detrend an X11SA series. As mentioned in the previous section, the HP filter can be seen as the minimization of an ad hoc function that attempts to balance fitting versus smoothing. It can also be seen as an optimal signal extraction filter in the UCARIMA model

$$x_t = m_t + c_t \quad (8.44)$$

$$\nabla^2 m_t = b_t \quad (8.45)$$

where c_t and b_t are mutually orthogonal white-noise variables, with variances σ_c^2 and σ_b^2; the standard application of the filter to quarterly series sets $\sigma_c^2 = 1600\,\sigma_b^2$. Algorithms to obtain the filter based on the minimization approach and on the Kalman filter estimation of the signal can be found in Danthine and Girardin (1989) and

8.12. SHORT-VERSUS LONG-TERM TRENDS; MEASURING ECONOMIC CYCLES   237

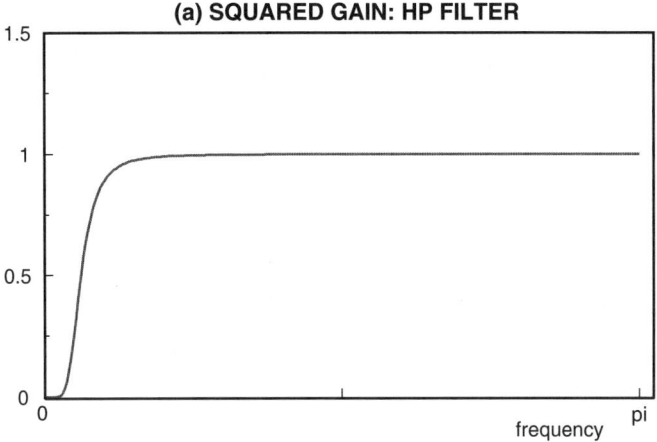

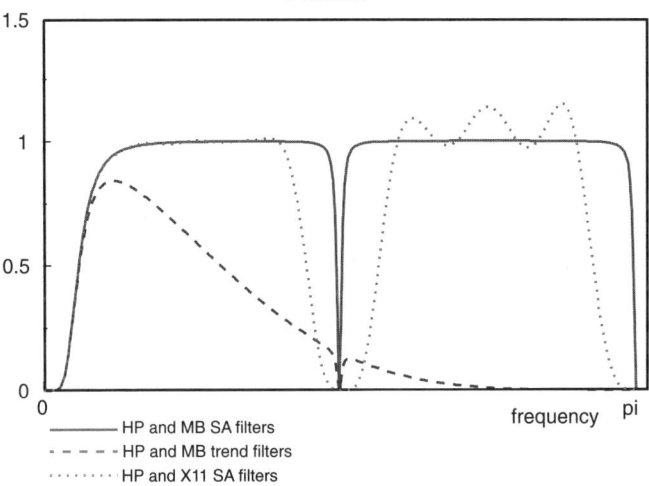

FIGURE 8.7 Squared gains of HP and SA Filters.

in Harvey and Jaeger (1993), respectively. An alternative algorithm estimates the signal through the WK filter. First, it is straightforward to find that (8.44) and (8.45) imply that the observed series follows the model $\nabla^2 x_t = \theta_H(B) a_t$, where $\theta_H = 1 - 1.7771B + .7994B^2$, and $\sigma_a^2 = 2000\sigma_b^2$. Denoting the HP filter to estimate the trend by $H_p(B,F)$ and applying expression (8.18), it is found that

$$H_p(B,F) = \frac{1}{2000} \frac{1}{\theta_H(B)} \frac{1}{\theta_H(F)}. \tag{8.46}$$

Trivially, the detrended series are obtained through the filter $H_c(B,F) = 1 - H_p(B,F)$. The squared gain of this last filter is displayed in Figure 8.7a. The filter is seen to remove the variation associated roughly with the interval $\omega \varepsilon [0, \pi/20]$, and hence will remove cycles with periods of 10 or more years. The numerical results obtained with the three algorithms are indistinguishable (Gómez 1999); the WK procedure is faster than the KF and considerably faster than the Danthine–Girardin procedure. The WK representation (8.46) is convenient for analytical discussion.

The HP filter to compute the cycle cannot be applied to the observed series, since the seasonality would be included in the cycle. It needs to be applied to either the SA series or the (short-term) trend. Therefore, in general, the two-step estimator of the cycle can be written as

$$\hat{c}_t = H_c(B,F)\,\nu(B,F)x_t = \eta(B,F)x_t \qquad (8.47)$$

where $H_c(B,F)$ denotes the HP filter; $\nu(B,F)$, the WK filter that provides the SA series or the trend; and $\eta(B,F)$, the convolution of the two. This last filter will be symmetric and centered, and using the model for x_t, given by (8.11), one can proceed with model-based analysis in a straightforward manner.

The squared gain of the $\eta(B,F)$ filter that estimates the cycle is given by the continuous line in Figure 8.7b when the SA series are used, and by the discontinuous line when the trend is considered. The dotted line in the figure displays the squared gain of the convolution of the X11 and HP filters. It is seen that the filter based on the trend is considerably more concentrated around the cyclical frequencies and ignores variation in the series of no cyclical interest. On the contrary, this variation would contaminate the cycle if the SA series is used as input.

As seen in expression (8.17), if the previous squared gain is applied to the spectrum of x_t, given in Figure 8.3b, the spectrum of the cycle estimator is obtained. Figure 8.8a displays the spectra obtained with the three inputs; the dotted line corresponds to the one based on the X11SA series, while the continuous and discontinuous lines correspond to the ones based on the model-based SA series and trend, respectively. They are seen to be similar in shape and the peak is associated, in the three cases, with a (roughly) 8-year cycle. The spectrum of the difference between the two cyclical components computed with the model-based SA series and trend is displayed in Figure 8.8b; it is close to a white-noise spectrum, and hence the cycle computed using the SA series is approximately equal to the cycle computed using the trend plus some additional noise.

Figure 8.9a compares the cycle estimates obtained with the three inputs (X11 was applied, in the X11ARIMA spirit, to the series extended at both ends with 3 years of forecasts and backcasts). The difference between using the X11SA series or the model-based SA one is seen to be minor. The difference between using the SA series or the trend cycle is, on the contrary, remarkable; the cycle estimator obtained from the trend is considerably smoother. During the 65 quarters considered, the cycle based on the SA series crosses the zero ordinate line 21 times. The cycle estimator based on the short-term trend behaves in a more sensible manner; it crosses the zero line only 7 times and cyclical periods are neatly defined. Figure 8.9b plots the MB short-term trend of Section 8.10 (equal to the seasonally adjusted and noise clean series) and the long-term trend obtained by applying the HP filter to the

8.12. SHORT-VERSUS LONG-TERM TRENDS; MEASURING ECONOMIC CYCLES

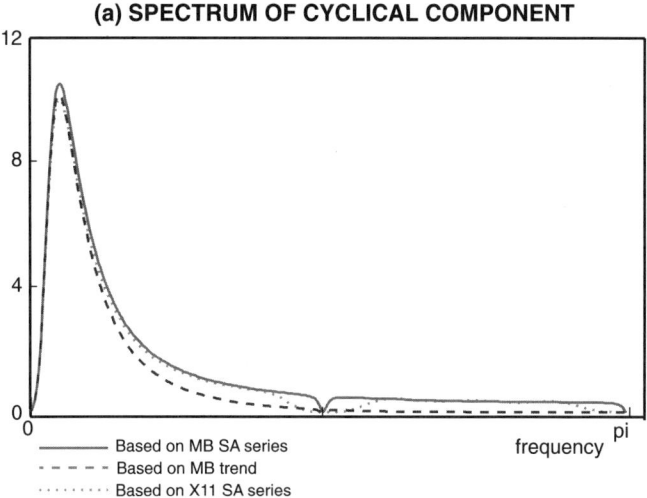

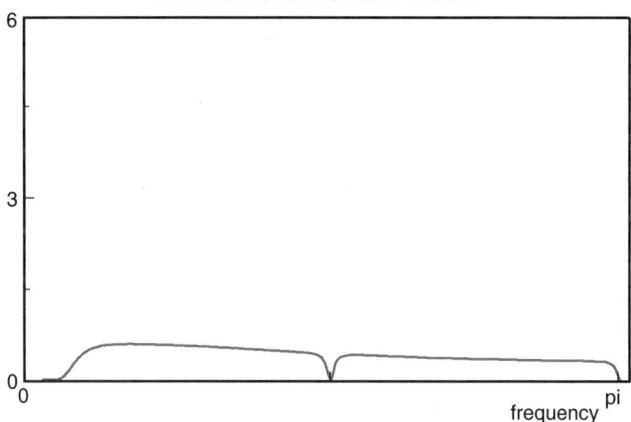

FIGURE 8.8 Spectra associated with the cyclical component.

previous short-term trend; the figure illustrates well the difference between the two and clearly indicates that the short-term trend is the signal of interest when looking at the quarter-to-quarter underlying growth of the series (i.e., the growth that results once the seasonal component and the noise have been removed). The long-term trend is of interest for a much larger horizon.

The model-based structure can be useful in more ways, as seen in Kaiser and Maravall (in press). But even if analysts using an ad hoc filter have no model for the component in mind, they will still worry about revisions in the estimator (implied by the two-sided structure of the filter). Because it considers a larger information set, the final estimator will be more accurate than the concurrent one, and the

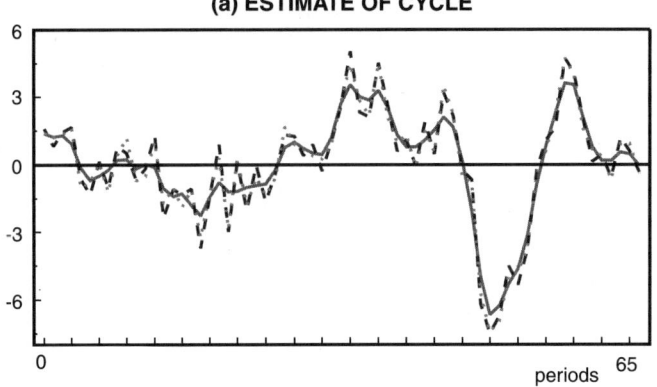
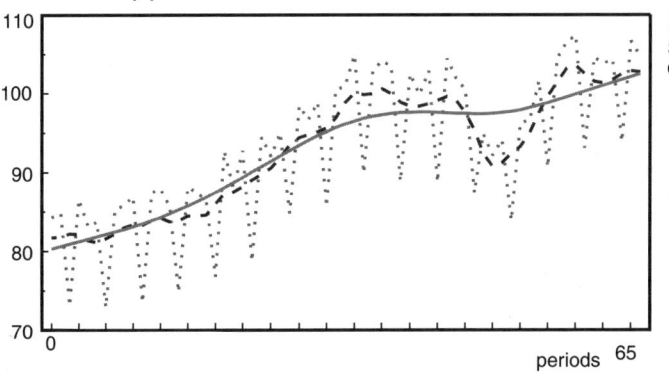
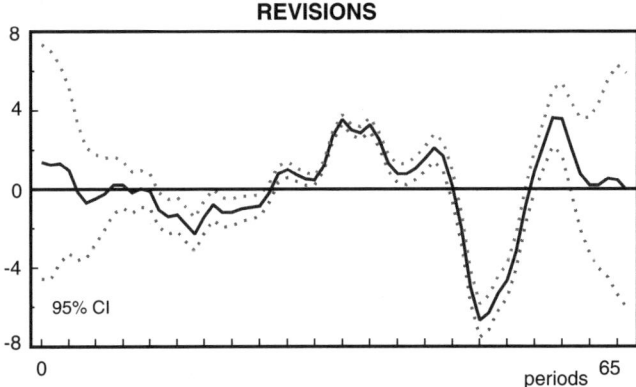

FIGURE 8.9 Trend and cycle estimated with different methods.

difference between the two estimators (i.e., the revision) can be considered an estimation error. Proceeding as in Section 8.8, and assuming that the HP filter is applied to the short-term trend, from (8.46), (8.18), and (8.11), expression (8.47) can be rewritten in terms of the observed series innovations, as

$$\hat{c}_t = k \frac{\theta_p(B)}{\theta_H(B)} \frac{\theta_p(F)(1-F^4)(1-F)}{\theta_H(F)\theta(F)} a_t = \delta(B,F) a_t \qquad (8.48)$$

where $k = .7994 k_p$. It is straightforward to see that the revision in the preliminary estimator $\hat{c}_{t-k|t}$ can then be expressed as

$$d_{k,t} = \hat{c}_t - \hat{c}_{t-k|t} = \sum_{j=1}^{\infty} \delta_{k+j} a_{t+j}$$

where invertibility of the denominator of (8.48) implies that the variance of $d_{k,t}$ can be computed using a finite number of terms. For the IPI example, the standard deviation of the revision error was computed for the estimator of the cycle based on the model-based trend. The 95% confidence intervals are shown in Figure 8.9c, from which two clear facts emerge: (1) even for a series with only 65 quarterly observations, historical estimation of the cycle is fairly precise; and (2) estimation for recent periods is unreliable. This poor performance is due mostly to the large revisions implied by the HP filter. One could exploit the model-based structure to obtain forecasts of the cycle (in a manner similar to that used in Section 8.9.1), but considering the size of the associated standard errors, these forecasts are of little interest.

REFERENCES

Akaike, H. and Ishiguro, M. (1980). *BAYSEA, a Bayesian Seasonal Adjustment Program*, Computer Science Monographs 13, The Institute of Statistical Mathematics, Tokyo.

Anderson, B. and Moore, J. (1979). *Optimal Filtering*. Prentice-Hall, Englewood Cliffs, NJ.

Ansley, C. F. and Kohn, R. (1985). Estimation, filtering and smoothing in state space models with incompletely specified initial conditions. *Ann. Stat.* **13**, 1286–1316.

Aoki, M. (1990). *State Space Modeling of Time Series*, (2nd ed.), Springer-Verlag, Berlin.

Bach, G. L., Cagan, P. D., Friedman, M., Hildreth, C. G., Modigliani, F., and Okun, A. (1976). *Improving the Monetary Aggregates: Report of the Advisory Committee on Monetary Statistics*. Board of Governors of the Federal Reserve System, Washington, DC.

Balchin, S. (1995). *A Description of the Seasonal Adjustment Methods X11, X11ARIMA, X12ARIMA, GLAS, STL, SEATS, STAMP and microCAPTAIN* (for the GSS Seasonal Adjustment Taskforce). Central Statistical Office.

Bank of England (1992). *Report of the Seasonal Adjustment Working Party*, Occasional Paper 2, Oct. 1992.

Bell, W. R. (1984). Signal extraction for nonstationary time series. *Ann. Stat.* **12**, 646–664.

Bell, W. R. (1995). Seasonal adjustment to facilitate forecasting. Arguments for not revising seasonally adjusted data. In *Proc. American Statistical Association*, Business and Economics Statistics Section.

Bell, W. R. and Hillmer, S. C. (1984). Issues involved with the seasonal adjustment of economic time series. *J. Business Econ. Stat.* **2**, 291–320.

Box, G. E. P. and Jenkins, G. M. (1976). *Time Series Analysis: Forecasting and Control*, Holden-Day, San Francisco.

Box, G. E. P., Hillmer, S. C., and Tiao, G. C. (1978). Analysis and modeling of seasonal time series. In A. Zellner (ed.), *Seasonal Analysis of Economic Time Series*, pp. 309–334. U.S. Dept. Commerce, Bureau of the Census, Washington, DC.

Box, G. E. P. and Tiao, G. C. (1975). Intervention analysis with applications to economic and environmental problems. *J. Am. Stat. Assoc.* **70**, 71–79.

Brockwell, P. and Davis, R. (1987). *Time Series: Theory and Methods*, Springer-Verlag, Berlin.

Bureau of the Census (1997). *X12-ARIMA Reference Manual, Beta Version*. Statistics Research Division, Bureau of the Census.

Burman, J. P. (1995). *Prophet: User Instructions and Software Description*. Applied Statistics Research Unit, Univ. of Kent.

Burman, J. P. (1980). Seasonal adjustment by signal extraction. *J. Roy. Stat. Soc. A* **143**, 321–337.

Burridge, P. and Wallis, K. F. (1988). Prediction theory for autoregressive moving average processes. *Econ. Rev.* **7**, 65–95.

Burridge, P. and Wallis, K. F. (1985). Calculating the variance of seasonally adjusted series. *J. Am. Stat. Assoc.* **80**, 541–552.

Burridge, P. and Wallis, K. F. (1984). Unobserved components models for seasonal adjustment filters. *J. Business Econ. Stat.* **2**, 350–359.

Canova, F. (1992). *Price Smoothing Policies: A Welfare Analysis*. Working Paper ECO No. 92/102, European Univ. Institute.

Cervantes (1605). *Las Aventuras del Ingenioso Hidalgo Don Quijote de la Mancha*, first part, Chapter XV.

Cleveland, W. S. (1972). *Analysis and Forecasting of Seasonal Time Series*, Ph.D. dissertation, Dept. Statistics, Univ. of Wisconsin—Madison.

Cleveland, W. P., Dunn, D. M., and Terpenning, I. J. (1978). SABL: A resistant seasonal adjustment procedure with graphical methods for interpretation and diagnosis. In A. Zellner (ed.), *Seasonal Analysis of Economic Time Series*, pp. 201–231. U.S. Dept. Commerce, Bureau of the Census, Washington, DC.

Cleveland, R. B., Cleveland, W. S., McRae, J. E., and Terpenning, I. J. (1990). STL: A seasonal-trend decomposition procedure based on lowess. *J. Offi. Stat.*, **6**, 3–73.

Cleveland, W. P. and Tiao, G. C. (1976). Decomposition of seasonal time series: A model for the X-11 program. *J. Am. Stat. Assoc.* **71**, 581–587.

Crafts, N. F. R., Leybourne, S. J., and Mills, T. C. (1989). Trends and cycles in British industrial production, 1700–1913. *J. Roy. Stat. Soc. A* **152**, 43–60.

Dagum, E. B. (1980). *The X11 Arima Seasonal Adjustment Method*. Statistics Canada, Catalogue 12-564E.

Danthine, J. P. and Girardin, M. (1989). Business cycles in Switzerland. A comparative study. *Eur. Econ. Rev.* **33**, 31–50.

De Jong, P. (1991). The diffuse Kalman filter. *Ann. Stat.* **19**, 1073–1083.

De Jong, P. and Chu-Chun-Lin, S. (1994). Fast likelihood evaluation and prediction for nonstationary state space models. *Biometrika* **81**, 133–142.

Den Butter, F. A. G. and Fase, M. M. G. (1991). *Seasonal Adjustment as a Practical Problem.* North-Holland, Amsterdam.

Engle, R. F. (1978). Estimating structural models of seasonality. In A. Zellner (ed.), *Seasonal Analysis of Economic Time Series*, pp. 281–297. U.S. Dept. Commerce, Bureau of the Census, Washington, DC.

Eurostat (1998a). *Seasonal Adjustment Methods: A Comparison.* Eurostat, Luxembourg, Sept. 1998.

Eurostat (1998b). *Eurostat Suggestions Concerning Seasonal Adjustment Policy.* SAM 98 Seminar, Bucharest, Oct. 1998.

Federal Reserve Board of Governors (1915). *First Annual Report*, Government Printing Office, Washington, DC.

Findley, D.F., Monsell, B.C., Bell, W.R., Otto, M.C., and Chen, S. (1998). New capabilities and methods of the X12ARIMA seasonal adjustment program (with discussion). *J. Business Econ. Stat.* **16**, 127–177.

Fiorentini, G. and Maravall, A. (1996). Unobserved components in arch models: An application to seasonal adjustment. *J. Forecasting* **15**, 175–201.

Fisher, B. (1995). *Decomposition of Time Series: Comparing Different Methods in Theory and Practice*, version 2.1, Luxembourg: Eurostat, April 1995.

Gersh, W. and Kitagawa, G. (1983). The prediction of time series with trends and seasonalities. *J. Business Econ. Stat.* **1**, 253–264.

Ghysels, E. (1994). On the periodic structure of the business cycle. *J. Business Econ. Stat.* **12**, 289–298.

Ghysels, E. (1993a). On the economics and econometrics of seasonality. *Advances in Econometrics. Sixth World Congress.* In C. A. Sims (ed.), Cambridge Univ. Press, Cambridge, UK.

Ghysels, E. (1993b). *Seasonality and Econometric Models*, special issue of the *J. Econ.* **55** (1–2).

Ghysels, E., Granger, C. W. J., and Siklos, P. L. (1996). "Is seasonal adjustment a linear or nonlinear data-filtering process? *J. Business Econ. Stat.* **14**, 374–397.

Ghysels, E. and Perron, P. (1993). The effect of seasonal adjustment filters on tests for a unit root. *J. Econ.* **55**, 57–98.

Gómez, V. (2000). The use of Butterworth filters for trend and cycle estimation in economic time series. Mimeo. Ministerio de Hacienda, Madrid.

Gómez, V. (1999). Three equivalent methods for filtering finite nonstationary time series. *J. Business Econ. Stat.* **17**, 109–116.

Gómez, V. and Maravall, A. (1996). *Programs TRAMO and SEATS; Instructions for the User.* Working Paper 9628, Servicio de Estudios, Banco de España.

Gómez, V. and Maravall, A. (1994). Estimation, prediction and interpolation for nonstationary series with the Kalman filter. *J. Am. Stat. Assoc.* **89**, 611–624.

Gómez, V., Maravall, A., and Peña, D. (1999). Missing observations in Arima models: Skipping approach versus additive outlier approach. *J. Econ.* **88**, 341–364.

Gourieroux, C. and Monfort, A. (1990). *Séries Temporelles et Modèles Dynamiques.* Economica, Paris.

Granger, C. W. J. (1966). The typical spectral shape of an economic variable. *Econometrica* **34**, 150–161.

Harrison, P. J. and Stevens, C. F. (1976). Bayesian forecasting. *J. Roy. Stat. Soc. B* **38**, 205–247.

Harvey, A. C. (1993). *Time Series Models*. Philip Allan, Deddington, UK.

Harvey, A. C. (1989). *Forecasting, Structural Time Series Models and the Kalman Filter*. Cambridge University Press, Cambridge, UK.

Harvey, A. C. and Jaeger, A. (1993). Detrending, stylized facts and the business cycle. *J. Appl. Econ.* **8**, 231–247.

Harvey, A. C. and Koopman, S. J. (1992). Diagnostic checking of unobserved components time series models. *J. Business Econ. Stat.* **10**, 377–390.

Harvey, A. C., Ruiz, E., and Sentana, E. (1992). Unobserved component time series models with ARCH disturbances. *J. Econ.* **52**, 129–157.

Harvey, A. C. and Todd, P. H. J. (1983). Forecasting economic time series with structural and Box–Jenkins models: A case study. *J. Business Econ. Stat.* **1**, 299–306.

Hillmer, S. C. (1985). Measures of variability for model-based seasonal adjustment procedures. *J. Business Econ. Stat.* **3**, 60–68.

Hillmer, S. C., Bell, W. R., and Tiao, G. C. (1983). Modeling considerations in the seasonal adjustment of economic time series. In A. Zellner (ed.), *Applied Time Series Analysis of Economic Data*, pp. 74–100. U.S. Dept. Commerce, Bureau of the Census, Washington, DC.

Hillmer, S. C. and Tiao, G. C. (1982). An Arima-model based approach to seasonal adjustment. *J. Am. Stat. Assoc.* **77**, 63–70.

Hodrick, R. and Prescott, E. (1980). *Post-War U.S. Business Cycles: An Empirical Investigation*. Carnegie Mellon Univ. manuscript.

Hylleberg, S. (ed.) (1992). *Modeling Seasonality*, Oxford University Press, Oxford, UK.

Jenkins, G. M. (1979). Practical experiences with modelling and forecasting time series. In O. D. Anderson (ed.), *Forecasting*, North-Holland, Amsterdam.

Kailath, T. (1976). *Lectures on Linear Least-Squares Estimation*. Springer-Verlag, New York.

Kaiser, R. and Maravall, A. (in press). Trend, seasonality and the business cycle; the Hodrick–Prescott filter revisited (mimeo). *Span. Econ. Rev. I*.

Kendall, M. (1976). *Time Series*, Griffin, London.

King, R. G. and Rebelo, S. T. (1993). Low frequency filtering and real business cycles. *J. Econ. Dyn. Contr.* **17**, 207–233.

Kitagawa, G. (1987). Non-Gaussian state space modeling of nonstationary time series. *J. Am. Stat. Assoc.* **82**, 1032–1063.

Kohn, R. and Ansley, C. F. (1987). Signal extraction for finite nonstationary time series. *Biometrika* **74**, 411–421.

Koopman, S. J., Harvey, A. C., Doornik, J. A., and Shephard, N. (1996). *Stamp: Structural Time Series Analyser, Modeller and Predictor*, Chapman and Hall, London.

Maravall, A. (1995). Unobserved components in economic time series. In H. Pesaran and M. Wickens. (eds.), *The Handbook of Applied Econometrics*, Vol. 1, Basil Blackwell, Oxford.

Maravall, A. (1994). Use and misuse of unobserved components in economic forecasting. *J. Forecasting* **13**, 157–178.

Maravall, A. (1993). Stochastic linear trends: Models and estimators. *J. Econ.* **56**, 5–37.

Maravall, A. (1989). On the dynamic structure of a seasonal component. *J. Econ. Dyn. Contr.* **13**, 81–91.

Maravall, A. (1988). The use of Arima models in unobserved components estimation. In W. Barnet, E. Berndt, and H. White (eds.), *Dynamic Econometric Modeling*. Cambridge Univ. Press, Cambridge, UK.

Maravall, A. (1987). On minimum mean squared error estimation of the noise in unobserved component models. *J. Business Econ. Stat.* **5**, 115–120.

Maravall, A. (1985). On structural time series models and the characterization of components. *J. Business Econ. Stat.* **3**(4), 350–355.

Maravall, A. (1983). Comment on "modelling considerations in the seasonal adjustment of economic data." In A. Zellner (ed.), *Applied Time Series Analysis of Economic Data*, U.S. Department of Commerce. Bureau of the Census.

Maravall, A. and Mathis, A. (1994). Encompassing univariate models in Multivariate Time Series: A case study. *J. Econ.* **61**, 197–233.

Maravall, A. and Planas, C. (in press). Estimation error and the specification of unobserved component models. Working Paper 9608, Servicio de Estudios, Banco de España. *J. Econ.*

Maravall, A. and Pierce, D. A. (1983). Preliminary-data error and monetary aggregate targeting. *J. Business Econ. Stat.* **1**, 179–186.

Miron, J. A. (1986). Financial panics, the seasonality of nominal interest rates and the founding of the Fed. *Am. Econ. Rev.* **76**, 125–140.

Moore, G. H., Box, G. E. P., Kaitz, H. B., Stephenson, J. A., and Zellner, A. (1981). *Seasonal Adjustment of the Monetary Aggregates: Report of the Committee of Experts on Seasonal Adjustment Techniques*, Board of Governors of the Federal Reserve System, Washington, DC.

Nelson, D. B. (1996). Asymptotic filtering theory for multivariate ARCH models. *J. Econ.* **71**, 1–47.

Nerlove, M., Grether, D. M., and Carvalho, J. L. (1979). *Analysis of Economic Time Series: A Synthesis*. Academic Press, New York.

Nourney, M. (1986). Umstellung der Zeitreihenanalyse. *Wirtschaft Statistik* **11**, 841–852.

Oppenheim, A. V. and Schaffer, R. W. (1989). *Discrete-Time Signal Processing*. Prentice-Hall, Englewood Cliffs, NJ.

Osborn, D. R. (1988). Seasonality and habit persistence in a life cycle model of consumption. *J. Appl. Econ.* **3**, 255–266.

Piccolo, D. and Vitale, C. (1981). *Metodi statistici per l'analisi economica*. Il Mulino, Bologna.

Pierce, D. A. (1980). Data revisions in moving average seasonal adjustment procedures. *J. Econ.* **14**, 95–114.

Pierce, D. A. (1979). Signal extraction error in nonstationary time series. *Ann. Stat.* **7**, 1303–1320.

Plosser, C. I. (1978). A time series analysis of seasonality in econometric models. In A. Zellner (ed.), *Seasonal Analysis of Economic Time Series*, pp. 365–397. U.S. Dept. Commerce, Bureau of the Census, Washington, DC.

Roberts, S. A. and Harrison, P. J. (1984). Parsimonious modelling and forecasting of seasonal time series. *Eur. J. Oper. Res.* **16**, 365–377.

SARA Committe (1998). The Results of the SARA Committe. ISTAT June 1998 Rome Conference.

Shephard, W. (1994). Local scale models: State space alternative to integrated GARCH processes. *J. Econ.* **60**, 181–202.

Shiskin, J., Young, A. H., and Musgrave, J. C. (1967). *The X11 Variant of the Census Method II Seasonal Adjustment Program.* Technical Paper 15, Bureau of the Census, Washington, DC.

Statistiches Bundesamt (1997). *Methodological Outline of the BV4 Decomposition Method* (mimeo).

Stephenson, J. A. and Farr, H. T. (1972). Seasonal adjustment of economic data by application of the general linear statistical model. *J. Am. Stat. Assoc.* **67**, 37–45.

Stock, J. H. and Watson, M. W. (1988). Variable trends in economic time series. *J. Econ. Persp.* **2**, 147–174.

Tiao, G. C. (1983). Study notes on Akaike's seasonal adjustment procedures. In A. Zellner (ed.), *Applied Time Series Analysis of Economic Data*, pp. 44–45. U.S. Dept. Commerce, Bureau of the Census, Washington, DC.

Tiao, G. C. and Hillmer, S. C. (1978). Some consideration of decomposition of a time series. *Biometrika* **65**, 497–502.

Wallis, K. F. (1974). Seasonal adjustment and relations between variables. *J. Am. Stat. Assoc.* **69**, 18–31.

Watson, M. W. (1986). Univariate detrending methods with stochastic trends. *J. Monetary Econ.* **18**, 49–75.

Whittle, P. (1963). *Prediction and Regulation by Linear Least-Squares Methods.* English Univ. Press, London.

Zellner, A. (ed.) (1983). *Applied Time Series Analysis of Economic Data*, Proceedings of a Bureau of the Census-NBER-ASA Conference, U.S. Dept. Commerce, Bureau of the Census, Washington, DC.

Zellner, A. (ed.) (1978). *Seasonal Analysis of Economic Time Series*, Proceedings of a Bureau of the Census-NBER-ASA Conference, U.S. Dept. Commerce, Bureau of the Census, Washington, DC.

PART II

Advanced Topics in Univariate Time Series

CHAPTER 9

Heteroscedastic Models

Ruey S. Tsay
University of Chicago

In options trading and in the foreign exchange rate market, volatility plays an important role. Here volatility means conditional variance of the underlying asset return. There are other measurements of volatility available in the literature. For instance, in interest rate studies, the "spread" between the long-term rate and short-term rate may be used to measure market variability. Indeed, the spread between 3-year Treasury notes and 3-month Treasury bills plays an important role in studying the term structure of U.S. interests. This non-uniqueness in definition is highly related to the fact that there is no direct measurement of volatility available. It also makes the evaluation of the accuracy of volatility forecasts difficult.

In options markets, if one accepts the idea that the prices are governed by a statistical model such as the Black–Scholes formula, then one can use the actual prices to obtain the "implied" volatility. But, this approach is often criticized for using a specific model, which in turn is based on some assumptions that might be hard to justify in practice. For instance, the well-known Black–Scholes formula is derived under the normality assumption. The implied volatility might be quite different from the actual volatility.

In this chapter, we define *volatility* as the conditional variance of an asset return and discuss econometric and statistical models available in the literature to model the evolution of volatility over time. The models discussed include the conditional heteroscedastic autoregressive (ARCH) model of Engle (1982), the generalized ARCH (GARCH) model of Bollerslev (1986), the exponential GARCH (EGARCH) model of Nelson (1991), the conditional heteroscedastic autoregressive moving-average (CHARMA) model of Tsay (1987), the random coefficient autoregressive (RCA) model of Nicholls and Quinn (1982), and the stochastic volatility (SV) models of Melino and Turnbull (1990), Harvey et al. (1994), and Jacquier et al. (1994). We shall also discuss advantages and weaknesses of each volatility model.

A Course in Time Series Analysis, Edited by Daniel Peña, George C. Tiao, and Ruey S. Tsay.
ISBN 0-471-36164-X. © 2001 John Wiley & Sons, Inc.

Although volatility is not directly measurable, it has some basic properties that are commonly seen in asset returns: (1) there are volatility clusters, that is, volatility may be high for certain time periods and low for other periods; (2) volatility evolves over time in a continuous manner, that is, there does not appear to have volatility jumps; (3) volatility does not diverge to infinity, that is, volatility varies within some fixed range—in statistical terms, this means volatility is stationary; and (4) volatility seems to react differently to a big positive return and a big negative return. These properties play an important role in the development of volatility models. Some of the volatility models were proposed specifically to correct the weaknesses of the existing models. For example, the EGARCH model was developed to capture the asymmetry between big "positive" and "negative" asset returns.

Let z_t be the return series of an asset, specifically, $z_t = \ln(p_t) - \ln(p_{t-1})$, where p_t is the price of the underlying asset at time t. If the asset is a stock with dividend payment d_t, then the return becomes $z_t = \ln(p_t + d_t) - \ln(p_{t-1})$. Other definitions of return can also be used. Treating z_t as a time series, it is informative to consider the conditional mean and conditional variance of z_t given F_{t-1}:

$$\mu_t = E(z_t|F_{t-1}), \qquad h_t = \text{Var}(z_t|F_{t-1}) = E[(z_t - \mu_t)^2|F_{t-1}] \qquad (9.1)$$

where F_{t-1} denotes the information set available at time $t-1$. Typically, F_{t-1} consists of all linear functions of the past returns. For simplicity, we assume that $\mu_t = 0$. In practice, if $\mu_t \neq 0$, then one should obtain a time series model for z_t and consider the mean-removed process $z_t^* = z_t - \mu_t$.

All volatility models mentioned above are models for h_t of (9.1). These models can be classified into two categories. The models in the first category use a deterministic function to govern the evolution of h_t, whereas those in the second category use a stochastic equation to describe h_t. The stochastic volatility models belong to the second category.

9.1. THE ARCH MODEL

The first model that provides a systematic framework for volatility modeling is the ARCH model of Engle (1982). The basic idea of ARCH models is that (1) the asset return z_t is serially uncorrelated, but dependent and (2) the dependence of z_t can be described by a simple quadratic function. Specifically, an ARCH(r) model assumes that

$$z_t = \sqrt{h_t}\epsilon_t, \qquad h_t = \alpha_0 + \alpha_1 z_{t-1}^2 + \cdots + \alpha_r z_{t-r}^2 \qquad (9.2)$$

where $\{\epsilon_t\}$ is a sequence of independent and identically distributed (iid) random variables with mean zero and variance 1, $\alpha_0 > 0$ and $\alpha_i \geq 0$ for $i > 0$. The coefficients α_i must satisfy some regularity condition to ensure that the unconditional variance of z_t is finite. In practice, ϵ_t is often assumed to follow the standard normal or a student-t distribution.

9.1. THE ARCH MODEL

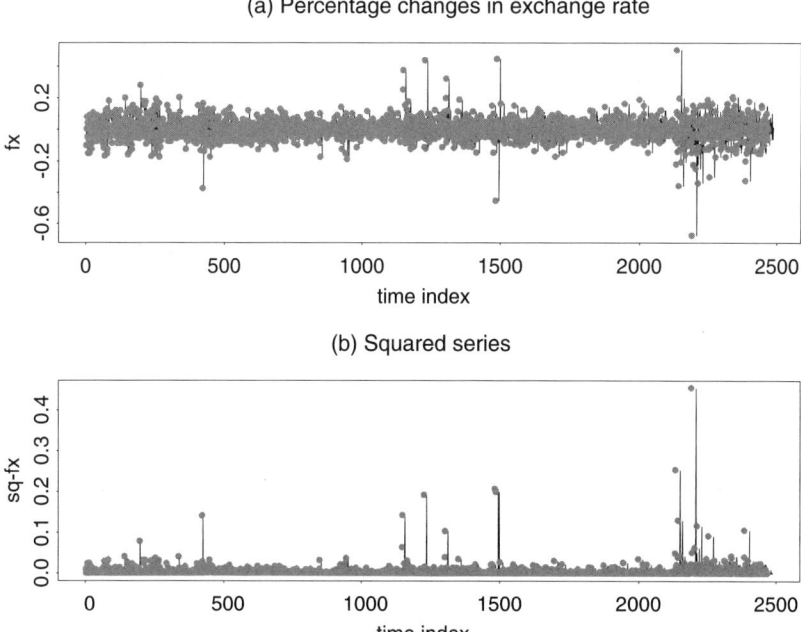

FIGURE 9.1 (a) Time plot of 10-min returns of exchange rate between the German deutsche mark and the U.S. dollar; (b) Squared returns.

From the structure of the model, it is seen that large past squared return $\{z_{t-i}^2\}_{i=1}^r$ imply a large conditional variance h_t for the return z_t. Consequently, z_t tends to assume a large value (in modulus). This means that, under the ARCH framework, large returns tend to be followed by another large return. Here I use the word "tend" because a large variance does not necessarily produce a large variate. It only says that the probability of obtaining a large variate is greater than that of a smaller variance. This feature is similar to the volatility clustering observed in asset returns.

To see the ARCH effect, Figure 9.1 shows the time plots of (1) the percentage changes in deutsche mark/U.S. dollar exchange rate measured in 10-min intervals from June 5, 1989 to June 19, 1989 for 2488 observations and (2) the squared series of the percentage changes. Some significant percentage changes occurred occasionally, but there exist certain stable periods. Figure 9.2a shows the sample autocorrelation function (acf) of the percentage change series. Clearly, the series has no serial correlation. Figure 9.2b shows the sample partial autocorrelation function (pacf) of the squared series of percentage changes. It is seen that there are some big spikes in the pacf. Such spikes suggest that the percentage changes are not independent and have some ARCH effects.

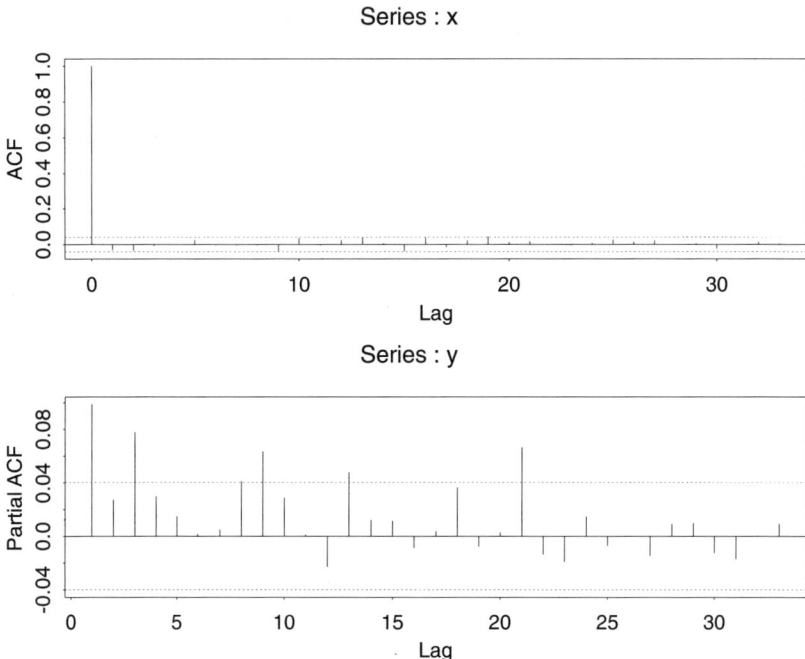

FIGURE 9.2 (a) Sample autocorrelation function of the return series of exchange rate and (b) sample partial autocorrelation function of the squared returns.

9.1.1. Some simple properties of ARCH models

To understand the ARCH models, it pays to study carefully the ARCH(1) model

$$z_t = \sqrt{h_t}\epsilon_t, \quad h_t = \alpha_0 + \alpha_1 z_{t-1}^2$$

where $\alpha_0 > 0$ and $\alpha_1 \geq 0$. First, the unconditional mean of z_t remains zero, because

$$E(z_t) = E[E(z_t \mid F_{t-1})] = 0.$$

Second, consider the unconditional variance of z_t:

$$\begin{aligned}\text{Var}(z_t) &= E(z_t^2) = E\big[E\big(z_t^2 \mid F_{t-1}\big)\big] \\ &= E\big(\alpha_0 + \alpha_1 z_{t-1}^2\big) = \alpha_0 + \alpha_1 E\big(z_{t-1}^2\big).\end{aligned}$$

Because z_t is a stationary process, $E(z_t^2) = E(z_{t-1}^2) = \text{Var}(z_t)$. Therefore, we have $\text{Var}(z_t) = \alpha_0 + \alpha_1 \text{Var}(z_t)$, and $\text{Var}(z_t) = \alpha_0/(1 - \alpha_1)$. Because variance must be positive, we need $0 \leq \alpha_1 < 1$. Third, in some cases, we need the higher-order moments of z_t to exist and, hence, α_1 must also satisfy further constraints. For instance,

9.1. THE ARCH MODEL

to study its tail behavior, we require that the fourth moment of z_t is finite. Under the normality assumption of ϵ_t in (9.2), we have

$$E(z_t^4 \mid F_{t-1}) = 3E(z_t^2 \mid F_{t-1}) = 3(\alpha_0 + \alpha_1 z_{t-1}^2)^2.$$

Therefore

$$E(z_t^4) = E[E(z_t^4 \mid F_{t-1})] = 3E(\alpha_0 + \alpha_1 z_{t-1}^2)^2 = 3E[\alpha_0^2 + 2\alpha_0\alpha_1 z_{t-1}^2 + \alpha_1^2 z_{t-1}^4]$$

from which, if z_t is fourth-order stationary with $m_4 = E(z_t^4)$, we have

$$m_4 = 3[\alpha_0^2 + 2\alpha_0\alpha_1 \operatorname{Var}(z_t) + \alpha_1^2 m_4]$$
$$= 3\alpha_0^2\left(1 + 2\frac{\alpha_1}{1 - \alpha_1}\right) + 3\alpha_1^2 m_4.$$

Therefore

$$m_4 = \frac{3\alpha_0^2(1 + \alpha_1)}{(1 - \alpha_1)(1 - 3\alpha_1^2)}.$$

This result has two important implications: (1) since the fourth moment is positive, we see that α_1 must also satisfy the condition $1 - 3\alpha_1^2 > 0$; that is, $0 \le \alpha_1^2 < \frac{1}{3}$; and (2) the unconditional kurtosis of z_t is

$$\frac{E(z_t^4)}{[\operatorname{Var}(z_t)]^2} = 3\frac{\alpha_0^2(1 + \alpha_1)}{(1 - \alpha_1)(1 - 3\alpha_1^2)} \times \frac{(1 - \alpha_1)^2}{\alpha_0^2} = 3\frac{1 - \alpha_1^2}{1 - 3\alpha_1^2} > 3.$$

Thus, the tail distribution of z_t is heavier than that of a normal distribution. In other words, the z_t under an ARCH(1) model is more likely than under normality to produce "outliers". This is in agreement with the empirical finding that outliers appear more often in asset returns than that implied by an iid sequence of normal random variates.

The above properties continue to hold for the general ARCH models, but the formulas become more complicated for higher-order ARCH models. The condition $\alpha_i \ge 0$ in equation (9.2) can be relaxed. It is a condition to ensure that the conditional variance h_t is positive for all t. In fact, a natural way to achieve positiveness in conditional variance is to rewrite an ARCH(r) model as

$$z_t = \sqrt{h_t}\epsilon_t, \quad h_t = \alpha_0 + Z'_{r,t-1}\Omega Z_{r,t-1} \quad (9.3)$$

where $Z_{r,t-1} = (z_{t-1}, \ldots, z_{t-r})'$ and Ω is an $r \times r$ non-negative definite matrix. The ARCH(r) model in (9.2) requires that Ω be diagonal. Thus, Engle's model uses a parsimonious approach to present a quadratic function. A simple way to achieve equation (9.3) is to employ a random coefficient model for z_t; see the CHARMA and RCA models discussed later.

9.1.2. Weaknesses of ARCH models

The advantages of ARCH models include properties discussed in the previous subsection. The model also has several weaknesses:

1. The model treats "positive" and "negative" returns in the same manner, because it depends on the square of the previous returns. In practice, it is well known that for financial time series the prices respond differently to positive and negative returns.
2. The ARCH model is rather restrictive. For instance, for the ARCH(1) model α_1^2 must be between $[0, \frac{1}{3}]$ if the series is to have a finite fourth moment. The constraint is even stronger for higher-order ARCH models.
3. The ARCH model does not provide any new insight for understanding financial time series. They only provide a mechanical way to describe the behavior of the conditional variance. It says nothing about what causes such behavior to occur.
4. ARCH models often over-predict the volatility, because they respond slowly to isolated large shocks to the return series.

9.1.3. Building ARCH models

A simple way to build an ARCH model consists of three steps. First, an ARIMA model is built for the observed time series to remove any serial correlations in the data. For most asset return series, this step amounts to remove the sample mean from the data if the sample mean is statistically significant. For simplicity, we continue to denote the mean-adjusted series by z_t. Then, examine the squared series z_t^2 to check for conditional heteroscedasticity. Two tests are available here. The first test is to check the usual Ljung–Box statistics of z_t^2; see McLeod and Li (1983). The second test for conditional heteroscedasticity is the Lagrange multiplier test of Engle (1982). This test is equivalent to the usual F statistic for testing $\alpha_i = 0$ $(i = 1, \ldots, k)$ in the linear regression

$$z_t^2 = \alpha_0 + \alpha_1 z_{t-1}^2 + \cdots + \alpha_k z_{t-k}^2 + e_t, \quad t = k+1, \ldots, n$$

where e_t denotes the error term, k is a prespecified positive integer, and n is the sample size. Let $SSR_0 = \sum_{t=k+1}^{n}(z_t^2 - \hat{\mu})^2$, where $\hat{\mu}$ is the sample mean of z_t^2, and $SSR_1 = \sum_{t=k+1}^{n}\hat{e}_t^2$, where $\hat{e}_t$ is the least-squares residuals of the above linear regression. Then, we have

$$F = \frac{(SSR_0 - SSR_1)/k}{SSR_1/(n - 2k - 1)}$$

which is asymptotically distributed as a chi-square distribution with k degrees of freedom. If the test is statistically significant, then conditional heteroscedasticity is detected. The third step of modeling is to use the pacf of z_t^2 to determine the ARCH order and to perform maximum likelihood estimation of the specified model. Using

9.1. THE ARCH MODEL

a pacf of z_t^2 to select the ARCH order can be justified as follows. From the model in (9.2), we have

$$h_t = \alpha_0 + \alpha_1 z_{t-1}^2 + \cdots + \alpha_r z_{t-r}^2.$$

For a given sample, z_t^2 is an unbiased estimate of h_t. Therefore, we expect that z_t^2 is related to $z_{t-1}^2, \ldots, z_{t-r}^2$ in a manner similar to that of an autoregressive model of order r. Note that a single z_t^2 is generally not a good estimate of h_t, but it can serve as an approximation that could be informative in specifying the order r.

Under the normality assumption, the likelihood function of an ARCH(r) model is

$$f(z_1, \ldots, z_n \mid \alpha) = f(z_n \mid F_{n-1}) f(z_{n-1} \mid F_{n-2}) \cdots f(z_{r+1} \mid F_r) f(z_1, \ldots, z_r \mid \alpha)$$

$$= \prod_{t=r+1}^{n} \frac{1}{\sqrt{2\pi h_t}} \exp\left[-\frac{z_t^2}{2h_t}\right] \times f(z_1, \ldots, z_r \mid \alpha)$$

where the marginal density function $f(z_1, \ldots, z_r \mid \alpha)$ is rather complicated. For simplicity, one may drop the last term from this likelihood function, especially when the sample size is sufficiently large. This results in a conditional likelihood function as

$$f(z_{r+1}, \ldots, z_n \mid \alpha, Z_r) = \prod_{t=r+1}^{n} \frac{1}{\sqrt{2\pi h_t}} \exp\left[-\frac{z_t^2}{2h_t}\right]$$

where h_t can be evaluated starting with h_i as the sample variance of z_t for $i \leq r$. We refer to estimates obtained by maximizing the likelihood function shown above as the conditional maximum likelihood estimates (MLEs).

9.1.4. An illustrative example

In this subsection, we analyze the percentage changes of exchange rate between the deutsche mark and dollar in 10-min intervals. The data are shown in Figure 9.1a. As shown in Figure 9.2a, the series has no serial correlations. However, pacf of the squared series z_t^2 shows some big spikes, especially at lags 1 and 3. There are some large pacf at higher lags, but the lower-order lags tend to be more important. Following the procedure discussed in the previous subsection, we can specify an ARCH(3) model for the series. Using the RATS program with conditional MLE, we obtain a fitted model as

$$h_t = .22 \times 10^{-6} + .328 z_{t-1}^2 + .073 z_{t-2}^2 + .103 z_{t-3}^2$$

where all the estimates are statistically significant at the 5% significant level, and the standard errors of the parameters are 0.46×10^{-8}, 0.0162, 0.0160, and 0.0147, respectively.

9.2. THE GARCH MODEL

The ARCH model has another disadvantage as it often requires many parameters to adequately describe the evolution of volatility of an asset return. For instance, for the monthly return series of S&P 500 index, one needs an ARCH(9) model for the volatility. An alternative model must be sought. One such an alternative is called the generalized ARCH (GARCH) model of Bollerslev (1986). A time series z_t follows a pure GARCH(r, s) model if $\mu_t = 0$ and

$$z_t = \sqrt{h_t}\epsilon_t, \quad h_t = \alpha_0 + \sum_{i=1}^{r}\alpha_i z_{t-i}^2 + \sum_{j=1}^{s}\beta_j h_{t-j} \tag{9.4}$$

where, again, $\{\epsilon_t\}$ is a sequence of iid random variables with mean 0 and variance 1.0, $\alpha_0 > 0$, $\alpha_i \geq 0$, $\beta_j \geq 0$, and $\sum_{i=1}^{\max(r,s)}(\alpha_i + \beta_i) < 1$. Here it is understood that $\alpha_i = 0$ for $i > r$ and $\beta_j = 0$ for $j > s$. The latter constraint on $\alpha_i + \beta_j$ implies that the unconditional variance of z_t is finite whereas its conditional variance h_t evolves over time. In practice, ϵ_t is often assumed to be a standard normal or student t distribution. Equation (9.4) reduces to a pure ARCH(r) model if $s = 0$.

It is easier to understand properties of GARCH models by using the following representation. Let $\eta_t = z_t^2 - h_t$ so that $h_t = z_t^2 - \eta_t$. By plugging $h_{t-i} = z_{t-i}^2 - \eta_{t-i}$ ($i = 0, \ldots, s$) into equation (9.4), we can rewrite the GARCH model as

$$z_t^2 = \alpha_0 + \sum_{i=1}^{\max(r,s)}(\alpha_i + \beta_i)z_{t-i}^2 + \eta_t - \sum_{j=1}^{s}\beta_j \eta_{t-j}. \tag{9.5}$$

It is easy to check that $\{\eta_t\}$ is a martingale difference series, that is, $E(\eta_t) = 0$ and $\text{Cov}(\eta_t, \eta_{t-j}) = 0$ for $j \geq 1$. However, $\{\eta_t\}$ is in general not an iid sequence. Equation (9.5) is an ARMA form for the squared series z_t^2. Thus, a GARCH model can be regarded as an application of the ARMA idea to the squared series z_t^2. It is then clear that

$$E(z_t^2) = \frac{\alpha_0}{1 - \sum_{i=1}^{\max(r,s)}(\alpha_i + \beta_i)}$$

provided that the denominator of the above fraction is positive.

The strengths and weaknesses of GARCH models can easily be seen by focusing on the simplest GARCH(1,1) model with

$$h_t = \alpha_0 + \alpha_1 z_{t-1}^2 + \beta_1 h_{t-1}, \quad 0 \leq \alpha_1, \beta_1 \leq 1, (\alpha_1 + \beta_1) < 1. \tag{9.6}$$

First, a large z_{t-1}^2 or h_{t-1} gives rise to a large h_t. This means that a large z_{t-1}^2 tends to be followed by another large z_t^2, creating, again, the well-known behavior of volatility

9.2. THE GARCH MODEL

clustering in the financial time series. Second, it can be shown that if $1 - 2\alpha_1^2 - (\alpha_1 + \beta_1)^2 > 0$, then

$$\frac{E(z_t^4)}{[E(z_t^2)]^2} = \frac{3[1-(\alpha_1+\beta_1)^2]}{1-(\alpha_1+\beta_1)^2 - 2\alpha_1^2} > 3.$$

Consequently, similar to ARCH models, the tailed distribution of a GARCH(1,1) process is heavier than that of a normal distribution. Third, the model provides a simple parametric form that can be used to describe the evolution of volatility.

The literature on GARCH models are enormous [see Bollerslev et al. (1992, 1994)], and the references cited therein. The model encounters the same weaknesses as the ARCH model. For instance, it responds equally to big positive and negative innovations. In addition, recent empirical studies of high-frequency financial time series indicate that the tailed behavior of GARCH models remains too short, even using student t innovations.

9.2.1. An illustrative example

The identification of GARCH models in practice is not simple. Only lower-order GARCH models are used in most applications, say, GARCH(1,1), GARCH(2,1), and GARCH(1,2) models. The estimation of GARCH models can be done in the same ways as ARCH models. We use the RATS program to performance estimation.

Example 9.1. In this example, we consider the monthly excess returns of S&P 500 index for 792 observations, starting from 1926 (see Fig. 9.3). Denote the return series by z_t. Figure 9.4 shows the sample acf of z_t and the sample pacf of z_t^2. The z_t series has some serial correlations at lags 1 and 3, but the key feature is that the pacf of z_t^2 shows strong linear dependence. If an MA(3) model is entertained, we obtain

$$z_t = \beta_0 + a_t - \theta_1 a_{t-1} - \theta_3 a_{t-3}$$

for the series. However, due to the program constraint in RATS, we shall use instead an AR(3) model

$$z_t = \phi_1 z_{t-1} + \phi_2 z_{t-2} + \phi_3 z_{t-3} + \beta_0 + a_t.$$

The fitted AR(3) model is

$$z_t = .088 z_{t-1} - .023 z_{t-2} - .123 z_{t-3} + .0066 + a_t, \quad \hat{\sigma}_a^2 = .00333. \tag{9.7}$$

For the GARCH effects, we shall use the GARCH(1,1) model

$$h_t = \alpha_0 + \beta_1 h_{t-1} + \alpha_1 a_{t-1}^2.$$

258 HETEROSCEDASTIC MODELS

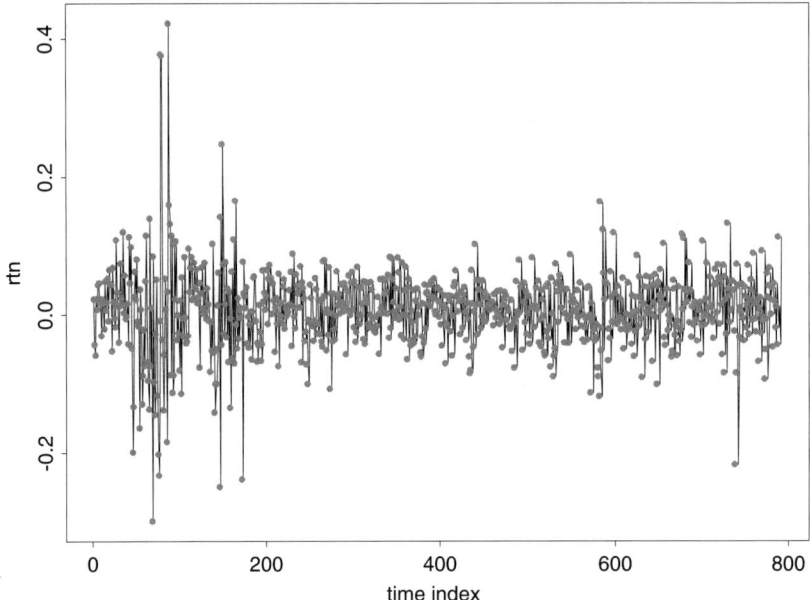

FIGURE 9.3 Time series plot of monthly S&P 500 excess returns.

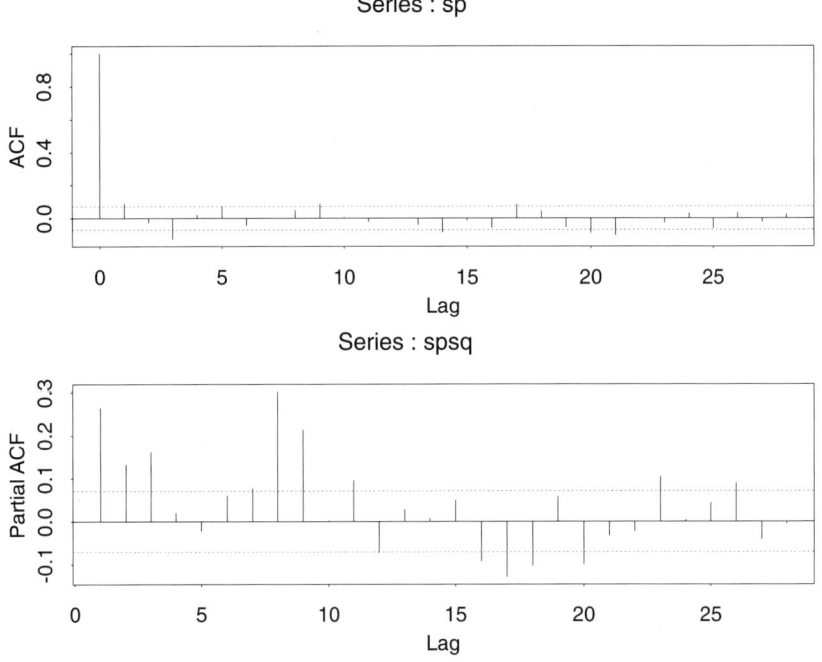

FIGURE 9.4 (a) Sample acf of the monthly excess returns of S&P 500 index; (b) sample pacf of the squared monthly excess returns.

9.2. THE GARCH MODEL

A joint estimation of the AR(3)-GARCH(1,1) model gives

$$z_t = .021 z_{t-1} - .034 z_{t-2} - .013 z_{t-3} + .0085 + a_t$$
$$h_t = .000099 + .8476 h_{t-1} + .1219 a_{t-1}^2.$$

From the 2nd equation, the implied unconditional variance of a_t is

$$\frac{.000099}{1 - .8476 - .1219} = .00325$$

which is very close to that of equation (9.7). However, t ratios of the parameters in the first equation suggest that all AR coefficients are insignificant at the 5% level. Therefore, we refine the model by dropping all AR coefficients. The refined model is

$$z_t = .0083 + a_t$$
$$h_t = .00010 + .8470 h_{t-1} + .1221 a_{t-1}^2.$$

The standard error of the parameter in the mean equation is .0015 whereas those of parameters in the 2nd equation are .00002, .0190 and .0201, respectively. The unconditional variance of a_t is $\frac{0.0001}{1-.847-.1221} = .00326$. This is a simple stationary GARCH(1,1) model. Note that the fitted model shows $\hat{\alpha}_1 + \hat{\beta}_1 = .9691$, which is close to 1. This phenomenon is commonly observed in practice and it leads to imposing the constraint $\alpha_1 + \beta_1 = 1$ in a GARCH(1,1) model, resulting in an integrated GARCH (or IGARCH) model.

Finally, to forecast the volatility of monthly excess returns of S&P 500 index, we can use the second equation. For instance

$$h_{t+1} = .0001 + .847 h_t + .1221 a_t^2$$

where a_t is the residual of the first equation and h_t is obtained from the 2nd equation, starting with $h_0 = 0$.

9.2.2. Remarks

An IGARCH(1,1) model can be written as

$$z_t = \sqrt{h_t} \epsilon_t, \quad h_t = \alpha_0 + \beta_1 h_{t-1} + (1 - \beta_1) z_{t-1}^2$$

where $\{\epsilon_t\}$ is defined as before and $1 > \beta_1 > 0$. For the monthly excess returns of

S&P 500 index, an estimated IGARCH(1,1) model is

$$z_t = .0067 + a_t, \quad h_t = .000119 + .8059 h_{t-1} + .1941 a_{t-1}^2$$

where the standard errors of the estimates are .0017, .000013, and .0144, respectively. The parameter estimates are close to those of the GARCH(1,1) model shown above, but there is a major difference between the two models. The unconditional variance of a_t, hence of z_t, is not defined under the IGARCH(1,1) model. This seems hard to justify for an excess return series.

9.3. THE EXPONENTIAL GARCH MODEL

To overcome some weaknesses of GARCH models in handling financial time series, Nelson (1991) proposes the exponential GARCH (EGARCH) model. In particular, to allow for asymmetric effects between positive and negative asset returns, he considers the weighted innovation

$$g(\epsilon_t) = \theta \epsilon_t + \gamma [|\epsilon_t| - E(|\epsilon_t|)] \tag{9.8}$$

where θ and γ are real constants. Both ϵ_t and $|\epsilon_t| - E(|\epsilon_t|)$ are zero-mean iid sequences with continuous distributions. Thus, $E[g(\epsilon_t)] = 0$. The asymmetry of $g(\epsilon_t)$ can easily be seen by rewriting it as

$$g(\epsilon_t) = \begin{cases} (\theta + \gamma)\epsilon_t - \gamma E(|\epsilon_t|) & \text{if } \epsilon_t \geq 0 \\ (\theta - \gamma)\epsilon_t - \gamma E(|\epsilon_t|) & \text{if } \epsilon_t < 0. \end{cases}$$

An EGARCH(r, s) model can then be written as

$$z_t = \sqrt{h_t} \epsilon_t, \quad \ln(h_t) = \alpha_0 + \frac{1 + \beta_1 B + \cdots + \beta_s B^s}{1 - \alpha_1 B - \cdots - \alpha_r B^r} g(\epsilon_{t-1}) \tag{9.9}$$

where α_0 is a constant, B is the back-shift (or lag) operator such that $B g(\epsilon_t) = g(\epsilon_{t-1})$ and $1 + \beta_1 B + \cdots + \beta_s B^s$ and $1 - \alpha_1 B - \cdots - \alpha_r B^r$ are polynomials with all zeros outside the unit circle and have no common factors. Again, equation (9.9) uses the usual ARMA parameterization to describe the evolution of the conditional variance of z_t, and some properties of the EGARCH model can be obtained in a similar manner as that of GARCH models. For instance, the unconditional mean of $\ln(h_t)$ is $\alpha_0 / (1 - \sum_{i=1}^r \alpha_i)$. However, it differs from the GARCH model in several ways. First, it uses logged conditional variance to relax the positiveness constraint of model coefficients. Also, the use of $g(\epsilon_t)$ enables the model to respond asymmetrically to positive and negative lagged values of z_t. Some further properties of the EGARCH model can be found in Nelson (1991).

9.3. THE EXPONENTIAL GARCH MODEL

To better understand the EGARCH model, let us consider the simple EGARCH(1,0) model

$$z_t = \sqrt{h_t}\epsilon_t, \quad (1-\alpha B)\ln(h_t) = \alpha_0 + g(\epsilon_{t-1}) \tag{9.10}$$

where ϵ_t's are iid standard normal and the subscript of α is omitted. In this case, $E(|\epsilon_t|) = \sqrt{2/\pi}$ and the model for $\ln(h_t)$ becomes

$$(1-\alpha B)\ln(h_t) = \begin{cases} \alpha_* + (\theta+\gamma)\epsilon_{t-1} & \text{if } \epsilon_{t-1} \geq 0 \\ \alpha_* + (\theta-\gamma)\epsilon_{t-1} & \text{if } \epsilon_{t-1} < 0 \end{cases} \tag{9.11}$$

where $\alpha_* = \alpha_0 - \sqrt{(2/\pi)}\gamma$. This is a nonlinear function similar to that of the threshold autoregressive model (TAR) of Tong (1978, 1990). It suffices to say that for this simple EGARCH model the conditional variance evolves in a nonlinear manner depending on the sign of z_{t-1}. Specifically, we have

$$h_t = h_{t-1}^\alpha \exp(\alpha_*) \begin{cases} \exp\left[(\theta+\gamma)\frac{z_{t-1}}{\sqrt{h_{t-1}}}\right] & \text{if } z_{t-1} \geq 0, \\ \exp\left[(\theta-\gamma)\frac{z_{t-1}}{\sqrt{h_{t-1}}}\right] & \text{if } z_{t-1} < 0. \end{cases}$$

The coefficients $(\theta+\gamma)$ and $(\theta-\gamma)$ show the asymmetry in response to positive and negative z_{t-1}. Cao and Tsay (1992) use nonlinear models, including EGARCH models, to obtain multi-step-ahead volatility forecasts.

9.3.1. An illustrative example

Nelson (1991) applies an EGARCH model to the daily excess returns of the value-weighted market index from the Center for Research of Security Prices for July 1962 to December 1987. The excess returns are obtained by removing monthly Treasury bill returns from the value-weighted index returns, assuming that the Treasury bill return was constant for each calendar day within a given month. There are 6408 observations. Denote the excess return by z_t. The model used is as follows:

$$z_t = \phi_0 + \phi_1 z_{t-1} + ch_t + a_t \tag{9.12}$$

$$\ln(h_t) = \alpha_0 + \ln(1+N_t w) + \frac{1+\beta B}{1-\alpha_1 B - \alpha_2 B^2} g(\epsilon_{t-1})$$

where h_t is the conditional variance of a_t given F_{t-1}, N_t is the number of nontrading days between trading days $t-1$ and t, α_0 and w are real parameters, and ϵ_t follows a generalized error distribution with probability density function

$$f(x) = \frac{v\exp[-(1/2)|x/\lambda|^v]}{\lambda 2^{(1+1/v)}\Gamma(1/v)}, \quad -\infty < x < \infty, \quad 0 < v \leq \infty$$

TABLE 9.1. Estimated AR(1)-EGARCH(2,1) Model for Daily Excess Returns of Value-weighted Market Index of CRSP (July 1962–Dec. 1987)

Parameter	α_0	w	γ	α_1	α_2	β	θ	ϕ_0	ϕ_1	c	v
Estimate	-10.06	.183	.156	1.929	$-.929$	$-.978$	$-.118$	$3.5 \cdot 10^{-4}$	.205	-3.361	1.576
Error		.346	.028	.013	.015	.015	.006	.009	$9.9 \cdot 10^{-5}$	.012	2.026	.032

where $\Gamma(.)$ is the gamma function and

$$\lambda = \left[\frac{2^{(-2/v)} \Gamma(1/v)}{\Gamma(3/v)} \right]^{1/2}.$$

The parameter c in (9.12) is called the *risk premium parameter*. Table 9.1 gives the parameter estimates and their standard errors. The first equation of model (9.12) has two features that are of interest: (1) it uses an AR(1) model to take care of possible serial correlations in the excess returns and (2) it uses the volatility h_t as a regressor to account for risk premium. This latter feature is related to the ARCH-M model, where "M" denotes that the mean of z_t depends on its volatility.

9.4. THE CHARMA MODEL

Many other models have been proposed in the literature to describe the evolution of conditional variance h_t in (9.1). We mention a model called *conditional heteroskedastic ARMA* (CHARMA) model that uses random coefficients to produce conditional heteroscedasticity; see Tsay (1987). The CHARMA model is not the same as the GARCH model, but the two models have similar second-order conditional properties. The CHARMA model can be generalized to the multivariate case in a rather parsimonious manner. A simple CHARMA model is defined as

$$\phi(B)(z_t - \mu) = \theta(B)a_t, \qquad \delta_t(B)a_t = \eta_t \qquad (9.13)$$

where $\phi(B)$ and $\theta(B)$ are the AR and MA polynomials of the usual stationary ARMA model, μ is the mean of z_t, η_t's are iid $N(0, \sigma_\eta^2)$ and $\delta_t(B) = 1 - \delta_{1,t} B - \cdots - \delta_{r,t} B^r$ is a purely random coefficient polynomial in B. The random coefficient vector $\delta_t = (\delta_{1,t}, \ldots, \delta_{r,t})'$ is a sequence of iid random vectors with mean zero and nonnegative definite covariance matrix Σ. In addition, $\{\delta_t\}$ is independent of $\{\eta_t\}$. For $r > 0$, the conditional variance of a_t in (9.13) is

$$h_t = \sigma_\eta^2 + (a_{t-1}, \ldots, a_{t-r}) \Sigma (a_{t-1}, \ldots, a_{t-r})',$$

which is equivalent to that of an ARCH(r) model if Σ is a diagonal matrix. Because Σ is a covariance matrix, it is nonnegative definite and, hence, $h_t \geq \sigma_\eta^2 > 0$. For financial time series, it is common to see that $\phi(B) = \theta(B) = 1$ so that $z_t = \mu + a_t$. An obvious difference between ARCH and CHARMA models is that the latter uses cross-products of the lagged values of z_t in the variance equation. The cross-product terms might be useful in some applications. For example, in modeling asset returns, the cross-product terms denote interations between previous returns. However, the number of cross-product terms increases quickly and some constraints are often needed to keep the model simple. From a theoretical point of view, higher-order properties of CHARMA models are harder than those of GARCH models, because it is harder to handle random coefficients than constant coefficients.

For illustration, we employ the CHARMA model

$$z_t = \phi_0 + a_t, \quad (1 - \delta_{1t}B - \delta_{2t}B^2)a_t = \eta_t$$

for the monthly excess returns of S&P 500 index used before in GARCH(1,1) study. The fitted model is

$$z_t = .00635 + a_t, \quad h_t = .00179 + (a_{t-1}, a_{t-2})\hat{\Sigma}(a_{t-1}, a_{t-2})'$$

where

$$\hat{\Sigma} = \begin{bmatrix} .1417(.0333) & -.0594(.0365) \\ -.0594(.0365) & .3081(.0340) \end{bmatrix},$$

where the numbers in parentheses are standard errors. The cross-product term of $\hat{\Sigma}$ has a t ratio of -1.63, which is marginally significant at the 10% level. If we refine the model to

$$z_t = \phi_0 + a_t, \quad (1 - \delta_{1t}B - \delta_{2t}B^2 - \delta_{3t}B^3)a_t = \eta_t$$

but assume that δ_{3t} is uncorrelated with $(\delta_{1t}, \delta_{2t})$, then we obtain the fitted model

$$z_t = .0068 + a_t, \quad h_t = .00136 + (a_{t-1}, a_{t-2}, a_{t-3})\hat{\Sigma}(a_{t-1}, a_{t-2}, a_{t-3})'$$

where the elements of $\hat{\Sigma}$ and their standard errors, shown in parentheses, are

$$\hat{\Sigma} = \begin{bmatrix} .1212(.0355) & -.0622(.0283) & 0 \\ -.0622(.0283) & .1913(.0254) & 0 \\ 0 & 0 & .2988(.0420) \end{bmatrix}.$$

All the estimates are now statistically significant at the 5% level.

9.5. RANDOM COEFFICIENT AUTOREGRESSIVE (RCA) MODEL

In the literature, the RCA model is introduced to account for variability among different subjects under study, similar to panel data analysis in econometrics and hierachical models in statistics. We classify the RCA model as a conditional heteroskedastic model, but historically it is used to obtain a better description of the conditional mean equation of the process by allowing for the parameters to evolve over time. A time series x_t is said to follow a RCA(p) model if it satisfies

$$z_t = c + \sum_{i=1}^{p}(\phi_i + \delta_{it})z_{t-i} + a_t \qquad (9.14)$$

where p is a positive integer, $\{\delta_t\} = \{(\delta_{1t}, \ldots, \delta_{pt})'\}$ is a sequence of independent random vectors with mean zero and covariance matrix Σ_δ, and $\{\delta_t\}$ is independent of $\{a_t\}$. See Nicholls and Quinn (1982) for further discussions of the model. The conditional mean and variance of the RCA model in (9.14) are

$$\mu_t = E(z_t \mid F_{t-1}) = \sum_{i=1}^{p} \phi_i z_{t-i}$$

$$\sigma_t^2 = h_t = \sigma_a^2 + (z_{t-1}, \ldots, z_{t-p})\Sigma_\delta(z_{t-1}, \ldots, z_{t-p})'$$

which is similar to that of a CHARMA model.

9.6. STOCHASTIC VOLATILITY MODEL

An alternative approach to describe the evolution of volatility is to introduce an innovation to the conditional variance equation of z_t. (Melino and Turnbull 1990, Harvey et al. 1994, Jacquier et al. 1994). The resulting model is referred to as a *stochastic volatility* (SV) model. Similar to EGARCH models, to ensure positiveness of the conditional variance, SV models use $\ln(h_t)$ instead of h_t. A simple SV model is defined as

$$z_t = \sqrt{h_t}\epsilon_t, \qquad (1 - \alpha_1 B - \cdots - \alpha_r B^r)\ln(h_t) = \alpha_0 + v_t \qquad (9.15)$$

where ϵ_t's are iid $N(0, 1)$, v_t's are iid $N(0, \sigma_v^2)$, $\{\epsilon_t\}$ and $\{v_t\}$ are independent, α_0 is a constant, and all zeros of the polynomial $1 - \sum_{i=1}^{r} \alpha_i B^i$ are outside the unit circle. Introducing the innovation v_t makes the SV model more flexible in describing the evolution of h_t, but it also increases the difficulty in parameter estimation. A quasi-likelihood method with Kalman filterting or Monte Carlo method is needed to estimate a SV model. Jacquier et al. (1994) provide some comparison of estimation results between quasi-likelihood and Monte Carlo Markov chain methods.

The appendix of Jacquier et al. (1994) provide some properties of the SV model when $r = 1$. For instance, with $r = 1$, we have

$$\ln(h_t) \sim N\left(\frac{\alpha_0}{1 - \alpha_1}, \frac{\sigma_v^2}{1 - \alpha_1^2}\right) \equiv N(\mu_h, \sigma_h^2)$$

and $E(z_t^2) = \exp[\mu_h + 1/(2\sigma_h^2)]$, $E(z_t^4) = 3\exp[2\mu_h^2 + 2\sigma_h^2]$, and $\text{corr}(z_t^2, z_{t-i}^2) = [\exp(\sigma_h^2 \alpha_1^i) - 1]/[3\exp(\sigma_h^2) - 1]$. Limited experience shows that SV models often provided improved fit in finite samples, but their contributions in forecasting volatility received mixed results.

9.7. LONG-MEMORY STOCHASTIC VOLATILITY MODEL

More recently, the SV model is further extended to allow for long memory in volatility, using the idea of fractional difference. A process is said to have long memory if its autocorrelation function decays at a hyperbolic, instead of an exponential, rate as the lag increases. The extension to long memory in volatility study is motivated by the fact that the autocorrelation function of the squared or absolute-value series of an asset return often decays slowly, even though the return series itself has no serial correlations; see Ding et al. (1993). A simple long-memory stochastic volatility (LMSV) model can be written as

$$z_t = \sqrt{h_t}\epsilon_t, \quad \sqrt{h_t} = \sigma \exp\left(\frac{u_t}{2}\right), \quad (1-B)^d u_t = \eta_t \qquad (9.16)$$

where $\sigma > 0$, ϵ_t's are iid $N(0, 1)$, η_t's are iid $N(0, \sigma_\eta^2)$ and independent of ϵ_t, and $0 < d < 0.5$. For such a model, we have

$$\begin{aligned}\ln(z_t^2) &= \ln(\sigma^2) + u_t + \ln(\epsilon_t^2) \\ &= \left[\ln(\sigma^2) + E(\ln\epsilon_t^2)\right] + u_t + \left[\ln(\epsilon_t^2) - E(\ln\epsilon_t^2)\right] \\ &\equiv \mu + u_t + e_t.\end{aligned}$$

Thus, the $\ln(z_t^2)$ series is a Gaussian long-memory signal plus a non-Gaussian white noise (Breidt et al. 1998). For applications, Ray and Tsay (2000) studied common long-memory components in daily stock volatilities of groups of companies classified by various characteristics. They found that companies in the same industrial or business sector tend to have more common long-memory components, such as large U.S. national banks and financial institutions.

REFERENCES

Bollerslev, T. (1986). Generalized autoregressive conditional heteroskedasticity. *J. Econo.* **31**, 307–327.

Bollerslev, T., Chou, R. Y., and Kroner, K. F. (1992). ARCH modeling in fiance. *J. Econo.* **52** 5–59.

Bollerslev, T., Engle, R. F., and Nelson, D. B. (1994). ARCH model. In R. F. Engle and D. C. McFadden (eds.), *Handbook of Econometrics,* Vol. IV, pp. 2959–3038. Elsevier Science, Amsterdam.

Breidt, F. J., Crato, N., and de Lima, P. (1998). On the detection and estimation of long memory in stochastic volatility. *J. Econo.* **83**, 325–348.

Cao, C. and Tsay, R. S. (1992). Nonlinear time series analysis of stock volatilities. *J. Appl. Econo.* **7**, 165–185.

Ding, Z., Granger, C. W. J., and Engle, R. F. (1993). A long memory property of stock returns and a new model. *J. Empi. Fina.* **1**, 83–106.

Engle, R. F. (1982). Autoregressive conditional heteroscedasticity with estimates of the variance of United Kingdom inflations. *Econometrica* **50**, 987–1007.

Harvey, A. C., Ruiz, E., and Shephard, N. (1994). Multivariate stochastic variance models. *Rev. Econ. Stud.* **61**, 247–264.

Jacquier, E., Polson, N. G., and Rossi, P. (1994). Bayesian analysis of stochastic volatility models (with discussion). *J. Business Econ. Stat.* **12**, 371–417.

McLeod, A. I. and Li, W. K. (1983). Diagnostic checking ARMA time series models using squared-residual autocorrelations. *J. Time Ser. Anal.* **4**, 269–273.

Melino, A. and Turnbull, S. M. (1990). Pricing foreign currency options with stochastic volatility. *J. Econ.* **45**, 239–265.

Nelson, D. B. (1991). Conditional heteroskedasticity in asset returns: A new approach. *Econometrica* **59**, 347–370.

Nicholls, D. F. and Quinn, B. G. (1982). *Random Coefficient Autoregressive Models: An Introduction*, Lecture Notes in Statistics, 11. Springer-Verlag, New York.

Ray, B. K. and Tsay, R. S. (in press). Long-range dependence in daily stock volatilities. *J. Business Econ. Stat.*

Tong, H. (1978). On a threshold model. In C. H. Chen (ed.), *Pattern Recognition and Signal Processing*. Sijhoff & Noordhoff, Amsterdam.

Tong, H. (1990). *Non-Linear Time Series: A Dynamical System Approach*, Oxford Univ. Press, Oxford, UK.

Tsay, R. S. (1987). Conditional heteroskedastic time series models. *J. Am. Stat. Assoc.* **82**, 590–604.

CHAPTER 10

Nonlinear Time Series Models: Testing and Applications

Ruey S. Tsay
University of Chicago

10.1. INTRODUCTION

Nonlinear time series analysis has gained much attention in recent years, due primarily to the fact that linear time series models have encountered various limitations in real applications and modern computers have provided advanced computational power which makes possible the nonlinear analysis. In addition, the development in nonparametric regression has established a solid foundation for nonlinear time series analysis.

Many nonlinear time series models have been introduced in the literature and shown to be useful in some applications. Consider parametric models. Granger and Andersen (1978) introduced bilinear models. Tong (1978, 1990) proposed the threshold autoregressive model and demonstrated that the model is capable of describing the asymmetric limit cycle of the annual sunspot number. Haggan and Ozaki (1981) considered the exponential autoregressive model and showed that the model is useful in modeling sound vibration. Priestley (1980) considered state-dependent models as a general framework for nonlinear analysis. Hamilton (1989) proposed Markov switching models to model the business cycles of macroeconomic time series. The above models employ explicit parametric forms that can, at best, be regarded as rough approximations to the underlying nonlinear characteristics of interest. It is usually hard to justify a priori the appropriateness of such an explicit model in real applications. To overcome this justification problem and to make use of recent developments in nonparametric regression, researchers in nonlinear time series analysis begin to explore the possibility of using data-driven methods such as nonparametric density estimation to identify the underlying characteristics of a time series. For example,

---

A Course in Time Series Analysis, Edited by Daniel Peña, George C. Tiao, and Ruey S. Tsay.
ISBN 0-471-36164-X. © 2001 John Wiley & Sons, Inc.

Robinson (1983) investigates asymptotic properties of nonparametric density estimation for time series data; Auestad and Tjøstheim (1990) apply a multivariate kernel smoothing method to estimate the conditional mean and conditional variance of a nonlinear autoregression; Lewis and Stevens (1991) use the multivariate adaptive regression splines (MARS) of Friedman (1991) to build adaptive spline threshold autoregressive models; Chen and Tsay (1993a) employ an arranged local regression procedure to construct functional-coefficient autoregressive models; Chen and Tsay (1993b) use nonparametric techniques to build nonlinear additive autoregressive models.

In the econometric literature, Engle (1982) proposes the conditional heteroscedastic autoregressive (ARCH) model to capture the serial dependence in conditional variance of a time series. The model has attracted much attention since 1985 or so, and several generalizations of the model are available in the literature, such as the generalized ARCH (GARCH) model of Bollerslev (1986), the exponential GARCH model of Nelson (1991), and the conditional heteroscedastic autoregressive moving-average model of Tsay (1987). These models have been discussed in Chapter 9.

Even confined to nonlinearity in the conditional expectation, it is impossible to review and summarize available results in a single chapter. The goal of this chapter is, therefore, to focus on nonlinear models that I have used. In Section 9.2, we briefly review some nonlinearity tests. In Section 9.3, we focus on the threshold autoregressive models with some applications. Finally, we briefly discuss advantages and disadvantages of some nonlinear models in Section 9.4.

10.2. NONLINEARITY TESTS

On the basis of available results in the literature both in real data analysis and in simulation study (e.g., Chan and Tong 1986b, Luukkonen et al. 1988), one can draw some conclusions concerning the performance of existing nonlinearity tests: (1) the idea of Lagrange multiplier tests seems to be powerful in detecting finite-order nonlinearity such as nonlinearity involving quadratic terms, (2) the idea of arranged autoregression is useful in spotting threshold nonlinearity, and (3) a test that uses the ideas (1) and (2) separately seems to suffer from power loss in detecting some types of nonlinear models. Consequently, it appears that we should combine ideas (1) and (2) in testing nonlinearity of a univariate time series. Such a combined test not only can overcome the weaknesses but is also able to retain the advantages of the individual tests. Motivated by this observation, we suggest next a procedure for a nonlinearity test in time series analysis that uses added variables to detect nonlinearity of bilinear (BI), exponential autoregressive (EXPAR), and smooth threshold autoregressive (STAR) models and employs arranged autoregression to detect threshold nonlinearity (see Tsay 1991).

10.2.1. The test

Consider an autoregression of order m

$$z_t = \phi_0 + \phi_1 z_{t-1} + \cdots + \phi_m z_{t-m} + a_t, \quad t = 1, 2, \ldots, n \quad (10.1)$$

10.2. NONLINEARITY TESTS

where $\{a_t\}$ is a sequence of martingale difference with mean zero and variance $\sigma_a^2 > 0$. It is well known that the ordinary least-squares estimates $\hat{\phi}_i$ are consistent for ϕ_i if z_t is an AR(p) process such that $p \leq m$ and the innovation process satisfies $E(|a_t|^\delta) < \infty$ for some $\delta > 2$, see Lai and Wei (1982). Therefore, the associated residual $\{\hat{a}_t\}$ is asymptotically a white-noise process if z_t is a linear AR(p) process. On the other hand, if z_t is bilinear, then $\hat{a}_t$ is related to $Y_{t-i}a_{t-j}$ for some i and j. Consequently, to detect the possibility of bilinearity in z_t one may apply the technique of added variables to the autoregression (10.1) with some suitably chosen variables such as $\{z_{t-i}\hat{a}_{t-i}\}$ and $\{\hat{a}_{t-i}\hat{a}_{t-i-1}\}$ for $i = 1, \ldots, m$. The same idea applies to the EXPAR and STAR models. More specifically, for the EXPAR model, we consider the added variables $z_{t-i}\exp(-z_{t-1}^2/\gamma)$ where γ is a normalization constant, such as $\gamma = \max\{|z_{t-1}|\}$. For the STAR model of Chan and Tong (1986a) with delay parameter d, we use the added variables $G(y_{t-d})$ and $z_{t-i}G(y_{t-d})$ where $y_{t-d} = (z_{t-d} - \bar{z}_d)/S_d$ with $\bar{z}_d$ and S_d the sample mean and standard deviation of z_{t-d}, respectively, and $G(.)$ is the cumulative distribution function (cdf) of the standard normal random variable.

Consider next the self-exciting threshold autoregressive (SETAR) models of Tong (1978). Since the models are piecewise linear in the domain of the threshold variable z_{t-d}, the traditional way of fitting an AR(m) model is not useful, because the estimates $\hat{\phi}_i$'s tend to show substantial fluctuation as data from different regimes are mixed together. To overcome this difficulty, the idea of arranged autoregression is useful. Roughly speaking, in an arranged autoregression the observed values of the "dependent variable" and the associated "design matrix" are sorted according to the values of the threshold variable. By so doing, we effectively transform a SETAR model into a linear regression model with model changes at the threshold values. This makes the technique of sequential estimation useful. In particular, the (normalized) predictive residuals can be used to detect the threshold nonlinearity. For instance, Petruccelli and Davies (1986) use normalized predictive residuals to derive a CUSUM test, and Tsay (1989) employs the predictive residuals to obtain an F test for threshold nonlinearity. We refer to this F test as a TAR-F test.

Putting the above two ideas together, we consider the following procedure for testing nonlinearity of a univariate time series:

1. For a given delay parameter d, fit recursively an arranged autoregression of order m to z_t, and calculate the normalized predictive residuals $\tilde{a}_t$ for $t = b+1, \ldots, n$, where b is chosen so that the $X'X$ matrix involved in the initial estimation is invertible.
2. Regress $\tilde{a}_t$ on the regressors $\{1, z_{t-1}, \ldots, z_{t-m}\}$, $\{z_{t-i}\tilde{a}_{t-i}, \tilde{a}_{t-i}\tilde{a}_{t-i-1} \mid 1 \leq i \leq m\}$, and $\{z_{t-1}\exp(-z_{t-1}^2/\gamma), G(y_{t-d}), z_{t-1}G(y_{t-d})\}$, where γ, y_{t-d} and $G(.)$ are defined as before, and compute the associated F statistic $\tilde{F}$.

If z_t is a stationary linear AR(p) process of order $p \leq m$, $\tilde{F}$ follows asymptotically a F distribution with degrees of freedom $3(m+1)$ and $n - b - 3(m+1)$. This result can be established along the same lines as in Tsay (1989).

Some remarks on the proposed testing procedure are in order. First, like many Lagrange multiplier tests, the selection of the added variables is somewhat arbitrary.

For example, we use only one added variable specifically for EXPAR models and two for STAR models. We believe that these three variables should be sufficient for reasonable EXPAR and STAR models because the second-order terms used can also detect certain nonlinearity of EXPAR and STAR models. In applications, one may choose the added variables based on the substantive information of the process under study. Also, other cumulative distribution functions can be used in lieu of the cdf of the standard normal random variable. Second, the selection of order m can be done in various ways such as via the Akaike information criterion [AIC (Akaike 1974)] or via an inspection of the sample partial autocorrelation function. Third, the number of observations b used to start the recursive estimation may depend on the order m and the sample size n. Fourth, the recursive estimation can be done via various algorithms such as the recursive least-squares method and the Kalman filter. The Kalman filter appears to be preferable when there are missing observations in the data (e.g., Tong and Yeung 1991). Fifth, the normalization constant γ is not critical so long as the resulting exponents are not too large for most of the data points. Finally, when the delay parameter is unknown, one may apply the test to some predetermined values of d.

10.2.2. Comparison and application

We now apply the test of the previous subsection to various real and simulated data so that its performance can be compared with other tests. This comparison serves several purposes. First, it is intended to show that the proposed test can, indeed, detect nonlinearity of various models such as BI, EXPAR, STAR, and SETAR. Second, for a given alternative nonlinear models, it shows that the proposed test performs well as compared with other existing tests that are known to work well. Third, it illustrates the application of the new test to real data.

Simulation

All the simulation results reported are based on 1000 replications each with 100 observations. Also, the AR order m is selected by AIC among $\{1,2,3,4\}$, $b = 10 + m$, and the delay parameter $d = 1$. For each realization of a given model, we generated 3100 data points with zero starting values, that is, setting z_t and a_t, the innovation, equal to zero for $t \leq 0$; but only the last 100 points were used as observations. The a_t's are standard normal random variates obtained from the RNNOR subroutine of the IMSL package.

Tables 10.1–10.5 give the empirical frequencies of rejecting a linear time series when the generating models are BI, EXPAR, logistic STAR, SETAR, and concurrent nonlinear, respectively. By *concurrent nonlinear models,* we meant models involving cross-products of the innovation a_t. The nonlinearity tests used in the simulation are the original F test (Ori-F) of Tsay (1986), the augmented F test (Aug-F) of Luukkonen et al. (1988), the TAR-F test, the CUSUM test, and the proposed new F test (New-F). Notice that Ori-F and Aug-F are based on least-squares estimates of the full data set whereas the remaining tests are based on recursive estimates of an arranged autoregression of order m. From the results we make the following

10.2. NONLINEARITY TESTS

TABLE 10.1. Empirical Frequencies of Rejecting a Linear Model Based on 5% and 10% Critical Values[a]

Model	β	CV[b]	Ori-F	Aug-F	TAR-F	CUSUM	New-F
a	−.6	5%	872	980	987	391	976
		10%	906	987	994	497	991
a	0.	5%	50	53	52	61	44
		10%	100	106	98	113	83
a	0.6	5%	859	970	924	949	968
		10%	898	982	953	973	991
b	−.6	5%	471	913	791	931	780
		10%	575	951	872	960	865

[a] The generating models are bilinear given by (a) $z_t = 0.5z_{t-1} + \beta z_{t-1}a_{t-1} + a_t$; (b) $z_t = a_t + 0.5a_{t-1} + \beta a_{t-1}^2$.
[b] Critical value.

TABLE 10.2. Empirical Frequencies of Rejecting a Linear Model Based on 5% and 10% Critical Values[a]

Φ	β	CV[b]	Ori-F	Aug-F	TAR-F	CUSUM	New-F
0.3	10.0	5%	126	283	269	826	999
		10%	203	422	367	951	999
0.3	20.0	5%	196	395	208	903	991
		10%	267	506	277	956	993
0.3	100.0	5%	90	189	183	976	784
		10%	115	258	244	984	833

[a] The generating models are exponential AR given by $z_t = [\phi + \beta \exp(-z_{t-1}^2)]z_{t-1} + a_t$.
[b] Critical value.

TABLE 10.3. Empirical Frequencies of Rejecting a Linear Model Based on 5% and 10% Critical Values[a]

β_0	β_1	α	CV[b]	Ori-F	Aug-F	TAR-F	CUSUM	New-F
−4.0	−.4	2.0	5%	620	886	338	374	566
			10%	722	934	473	497	696
−2.0	0.	2.0	5%	78	496	191	326	373
			10%	152	664	293	479	644
2.0	−.4	2.0	5%	736	675	738	594	501
			10%	830	783	821	696	642
0.0	0.	2.0	5%	46	43	46	51	51
			10%	79	89	96	97	99

[a] The generating models are logistic STAR given by $z_t = 1.0 - 0.5z_{t-1} + (\beta_0 + \beta_1 z_{t-1})G(\alpha z_{t-1}) + a_t$ with $G(z) = \frac{\exp(z)}{[1+\exp(z)]}$.
[b] Critical value.

TABLE 10.4. Empirical Frequencies of Rejecting a Linear Model Based on 5% and 10% Critical Values[a]

Φ_0	Φ_1	β_0	β_1	ω	CV[b]	Ori-F	Aug-F	TAR-F	CUSUM	New-F
1.0	−.5	−1.0	−.5	0.0	5%	62	567	121	275	461
					10%	119	680	209	397	607
2.0	0.5	0.5	−.4	1.0	5%	931	985	983	978	989
					10%	962	993	990	994	998
0.0	0.5	0.0	0.5	0.0	5%	47	45	35	43	37
					10%	89	98	69	94	88
0.0	0.5	0.0	−.5	0.0	5%	53	136	560	199	412
					10%	103	230	679	302	557

[a] The generating models are SETAR given by $z_t = \begin{cases} \phi_0 + \phi_1 z_{t-1} + a_t & \text{if } z_{t-1} \leq \omega \\ \beta_0 + \beta_1 z_{t-1} + a_t & \text{if } z_{t-1} < \omega. \end{cases}$
[b] Critical value.

observations:

1. As expected, the New-F test appears to work well for all the cases considered. On the other hand, each of the other tests shows certain weakness. For example, Table 10.2 shows that TAR-F test is not powerful in detecting EXPAR models. This is in agreement with the finding of Luukkonen et al. (1988).
2. The Aug-F, TAR-F and New-F tests all have good power in detecting bilinear nonlinearity.
3. The CUSUM and New-F test work well for the EXPAR alternatives.
4. The nonlinearity of logistic STAR models employed is relatively hard to detect (see Table 10.3). This is true for all the tests considered.
5. The Aug-F test seems to work well when the nonlinearity is caused mainly by the difference in the constant terms (see row 1 of Tables 10.3 and 10.4). However, the test has relatively low power when the nonlinearity is not caused by constant terms (see the last row of Table 10.4).
6. All the tests seem to have reasonable type I errors (see the case of linear models in Tables 10.1, 10.3, and 10.4).

TABLE 10.5. Empirical Frequencies of Rejecting a Linear Model Based on 5% and 10% Critical Values[a]

Model	CV[b]	Ori-F	Aug-F	TAR-F	CUSUM	New-F
a	5%	211	209	76	135	216
	10%	306	295	138	234	288
b	5%	239	308	453	154	447
	10%	331	403	554	213	537

[a] The generating models are concurrent nonlinear given by (a) $z_t = a_t + 0.5a_{t-1} - 0.6a_t a_{t-1}$; (b) $z_t = 0.5z_{t-1} - 0.6z_{t-1}a_t + a_t$.
[b] Critical value.

10.2. NONLINEARITY TESTS

TABLE 10.6. P values of Nonlinearity Tests on Real Data[a]

						$d=1$			$d=2$			$d=3$ or 8		
Data	TR	n	p	OF	AF	TF	CU	NF	TF	CU	NF	TF	CU	NF
Sunspot	Raw	280	11	.000	.000	.000	.000	.002	.000	.607	.000	.000	.573	.000
Lynx	Log	114	9	.003	.001	.015	.107	.062	.012	.000	.085	.041	.000	.015
Lynx	Raw	114	3	.000	.000	.002	.007	.000	.000	.009	.000	.000	.016	.000
Blowfly	Log	159	3	.000	.000	.000	.008	.002	.009	.018	.035	.000	.000	.000
Blowfly	Raw	159	2	.000	.000	.000	.064	.000	.000	.396	.006	.000	.000	.000
Series A	Raw	197	7	.828	.953	.455	.441	.244	.366	.597	.938	.746	.835	.504
Series B	Diff.	368	1	.003	.001	.007	.039	.038	.842	.925	.861	.000	.750	.001
Series C	Raw	226	2	.700	.019	.983	.869	.455	.890	.976	.000	.747	.996	.007

[a] In this table, OF, AF, TF, CU, and NF denote Ori-F, Aug-F, TAR-F, CUSUM, and New-F tests, respectively; TR stands for transformation; and ".000" indicates that the corresponding p-value is less than .001. The delay parameter $d=8$ is used for the blowfly series.

7. All the tests have relatively low power in detecting concurrent nonlinearity, which suggests that further investigation is needed in order to handle this type of nonlinearity.

Applications

We now apply the tests discussed earlier as well as the bispectrum test of Hinich (1982) and the DBS test of Brock et al. (1987) to some data sets that have been widely analyzed in the literature. Since the delay parameter is often unknown in applications, the set {1,2,3} or {1,2,8} was used as the possible values for d. These values have been used in the literature for the processes employed. Also $b = [n/10] + p$ with p the AR order used and $[h]$ the integer part of h. Table 10.6 gives the results of the tests. There ".000" denotes that the corresponding p value is less than .001. The data employed are (1) the annual sunspot series from 1700 to 1979, (2) the Canadian lynx series, (3) the observations from $t = 48$ to 206 of the blowfly population data used in Tong (1983) and Tsay (1988), and (4) series A–C of Box and Jenkins (1976). From the table we conclude the following:

1. The results of the CUSUM test depend very much on the threshold variable z_{t-d}. Consider, for instance, the sunspot series. The CUSUM test suggests linearity for $d = 2$ or 3, whereas the other tests indicate nonlinearity.
2. All the F tests suggest that series A is linear, whereas sunspot, lynx, and blowfly series are nonlinear.
3. For the first difference of series B, the use of delay parameter $d = 2$ fails to detect any nonlinearity. This is conceivable given the fact that the differenced series of a stock price is close to white noise.
4. The Aug-F test and New-F test with $d = 2$ or 3 seem to suggest some nonlinearity in series C whereas all of the other tests suggest linearity. We interpret this as an indication that the nonlinearity is caused either by difference in

TABLE 10.7. Results of BDS and Bispectrum Tests on Real Data Using Asymptotic 5% Critical Values[a]

Data	TR	n	p	BDS Test		Bispectrum	
				$m=2$	$m=3$	Normal	Linear
Sunspot	Raw	280	11	NL	NL	NG	L
Lynx	Log	114	9	L	NL	NG	L
Lynx	Raw	114	3	NL	NL	NG	L
Blowfly	Log	159	3	L	L	NG	NL
Blowfly	Raw	159	2	NL	NL	G	L
Series A	Raw	197	7	NL	L	G	L
Series B	Diff.	368	1	NL	NL	NG	NL
Series C	Raw	226	2	NL	NL	G	L

[a] The parameters used in this table were given in the text; L and NL denote linear and nonlinear, respectively, G and NG denote normality and nonnormality, respectively, and TR stands for transformation.

constant terms or by some STAR-type structure in the series, because both Aug-F and New-F tests are more sensitive to these two types of nonlinearity (see the simulation results of Tables 10.3 and 10.4). In fact, the outlier and level shift techniques of Chang et al. (1988) suggest that there are two level shifts at $t = 58$ and $t = 61$, respectively, and an innovational outlier at $t = 60$. After adjusting for these disturbances, all the tests fail to detect nonlinearity at the 5% level.

Table 10.7 gives the results of the Bispectrum and BDS tests by using 5% asymptotic critical values. For the Bispectrum test, the smoothing parameter M was determined by $\max\{10, [\sqrt{n} - 1]\}$, where n is the sample size, and the 80% fractile test was used. These values were used based on Hinich's suggestion. For the DBS test, each data set was properly filtered by fitting a linear AR(p) model before applying the test. The parameter m was 2 or 3, and ϵ was set equal to one standard deviation of the prefiltered process. The simulation results of Hsieh and LeBaron (1988) suggest that these choices often give the best performance of the test. From the table, the BDS test tends to suggest nonlinearity whereas the bispectrum test indicates non-normality over nonlinearity for several series.

More recent development in testing for threshold nonlinearity can be found in Hansen (1996) and the references cited therein.

10.3. THE TAR MODEL

Next, we consider a specific nonlinear model with interesting applications. A TAR model is a piecewise linear autoregressive model. However, it is piecewise linear in the space of the threshold variable, not piecewise linear in time. See Tiao and Tsay (1994) for further details. A simple TAR model of a time series z_t with threshold

10.3. THE TAR MODEL

variable z_{t-d} is defined as follows, where d is a positive integer and is referred to as the *threshold lag*. Partition the space of z_{t-d}, the real line, by

$$-\infty = r_0 < r_1 < \cdots < r_g < r_{g+1} = \infty,$$

where the r_h's are referred to as the *thresholds* and g is a *positive integer*. Then, an TAR model of order p for z_t is defined as

$$z_t = \phi_0^{(h)} + \phi_1^{(h)} z_{t-1} + \cdots + \phi_p^{(h)} z_{t-p} + \epsilon_{h,t} \quad \text{for } r_{h-1} \le z_{t-d} < r_h, \quad h = 1, \ldots, g \tag{10.2}$$

where the $\phi_i^{(h)}$'s are real numbers and $\{\epsilon_{h,t}\}$ is a sequence of independently and identically distributed Gaussian random variates with mean zero and variance σ_h^2. For model (10.2), $\{\epsilon_{h,t}\}$ and $\{\epsilon_{k,t}\}$ are independent if $h \ne k$. The partition $r_{h-1} \le y_{t-d} < r_h$ is referred to as the *h regime* of the model. This model was first proposed in Tong (1978) and Tong and Lim (1980). It has gained some popularity because the model is capable of producing limit cycle, time irreversibility, and asymmetric behavior of a time series. For example, Potter (1995) applies the model to U.S. quarterly real GNP and finds that the threshold nonlinearity of the series cannot be completely explained by oil shocks, political business cycle, and discrete changes in Federal Reserve policy; Geweke and Terui (1991) consider a Bayesian analysis of the model.

It is interesting to note that the stationarity of z_t in (10.2) does not require all of the zeros of the polynomial $\phi^{(h)}(B) = 1 - \phi_1^{(h)} B - \cdots - \phi_p^{(h)} B^h$ to be outside of the unit circle. On the contrary, limit cycles of an TAR model tend to result from certain alternations between explosive and contractive regimes. For necessary and sufficient conditions of stationarity for simple TAR models, see Petruccelli and Woolford (1984) and Chen and Tsay (1991).

The TAR model in (10.2) is a self-exciting model because the model uses its own lagged value z_{t-d} as the threshold variable. A general TAR model in fact can use other variable as threshold or use more than one threshold variables. Such generalizations are straightforward.

10.3.1. U.S. real GNP

To illustrate the usefulness of TAR models, we consider the series $\{Z_t\}$ of quarterly U.S. real GNP (in 1982 dollars) from the first quarter of 1947 to the first quarter of 1991, a total of 177 observations. The data are obtained from the Citibase database and are seasonally adjusted. In our analysis, we focus on the growth rate, namely

$$z_t = \log(Z_t) - \log(Z_{t-1})$$

so that there are 176 data points in the series. For simplicity, we shall not consider the minor seasonal behavior of the data commonly encountered in seasonally adjusted series published by the government. The growth series is shown in Figure 10.1.

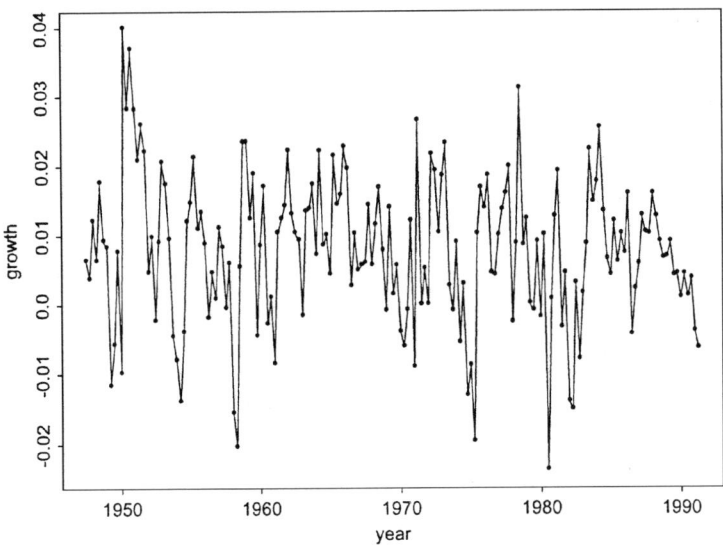

FIGURE 10.1 Growth of U.S. quarterly real GNP from 1947.II to 1991.I. The original GNP data were seasonally adjusted and in 1982 million dollars.

For linear models, it is easily obtained that the AR(2) model

$$z_t = .0041 + .33 z_{t-1} + .13 z_{t-2} + \epsilon_t \tag{10.3}$$

fits the data well, where the standard errors of the parameters are .001, .075, and .076, respectively, and the residual standard deviation is .00986. The residuals of model (10.3) give the Box–Ljung statistic $Q(12) = 10.1$, indicating no serial correlations in the residuals. There are two possible outliers at $t = 12$ and $t = 133$. The magnitudes of these two possible outliers, however, are not substantial.

In what follows, we adopt the TAR modeling approach of Tsay (1989) to specify a model for z_t. To specify tentatively the threshold lag d, Table 10.8 gives the results of a threshold nonlinearity test based on arranged AR(2) autoregressions with possible threshold $d \in \{1, \ldots, 6\}$. For each value of d, the data are arranged according to the order of z_{t-d}. Then, predictive residuals from the arranged autoregression are regressed against the predictor variables, giving rise to an asymptotic F test for independence between the residuals and the predictors that would be consistent with

TABLE 10.8. Threshold Nonlinearity Tests of U.S. Real GNP

d	1	2	3	4	5	6
F test	0.37	3.16	2.55	2.65	1.70	1.80
p value	.778	.026	.058	.051	.169	.150

10.3. THE TAR MODEL

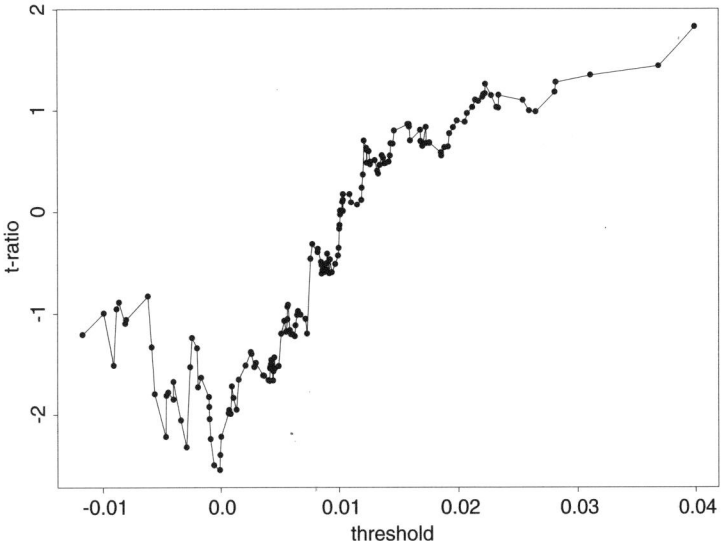

FIGURE 10.2 T-ratio of autoregressive coefficient at lag 2 versus threshold for the growth series of U.S. quarterly real GNP: 47.II-91.I.

linearity. From the table, it is seen that the linear model hypothesis seems untenable and that $d = 2$ is reasonable for the series as the corresponding p value is the smallest.

To determine the number of regimes and the threshold values r_h's, Figure 10.2 shows the sequential t ratio of lag 2 AR estimate plotted against the threshold variable z_{t-d} in an arranged autoregression of order 2 and $d = 2$. Major changes in the slope of the t ratios suggest regime partitions. Thus, this plot indicates that the data can be partitioned into two regimes with a threshold at $z_{t-2} = 0.0$. Therefore, we tentatively specify an TAR(2) model with two regimes separated by this threshold. Alternatively, Tsay (1998) uses information criteria such as the Akaike information criterion or the Bayesian information criterion (Schwarz 1978) to select the thresholds of a TAR model. With such a specification, we obtained the model

$$z_t = \begin{cases} \phi_0^{(1)} + \phi_1^{(1)} z_{t-1} + \phi_2^{(1)} z_{t-2} + \epsilon_{1,t} & \text{if } z_{t-2} \leq 0.0 \\ \phi_0^{(2)} + \phi_1^{(2)} z_{t-1} + \phi_2^{(2)} z_{t-2} + \epsilon_{2,t} & \text{if } z_{t-2} > 0.0, \end{cases} \quad (10.4)$$

where the numbers of observations are 37 and 137, respectively, and the parameters are

h	$\phi_0^{(h)}$(std)	$\phi_1^{(h)}$(std)	$\phi_2^{(h)}$(std)	σ_h
1	$-.0039(.0033)$	$.44(.18)$	$-.79(.33)$	$.0120$
2	$.0038(.0014)$	$.31(.08)$	$.20(.11)$	$.0087$

The TAR model in (10.4) fits the data very well. The normalized residuals, by taking into account the difference in residual standard deviations of the two regimes, give $Q(12) = 5.8$. This model is close to that of Potter (1995), who uses a slightly different data span and includes an additional term z_{t-5} in the model, which presumably is caused by the minor seasonal behavior mentioned earlier.

The most striking feature of the TAR(2) model in (10.4) is that, by treating a negative growth in GNP as "contraction" and a positive growth as "expansion," we see that the economy behaves differently after "contraction" and "expansion." For example, the AR polynomial of the first regime has a pair of complex roots, indicating some cyclical behavior of the GNP after a contraction. On the other hand, the AR polynomial of the second regime has two real roots, showing that the economy tends to decay exponentially to some mean level after an expansion.

To facilitate further discussion and to gain insight into the structure of the GNP data, we refine the TAR model in (10.4) by incorporating the relative size of z_{t-1} with respect to z_{t-2}, the threshold variable of the model. More specifically, we generalize the TAR(2) model in (10.4) to a four-regime TAR model with regimes shown in Figure 10.3. These four regimes have straightforward meaning:

- Regime I: $z_{t-1} \leq z_{t-2} \leq 0$. This regime denotes a recession period in which the economy changed from contraction to an even worse one.
- Regime II: $z_{t-1} > z_{t-2}$ but $z_{t-2} \leq 0$. Here the economy was in contraction, but improving.
- Regime III: $z_{t-1} \leq z_{t-2}$ but $z_{t-2} > 0$. This regime corresponds to a period in which the economy was reasonable, but declining.

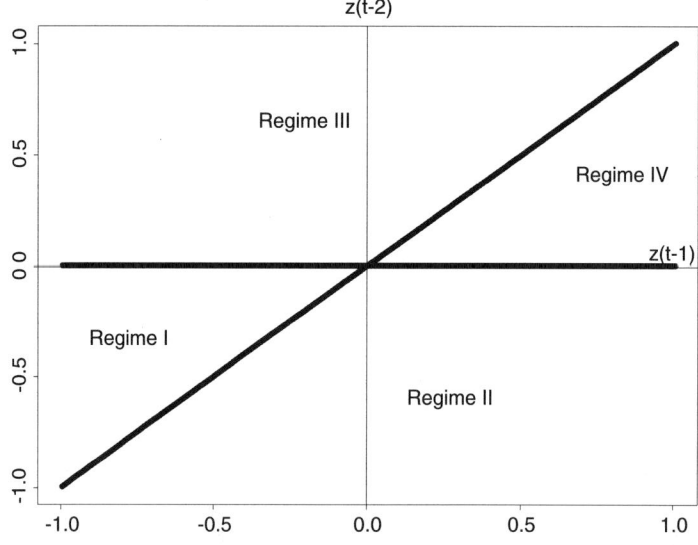

FIGURE 10.3 Regimes of a TAR model for the growth series of U.S. quarterly real GNP: 47.II-91.I.

10.3. THE TAR MODEL

- Regime IV: $z_{t-1} > z_{t-2} > 0$. This is an expansion period in which the economy was reasonable and became stronger.

The resulting TAR model is

$$z_t = \begin{cases} -.015 - 1.076 z_{t-1} + \epsilon_{1,t} & \text{regime I} \\ -.006 + .630 z_{t-1} - .756 z_{t-2} + \epsilon_{2,t} & \text{regime II} \\ .006 + .438 z_{t-1} + \epsilon_{3,t} & \text{regime III} \\ .004 + .443 z_{t-1} + \epsilon_{4,t} & \text{regime IV} \end{cases} \quad (10.5)$$

where all the parameters have a t ratio greater than 2 in modulus except for the constant terms in regimes II and IV, for which the t ratios are -1.35 and 1.32, respectively. The residual standard deviations are $\sigma_1 = .0062$, $\sigma_2 = .0132$, $\sigma_3 = .0094$, and $\sigma_4 = .0082$. The numbers of observations in each regime are 6, 31, 79, and 58, respectively. It is gratifying to see that there were only six cases in the recession period, regime I. Furthermore, it was even more reassuring to see the negatively explosive nature of the model in regime I, which indicates that the economy usually recovers quickly from the recession period. In fact, there were only three occasions in which we had more than two consecutive negative growth in quarterly real GNP during the entire data span.

The model of regime II is also interesting. Since $z_{t-1} > z_{t-2}$, $z_{t-2} < 0$ and the constant term is not statistically significant at the usual 5% critical value, the model tends to have positive conditional means, suggesting that the economy is more likely to grow continuously out of recession once a recovery has started. The two models in regimes III and IV are relatively close. They indicate a general positive mean level for the U.S. GNP during expansion periods. Also, with an autoregressive coefficient of about .44, these two models imply an average growth rate of approximately between 2.9 and 4.3% per year.

10.3.2. Postsample forecasts and discussion

To compare the linear and the TAR models in (10.3)–(10.5) for describing the GNP growth, we consider out-of-sample forecasts of these three models. The comparison is based on mean-squared errors of forecasts and is performed according to the following procedure:

1. Consider the 60 subseries $(z_1, \ldots, z_\ell)$ for $\ell = 104, \ldots, 163$ of the data.
2. For each subseries, estimate the parameters and compute out-of-sample forecasts of 1 to 12 steps ahead and the associated forecast errors. For the TAR models, multi-step ahead forecasts are obtained via simulation of 2000 realizations at each step. The mean of these 2000 realizations is treated as a point forecast.
3. For 1–12-step-ahead forecasts, compute the mean-squared errors of forecast according the regimes of the TAR models.

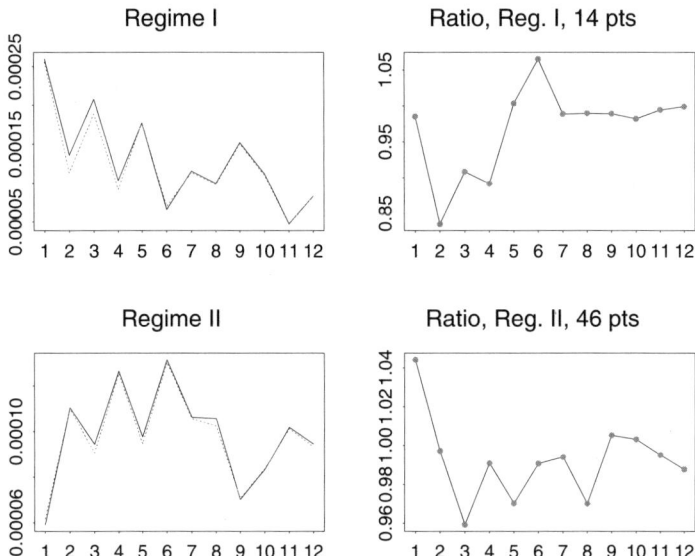

FIGURE 10.4 Mean-squared error of out-of-sample forecasts for a two-regime TAR model and linear AR(2) model.

Figure 10.4 shows the comparison between the two-regime TAR model in (10.4) and the linear AR(2) model in (10.3), where both the mean-squared errors of forecasts and their ratio (TAR/linear) are given. For mean-squared errors the solid line corresponds to the linear AR(2) model. From the plot, it is seen that the two-regime TAR model performs better than the linear model, especially in Regime 1 and the cases of 2–4-step-ahead forecasts. The gain of TAR model in the second regime is rather small. Figure 10.5 gives the corresponding comparison between the four-regime TAR model in (10.5) and the linear model in (10.3). From the plots, it is clear that the TAR model outperforms the linear model by a substantial margin in short-term forecasts at regimes I and II. The ratio of the 1-step-ahead forecast in regime I is as low as .25. This result shows strongly that the TAR model describes the dynamic behavior of the real GNP much better than does the linear model when the U.S. economy is declining.

Discussion
The comparisons presented above suggest that the TAR models, by their piecewise linear nature, can capture the asymmetric behavior of the real GNP during recession and expansion periods whereas the linear AR model can reflect properties of only the "majority" of the data. In this particular instance, the case of $z_{t-2} > 0$ has most of the data points (137 out of 176) so that the linear model built is close to that of the second regime of the TAR(2) model in (10.4). Consequently, the TAR model can gain forecast accuracy when the GNP is in the first regime. This also appears to be the reason underlying the success of the 4-regime TAR model in (10.5). Such

10.3. THE TAR MODEL

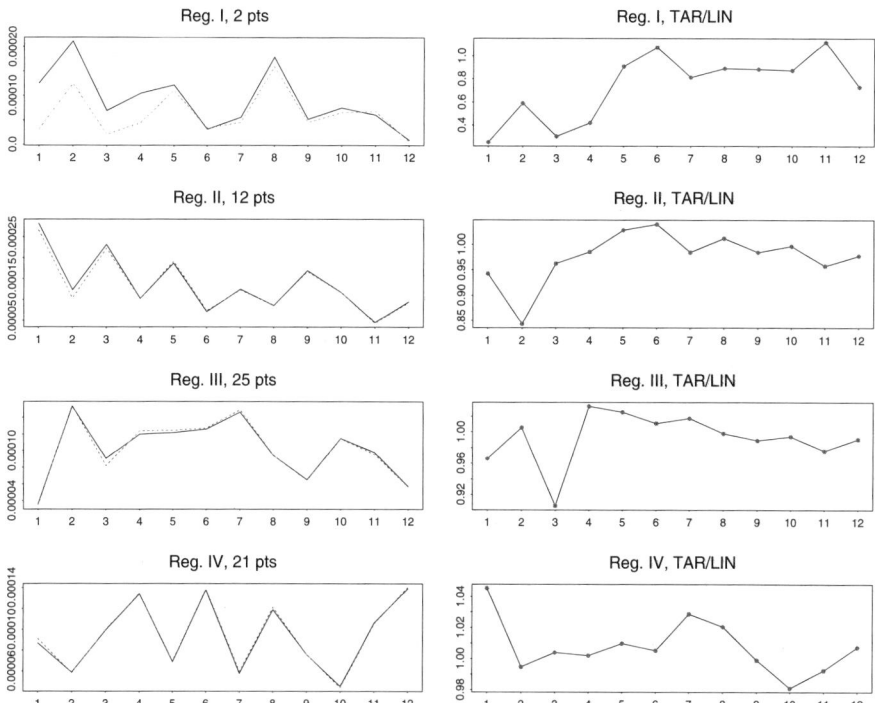

FIGURE 10.5 Mean squared error of out-of-sample forecasts for a four-regime TAR model and linear AR(2) model.

a conclusion appears to be obvious; nevertheless, it highlights the contribution of nonlinear time series analysis. Further, the success of TAR models in describing the real GNP data indicates clearly that the economy does not follow a simple linear model. The explosive model in regime I of model (10.5) provides convincing evidence that there is asymmetry in the economy. This result has various implications:

1. By considering the implied predictive distributions (or simply the conditional distributions) of the TAR model in (10.5), we see that the probability of growing out of a recession is different from that of getting into one. Thus, the model via a different route supports the finding of Hamilton (1989), who uses a Markov switching model to show that the transition probability from recession to expansion is different from that from expansion to recession.
2. The regime partition of the TAR model in (10.5) enables one to use it to provide probability assessment of turning points in the U.S. economy. This point might be of interest in its own right.
3. Since only the model of regime II in model (10.5) has a pair of complex roots, the model suggests that business cycles, if any, in quarterly real U.S. GNP are rather weak.

4. The TAR result obtained in this chapter is in agreement with Diebold and Rudebusch (1990), who investigate duration dependence in the American business cycle.
5. Our analysis enhances the finding of Potter (1995) that linear models can easily overlook certain dynamic behavior of an economic time series.
6. It is interesting to see that all except one of the six points in regime I of the four-regime TAR model are in the second quarter. The fact that those data in the second quarter behaved differently from the others might explain the result obtained in Ghysels (1991).
7. Finally, TAR models similar to those used in this chapter were employed more recently by Martinez and Espasa (1998) to compare behavior of quarterly gross demostic product (GDP) between the United States and Spain with interesting and insightful results.

10.4. CONCLUDING REMARKS

In this chapter, we considered a general test procedure for detecting nonlinearity in a univariate time series and demonstrated nonlinear application by using the threshold autoregressive models. As mentioned in the introduction, many other nonlienar models are available. See Tong (1990) for a good summary of nonlinear models. Here we shall discuss advantages and disadvantages of some nonlinear models. First, the GARCH models and stochastic volatility models are widely used to describe the evolution of conditional variance of a security return. See Bollerslev et al. (1992) for a review of GARCH models and Jacquier et al. (1994) for stochastic volatility models. These models can produce the heavy tails commonly seen in financial time series. The stochastic volatility models often provide better fit in applications. However, no definite conclusions can be drawn between the two classes of model in out-of-sample forecasting comparison. A somewhat unsatisfactory feature of these models is that a fitted model may depend heavily on a small number of data points. Second, bilinear models are basically symmetric models in the sense that they cannot produce asymmetric limit cycles. The threshold models, on the other hand, are capable of modeling asymmetric business cycles. The Markov switching models of Hamilton (1989) can also produce asymmpetric cycles. However, maximum likelihood estimation of the Markov swiching models can be difficult. My own experience shows that in most applications a three-state switching model is needed. The probability transition matrix then involves many parameters, making the model even harder to estimate. Third, there is substantial interest in using smooth transition autoregressive (STAR) models, especially among econometricians. These models have a continuous conditional expectation and can also describe asymmetric limit cycle (Granger and Teräsvirta 1993). However, it is hard to obtain efficient estimates of the transition parameters. Most, if not all, examples shown in the literature have large standard errors for the estimated transition parameters. The threshold model can be modified to have a continuous expectation. The modification raises some interesting features between discontinuous and continuous threshold models. See Chan (1993) and Chan and Tsay (1998) for asymptotic properties of conditional least-squares estimators of the threshold. Fourth, the state-dependent model of Priestley (1980) is general, but not easy to use in practice.

ACKNOWLEDGMENT

This research is supported by the National Science Foundation and the Graduate School of Business, University of Chicago.

REFERENCES

Akaike, H. (1974). A new look at statistical model identification. *IEEE Trans. Auto. Cont.* **AC-19**, 716–722.

Auestad, B. and Tjøstheim, D. (1990). Identification of nonlinear time series: First order characterization and order determination. *Biometrika* **77**, 669–687.

Bollerslev, T. (1986). Generalized autoregressive conditional heteroscedasticity. *J. Econ.* **31**, 307–327.

Bollerslev, T. Chou, R., and Kroner, K. F. (1992). ARCH modeling in finance: A review. *J. Econ.* **52**, 5–59.

Box, G. E. P. and Jenkins, G. M. (1976). *Time Series Analysis, Forecasting and Control*. Holden-Day, San Francisco.

Brock, W., Dechert, D., and Scheinkman, J. (1987). *A Test for Independence Based on the Correlation Dimension*. Working Paper, Social Systems Research Institute, Univ. Wisconsin–Madison.

Chan, K. S. (1993). Consistency and limiting distribution of the least squares estimator of a threshold autoregressive model. *Ann. Stat.* **21**, 520–533.

Chan, K. S. and Tong, H. (1986a). On estimating thresholds in autoregressive models. *J. Time Ser. Anal.* **7**, 179–190.

Chan, W. S. and Tong, H. (1986b). On tests for nonlinearity in time series analysis. *J. Forecasting* **5**, 217–228.

Chang, I., Tiao, G. C., and Chen, C. (1988). Estimation of time series parameters in the presence of outliers. *Technometrics* **30**, 193–204.

Chan, K. S. and Tsay, R. S. (1998). Limiting properties of the conditional least squares estimator of a continuous TAR model. *Biometrika* **85**, 413–426.

Chen, R. and Tsay, R. S. (1991). On the ergodicity of TAR(1) processes. *Ann. Appl. Prob.* **1**, 613–634.

Chen, R. and Tsay, R. S. (1993a). Functional-coefficient autoregressive models. *J. Am. Stat. Assoc.* **88**, 298–308.

Chen, R. and Tsay, R. S. (1993b). Nonlinear additive ARX models. *J. Am. Stat. Assoc.* **88**, 955–967.

Diebold, F. X. and Rudebusch, G. D. (1990). A nonparametric investigation of duration dependence in the American business cycle. *J. Polit. Econ.* **98**, 596–616.

Engle, R. (1992). Autoregressive conditional heteroscedasticity with estimates of the variance of UK inflation. *Econometrica* **50**, 987–1008.

Friedman, L. H. (1991). Multivariate adaptive regression splines (with discussions). *Ann. Stat.* **19**, 1–141.

Geweke, J. and Terui, N. (1993). Bayesian threshold autoregressive models for nonlinear time series. *J. Time Ser. Anal.* **14**, 441–454.

Granger, C. W. J. and Andersen, A. P. (1978). *An Introduction to Bilinear Time Series Models*. Vandenhoek & Ruprecht, Gottingen.

Granger, C. W. J. and Teräsvirta, T. (1993). *Modeling Nonlinear Economic Relationships*, Oxford Univ. Press, Oxford.

Ghysels, E. (1991). *Are Business Cycle Turning Points Uniformly Distributed Throughout the Year?* Working Paper, Univ. Montreal.

Haggan, V. and Ozaki, T. (1981). Modeling nonlinear vibrations using an amplitude-dependent autoregressive time series model. *Biometrika* **68**, 189–196.

Hamilton, J. D. (1989). A new approach to the economic analysis of nonstationary time series and the business cycle. *Econometrica* **57**, 357–384.

Hansen, B. E. (1996). Inference when a nuisance parameter is not identified under the alternative. *Econometrica* **64**, 413–430.

Hinich, M. J. (1982). Testing for Gaussianity and linearity of a stationary time series. *J. Time Ser. Anal.* **3**, 169–176.

Hsieh, D. A. and LeBaron, B. (1988). *Finite Sample Properties of the BDS Statistic*. Working Paper, Graduate School of Business, Univ. Chicago.

Jacquier, E., Polson, N. G., and Rossi, P. E. (1994). Bayesian analysis of stochastic volatility models. *J. Business Econ. Stat.* **12**, 371–417.

Jones, D. A. (1978). Nonlinear autoregressive processes. *Proc. Roy. Soc. Lond. A* **360**, 71–95.

Keenan, D. M. (1985). A Tukey nonadditivity-type test for time series nonlinearity. *Biometrika* **72**, 39–44.

Lai, T. L. and Wei, C. Z. (1982). Least squares estimates in stochastic regression models with applications to identification and control of dynamic systems. *Ann. Stat.* **10**, 154–166.

Lewis, P. A. W. and Stevens, J. G. (1991). Nonlinear modeling of time series using multivariate adaptive regression splines (MARS). *J. Am. Stat. Assoc.* **86**, 864–877.

Luukkonen, R., Saikkonen, P., and Terasvirta, T. (1988). Testing linearity against smooth transition autoregressive models. *Biometrika* **75**, 491–499.

Maravall, A. (1983). An application of nonlinear time series forecasting. *J. Business Econ. Stat.* **1**, 66–74.

Martinez, J. M. and Espasa, A. (1998). *Modeling Nonlinearities in GDP: Some Differences between U.S. and Spanish Data*. Working Paper, Dept. Estadistica y Econometria, Univ. Carlos III de Madrid, Spain.

McLeod, A. I., and Li, W. K. (1983). Diagnostic checking ARMA time series models using squared-residual autocorrelations. *J. Time Ser. Anal.* **4**, 269–273.

Nelson, D. (1991). Conditional heteroscedasticity in asset returns: A new approach. *Econometrica* **59**, 347–370.

Petruccelli, J. and Davies, N. (1986). A portmanteau test for self-exciting threshold autoregressive-type nonlinearity in time series. *Biometrika* **73**, 687–694.

Petruccelli, J. and Woolford, S. W. (1984). A threshold AR(1) model. *J. Appl. Prob.* **21**, 270–286.

Potter, S. M. (1995). A nonlinear approach to U.S. GNP. *J. Appl. Econ.* **10**, 109–125.

Priestley, M. B. (1980). State dependent models: A general approach to nonlinear time series analysis. *J. Time Ser. Anal.* **1**, 57–71.

Robinson, P. M. (1983). Non-parametric estimation for time series models. *J. Time Ser. Anal.* **4**, 185–208.

Schwarz, G. (1978). Estimating the dimension of a model. *Ann. Stat.* **6**, 461–464.

Tiao, G. C. and Tsay, R. S. (1994). Some advances in nonlinear and adaptive modeling in time series. *J. Forecasting* **13**, 109–131.

Tong, H. (1978). On a threshold model. In C. H. Chen (ed.), *Pattern Recognition and Signal Processing*. Sijhoff & Noordhoff, Amsterdam.

Tong, H. (1983). *Threshold Models in Nonlinear Time Series Analysis*, Lecture Notes in Statistics, Vol. 21. Springer-Verlag, New York.

Tong, H. (1990). *Nonlinear Time Series: A Dynamical System Approach*, Oxford Univ. Press, London.

Tong, H. and Lim, K. S. (1980). Threshold autoregression, limit cycles, and cyclical data (with discussion). *J. Roy. Statist. Soc.* **B42**, 45–292.

Tong, H. and Yeung, I. (1991). On tests for self-exciting threshold autoregressive-type nonlinearity in partially observed time series. *Appl. Stat.* **40**, 43–62.

Tsay, R. S. (1986). Nonlinearity tests for time series. *Biometrika* **73**, 461–466.

Tsay, R. S. (1987). Conditional heteroscedastic time series models. *J. Am. Stat. Assoc.* **82**, 590–604.

Tsay, R. S. (1988). Nonlinear time series analysis of blowfly population. *J. Time Ser. Anal.* **9**, 247–263.

Tsay, R. S. (1989). Testing and modeling threshold autoregressive processes. *J. Am. Stat. Assoc.* **84**, 231–240.

Tsay, R. S. (1991). Detecting and modeling nonlinearity in univariate time series analysis. *Statistica Sinica* **1**, 431–451.

Tsay, R. S. (1998). Testing and modeling multivariate threshold models. *J. Am. Stat. Assoc.* **93**, 1188–1202.

CHAPTER 11

Bayesian Time Series Analysis

Ruey S. Tsay
University of Chicago

11.1. INTRODUCTION

Estimation and model selection are two main components of time series analysis. There are many results available in the literature that concern parameter estimation of a given model and model selection within a specified class of models. For example, the exact maximum likelihood method was widely investigated in the 1980s for autoregressive moving-average (ARMA) models (e.g., Ansley 1979, Jones 1980, Hillmer and Tiao 1979), and the conditional least-squares approach was proposed for nonlinear models (e.g., Tong 1990 and references cited therein). For model selection, some popular model selection criteria such as AIC and its variants are commonly used to select the order of an autoregressive process or a threshold autoregressive models (see, e.g., Priestley 1981, Brockwell and Davis 1991, Tong 1990).

There is, however, no unified approach or program that can be used to estimate most of the linear and nonlinear models considered in the literature. For example, special packages are needed to apply bilinear models, threshold models or Markov switching models. Furthermore, there is little discussion of model selection across different classes of nonlinear models. Much work on model selection in the literature focuses on nested models for which the traditional maximum likelihood ratio tests or Largange multiplier tests or information criterion functions apply. For non-nested models, model discrimination becomes much more involved, especially when the competing models are nonlinear. In the time series literature, Li (1993) adopts the idea of separate families of hypotheses of Cox (1962) and proposes a test statistic for discriminating bilinear and threshold models. The test statistic has an asymptotic chi-squared distribution with one degree of freedom. However, Li's test is closely

---

A Course in Time Series Analysis, Edited by Daniel Peña, George C. Tiao, and Ruey S. Tsay.
ISBN 0-471-36164-X. © 2001 John Wiley & Sons, Inc.

11.1. INTRODUCTION

related to the method of selecting a model with smaller residual variance and is not applicable to other nonlinear models.

The purpose of this chapter is to consider a unified Bayesian approach that can be used to estimate most of the univariate time series models available in the literature and to select an appropriate model for a time series when the candidate models may be nonnested, nonlinear (Chen et al. 1997). More specifically, our objective is to consider an approach that is widely applicable in univariate time series analysis. The models can be linear or nonlinear, and the approach can discriminate between nonnested nonlinear models. The approach used is based on Gibbs sampling and requires some prior specification. In particular, our approach to model selection allows each observation to select one of the candidate models. The key prior specification here is the probability that an individual observation is generated by a given model given that both observations adjacent in time are generated by that same model [see equation (11.7).] Sensitivity analysis of prior specification will be discussed later.

Because the approach considered uses Gibbs sampling, it may require substantial computing time in some applications. Our goal is not to develop the most efficient approach for univariate time series analysis, but a unified approach that is applicable to most parametric models. In a given application where the entertained models are specified, it is often possible to reduce the computing time by some special algorithm or theoretical derivation. However we shall not focus on those special issues.

Fully Bayesian analysis of time series data have been considered in Monahan (1983), West and Harrison (1986), and some chapters in Spall (1988). Many of the analyses can only entertain simple models in real applications because they require prohibitive computation. This limitation is largely overcome by using the Gibbs sampler. The main differences between this chapter and the abovementioned Bayesian analyses include that (1) we use a general framework consisting of a mixture of several models, (2) we treat initial conditions of a time series as parameters so that the analysis is applicable to unit-root nonstationary series as well as nonlinear series, and (3) we make use of the recent developments in Markov chain Monte Carlo (MCMC) methods so that the computation is greatly simplified. As such the Bayesian analysis considered is very general and widely applicable.

The chapter is organized as follows. In Section 11.2, we give the general framework of models considered in this chapter and show that many time series models considered in the literature are special cases of our model. Section 11.3 considers model estimation via the Gibbs sampling. In particular, we treat starting values of the time series and the innovational series as parameters and consider the conditional likelihood function of a parameter given the others. We also discuss methods for implementing the Gibbs sampler when the parameter under study is nonlinear, such as the moving-average parameters in an ARMA model. Consequently, the Bayesian analysis considered can handle nonlinear models as well as unit-root nonstationary models. Section 11.4 is devoted to model discrimination. Here a simple switching framework is used in which the competing non-nested nonlinear models become submodels of a mixture. Under this framework, each individual observation can select its own model from the mixture. The posterior probability that particular observations are associated with a particular model can then be used to select an appropriate model for the whole series. Advantages of the this model selection method are discussed. This idea for model discrimination was used in McCulloch and Tsay (1994), George et al. (1996), and Chen et al. (1997). Finally, we consider some simulated and real examples in Section 11.5.

11.2. A GENERAL UNIVARIATE TIME SERIES MODEL

The model considered in this chapter is

$$z_t = f(z_{t-1}, \ldots, z_{t-p}; a_{t-1}, \ldots, a_{t-q}; \boldsymbol{\beta}_f) + a_t$$
$$a_t = g_t \epsilon_t \quad (11.1)$$
$$g_t = g(z_{t-1}, \ldots, z_{t-u}; a_{t-1}, \ldots, a_{t-v}; g_{t-1}, \ldots, g_{t-w}; \boldsymbol{\beta}_g)$$

where z_t is a univariate time series; $f(.)$ and $g(.)$ are two known functions with finite-dimensional parameter vectors $\boldsymbol{\beta}_f$ and $\boldsymbol{\beta}_g$, respectively; p, q, u, v, and w are nonnegative integers; and $\{\epsilon_i\}$ is a sequence of independent and identically distributed random variables with mean zero and variance one. The function $g(.)$ is assumed to be positive; it governs the evolution of the volatility (i.e., conditional variance) of the innovational series a_t. For simplicity, we focus on the case that ϵ_t's are standard normal random variables, that is, that a_t is conditionally normal. However, it is easily seen that ϵ_t can be any continuous random variables with a well defined density function.

Model (11.1) is a general model, because it encompasses many commonly used models in the literature. Some specific examples are

1. If $g(.) = \beta_1$, which is a positive constant, and $f(.) = \sum_{i=1}^{p} \phi_i z_{t-i} - \sum_{i=1}^{q} \theta_i a_{t-i}$, then model (11.1) reduces to the well-known ARMA of Box et al. (1994).
2. If $f(.) = 0$ and $g^2(.) = \gamma_0 + \sum_{i=1}^{q} \gamma_i a_{t-i}^2$, where $\gamma_0 > 0$ and $\gamma_i \geq 0$, then the model becomes the well-known conditional autoregressive heteroscedastic (ARCH) model of Engle (1982). The ARCH model and its variants are widely used in finance to model the volatility of a security return.
3. If $f(.) = 0$ and $g^2(.) = \gamma_0 + \sum_{i=1}^{v} \gamma_i a_{t-i}^2 + \sum_{i=1}^{w} \lambda_i g_{t-l}^2$, where $\gamma_0 > 0$, $\gamma_i \geq 0$ and $\lambda_i \geq 0$, then we have the generalized ARCH (GARCH) model of Bollerslev (1986).
4. If $f(.) = 0$ and $g(.) = \exp(\gamma_0 + \sum_{i=1}^{u} \beta_i z_{t-i} + \sum_{j=1}^{v} \gamma_j a_{t-j})$, then model (11.1) becomes a stochastic volatility model in which the conditional variance of the series is related to past observations and past innovations. This model is similar to that in Tsay (1987) and can be extended to include models that allow for asymmetric responses to positive and negative innovations.
5. If $g(.) = \beta_1 > 0$, a constant, and $f(.) = \sum_{i=1}^{p} \phi_i z_{t-i} - \sum_{i=1}^{q} \theta_i a_{t-i} + \sum_{i=1}^{p} \sum_{j=1}^{q} \beta_{ij} z_{t-i} a_{t-j}$, then model (11.1) becomes the bilinear model of Granger and Andersen (1978) and Subba Rao (1981).

6. If $f(.) = \phi_0^{(i)} + \sum_{j=1}^{p} \phi_j^{(i)} z_{t-i}$ and $g(.) = \sigma^{(i)} > 0$ for $r_{i-1} \le z_{t-d} \le r_i$, where d is a positive integer and r_i's are real numbers satisfying $-\infty = r_0 < r_1 < \cdots < r_k = \infty$, then model (11.1) becomes the threshold autoregressive (TAR) model of Tong (1978, 1990).

Model (11.1) also provides a framework to combine different time series models. For example, if $f(.) = \phi_0 + \phi_1 z_{t-1} - \theta_1 a_{t-1}$ and $g(.) = \omega_0 + \omega_1 a_{t-1} > 0$ almost surely, then z_t is an ARMA process with a concurrent bilinear innovation. Such an innovational series also shows stochastic volatility as that of an ARCH model. In Section 11.4, we use model (11.1) to develop a switching model for model discrimination of nonnested nonlinear models.

11.3. ESTIMATION

In this section, we discuss a general approach to parameter estimation of model (11.1). The approach is Bayesian and makes use of the Gibbs sampling. In particular, we assume that the time series z_t starts at time $t = 1$ with unknown starting values, lagged innovations, and lagged g values. We treat these initial values as unknown parameters of the model and estimate them jointly with other parameters. This marks a major difference between the approach and many existing estimation methods, because those existing methods assume either the starting values are zero or the process under study is stationary (see, e.g., Brockwell and Davis 1991). The idea of treating starting values and innovations of a time series process as unknown parameters has been used previously in the literature primarily for ARMA models. For example, in exact likelihood estimation of ARMA models, those starting values and innovations are estimated by using the dynamic structure of the data. Chen et al. (1997) apply the idea to nonlinear models and Li and Tsay (1998), to multivariate ARMA models.

Consider model (11.1). Let $p^* = \max\{p, u\}$, $q^* = \max\{q, v\}$, $\mathbf{z}_0 = (z_{-p^*+1}, z_{-p^*+2}, \ldots, z_0)'$ be the starting values of z_t, $\mathbf{a}_0 = (a_{-q^*+1}, a_{-q^*+2}, \ldots, a_0)'$ be the starting innovations, and $\mathbf{g}_0 = (g_{-w+1}, \ldots, g_0)'$ the starting lagged g values. Finally, let $\widehat{\ } = (\mathbf{z}_0', \mathbf{a}_0', \mathbf{g}_0', \boldsymbol{\beta}_f', \boldsymbol{\beta}_g')'$ be the set of all parameters of model (11.1). For n observations $\{z_t\}_{t=1}^n$, let $\mathbf{Z}_t = (z_1, \ldots, z_t)'$. It is easily seen that the conditional mean and variance of z_t given $\mathbf{Z}_{t-1}$ and $\widehat{\ }$ are

$$E(z_t \mid \mathbf{Z}_{t-1}, \widehat{\ }) = f(z_{t-1}, \ldots, z_{t-p}; a_{t-1}, \ldots, a_{t-q}) \equiv f_t$$
$$\text{Var}(z_t \mid \mathbf{Z}_{t-1}, \widehat{\ }) = g^2(z_{t-1}, \ldots, z_{t-u}; a_{t-1}, \ldots, a_{t-v}; g_{t-1}, \ldots, g_{t-w}) \equiv g_t^2.$$

Therefore, the log-likelihood function of the data can be written as

$$L(\mathbf{Z}_n, \widehat{\ }) = \sum_{t=1}^{n} \ln p(z_t \mid \mathbf{Z}_{t-1}, \widehat{\ }),$$

which under normality becomes

$$L(Z_n, \frown) = \frac{-1}{2} \sum_{t=1}^{n} \left[\ln\left(2\pi g_t^2\right) + \frac{(z_t - f_t)^2}{g_t^2} \right].$$

Given prior distribution $p(\frown)$, the log of the joint posterior distribution function for the model is

$$\ell(\Omega \mid Z_n) \propto \ln[p(\frown)] - \frac{1}{2} \sum_{t=1}^{n} \left[\ln\left(2\pi g_t^2\right) + \frac{(z_t - f_t)^2}{g_t^2} \right]. \qquad (11.2)$$

The ability to evaluate this posterior function plays a key role in our Bayesian approach. For the general model in (11.1), this posterior function involves many parameters and might be difficult to handle. Some methods are available in the literature to overcome this difficulty, especially when special cases of model (11.1) are entertained. For example, the Kalman filter can be used to evaluate this posterior function recursively for linear Gaussian ARMA models with a flat prior (Jones 1980). The EM algorithm can be used if model (11.1) is in the form of a component model (Shumway and Stoffer 1982). More recently, the Gibbs sampler has been shown to be useful in obtaining the joint posterior distribution of $\frown$ for some time series models. For example, the Gibbs sampler with the Metropolis algorithm is found to be useful in modeling linear Gaussian ARMA models with conditionally conjugate priors. Here the Metropolis algorithm is used primarily to handle nonlinear parameters for which no closed-form formulas are available to simplify the Gibbs draw. In Carlin et al. (1992a), the Gibbs sampler in conjunction with scale mixtures of normal distributions was used to analyze nonlinear state-space models. An advantage of the Gibbs sampler is that the joint posterior distribution of the model parameters in (11.2) can be obtained iteratively by using lower-dimensional conditional posterior distributions. As a special case, one may consider all one-dimensional conditional posterior distributions in implementing the sampler. The one-dimensional posterior distributions obtained from (11.2) are easy to evaluate. Another advantage of the Gibbs sampler is that only conditional prior specification is needed. Other Bayesian analyses of time series models using Markov chain methods include those by Marriott et al. (1996), Chib and Greenberg (1994), and Li and Tsay (1998), among many others.

In this chapter, we also use the Gibbs sampler. However, we shall not use the Metropolis algorithm to handle nonlinear parameters. Instead, we follow the work of Chen et al. (1997) and employ the griddy Gibbs approach of Tanner (1991) for those parameters that do not have closed-form formulas to facilitate the Gibbs draws. Advantages of the griddy Gibbs include simplicity and wide applicability. The computational burden of the griddy Gibbs, however, may be heavy. The Gibbs sampling and griddy Gibbs used are given below.

11.3.1. Gibbs sampling

For simplicity, we consider the case of three parameters $(\theta_1, \theta_2, \theta_3)$ and assume that subroutines are available to draw samples from the three full conditional posterior distributions

$$f_1(\theta_1 \mid \theta_2, \theta_2, \mathbf{Z}), \qquad f_2(\theta_2 \mid \theta_3, \theta_1, \mathbf{Z}), \qquad f_3(\theta_3 \mid \theta_1, \theta_2, \mathbf{Z}) \qquad (11.3)$$

where $\mathbf{Z} = (z_1, z_2, \ldots, z_n)'$ denotes the vector of n consecutive observations observed at equally spaced time-intervals. In our applications, we use mainly well-known distributions so that the random drawings were carried out by subroutines in the NAG package. This, however, is not a necessity for the general use of the sampler.

The Gibbs sampler employed in this chapter works as follows:

1. Consider an arbitrary set of starting values for the three parameters, say, $\theta_{10}, \theta_{20}, \theta_{30}$.
2. Generate M sets of random observations drawn iteratively from the full conditional posterior distributions in (11.3). More specifically, the first set of random observations $\{\theta_{11}, \theta_{21}, \theta_{31}\}$ is obtained as follows:

   θ_{11} is drawn from $f_1(\theta_1 \mid \theta_{20}, \theta_{30}, \mathbf{Z})$
   θ_{21} is drawn from $f_2(\theta_2 \mid \theta_{30}, \theta_{11}, \mathbf{Z})$
   θ_{31} is drawn from $f_3(\theta_3 \mid \theta_{11}, \theta_{21}, \mathbf{Z})$.

   The second set of random observations $\{\theta_{12}, \theta_{22}, \theta_{32}\}$ is obtained as follows:

   θ_{12} is drawn from $f_1(\theta_1 \mid \theta_{21}, \theta_{31}, \mathbf{Z})$
   θ_{22} is drawn from $f_2(\theta_2 \mid \theta_{31}, \theta_{12}, \mathbf{Z})$
   θ_{32} is drawn from $f_3(\theta_3 \mid \theta_{12}, \theta_{22}, \mathbf{Z})$

   and so on.
3. Generate another N sets of random observations as in step 2 above to form a random sample of size N for $(\theta_1, \theta_2, \theta_3)$. Denote the random sample by

$$\left(\theta_1^{(1)}, \theta_2^{(1)}, \theta_3^{(1)}\right), \left(\theta_1^{(2)}, \theta_2^{(2)}, \theta_3^{(2)}\right), \ldots, \left(\theta_1^{(N)}, \theta_2^{(N)}, \theta_3^{(N)}\right). \qquad (11.4)$$

4. Estimate the posterior marginals from the random sample in (11.4).

Tiao and Tsay (1994) use Gibbs sampling to perform a Bayesian analysis of random variance shift model. For properties of Gibbs sampling, see Tierney (1994), Raftery and Lewis (1992), and the references cited therein.

11.3.2. Griddy Gibbs

Let ω_i be the ith element of Θ and $[b_0, b_1]$ be the support of ω_i. In practice, the support is determined by properties of the entertained model. For example, if ω_i denotes the lag 1 coefficient of an AR(1) model, namely, $z_t = \omega_i z_{t-1} + \epsilon_t$, then $[b_0, b_1] = [-1, 1]$ so that the process is not explosive. The conditional posterior distribution function of ω_i given the data, all the other parameters and prior distribution $p(\omega_i)$ is

$$p(\omega_i \mid \mathbf{Z}_n, \Theta_{(i)}) \propto p(\omega_i \mid \Theta_{(i)}) \prod_{t=1}^{n} N(f_t, g_t^2) \qquad (11.5)$$

where $\Theta_{(i)}$ denotes all the parameters in Θ except ω_i and f_t and g_t are functions of ω_i. The griddy Gibbs draws a realization of ω_i by the following procedure:

- Select a grid of m points in the support $[b_0, b_1]$ or for a subset of $[b_0, b_1]$.
- For each grid point, evaluate the conditional posterior distribution function of ω_i in (11.5).
- Draw a random realization of ω_i from the selected grid based on the values of the conditional posterior distribution function.

From this procedure, it is clear that the actual value of the normalization constant of the conditional posterior function is not needed in implementing the griddy Gibbs. The prior distribution $p(\omega_i \mid \Theta_{(i)})$ may assume many forms depending on the substantive information of the problem under study. It is clear, however, that a uniform prior simplifies the computation involved.

11.3.3. An illustrative example

Let us consider in details the Gibbs sampler used for the following simple bilinear model

$$z_t = \phi_0 + \phi_1 z_{t-1} - \theta a_{t-1} + \beta z_{t-1} a_{t-1} + a_t, \qquad a_t = \sigma \epsilon_t$$

where $\sigma > 0$ is the standard deviation of the innovation series a_t. The parameters of this model are $\Theta = (\phi_0, \phi_1, \theta, \beta, z_0, a_0, \sigma)'$, where z_0 is the starting value of the series and a_0 is the starting innovation. The Gibbs samples of these parameters can be drawn as follows:

- The two AR coefficients ϕ_0 and ϕ_1 can be drawn easily because they are linear parameters and have a closed-form formula when conjugate prior is used. Specifically, conditional on the other parameters, we can express the two AR parameters in a linear regression setup in a manner similar to that of an MA(1) model. Let $m_1 = \theta - \beta z_0$, $x_{11} = 1$ and $x_{21} = z_0$. For $t > 1$, define recursively $m_t = (\theta - \beta z_{t-1})m_{t-1}$, $x_{1t} = 1 + (\theta - \beta z_{t-1})x_{1,t-1}$, and $x_{2t} = z_{t-1} + (\theta - \beta z_{t-1})x_{2,t-1}$.

11.3. ESTIMATION

Furthermore, define $z_t^* = z_t - m_t$. Then, we have

$$z_t^* = \phi_0 x_{1t} + \phi_1 x_{2t} + a_t, \quad t = 1, \ldots, n.$$

Therefore, ϕ_0 and ϕ_1 can be drawn jointly by using the usual result of Gibbs sampling for linear regression model with conjugate prior.

- The variance parameter σ^2 can also be drawn by using the usual technique, because conditional on other parameters σ^2 has an inverted chi-square distribution under the normality assumption and conjugate prior.
- The starting value z_0 can be drawn again by using results of linear regression analysis. Specifically, define $z_1^* = z_1 - \phi_0 + \theta a_0$, $x_1^* = \phi_1 + \beta a_0$, and $z_t^* = z_t - \phi_0 - \phi_1 z_{t-1} + (\theta - \beta z_{t-1})z_{t-1}^*$ and $x_t^* = (\theta - \beta z_{t-1})x_{t-1}^*$ for $t > 1$. Then, we have

$$z_t^* = x_t^* z_0 + a_t, \quad t = 1, \ldots, n$$

and the result of Gibbs sampler for simple linear regression applies.

- Similarly, the starting innovation a_0 can be drawn by using the result of linear regression analysis. Define $z_1^* = z_1 - \phi_0 - \phi_1 z_0$, $x_1^* = -\theta + \beta z_0$ and $z_t^* = z_t - \phi_0 - \phi_1 z_{t-1} + (\theta - \beta z_{t-1})z_{t-1}^*$ and $x_t^* = (\theta - \beta z_{t-1})x_{t-1}^*$ for $t > 1$. Then, we obtain

$$z_t^* = x_t^* a_0 + a_t, \quad t = 1, \ldots, n,$$

which is a simple linear regression.

- Finally, the MA coefficient θ and the bilinear parameter β are nonlinear, and there exist no closed-form formulas to simplify the Gibbs draw. One possible approach to overcome this difficulty is to use the Metropolis algorithm. In this chapter, we use the griddy Gibbs approach. As mentioned before, for these two parameters, the individual conditional posterior distribution functions can be evaluated easily over a grid of finite points. For the MA coefficient θ, the support is $[-1, 1]$ whereas that of the bilinear parameter β must satisfy the condition $\phi_1^2 + \sigma^2 \beta^2 < 1$. See Liu (1989) for the stationarity condition of the bilinear model.

Note that in theory all parameters can be drawn by using the griddy Gibbs. However, it is desirable to use closed-form formulas whenever available and to draw several parameters jointly whenever possible. Drawing one parameter at a time using the griddy Gibbs could result in slow convergence of the sampler.

In the illustration above, all techniques used are not limited to the bilinear model. On the contrary, the estimation procedure used is widely applicable in linear and nonlinear time series analysis. Only the closed-form formulas and the likelihood function need to be changed when other models are entertained.

A potential weakness in using the griddy Gibbs is the specification of parameter support. For simple models, one can use the theoretical properties of the model such as stationarity, invertibility, or existence of some moments to select the supports. However, for high-dimensional models, the interdependence of the parameters may complicate the specification. In our implementation of the griddy Gibbs, we use an iterative method. We start with a relatively wide interval for a given parameter and refine the interval after some Gibbs iterations. In practice, this means one needs to run the Gibbs sampler several times in order to obtain estimates of a model. Given the advance in computing facilities and the gains in understanding the series over the iterations, we believe that this is not a serious drawback for the estimation method used. Furthermore, when the number of parameters is large, one can start with a sparse grid in the initial Gibbs iterations to reduce the computation in refining the specification of parameter supports.

11.4. MODEL DISCRIMINATION

In this section we consider the problem of model selection in nonlinear time series analysis, especially when the competing models are not nested. Such a model-selection problem is important because many classes of nonlinear models have been proposed in the literature and there exists no simple method to effectively discriminate one class of models from another. For example, both the TAR and bilinear models have been used to analyze the annual sunspot data with proponents claiming better fit for their model (Tong 1990, Gabr and Subba Rao 1981). Another example is that many ARCH-type models have been used to describe and predict the volatility of the monthly S&P 500 excess returns, and there is no agreement on which model is most appropriate.

Our approach to model discrimination is to let individual observations make their own choice of model. Consider the case of two competing nonnested nonlinear models. We use these competing models to define the functions $f(.)$ and $g(.)$ of model (11.1) and introduce a simple switching scheme that allows each individual observation to select its own model. Thus, under the approach adopted, the two competing models become submodels of a mixed model, and each individual observation can select its own submodel. In real applications, the dynamic structure of a time series cannot change abruptly over time. A structural change tends to occur gradually over a period of time. Therefore, it is reasonable to assume that the model selection of individual observations evolves over time in a smooth fashion. This consideration leads us to employ a simple switching scheme to govern the model selection. The selection results of individual observations provide information about which submodel is more appropriate for the data. This information can then be used to make model selection.

The idea behind the mixed model is simple. We believe that the issue of model discrimination exists only when the two competing nonnested nonlinear models fit the data well; otherwise, the selection is clear. Consequently, a better way to discriminate between models is to let each individual observation select its own model. Moreover, it is conceivable that certain portions of the data fit one model nicely whereas the remaining data fit the other model better. In this situation, the mixed model considered appears to be more appropriate.

One can also treat the mixed-model approach as a generalization of the odds ratios commonly used in Bayesian inference. In computing an odds ratio, we assume that *all* of the data points belong to the candidate model. On the other hand, under the mixed model, observations can belong to different models. Thus, the method used provides another level of flexibility over the odds ratio.

The switching scheme for model discrimination has been used in McCulloch and Tsay (1994) to test for "trend stationarity" versus "difference stationarity" of a linear time series and in George et al. (1996) to distinguish between fixed-coefficient versus random-coefficient autoregressive models. However, these two papers use Markov switching and only consider linear models for which closed-form formulas are available. The approach of Chen et al. (1997) is much more general as it can handle a wide range of linear and nonlinear models. It improves the procedure by treating starting values and innovations as parameters. This improvement could be significant in applications because it relaxes the assumption that the starting values and innovations are either fixed or equal to their expectation. For nonstationary series to which most real-world time series belong, the unconditional expection of a series might not exist.

The probabilistic mechanism of the mixed model can give rise to a large number of possible submodel configurations; for a given time series of length n, the possible number of submodel configurations is 2^n. In applications, these configurations might require intensive computation in model estimation. However, as illustrated in George and McCulloch (1993), McCulloch and Tsay (1994), and George et al. (1996), this computational difficulty can be overcome by using Gibbs sampling.

11.4.1. A mixed model with switching

The framework used for discriminating between two competing models is the two-state switching model:

$$z_t = \begin{cases} f_{1,t} + a_{1t}, & a_{1t} = g_{1,t}\epsilon_t & \text{if } s_t = 1 \\ f_{2,t} + a_{2t}, & a_{2t} = g_{2,t}\epsilon_t & \text{if } s_t = 2 \end{cases} \quad (11.6)$$

where $f_{i,t}$, $g_{i,t}$ and a_{it} are defined as in (11.1) and $\{s_t\}$ is a sequence of states. The state switching is governed by

$$P(s_t = i \mid s_{t-1} = s_{t+1} = i) = \eta, \qquad P(s_t = i \mid s_{t-1} \neq s_{t+1}) = 0.5. \quad (11.7)$$

Thus the switching depends on the two nearest neighbors of the observation in time. Each observation has a conditional probability η to stay with the same model as its neighbors. When the two adjacent neighbors are in different states, the probability of model switch is neutral at .5. Because structural changes tend to occur gradually over a period of time, a large η seems to be more realistic in application. If $\eta = 1$, then all observations come from one of the two competing models. In this chapter, we use η

close to 1. We consider values of η in the interval [.95, .9999] to study the sensitivity of model selection with respect to the choice of η. Alternatively, one could put a hyperprior on η that has most of the probability mass on values close to 1. Including the large set of parameters s_t in model (11.6) makes it very flexible. In application strong prior information (large η) is needed to get reasonable results.

One way to appreciate the implication of η is to consider the independence case in which model change is independent over time. In this case, there are 8 possible model configurations for every three consecutive observations and the probability that all three observations belong to the same model is only $\eta/4$, which is less than .25. Thus, the chance of model change is substantial when η is not large. Of course, the independence assumption is an extreme case and is often unrealistic in application. Our discussion is meant only to justify the use of a large value for η.

There are many ways to describe the transition of model selection from one observation to another, ranging from independent Bernoulli trials to complicated dynamic mechanism. Our choice of (11.7) is based on several considerations. First, the transition is very flexible; it covers a wide range of possibilities by varying η. For example, $\eta = 0.5$ corresponds to independent Bernoulli trials with probability .5, and $\eta = 1$ implies that change can occur only when two neighbors belong to different models. Second, it is easy to use because the user only needs to specify a single parameter. In the traditional two-state Markov switching model, one needs to specify two parameters for the probability transition matrix. Third, the equation is intuitively appealing. It easily reflects the common sense of smooth model change. Fourth, the scheme can be extended to involve other neighboring systems, including two observations prior and after the observation. Such a specification would provide an alternative way to specify strong prior information that nearby observations are likely to come from the same model.

11.4.2. Implementation

Model selection is based on the posterior distribution of the parameters s_t in model (11.6). This posterior is computed in the obvious way by using Gibbs sampling and drawing the s_t's given the parameters of both models and then drawing the parameters of the individual models given the s_t values in a manner similar to that outlined in Section 11.3.

To use our method for model discrimination, we consider the following procedure:

1. For each submodel we specify prior distributions for the model parameters and then use all the data and the estimation method of Section 11.3 to obtain estimates (typically posterior means).
2. Choose a value for η and perform a Gibbs estimation of the mixed model in (11.6). Initial values for the submodel parameters are obtained from step 1. Step 1 also provides an initial choice of the support and grid for each parameter drawn using the griddy method.

11.5. EXAMPLES

3. Check the convergence of Gibbs sampler. Refine and iterate Gibbs sampler if necessary.
4. Use the posterior distribution of model selection of individual observations to make inference.

Once the posterior distribution of individual selection is available, one can make inference of model selection based on the objective of the analysis. For example, if the objective is forecasting, one may pay more attention to the model selection of observations close to the forecast origin. If the objective is the dynamic structure of the data, then posterior mean or median can be used for overall selection. It is conceivable that the data might not have sufficient information to distinguish one competing model from another. In this case, one might search for more data or for ways to further improve the model. It would be unwise to assume that a statistical method can always distinguish two competing models based on a finite sample of observations.

Finally, it is important in practice to study the sensitivity of model selection with respect to prior specification such as η and to check the convergence of the Gibbs sampler. By varying priors and the number of iterations and starting values of the Gibbs sampler, one can learn the stability of model selection.

11.5. EXAMPLES

We illustrate the unified approach to estimating and modeling time series by some simulated and real examples. For the simulated examples, we consider AR versus MA models and TAR versus bilinear models. We analyse two real data sets. We compare TAR and bilinear models for the annual sunspot numbers and ARCH(2) and GARCH(1,1) models for monthly excess returns of the S&P 500 stock market portfolio.

Example 11.1. Figure 11.1a shows a time plot of 300 observations generated from the model

$$z_t = \begin{cases} .8z_{t-1} + a_t & \text{if } t = 101, \ldots, 200 \\ a_t + .3a_{t-1} + .4a_{t-2} & \text{if } t = 1, \ldots, 100; 201, \ldots, 300 \end{cases}$$

where $a_t = .5\epsilon_t$, $a_0 = a_{-1} = 0$ and $z_0 = 0$. This is a mixed model with two change points at $t = 101$ and $t = 201$. The model change at $t = 201$ can be seen in Figure 11.1a, but that at $t = 101$ is not obvious. Our goal here is to illustrate the performance of the approach adopted in estimation and model selection. For the AR model, the parameter vector is $\hat{}_1 = (\phi_1, z_0, \sigma_1)'$, where ϕ_1 is the lag 1 AR coefficient and σ_1 is the standard deviation of the AR innovation. For the MA model, the parameter vector is $\hat{}_2 = (\theta_1, \theta_2, a_0, a_{-1}, \sigma_2)'$, where θ_i are MA coefficients, a_i are starting innovations and σ_2 denotes the standard deviation of the MA innovation.

298 BAYESIAN TIME SERIES ANALYSIS

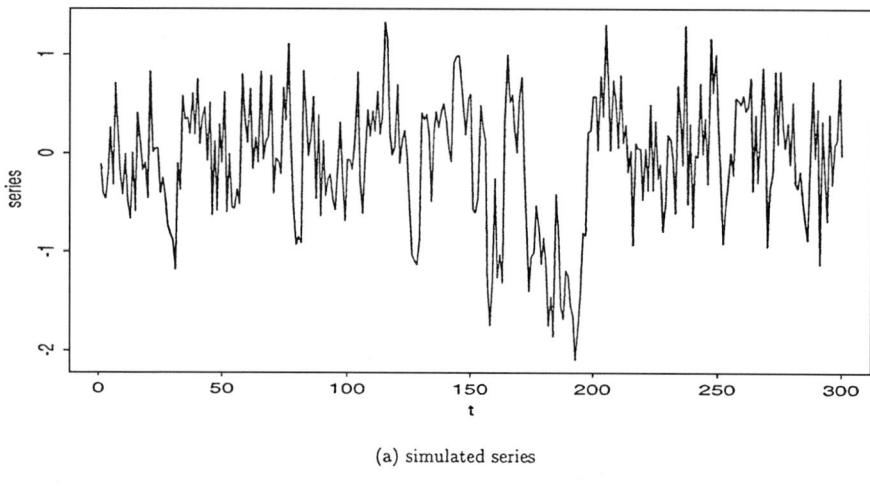

(a) simulated series

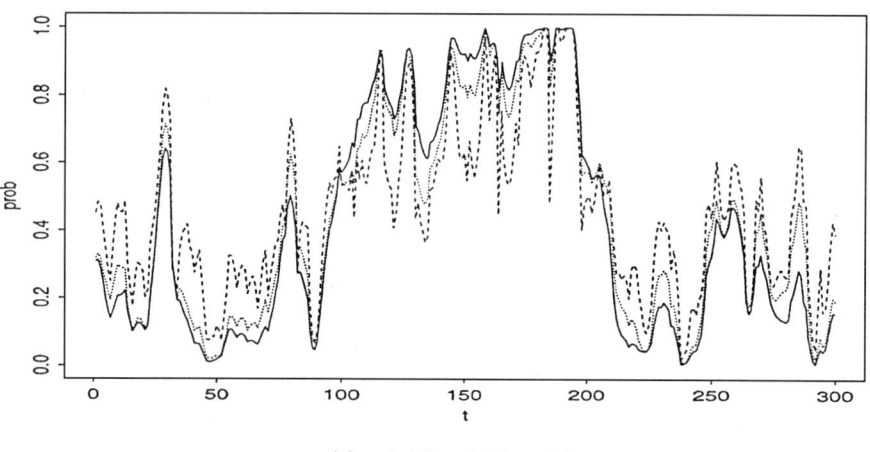

(b) probability of MA model

FIGURE 11.1 Time plot and posterior probabilities for Example 11.1.

Following the procedure in Section 11.4.2, we began with Gibbs samples for each model, assuming that all of the data belong to that model, to obtain initial parameter estimates and the initial parameter supports for the griddy Gibbs. The prior for any parameter drawn by griddy Gibbs is the uniform distribution over its interval support. The initial Gibbs samples used 300 iterations. Using results of the initial Gibbs samples and a given η for conditional switching probability, we ran 2500 Gibbs iterations to obtain posterior distribution of individual model selection. The estimated posterior probabilities are based only on the last 2000 iterations. This step of ignoring the first 500 Gibbs iterations was taken to reduce the effect of initial parameter specification. Figure 11.1b shows the posterior mean of selecting the MA model for each individual observation for $\eta = .95, .99, .995$, respectively. The solid line is for $\eta = .995$ and the dotted line for $\eta = .99$. The effect of η on model selection is seen from the three posterior probabilities. As expected, $\eta \geq .99$ works better and is preferred. For this example, it is seen that the adopted model selection method works reasonably well. It points out clearly the two change points and is able to identify the generating model. For estimation, the posterior distributions of the parameters are well behaved and are centered roughly around the true values.

Example 11.2. In this example, we generated 300 observations from the mixed model

$$z_t = \begin{cases} \begin{cases} .8z_{t-1} + a_{1t} & \text{if } z_{t-1} \geq 0 \\ -.8z_{t-1} + a_{1t} & \text{if } z_{t-1} < 0 \end{cases} & \text{if } t = 1, \ldots, 100; 201, \ldots, 300 \\ .5z_{t-1} + .2z_{t-1}a_{2,t-1} + a_{2t} & \text{if } t = 101, \ldots, 200 \end{cases}$$

where $z_0 = a_0 = 0$, $a_{1t} = 0.5\epsilon_t$, and $a_{2t} = 0.3\epsilon_t$. This is a mixture of TAR and bilinear models with two change points at $t = 101$ and $t = 201$. The TAR model has two regimes separated by the threshold variable z_{t-1} at threshold $r = 0$. In each regime, the model is AR(1). The bilinear model used contains a single bilinear term $.2z_{t-1}a_{t-1}$ and is referred to as a "diagonal" bilinear model. A special feature of such a bilinear model is that the mean of the series is nonzero, even though there is no constant term in the model. Properties of diagonal bilinear models are more complicated than those of nondiagonal bilinear models. (Guegan 1994).

The data of this example are shown in Figure 11.2a. Even a careful reading of the plot cannot reveal the two change points easily. In our analysis, we assume that the threshold variable z_{t-1} is known, but the threshold r is unknown. Thus, the parameter vector for the TAR submodel is $\hat{}_1 = (r, \phi_1^{(1)}, \phi_1^{(2)}, \sigma_1, z_0)'$ where r denotes the threshold, $\phi_1^{(i)}$ is the AR(1) coefficient of the ith regime, and σ_1 is the innovational standard deviation. For the bilinear submodel, the parameter vector is $\hat{}_2 = (\phi, \beta, \sigma_2, z_0, a_0)'$, where ϕ and β are the AR and bilinear coefficient, respectively; σ_2 denotes the standard deviation of innovations; and z_0 and a_0 denote the starting value and innovation, respectively. In sum, there are 10 parameters in the mixed model used for model selection.

Following the procedure of Section 11.4.2 and using essentially the same Gibbs steps and numbers of iterations as those of Example 11.1, we obtain the posterior probability of selecting the bilinear model for each observation. These probabilities are shown in Figure 11.2b for $\eta = .95, .99, .999$. It is seen that the procedure works well for the data, especially for $\eta = .999$. The two change points and the generating model are clearly identified. The result for $\eta = .95$ given by the dashed line in Figure 11.2b shows some model uncertainty at some patches of observations. This is reasonable because the prior probability for model change in this case is substantial.

Example 11.3. In this example, we generated 300 observations all from the TAR submodel of Example 11.2. However, we assume that the bilinear submodel is another

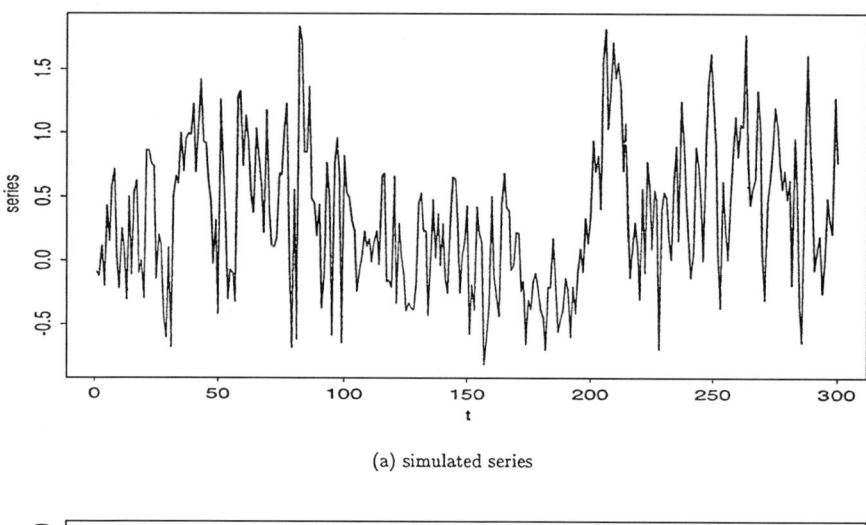

(a) simulated series

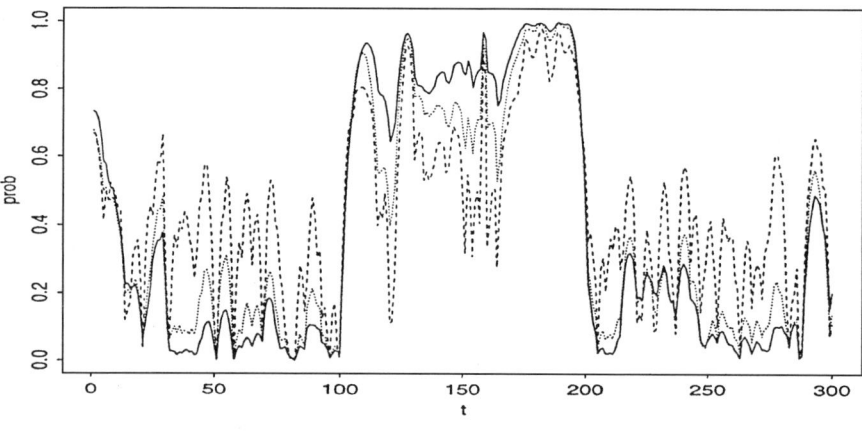

(b) probability of bilinear model

FIGURE 11.2 Time plot and posterior probabilities for Example 11.2.

competing model and apply the procedure to discriminate between these two models. Figure 11.3a shows the data, whereas Figure 11.3b gives the posterior probabilities of selecting the bilinear model by the individual observations. These probabilities were obtained by using the same starting values, the same Gibbs steps and iterations as those of Example 11.2, except that the probability of conditional model switching is set at $\eta = .99, .999,$ and $.9999$, respectively. From the probability plot, it is seen that the adopted procedure indeed selects the generating model for the data, especially when η is close to 1. The case of $\eta = .99$ shows some model uncertainty, even though only some isolated points have posterior probability greater than .5 for the bilinear model. This example thus shows that the prior specification of η should be close to 1 in applications, say, $\eta \geq .99$.

11.5. EXAMPLES

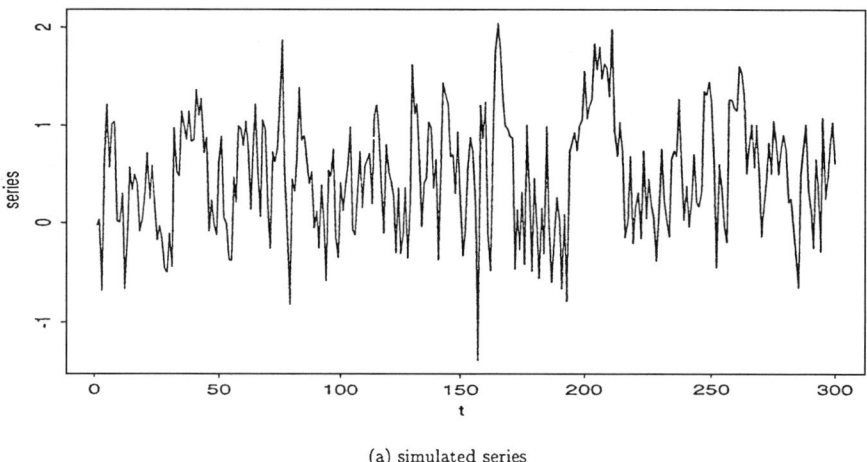

(a) simulated series

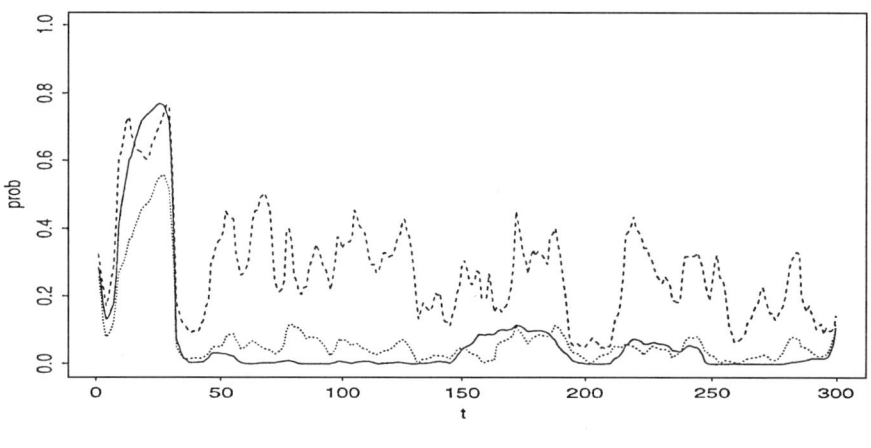

(b) probability of bilinear model

FIGURE 11.3 Time plot and posterior probabilities for Example 11.3.

Example 11.4. In this example, we consider the annual Wolf sunspot number from 1700 to 1979 for 280 observations. The data shown in Figure 11.4a are listed in Tong (1990) and have been widely used in nonlinear time series analysis. It is generally believed that this series is nonlinear, but there is no agreement on which nonlinear model is most appropriate for the data. When the subsample from 1700 to 1921 was used, Gabr and Subba Rao (1981) identified a bilinear model for the series whereas Tong (1990) specified a two-regime TAR model. Li (1993) applied a test statistic, which uses the idea of separate families of hypotheses of Cox (1962), to the subsample and concluded that the bilinear model of Gabr and Subba Rao is more appropriate. However, from a theoretical viewpoint, bilinear models do not possess the asymmetric feature between rise and fall of the cyclical pattern observed

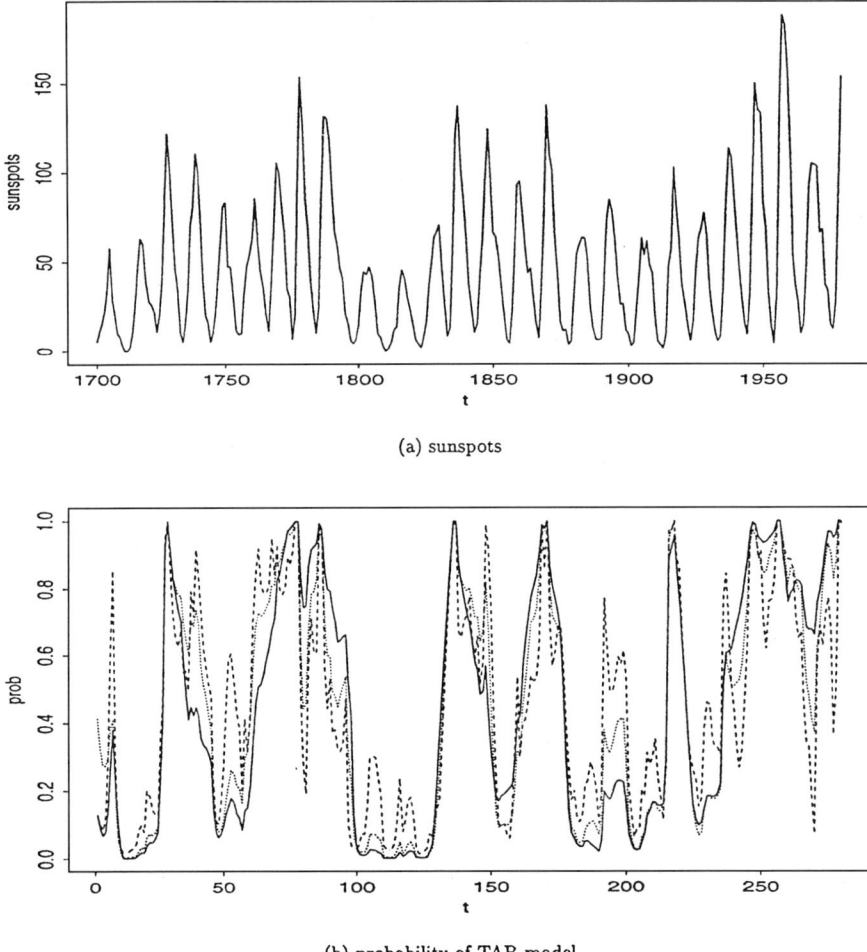

FIGURE 11.4 Time plot and posterior probabilities for annual sunspot numbers.

in the sunspot number. On the other hand, the TAR is capable of producing asymmetric cycle, but it has larger residual variance in the subsample. The issue of model selection remains.

Our analysis here is to apply the model discrimination procedure to the full sample, assuming that the bilinear model of Gabr and Subba Rao (1981) and the TAR of Tong (1990) as two competing models. The bilinear model considered assumes the form

$$z_t = \alpha_0 + \alpha_1 z_{t-1} + \alpha_2 z_{t-2} + \alpha_9 z_{t-9} + \beta_{21} z_{t-2} a_{t-1} + \beta_{81} z_{t-8} a_{t-1} + \beta_{13} z_{t-1} a_{t-3}$$
$$+ \beta_{43} z_{t-4} a_{t-3} + \beta_{16} z_{t-1} a_{t-6} + \beta_{24} z_{t-2} a_{t-4} + \beta_{32} z_{t-3} a_{t-2} + a_t \quad (11.8)$$

11.5. EXAMPLES

where $a_t = \sigma \epsilon_t$. Besides the 12 parameters shown in equation (11.8), this bilinear model also needs 9 starting values $z_0, z_{-1}, \ldots, z_{-8}$ and 6 starting innovations $a_0, \ldots, a_{-5}$. In total, estimation of this bilinear model considers 27 parameters some of which are highly nonlinear. As shown by the simpler bilinear example of Section 11.3, we can estimate this bilinear model via the Gibbs sampler considered. The Gibbs draws of the nonlinear parameters can be done by the griddy Gibbs.

The threshold model built by Tong (1990) is

$$z_t = \begin{cases} \phi_0^{(1)} + \sum_{i=1}^{3} \phi_i^{(1)} z_{t-i} + a_t^{(1)} & \text{if } z_{t-3} \leq r \\ \phi_0^{(2)} + \sum_{i=1}^{11} \phi_i^{(2)} z_{t-i} + a_t^{(2)} & \text{if } z_{t-3} > r \end{cases} \quad (11.9)$$

where $a_{it} = \sigma_i \epsilon_t$. Including innovational standard deviation, this TAR model contains 5 parameters in regime 1 and 13 parameters in regime 2. Counting the threshold r and 11 starting values $z_0, z_{-1}, \ldots, z_{-10}$, we are effectively estimating 30 parameters for the TAR model. Except for the threshold r, all of the parameters has closed-form formulas and can be drawn easily. Conditioned on other parameters, the threshold r becomes a change point of the data. Gibbs draws of r, therefore, can be done by either the griddy Gibbs or the method in Carlin et al. (1992a).

Again, we follow the procedure in Section 11.4.2 to carry out the model selection. Because of the large number of parameters involved, we used 3500 Gibbs iterations for this example, but discarded results of the first 500 iterations in computing the posterior probabilities. Figure 11.4b shows the posterior probabilities of selecting the TAR model by the individual observations, where the solid, dashed, and dotted lines are for $\eta = .99, .999$, and $.9999$, respectively. In our analysis, we carried out many Gibbs samples and found that the posterior probability plot is stable. From the plots, it is seen that the data do not strongly favor a single model. For certain periods, the TAR was preferred. But for other periods, the bilinear model was selected. It seems that the data are not sufficiently informative to discriminate between these two competing models. However, the TAR model appears to be the choice of model by the most recent observations. This is in good agreement with the results of forecasting comparison in Tong (1990, Section 7.3) who showed that the TAR model produced better out-of-sample forecasts of the sunspot numbers for the latter part of the data.

The fact that the data were not very informative in choosing a single model is understandable. First, the two competing models entertained contain many parameters, making them rather flexible and capable of providing good fit in finite samples. In this circumstance, one might need a large number of observations to distinguish one model from the other. Second, there exists the possibility that neither of the two competing models is appropriate for the data. This is evident in the posterior probability plot of Figure 11.4b where the TAR model was preferred when the sunspot number was high and the bilinear model was chosen when the sunspot number was low. In addition, our residual analysis shows that the normalized residuals of the mixed model has lag 1 serial correlation, even though the correlation is relatively weak.

In summary, the procedure used does not pinpoint a single model for the annual sunspot number from 1700 to 1979. However, it produces results that are reasonable

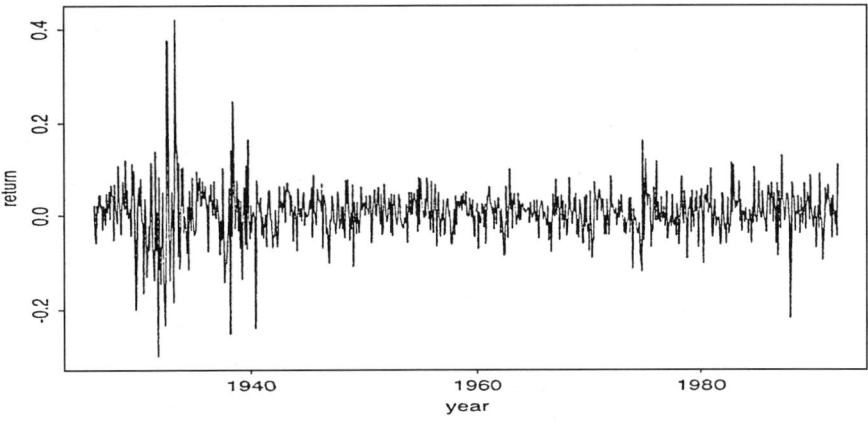

(a) Monthly SP500 excess returns

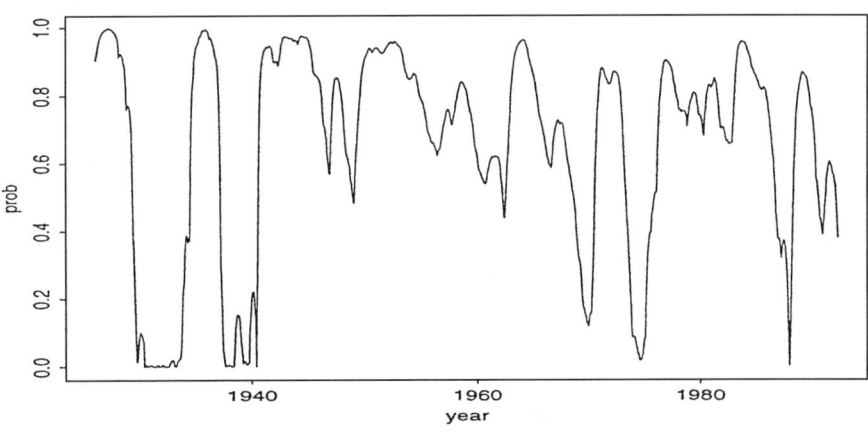

(b) probability of ARCH(2) model

FIGURE 11.5 Time plot and posterior probabilities for monthly S&P 500 returns.

and in agreement with those available in the literature. It would be unwise to expect that a model selection method can always select a single model based on a finite number of observations. One must take into consideration the possibility that there exists no true model for a real-world time series. When the data are not sufficiently informative, a good model-selection procedure should be able to reveal it. In this sense, the adopted model-discrimination method appears to be reasonable.

Example 11.5. Figure 11.5a is a time plot of monthly excess returns of the S&P 500 portfolio from January 1926 to December 1991 giving 792 observations. This series has been widely analyzed in volatility studies, but there is little agreement on what is the most appropriate model for the data. Our goal here is to compare between ARCH(2) and GARCH(1,1) models for the data. The model considered is

$$z_t = \begin{cases} \theta_0 + a_{1t}, & a_{1t} = g_{1,t}\epsilon_t, & g_{1,t}^2 = \gamma_0 + \gamma_1 a_{1,t-1}^2 + \gamma_2 a_{1,t-2}^2 & \text{if } s_t = 1 \\ \beta_0 + a_{2t}, & a_{2t} = g_{2,t}\epsilon_t, & g_{2,t}^2 = \alpha_0 + \alpha_1 g_{2,t-1}^2 + \beta_1 a_{2,t-1}^2 & \text{if } s_t = 2. \end{cases}$$

The state switching is governed by equation (11.7) with $\eta = .999$. Figure 11.5(b) plots the posterior probability of the ARCH(2) model based on the Gibbs sampler with 500 initial iterations and 4000 general iterations. The mean of the posterior probabilities is .65, so that the overall fit is slightly in favor of the ARCH(2) model. On the other hand, by comparing Figures 11.5a and 11.5b, the GARCH(1,1) model was selected by most observations that appear to be volatile. Thus, our result indicates that the evidence of GARCH(1,1) model reported in the literature is due largely to the few visibly volatile periods of the U.S. economy. While such a conclusion is understandable, the analysis used does highlight the influential periods for using GARCH(1,1) model. This shows that the adopted model discrimination procedure can be used to monitor the evolution of the time series under study.

In the estimation, we used various constraints to ensure that the two submodels have proper unconditional variances. For instance, we require $\alpha_0 > 0$, $0 \leq \alpha_1 + \beta_1 < 1$, $\alpha_1 \geq 0$, and $\beta_1 \geq 0$ so that the GARCH(1,1) model is not integrated. Such constraints are easy to implement under the adopted unified approach.

ACKNOWLEDGMENT

This research is supported in part by the National Science Foundation and the Graduate School of Business, University of Chicago.

REFERENCES

Ansley, C. F. (1979). An algorithm for the exact likelihood of a mixed autoregressive moving average process. *Biometrika* **66**, 59–65.

Bollerslev, T. (1986). Generalized autoregressive conditional heteroscedasticity. *J. Econo.* **31**, 307–327.

Box, G. E. P., Jenkins, G. M., and Reinsel, G. C. (1994). *Time Series Analysis: Forecasting and Control*, 3rd ed. Prentice-Hall, Englewood Cliffs, NJ.

Brockwell, P. J. and Davis, R. A. (1991). *Time Series: Theory and Methods*, 2nd ed. Springer-Verlag, New York.

Carlin, B. P., Gelfand, A. and Smith, A. F. M. (1992a). Hierarchical Bayesian analysis of change point problems. *Appl. Stat.* **41**, 389–405.

Carlin, B. P., Polson, N. G., and Stoffer, D. S. (1992b). A monte carlo approach to nonnormal and nonlinear state-space modeling. *J. Am. Stat. Assoc.* **87**, 493–500.

Chen, C. W. S., McCulloch, R. E., and Tsay, R. S. (1997). A unified approach to estimating and modeling linear and nonlinear time series. *Statistica Sinica* **7**, 451–472.

Chib, S. and Greenberg, E. (1994). Bayes inference in regression models with ARMA(p,q) errors. *J. Econo.* **64**, 183–206.

Cox, D. R. (1962). Further results on test of separate families of hypotheses. *J. Roy. Stat. Soc. B* **24**, 406–424.

Engle, R. F. (1982). Autoregressive conditional heteroscedasticity with estimates of the variance of United Kingdom inflation. *Econometrica* **50**, 987–1007.

Gabr, M. M. and Subba Rao, T. (1981). The estimation and prediction of subset bilinear time series models with applications. *J. Time Ser. Anal.* **2**, 155–171.

George, E. I. and McCulloch, R. E. (1993). Variable selection via Gibbs sampling. *J. Am. Stat. Assoc.* **88**, 881–889.

George, E. I., McCulloch, R. E., and Tsay, R. S. (1996). Two approaches to Bayesian model selection with applications. In D. A. Berry, K. M. Chaloner, and J. K. Geweke, (eds.), *Bayesian Analysis of Statistics and Econometrics*, Wiley, New York.

Granger, C. W. J. and Andersen, A. P. (1978). *Introduction to Bilinear Time Series Models*. Vandenhoeck & Ruprecht, Göttingen.

Guegan, D. (1994). *Series Chronologiques Nonlinear a Temps Discret*. Economica, Paris.

Hillmer, S. C. and Tiao, G. C. (1979). Likelihood function of stationary multiple autoregressive moving average models. *J. Am. Stat. Assoc.* **74**, 652–660.

Jones, R. H. (1980). Maximum likelihood fitting of ARMA models to time series with missing observations. *Technometrics* **22**, 389–395.

Li, H. and Tsay, R. S. (1998). A unified approach to identifying multivariate time series models. *Jo. Am. Stat. Assoc.* **93**, 770–782.

Li, W. K. (1993). A simple one degree of freedom test for non-linear time series model discrimination. *Statistica Sinica* **3**, 245–254.

Liu, J. (1989). A simple condition for the existence of some stationary bilinear time series. *J. Time Ser. Anal.* **10**, 33–39.

Marriott, J., Ravishanker, N., Gelfand, A., and Pai, J. (1996). Bayesian analysis of ARMA processes: complete sampling-based inference under exact likelihoods. In D. A. Berry, K. M. Chaloner, and J. K. Geweke (eds.), *Bayesian Analysis of Statistics and Econometrics*. Wiley, New York.

McCulloch, R. E. and Tsay, R. S. (1994). Bayesian inference of trend- and difference-stationarity. *Econo. Theory* **10**, 596–608.

Monahan, J. F. (1983). Fully Bayesian analysis of ARMA time series models. *J. Econ.* **21**, 307–331.

Priestley, M. B. (1981). *Spectral Analysis and Time Series*. Academic Press. Orlando, FL.

Raftery, A. E. and Lewis, S. M. (1992). One long run with diagnostics: implementation strategies for Markov chain Monte Carlo. *Stat. Sci.* **7**, 493–497.

Shumway, R. H. and Stoffer, D. S. (1982). An approach to time series smoothing and forecasting using the EM algorithm. *J. Time Ser. Anal.* **3**, 253–264.

Spall, J. C. (1988), *Bayesian Analysis of Time Series and Dynamic Models*. Marcel Dekker, New York.

Subba Rao, T. (1981). On the theory of bilinear time series models. *J. Roy. Stat. Soc. B* **43**, 244–255.

REFERENCES

Tanner, M. A. (1991). *Tools for Statistical Inference*. Springer-Verlag, New York.

Tiao, G. C. and Tsay, R. S. (1994). Some advances in nonlinear and adpative modeling in time series. *J. Forecasting* **13**, 109–131.

Tierney, L. (1994). Markov chains for exploring posterior distributions. *Ann. Stat.* **22**, 1701–1762.

Tong, H. (1978). On a threshold model. In C. H. Chen (ed.), *Pattern Recognition and Signal Processing*. Sijhoff & Noordhoff, Amsterdam.

Tong, H. (1990). *Nonlinear Time Series: A Dynamical System Approach*. Oxford Univ. Press, Oxford.

Tsay, R. S. (1987). Conditional heteroscedasticity in time series analysis. *J. Am. Stat. Assoc.* **82**, 590–604.

West, M. and Harrison, P. J. (1986). Monitoring and adaptation in Bayesian forecasting models. *J. Am. Stat. Assoc.* **81**, 741–750.

CHAPTER 12

Nonparametric Time Series Analysis: Nonparametric Regression, Locally Weighted Regression, Autoregression, and Quantile Regression

Siegfried Heiler
Universität Konstanz

12.1. INTRODUCTION

In this chapter we discuss the application of some nonparametric techniques to time series. There is indeed a long tradition in applying nonparametric methods in time series analysis, and this holds true not only for certain test situations, such as runs tests for randomness of a stochastic sequence, permutation tests, or certain rank tests.

In Chapter 2 of this book the periodogram is introduced. Although the periodogram is an asymptotically unbiased estimate of the spectral density of an underlying stationary process, it is well known that it is not consistent. Therefore, as early as the 1950s, smoothing the periodogram directly with a so-called spectral window or using a system of weights, according to a lag window with which the empirical autocovariances are multiplied in the calculation of the Fourier transform, was introduced. Quite a number of different windows were proposed, and with respect to the window width similar rules hold for achieving consistent estimates, such as the ones we will shortly discuss in the context of nonparametric regression later in this text. Nonparametric spectral estimation is extensively treated in many textbooks on time series analysis, to which the interested reader is referred. It will not be treated in this chapter.

A Course in Time Series Analysis, Edited by Daniel Peña, George C. Tiao, and Ruey S. Tsay.
ISBN 0-471-36164-X. © 2001 John Wiley & Sons, Inc.

Another area where nonparametric ideas have been applied for a long time is smoothing and decomposing seasonal time series. Local polynomial regression can be traced back to 1931 (Macaulay 1931). Fisher (1937) and Jones (1943) discussed a local least-squares fit under the side condition that a locally constant periodic function (for modeling seasonal fluctuations) be annihilated, and later Bongard (1960) developed a unified principle for treating the interior and the boundary part (with and without seasonal variations) of a time series derived from a local regression approach. These ideas will be taken up later again in Section 12.8, since they represent an attractive alternative to smoothing and seasonal decomposition procedures based on linear time series models.

The aim of this chapter is to present some basic concepts of nonparametric regression, including locally weighted regression with special emphasis on their application to time series. Nonparametric regression has now become an area with an abundance in new methodological proposals and developments. It is therefore impossible to give a comprehensive survey on the subject in a chapter of a general textbook on time series. We will hence concentrate on the basic ideas only. The reader interested in more details may be refered to a survey paper by Härdle, et al. (1997), where more specific areas, proposals, and further references can be found.

The ARMA model is a typical linear time series model. In Chapters 10, 11, and 13, we encounter specific nonlinear models. ARCH type models and their variants are also of a very specific nonlinear type to capture volatility phenomena. In contrast to that in nonparametric regression, no assumption is made about the form of the regression function. Only some smoothness conditions are required. The complexity of the model will be determined completely by the data. One lets the data speak for themselves, thereby avoiding subjectivity in selecting a specific parametric model. But the gain in flexibility has a price. One has to choose bandwidths. We return to this later. In addition, a higher complexity in the mathematical argumentation is involved. However, asymptotic considerations will not be discussed in detail in this chapter.

Because of their flexibility, nonparametric regression techniques may serve as a first step in the process of finding an adequate parametric model. If no model can be found that describes the underlying structure adequately, then the results of nonparametric estimation may be used directly for forecasting or describing the characteristics of the time series.

12.2. NONPARAMETRIC REGRESSION

Since forecasting is an important objective of many time series significant analyses, estimating the conditional distribution or some of its characteristics plays a considerable role. For point prediction the conditional mean or median is of particular interest. In order to obtain confidence or prediction intervals, estimates of conditional variances or conditional quantiles are needed. The latter are also of interest in studying volatility in financial time series. The first step is therefore to look at nonparametric estimation of densities and conditional densities. Let $x \in \mathbb{R}$ be a random variable

TABLE 12.1. Selected Kernel Functions

Name	Kernel
Uniform	$\frac{1}{2}\Pi_{[-1,1]}(u)$
Triangle	$(1-|u|)\Pi_{[-1,1]}(u)$
Epanechnikov	$\frac{3}{4}(1-u^2)\Pi_{[-1,1]}(u)$
Bisquare	$\frac{15}{16}(1-2u^2+u^4)\Pi_{[-1,1]}(u)$
Triweight	$\frac{35}{32}(1-3u^2+3u^4-u^6)\Pi_{[-1,1]}(u)$
Gaussian	$\frac{1}{\sqrt{2\pi}}\exp\left(-\frac{1}{2}u^2\right)$

whose distribution has a density f, and let $x_1, \ldots, x_n$ be a random sample from x. Then a *kernel density estimator* for f is given by

$$f_n(x) = \frac{1}{nh_n} \sum_{i=1}^{n} K\left(\frac{x_i - x}{h_n}\right). \qquad (12.1)$$

Here K is a so-called *kernel* function, which is a symmetric density assigning weights to the observations x_i that decrease with the distance between x and x_i. Some popular kernel functions are listed in Table 12.1 and exhibited in Fig. 12.1. The first five have the interval $[-1, 1]$ as support, whereas the Gaussian kernel has infinite support. The term h_n is the *bandwidth* which drives the size of the local neighborhood included in the estimation of f at x. The bandwidth depends on the sample size n and has to fulfil $h_n \to 0$ and $nh_n \to \infty$ for $n \to \infty$ as necessary condition for consistency. But for practical applications this asymptotic condition is not very helpful. A very small bandwidth will lead to a wiggly course of the estimated density, whereas a large bandwidth yields a smooth course but will possibly flatten out interesting details. Bandwidth selection will be dealt with in Section 12.7.

A k_n *nearest-neighbor* $(k_n - NN)$ *estimator* of f is obtained by substituting the fixed bandwidth h_n in (12.1) by the random variable $H_{n,k_n}(x)$ measuring the distance between x and the k_n nearest observation among the x_i, $i = 1, \ldots, n$. Nearest-neighbour estimators have the property that the number of observations used for the local approach is fixed. This is an advantage if the x space shows a greatly unbalanced design. On the other hand, the bias varies from point to point because of the variable local bandwidth.

For $x \in \mathbb{R}^p$ a kernel $K : \mathbb{R}^p \to \mathbb{R}$ is needed in (12.1). In this case either *product kernels*

$$K(u) = \prod_{j=1}^{d} K_j(u_j)$$

with kernels K_j and $K_j : \mathbb{R} \to \mathbb{R}$, bandwidth h_j in coordinate j, and $h_n = h_1 \cdots h_p$

12.2. NONPARAMETRIC REGRESSION

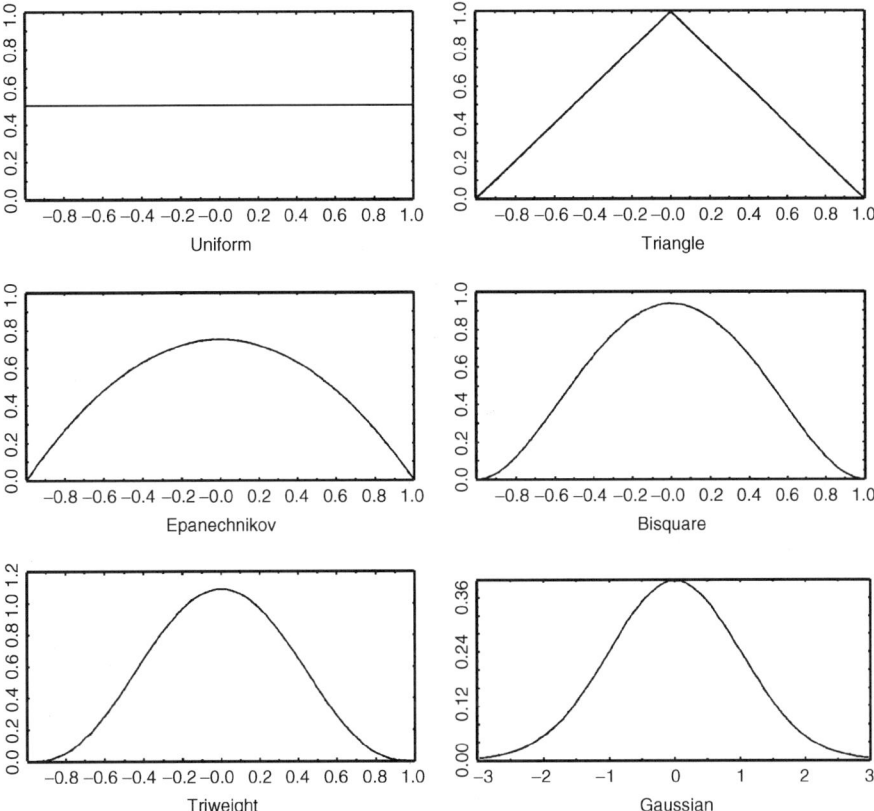

FIGURE 12.1 Some popular kernel functions in practice.

or *norm kernels*

$$K(u) = K(\|u\|)$$

with a suitable norm on $\mathbb{R}^p$ are used. In connection with time series applications frequently *product kernels* are applied,

$$f_n(x) = \frac{1}{n} \sum_{i=1}^{n} \prod_{j=1}^{p} \frac{1}{h_j} K_j \left(\frac{x_{ij} - x_j}{h_j} \right) \qquad (12.2)$$

and $h_j = \hat{\sigma}_j \cdot h$ with an estimated standard deviation in the jth coordinate is a popular choice for the bandwidths.

Now let (y, x) with $y \in \mathbb{R}, x \in \mathbb{R}^p$ be a random vector with joint density $f(y, x)$ and let $f_X(x)$ be the marginal density of x. Then the conditional density

$g(y|x) = f(y,x)/f_X(x)$ can be estimated by inserting a kernel density estimator or a corresponding nearest-neighborhood estimator in the nominator and denominator of $g(y|x)$. With the choice of a kernel function

$$K = \mathbb{R}^{p+1} \to \mathbb{R}, \, K(y,x) = K_1(y)K(x)$$

and bandwidths h_1 resp. (respectively) h we obtain the kernel estimator for the conditional density

$$g_n(y|x) = \frac{h_1^{-1} \sum_{i=1}^n K_1\left(\frac{y_i-y}{h_1}\right) K\left(\frac{x_i-x}{h}\right)}{\sum_{i=1}^n K\left(\frac{x_i-x}{h}\right)}. \tag{12.3}$$

An estimator for the *conditional mean* $m(x) = \int_{-\infty}^{\infty} y g(y|x) dy$ is obtained when we replace g in the integral by its estimator g_n. For K_1 a symmetric density this immediately yields

$$m_n(x) = \frac{\sum_{i=1}^n y_i K\left(\frac{x-x_i}{h}\right)}{\sum_{i=1}^n K\left(\frac{x-x_i}{h}\right)}. \tag{12.4}$$

This is the well-known *Nadaraya–Watson nonparametric regression estimator* (NW estimator), (Nadaraya 1964, Watson 1964). We see that it can be written as a weighted mean

$$m_n(x) = \sum_{i=1}^n y_i w_{n,i}(x; x_1, \ldots, x_n) \tag{12.5}$$

where the random weights depend on the point x and the random variables $x_1, \ldots, x_n$. Apart from conditional means, conditional quantiles are also of interest in various time series applications. Let

$$F(y|x) = \int_{-\infty}^{y} g(y|x) dy \tag{12.6}$$

denote the conditional distribution function of y given x. Then the conditional α quantile at x, $q_\alpha(x)$ is defined as

$$q_\alpha(x) = \inf\{y \in \mathbb{R} \mid F(y|x) \geq \alpha\}, \quad 0 < \alpha < 1. \tag{12.7}$$

If $g(\cdot|x)$ is strictly positive, then, of course, $q_\alpha(x)$ is the unique solution of $F(y|x) = \alpha$, that is, $q_\alpha(x) = F^{-1}(\alpha|x)$. One possible procedure for estimating q_α is to take the empirical α quantile of an estimator $F_n = (\cdot|x)$ according to (12.7).

Let $F_1(z) = \int_{-\infty}^{z} K_1(u) du$ be the distribution function pertaining to the kernel K_1. Then the estimated conditional distribution, obtained by integrating $g_n(\cdot|x)$ from

12.2. NONPARAMETRIC REGRESSION

$-\infty$ to y, is given by

$$F_n(y \mid x) = \frac{\sum_{i=1}^n K\left(\frac{x_i - x}{h}\right) F_1\left(\frac{y - y_i}{h_1}\right)}{\sum_{i=1}^n K\left(\frac{x_i - x}{h}\right)}. \tag{12.8}$$

Let us assume that K_1 has support $[-1, 1]$. Then we have

$$F_1\left(\frac{y - y_i}{h_1}\right) = \begin{cases} 1 & \text{for} \quad y_i \leq y - h_1 \\ 0 & \text{for} \quad y_i \geq y + h_1 \end{cases}$$

so that in this case

$$F_n(y \mid x) = \frac{1}{\sum_{i=1}^n K\left(\frac{x_i - x}{h}\right)} \Bigg\{ \sum_{i=1}^n \mathbf{1}_{(-\infty, y - h_1]}(y_i) K\left(\frac{x_i - x}{h}\right) \\ + \sum_{i=1}^n \mathbf{1}_{(y - h_1, y + h_1)}(y_i) F_1\left(\frac{y - y_i}{h_1}\right) K\left(\frac{x_i - x}{h}\right) \Bigg\}. \tag{12.9}$$

One can see that the estimation contains only observations in the regressor space laying in a band around x. The first sum on the RHS includes observations, whose y values are less than or equal to $y - h_1$. The second sum contains observations with y_i values in a neighborhood of y. In contrast to a usual empirical distribution function here, observations greater than y obtain a positive weight.

Of particular interest may be the median regression function $q_{1/2}$ for asymmetric distributions as an alternative to ordinary regression based on the mean. Another interesting application may be the estimation of $q_{\alpha/2}$ and $q_{1-\alpha/2}$ in order to get predictive intervals. These can be compared with intervals obtained from parametric models, which lack the possibility to evaluate the bias due to misspecification of the model.

Taking some boundary corrections into account, for a not-too-unbalanced design the second sum in (12.9) can be approximated by $\sum_{i=1}^n \mathbf{1}_{(y - h_1, y]} K[(x_i - x)/h]$, so that the conditional distribution function is estimated by

$$\tilde{F}_n(y \mid x) = \frac{\sum_{i=1}^n \mathbf{1}_{(-\infty, y]}(y_i) K\left(\frac{x_i - x}{h}\right)}{\sum_{i=1}^n K\left(\frac{x_i - x}{h}\right)}. \tag{12.10}$$

This estimator was for $x \in \mathbb{R}$ considered by Horvath and Yandell (1988), who proved asymptotic results for the iid case. Abberger (1996) derives from (12.10) the empirical quantile function

$$q_{n,\alpha}(x) = \inf\{y \in \mathbb{R} \mid \tilde{F}_n(y \mid x) \geq \alpha\}, \qquad 0 < \alpha < 1 \tag{12.11}$$

and investigates the behavior of $\tilde{F}_n$ and $q_{n,\alpha}$ in applications to stationary time series.

12.3. KERNEL ESTIMATION IN TIME SERIES

When a kernel or nearest-neighbor (NN) estimator is applied to dependent data, as is the case in time series, it is affected only by the dependence among the observations in a small window and not by that between all data. This fact reduces the dependence between the estimates, so that many of the techniques developed for independent data can be applied in these cases as well. This fact was called "the whitening by windowing principle" by Hart (1996). A typical situation for an application to a time series $\{z_t\}$ is that the regressor vector x consists of past time series values

$$x_t = (z_{t-1}, \ldots, z_{t-p}), \tag{12.12}$$

which leads to the very general nonparametric autoregression model

$$z_t = m(z_{t-1}, \ldots, z_{t-p}) + a_t, \quad t = p+1, p+2, \ldots \tag{12.13}$$

with $\{a_t\}$ a white-noise sequence. Of course x_t might also include time series values of other predictive variables such as leading indicators.

An indispensable requirement for proving asymptotic properties of kernel estimates in this and related situations is that the underlying processes are stationary. Another condition is that the memory of these underlying processes decreases with distance between events and that the rate of decay can be estimated from above by so-called *mixing conditions*. So-called strong mixing conditions are used by Robinson (1983, 1986). Collomb (1984, 1985) worked with so-called ϕ or uniform mixing conditions. We will not present these fairly complicated asymptotic considerations here. But we would like to remark that these mixing conditions are hard to check in practice.

In Chapter 3 we encountered the linear autoregressive model $z_t = \phi_1 z_{t-1} + \cdots + \phi_p z_{t-p} + a_t$, and in Chapter 10 threshold autoregression is discussed where the autoregressive parameters vary according to some threshold variable. In contrast to these examples, the model (12.13) is much more general and flexible, and its estimation may lead to insights that can be helpful in choosing an appropriate parametric (possibly nonlinear) model afterward.

For $x \subset \mathbb{R}^p$, x_t as in (12.12) and weights

$$w_{n,t} = \frac{K\left(\frac{x_t - x}{h}\right)}{\sum_{s=p+1}^{n} K\left(\frac{x_s - x}{h}\right)}$$

the Nadaraya-Watson estimator in model (12.13) is given by

$$m_n(x) = \sum_{s=p+1}^{n} z_t w_{n,t}(x). \tag{12.14}$$

12.3. KERNEL ESTIMATION IN TIME SERIES

For x equal to the last observed pattern, $x = (z_n, z_{n-1}, \ldots, z_{n-p+1})'$ this provides a one-step-ahead predictor for z_{n+1} that allows a very intuitive interpretation. Given the course of the time series observed over the last p instants, the predictor is a weighted mean of all those time series values in the past, which followed a course pattern that is similar to the last observed one. The weights depend on how close the pattern observed in the past comes to the pattern given by $(x_n, \ldots, x_{n-p+1})'$.

A *k-step-ahead predictor* is given if z_t in (12.14) is replaced by z_{t-k+1}:

$$m_{n,k} = \sum_{t=p+1}^{n-k+1} z_{t+k-1} w_{n,t}(x), \qquad k = 1, 2, \ldots. \tag{12.15}$$

This predictor does not use the variables $z_{n+1}, \ldots, z_{n+k}$, which are unknown, but may contain information about the conditional expectation $E(z_{n+k} \,|\, (z_n, \ldots, z_{n-p+1})')$. They might be replaced by estimates in a multistep procedure that consists in a succession of one-step-ahead forecasts. This procedure can lead to a smaller mean-squared error than the multistep procedure (12.15). For a different proposal, see Chen (1996).

Up to now we have considered only the autoregressive case where the regressor vector contains past time series values. The case of vector autoregression, where for each individual (scalar) time series past values of related time series or leading indicators are also included in the regression vector, can be treated in a similar way as nonparametric autoregression, although the number of components in x is restricted due to the *curse of dimensionality*, which we discuss later.

If the regressor vector $x_t = (z_{t-1}, \ldots, z_{t-p})'$ is used in estimating conditional distribution functions and conditional quantiles, such as in (12.10) and (12.11), then we arrive at *quantile autoregression*. The median autoregression $q_{n,1/2}$ may serve as an alternative to the mean autoregression (12.14). In financial data one is often interested in the behavior of quantiles in the tails. For instance, the *value at risk* of a certain asset is measured by looking at low quantiles ($\alpha = 0.01$ or $\alpha = 0.05$) of the conditional distribution of the corresponding series of returns.

Abberger (1996) applied quantile autoregression to time series of daily stock returns. In order to assess such models, forecast error cannot serve as a criterion, since quantiles are not observable. Abberger proposed the criterion

$$\xi_\alpha = \frac{1 - \sum_{t=1}^{n} \rho_\alpha(z_t - q_\alpha(x_t))}{\sum_{t=1}^{n} \rho_\alpha(z_t - q_\alpha)} \tag{12.16}$$

where

$$\rho_\alpha(u) = \alpha \mathbf{1}_{[0,\infty)}(u) u + (\alpha - 1) \mathbf{1}_{(-\infty,0)}(u) u \tag{12.17}$$

is the loss function introduced by Koenker and Bassett (1978) in their seminal paper on quantile regression and q_α is the unconditional α-quantile of the corresponding distribution.

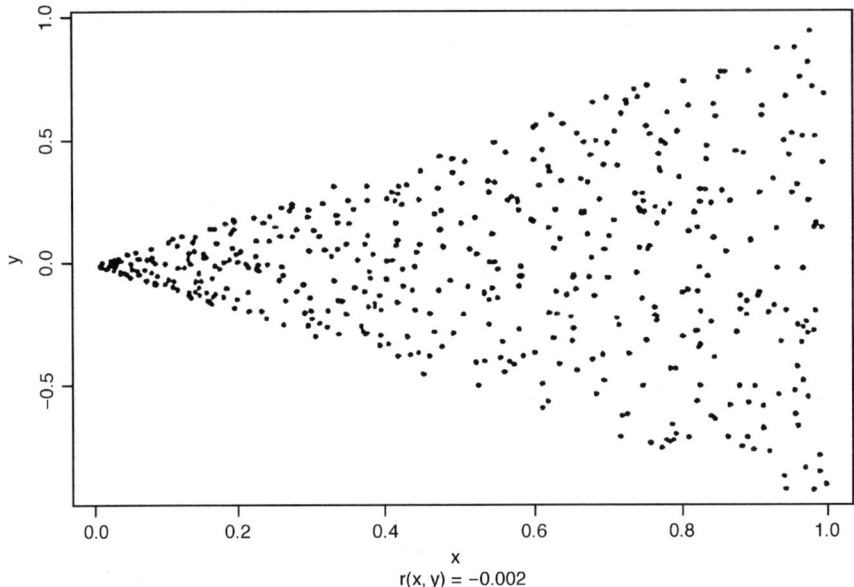

FIGURE 12.2 Simulated heteroskedastic data, $n=500$.

The term ξ_α is constructed according to the R^2 criterion in ordinary regression. It assumes values between zero and one, where $\xi_\alpha = 0$ if $q_\alpha(x_t) = q_\alpha$ for all x_t and $\xi_\alpha = 1$ if $z_t = q_\alpha(x_t)$ for all t and all α, that is, if the distribution of $\{z \mid x\}$ is a one-point distribution. Figure 12.2 and Table 12.2 illustrate the behavior of ξ_α with a simulated conical data set of 500 observations.

The observations are heteroscedastic and have mean zero. The correlation between x and y is -0.002. In Table 12.2 empirical ξ_α values for different α are exhibited. They are calculated by replacing in (12.16) $q_\alpha(x_t)$ by its kernel estimator $q_{n,\alpha}(x_t)$ and q_α by the empirical unconditional quantile of the first $t - 1$ data values $z_1, \ldots, z_{t-1}$. The latter can be interpreted as a naive forecast of $q_\alpha(x_t)$.

The findings of Abberger (1996, 1997) for several German stock returns were ξ_α values close to zero for the median and increasing in a U-shaped form toward the boundary areas around $\alpha = .01$ resp. $\alpha - .99$.

In Chapter 9 ARCH and GARCH models are introduced. They represent a very specific kind of parametric modeling for studying the phenomenon of volatility. A flexible alternative to the combination of an ARMA model with ARCH or GARCH residuals is given by the *conditional heteroscedastic autoregressive nonlinear* (CHARN-)

TABLE 12.2. ξ_α **Values for the Data in Figure 12.2**

α	0.01	0.05	0.10	0.25	0.50	0.75	0.90	0.95	0.99
ξ_α	0.43	0.36	0.27	0.10	0.01	0.11	0.26	0.34	0.41

12.3. KERNEL ESTIMATION IN TIME SERIES

model

$$z_t = m(x_t) + \sigma(x_t)\xi_t \tag{12.18}$$

studied by Härdle and Yang (1996) or Härdle et al. (in press). Here $x_t = (z_{t-1}, \ldots, z_{t-p})'$ is again the autoregressive vector (12.12), ξ_t is a random variable with mean zero and variance one; $\sigma^2(x)$ is called the *volatility function*. Given an estimator for m, e. g. the NW estimator m_n according to (12.14), it was suggested that $\sigma^2(x)$ can be estimated by

$$\sigma_n^2(x_t) = g_n(x_t) - m_n^2(x_t) \tag{12.19}$$

where

$$g_n(x) = \frac{\sum_{t=1}^n K\left(\frac{x_t-x}{h}\right) z_t^2}{\sum_{t=1}^n K\left(\frac{x_t-x}{h}\right)} = \sum_{t=1}^n z_t^2 w_{n,t}(x). \tag{12.20}$$

Since the estimator (12.19) is based on a difference, a negative variance estimator may occasionally result. This can be avoided if the volatility function is estimated on the basis of residuals. See (12.51), the discussion in text surrounding it, and Feng and Heiler (1998a).

In the context of time series analysis not only past values of the time series itself or of related series may occur as regressor variables, but also the time index itself, in which case $x_t = t$, or some functions of the time index like polynomials or trigonometric functions. This leads to smoothing approaches. In the case $m(x_t) = m(t)$ the NW estimator at t consists in a weighted mean of the time series values in a neighborhood $[t-h, t+h]$ of z_t with nonrandom weights. Polynomials and trigonometric functions in t are used in decomposing a seasonal time series into trend-cyclical and seasonal components according to an unobserved components model. This application will be studied in Section 12.8 after the discussion of locally weighted regression.

In the area of quantile estimation the regressor $x_t = t$ leads to quantile smoothing. This technique was used by Abberger (1996, 1997) in order to compare the results of a nonparametric procedure for stock returns with those of a GARCH model, evaluated with an *S–Plus* package under the standard assumption of an underlying Gaussian distribution. As an example, we take daily discrete DAX returns, defined as $z_t = (price_t - price_{t-1})/price_{t-1}$, exhibited in Figure 12.3.

Since the Gaussian distribution is completely determined by mean and variance, conditional quantiles can easily be calculated from the outcomes of the GARCH model estimation. The results are depicted in Figures 12.4 and 12.5 for the lower and upper quartiles and for the .1 and .9 quantiles, respectively. Two messages can be learned from the results. The first is that the asymmetric behavior of volatility, which is revealed by the nonparametric approach, will remain completely hidden by the choice of a wrong parametric model that is being offered as the default option by the package. In the presented example, which is not untypical for stock returns, volatility

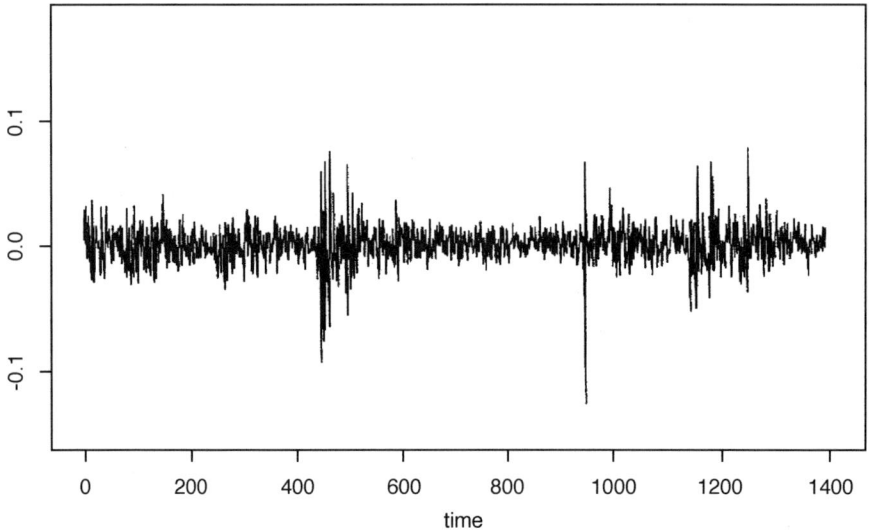

FIGURE 12.3 Time series of daily DAX returns from Jan. 2, 1986 to Aug. 13, 1991.

is a phenomenon that has mainly to do with movements in the lower tails of the conditional distributions. The second finding in the figures is that kernel smoothing is very robust toward aberrant and erratic observations in the course of the time series, whereas GARCH models react very sensitively to them.

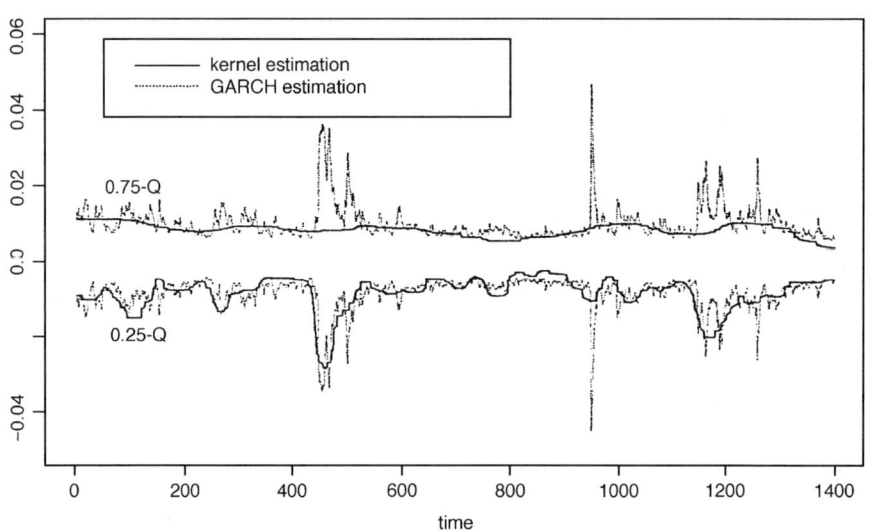

FIGURE 12.4 Estimation of 0.25 and 0.75 quantiles of daily DAX returns.

12.4. PROBLEMS OF SIMPLE KERNEL ESTIMATION AND RESTRICTED APPROACHES

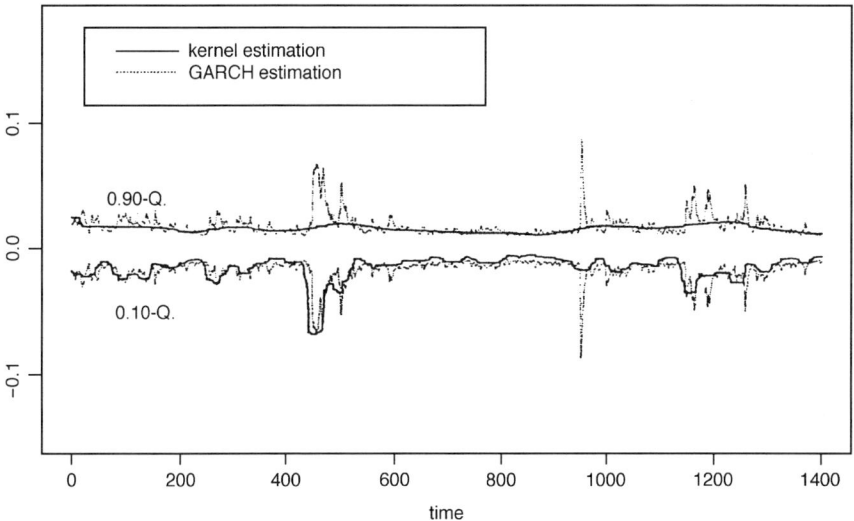

FIGURE 12.5 Estimation of 0.10- and 0.90-quantiles of daily DAX returns.

12.4. PROBLEMS OF SIMPLE KERNEL ESTIMATION AND RESTRICTED APPROACHES

The nonparametric approaches we have treated so far suffer from two drawbacks. One is the so-called curse of dimensionality; the other is increased bias in cases of a highly clustered design density and particularly at the boundaries of the x space. *Curse of dimensionality* describes the fact that in higher-dimensional regression problems the subspace of $\mathbb{R}^{p+1}$ spanned by the data is rather empty, that is, there are only few observations in the neighborhood of a point $x \in \mathbb{R}^p$. In practice, this happens to be the case already for $p > 2$.

Several proposals have been made to cope with the curse-of-dimensionality problem. We will describe only two of them very shortly. The first consists in decomposing $\mathbb{R}^p$ into a class of J disjoint course patterns, A_j, $j = 1, \ldots, J$, with the aid of a non-hierarchical cluster analysis. These J disjoint sets serve then as the states of a homogeneous Markov chain. In the model

$$m(x_t) = \mathrm{E}[z_t \mid x_t \in A_j] \text{ for } x_t \in A_j, j = 1, \ldots, J$$

where x_t is the autoregressive vector (12.12), m is estimated by

$$m_n(x_t) = N_j^{-1} \sum_{s=1}^{n} z_s \mathbf{1}_{A_j}(x_s)$$

where N_j is the number of course patterns of length p from the time series in A_j. Here the estimator is an unweighted mean of all values following courses in pattern class A_j. Markov chain models of this type were first used by Yakowitz (1979b) for analyzing time series of water runoff in rivers. Asymptotic properties for this type of model are discussed by Collomb (1980, 1983).

Gourieroux and Monfort (1992) examined a corresponding model for economic time series by incorporating volatility. They called their model

$$z_t = \sum_{j=1}^{J} \alpha_j \mathbf{1}_{A_j}(x_t) + \sum_{j=1}^{J} \beta_j \mathbf{1}_{A_j}(x_t)\xi_t$$

a *qualitative threshold ARCH model*. For further discussion of Markov chain models we also refer to Chapter 12.

Another proposal in order to cope with the curse of dimensionality is given by the so-called generalized additive models, studied by Hastie and Tibshirani (1990), which are defined as

$$z_t = m_0 + \sum_{j=1}^{p} m_j(z_{t-i_j}) + a_t.$$

The components m_j are again of a general form. For estimation, the so-called backfitting algorithms such as the *alternating conditional expectation algorithm* (ACE) of Breiman and Friedman (1985) or the BRUTO algorithm of Hastie and Tibshirani (1990) may be used. The main idea of backfitting goes as follows. In the model above, $E[z_t - m_0 - \sum_{j \neq k} m_j(z_{t-i_j})] = m_k(z_{t-i_k})$. Hence the variable in square brackets can be used to obtain a nonparametric estimate for $m_k(z_{t-i_k})$. Of course, the other m_j are unknown as well, so that the estimation procedure has to be iterated until all the $m_{n,j}$ converge. For a more detailed study of generalized additive models, the reader is referred to the book by Hastie and Tibshirani as well as to the two interesting papers by Chen and Tsay (1993) (1993a, 1993b). For further discussion and other approaches, see also Härdle et al. (1997). Quite a few proposals can be found in the literature dealing with the bias problem of NW estimators close to the boundary and in cases of an unbalanced design in the x space. Gasser and Müller (1979, 1984) suggested for the case $p = 1$ a system of variable weights, Gasser et al. (1985) developed asymmetric boundary kernels, and Messer and Goldstein (1993) suggested variable kernels that automatically become deformed and thus reduce the bias in the boundary area.

Yang (1981) and Stute (1984) suggested a symmetrized k-NN estimator, and Michels (1992) proposed boundary kernels for bias reduction that can be carried over to the case $p > 1$. We do not discuss the abovementioned proposals in more detail since the disadvantages mentioned earlier can be repaired by using locally weighted regression.

12.5. LOCALLY WEIGHTED REGRESSION

Locally weighted respectively local polynomial regression was introduced into the statistical literature by Stone (1977) and Cleveland (1979). The statistical properties were investigated since then in papers by Tsybakov (1986), Fan (1993), Fan and Gijbels (1992, 1995), Ruppert and Wand (1994), and many others. A detailed description may be found in the book of Fan and Gijbels (1996).

For the sake of simplicity we start with the assumption that the regressor x is a scalar. For a better understanding we regard the data as being generated by a location-scale model

$$y = m(x) + \sigma(x)\xi \tag{12.21}$$

akin to the one considered in (12.18), where the ξ are independent with $E(\xi) = 0$, $Var(\xi) = 1$ and $m(x_0) = E(y \mid x = x_0)$. m is assumed to be smooth in the sense that the $(p+1)$th derivative exists at x_0, so that it can be expanded in a Taylor series around x_0

$$m(x) = m(x_0) + (x - x_0)m'(x_0) + \cdots + (x - x_0)^r \frac{m^{(r)}(x_0)}{r!} + R_r(x) \tag{12.22}$$

with the remainder term

$$R_r(x) = (x - x_0)^{r+1} m^{(r+1)}(x_0 + \theta(x - x_0))/(r-1)!, \; 0 < \theta < 1. \tag{12.23}$$

With

$$\beta_j(x_0) = \frac{m^{(j)}(x_0)}{j!}, \quad j = 0, 1, \ldots, r \tag{12.24}$$

we arrive at a local polynomial representation for m:

$$m(x) \approx \sum_{j=0}^{r} \beta_j(x_0)(x - x_0)^j. \tag{12.25}$$

This approach motivates the nonparametric estimation of m as a local polynomial by solving the least-squares problem:

$$\min_{\beta \in \mathbb{R}^{r+1}} \left\{ \sum_{i=1}^{n} \left[y_i - \sum_{j=0}^{r} (x_i - x)^j \beta_j \right]^2 K\left(\frac{x_i - x}{h}\right) \right\}.$$

With the design matrix X_x having the n rows $[1, x_i - x, \ldots, (x_i - x)^r]$, the diagonal weight matrix $W_x = \text{diag}\{K[(x_i - x)/h]\}$ and the vector $y = (y_1, \ldots, y_n)'$, the solutions at x is given by

$$\hat{\beta}(x) = (X'_x W_x X_x)^{-1} X'_x W_x y, \qquad (12.26)$$

and where e_j is the jth unit vector in $\mathbb{R}^{r+1}$. We see immediately that

$$\hat{m}(x) = \hat{\beta}_0 = e'_1 (X'_x W_x X_x)^{-1} X'_x W_x y \qquad (12.27)$$

and that with

$$\hat{m}^{(j)}(x) = \hat{\beta}_j(x) j! = j! e'_{j+1} (X_x W_x X_x)^{-1} X'_x W_x y, \qquad j = 1, \ldots, r \qquad (12.28)$$

an estimator for the jth derivative of m is given.

The case $r = 0$ yields the Nadaraya–Watson estimator (12.14). Let $u = (r_r(x_1))_{i=1}^n$ be the residual vector containing the remainder terms according to (12.23) at the data points. Then the conditional bias of $\hat{\beta}(x)$ is given by

$$B(\hat{\beta}(x)) = (X'_x W_x X_x)^{-1} X'_x W_x u$$

and with $\Sigma_x = W(x)^2 \text{diag}(\sigma^2(x_i))$, its conditional covariance matrix is

$$\text{Var}(\hat{\beta}(x)) = (X'_x W_x X_x)^{-1} (X'_x \Sigma_x X_x)(X'_x W_x X_x)^{-1}.$$

These last two expressions cannot be used directly since they contain the unknown vector u of remainder terms and the unknown diagonal matrix Σ_x.

A first-order asymptotic expansion of the variance and the bias term uses the moments of K and K^2, denoted by

$$\mu_j = \int u^j K(u) du \qquad \text{and} \qquad v_j = \int u^j K^2(u) du,$$

which are contained in the matrices

$$S = (\mu_{j+l})_{0 \le j,l \le r}, \qquad \tilde{S} = (\mu_{j+l+1})_{0 \le j,l \le r}, \qquad S^* = (v_{j-l})_{0 \le j,l \le r}$$

and the vectors $c_r = (\mu_{r+1}, \ldots, \mu_{2r+1})$, $\tilde{c}_r = (\mu_{r+2}, \ldots, \mu_{2r+2})$. For an iid sample $(y_1, x_1), \ldots, (y_n, x_n)$ with the marginal density $f(x) > 0$ and with $f, m^{(r+1)}$ and σ^2 continous in a neighborhood of x, we obtain for $h \longrightarrow 0$ and $nh_n \longrightarrow \infty$ the asymptotic conditional variance

$$\text{Var}(\hat{m}^{(j)}(x)) = e'_{j+1} S^{-1} S^* S^{-1} e_{j+1} \frac{(j!)^2 \sigma^2(x)}{f(x) n h^{1+2j}} + o_p \left(\frac{1}{n h^{1+2j}} \right). \qquad (12.29)$$

12.5. LOCALLY WEIGHTED REGRESSION

For the asymptotic conditional bias we have to distinguish between the cases where $r - j$ is odd and where $r - j$ is even. For $r - j$ odd, we have

$$\text{Bias}\big(\hat{m}^{(j)}(x)\big) = e'_{j+1} S^{-1} c_r \frac{j!}{(r+1)!} m^{(r+1)}(x) h^{r+1-j} + o_p(h^{r+1-j}). \quad (12.30)$$

For $(r - j)$ even, the asymptotic bias is

$$\text{Bias}\big(\hat{m}^{(j)}(x)\big) = e'_{j+1} S^{-1} \tilde{c}_r \frac{j!}{(r+2)!} \cdot$$

$$\left\{ m^{(r+2)}(x) + (r+2) m^{(r+1)}(x) \frac{f'(x)}{f(x)} \right\} h^{r+2-j} + o_p(h^{r+2-j}) \quad (12.31)$$

provided that f' and $m^{(r+2)}$ are continuous in a neighborhood of x and $nh^3 \longrightarrow \infty$. As a very interesting fact, we notice the difference in asymptotic bias between $r - j$ odd and $r - j$ even. For instance, we have for the NW estimator ($r = 0, j = 0$)

$$B(m_n(x)) = h^2 \left[\frac{m''(x)}{2} + \frac{m' f'(x)}{f(x)} \right] \mu_2 + o_p(h^2),$$

whereas for the local linear approach we obtain

$$B(\hat{m}(x)) = \frac{h^2 m''(x) \mu_2}{2} + o_p(h^2).$$

We see that the bias of the local linear estimator has a simpler structure. The linear term in the bias expansion vanishes, whereas the expression for the variance is the same in both cases and given by $v_0 \sigma^2(x)/nh$. The bias of the NW estimator does not only depend on m', but also on the score function $-f'/f$. This is the reason why an unbalanced design leads to an increased bias.

Similar considerations hold for higher-order polynomials. In practice, this means that for estimating m it is sufficient to consider $r = 1$ or $r = 3$, and for m' only $r = 2$ or $r = 4$ should be considered. In many applications $r = j + 1$ suffices. Fitting a higher-order polynomial will possibly reduce the bias, but on the other hand the variance will increase since more parameters have to be estimated locally.

If the regressor x is a vector rather than a scalar, in most cases a local linear approach is chosen since in this case the step from $r = 1$ to $r = 3$ leads to a strong increase of parameters to be estimated locally, which entails an unacceptable increase in variance. Since

$$\hat{\beta}_j(x) = e'_{j+1} \hat{\beta} = e'_{j+1} (X'_x W_x X_x)^{-1} X'_x W_x y = \sum_{i=1}^{n} w^j_{ni} \left(\frac{x_i - x}{h} \right) y_i \quad (12.32)$$

for estimating $\beta_j(x) = m^{(j)}(x)/j!$, we have a similar expression as a weighted mean such as that for the NW estimator (12.14). The weights depend on the observations x_i and on the location of x in the design space.

It can be seen easily that the weights $w_{ni}^j(u_t) = w_{ni}^j[(x_i - x)/nh]$ satisfy the discrete moment conditions

$$\sum_{i=1}^{n}(x_i - x)^q w_{ni}^j\left(\frac{x_i - x}{h}\right) = \delta_{jq} \quad \text{with } 0 \le j, q \le r.$$

As a consequence of this, the sample bias for estimating a polynomial with degree less than or equal to r is zero.

The variance of $\hat{m}^{(j)}(x)$ is given by

$$\text{Var}\left(\hat{m}^{(j)}(x)\right) = \sum_{i=1}^{n} w_{ni}^j\left(\frac{x_i - x}{h}\right)^2 \sigma^2(x_i).$$

The kernel with the weights $w_{ni}^j(u_t)$ is called the *active kernel*.

A first-order approximation to the w_{ni}^j is given if $(X_x' W_x X_x)$ is replaced by the kernel moments matrix S. The corresponding kernel

$$\tilde{K}^{(j)}(u) = e_{j+1}' S^{-1}(1, u, \ldots, u^r)' K(u) \tag{12.33}$$

is called the *equivalent kernel*. It satisfies the corresponding moment conditions

$$\int u^q \tilde{K}(j)(u) du = \delta_{jq} \quad 0 \le j, q \le r. \tag{12.34}$$

For instance, for the case $r = 1$, $j = 0$, we have $\tilde{K}(u) = K(u)$, and for $r = 2$, $j = 1$ (estimation of m'), $\tilde{K}^{(1)}(u) = \mu_2^{-1} u K(u)$. This means that for estimating m itself in the interior of the x space the effective kernel is equal to the chosen symmetric kernel function itself, whereas for estimating the first derivative, $\tilde{K}^{(1)}$ is a skew function. As a general result, $\tilde{K}^{(j)}$ is symmetric for j even and skew for j odd.

In terms of equivalent kernels, the asymptotic conditional variance and the asymptotic conditional bias (for $r - j$ odd) are

$$\text{Var}\left(\hat{m}^{(j)}(x)\right) = \frac{(j!)^2 \sigma^2(x)}{f(x) n h^{1+2j}} \int \tilde{K}^{(j)2}(u) du + o_p(nh^{-1-2j}), \tag{12.35}$$

$$\text{Bias}\left(\hat{m}^{(j)}(x)\right) = \frac{j!}{(r+1)!} m^{(r+1)}(x) h^{r+1-j} \int u^{r+1} \tilde{K}^{(j)}(u) du + o_p(h^{-r-1+j}). \tag{12.36}$$

The major advantage of local polynomial regression over other smoothing methods

12.5. LOCALLY WEIGHTED REGRESSION

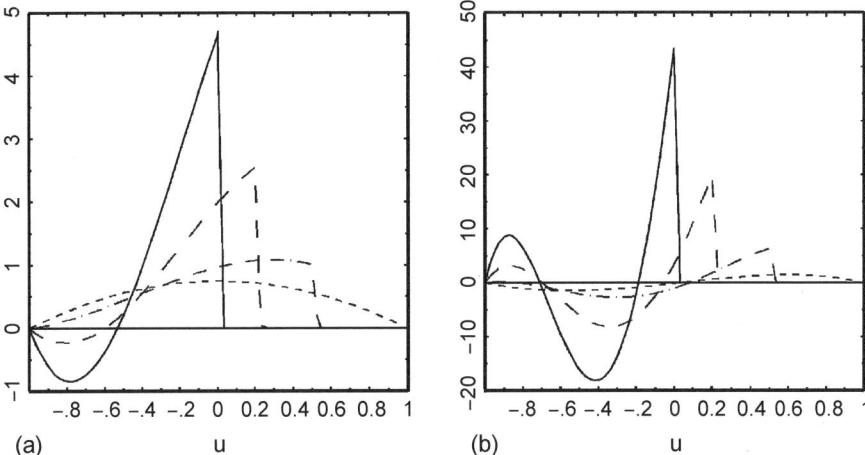

FIGURE 12.6 Active kernels derived from the Epanechnikov kernel with $nh = 30$ at the right boundary for (a) $r = 1$, $j = 0$ and (b) $r = 2$, $j = 1$. Estimation at interior points (short dashes), at $x = x^* - 15$ (dashes and points), at $x^* - 6$ (long dashes) and at the boundary point x^* (solid line).

consists in the automatic adaptation of the active resp. equivalent kernel to the estimation situation in the boundary area. If x is scalar and $x_* = \min(x_i)$, $x^* = \max(x_i)$, then for a given bandwith h the interior of the x space is given by all observations in the interval $[x_* + h, x^* - h]$. For all x in this interval the equivalent kernels $\tilde{K}^{(j)}$ have the abovementioned symmetry resp. asymmetry property. In the left boundary part $[x_*, x_* + h]$ the number of left neighbors in a local neighborhood of a point x will be small compared to the number of right neighbors, and for $x = x_*$, we have only right neighbors. Corresponding considerations hold for the right boundary part $[x^* - h, x^*]$. For $x \in \mathbb{R}^p$, $(p > 1)$ the boundary area will often cover an important part of the whole design space. For $(r - j)$ odd, the active resp. equivalent kernels automatically adapt to the skew data situation in the boundary area. The situation in the right boundary area is illustrated in Fig. 12.6 for the Epanechnikov kernel $K(u) = \frac{3}{4}(1 - u^2)_+$ for a local linear estimation of $m(r = 1, j = 0)$ and a local quadratic estimation of $m'(r = 2, j = 1)$.

We see how the weighting systems become deformed toward the boundary. The pictures for the left boundary area are symmetric with those in Figure 12.6. Since the size of the local neighborhood shrinks toward the boundary, the bias part of the mean-squared error (MSE) will be lower in the boundary area than in the interior. On the other hand, the variance part will increase since fewer observations are included in the local estimation and also because of the increasing deformation of the weighting system toward the boundary. Usually, the increase in variance overcompensates for the reduction of the bias, particularly if m'' remains roughly the same in the boundary area. As a conseqence, the MSE will increase toward the boundary. The increase will be even more pronounced for higher-order polynomials.

For $x \in \mathbb{R}^p$ the local linear fit is given as the solution of the least-squares criterion

$$\sum_{i=1}^{n}[y_i - \beta_0 - \beta'(x_i - x)]^2 K\left(\frac{x_i - x}{h}\right)$$

where K is a p–variate kernel. With the design matrix X_x with rows $(1, (x_{i1} - x_1), \ldots, (x_{ip} - x_p))$ the solution has the same form as in (12.27). Let K be a product kernel composed of the same univariate kernel and bandwidth h in each coordinate, and let $H_m(x)$ be the Hessian matrix of the second derivatives of m. Then we get an asymptotic expression for the variance and the bias in the interior (Ruppert and Wand, 1994)

$$\text{Var}(\hat{m}(x)) = \frac{v_0 \sigma^2(x)}{f(x)nh^p} + o_p(nh^p) \tag{12.37}$$

and

$$\text{Bias}(\hat{m}(x)) = \frac{h^2}{2} \mu_2 tr\{H_m(x)\} + o_p(ph^2). \tag{12.38}$$

These considerations about the advantage of a local linear approach compared to the local constant estimation, about its design adaptation property and its automatic boundary adaptation, hold for the multivariate case in a similar way.

Up to now we have considered local least-squares regression to estimate the mean function m. But the idea of locally weighted regression turns out to be a very versatile tool for estimation in a variety of situations.

Yu and Jones (1998) consider the estimation of the conditional distribution function $F(y \,|\, x)$. Let $F_1(u) = \int_{-\infty}^{u} K_1(v)dv$ be the distribution function pertaining to a symmetric kernel density K_1, and let h_2 be a bandwidth. Yu and Jones consider a local linear approach for $F(y \,|\, x)$ that is motivated by the approximations

$$E\left[F_1\left(\frac{y - y_0}{h_2}\right) \,\Big|\, x_0\right] \approx F(y_0 \,|\, x_0)$$

and

$$F(y_0 \,|\, x_0) \approx F(y_0 \,|\, x) + \dot{F}(y_0 \,|\, x)(x - x_0) = \beta_0 + \beta_1'(x - x_0)$$

where $\dot{F}(y_0 \,|\, x) = \partial F(y_0 \,|\, x)/\partial x$. This suggests the least-squares approach

$$\sum_{i=1}^{n}\left[F_1\left(\frac{y_i - y}{h_2}\right) - \beta_0 - \beta'(x_i - x)\right]^2 K\left(\frac{x_i - x}{h_1}\right)$$

12.5. LOCALLY WEIGHTED REGRESSION

where K is a second kernel with bandwidth h_1. The solution

$$\tilde{F}_{h_1,h_2}(y \mid x) = \hat{\beta}_0 = e'_1(X'_x W_x X_x)^{-1} X'_x W_x \tilde{y} \tag{12.39}$$

with $\tilde{y} = (F_1[(y_1 - y)/h_2], \ldots, F_1[(y_n - y)/h_2])'$ is called a *local linear double-kernel smoothing* by the authors. The estimator is continuous and has zero as left boundary value (for $y \to -\infty$) and 1 as right boundary value. It can happen that the estimator ranges outside $[0, 1]$. But this does not, as the authors say, create problems in estimating q_α by

$$\tilde{q}_\alpha(x) = \tilde{F}^{-1}_{h_1,h_2}(\alpha \mid x).$$

This estimator involves the problem that two bandwidths h_1 and h_2 have to be chosen. For a possible procedure with $h_2 < h_1$ we refer the reader to the paper by Fan et al. (1996) considering a related idea for estimating the conditional density itself.

$$E\left[\frac{1}{h_2} K_1\left(\frac{y - y_0}{h_2}\right)\right] \approx g(y_0 \mid x) + \dot{g}(y_0 \mid x)(x - x_0)$$

$$= \beta_0 + \beta'(x - x_0)$$

where $\dot{g}(y \mid x) = \partial g(y \mid x)/\partial x$, leads to the least squares criterion

$$\sum_{i=1}^{n}\left[\frac{1}{h_2} K_1\left(\frac{y_i - y}{h_2}\right) - \beta_0 - \beta'(x - x_0)\right]^2 K\left(\frac{x_i - x}{h_1}\right) \tag{12.40}$$

with the solution $\hat{g}(y \mid x) = \hat{\beta}_0$ as in (12.39), where now the vector $\tilde{y}$ is

$$\tilde{y} = \frac{1}{h_2}\left(K_1\left(\frac{y_1 - y}{h_2}\right), \ldots, K_1\left(\frac{y_n - y}{h_2}\right)\right)'.$$

The local constant approach leads to the traditional estimator (12.3). Fan et al. also consider the case of a local quadratic approach for estimating the first derivative. We will not pursue this case further here, since for the quadratic term $p(p + 1)/2$ more parameters have to be estimated.

In all local regression approaches we used so far, we used the least-squares criterion. Let us now look at cases where instead of the square function, another convex loss function, $\rho : \mathbb{R} \to \mathbb{R}$, is used, which has a unique minimum at zero, and let $m_\rho(x) = \operatorname{argmin}_{\beta_0} E[\rho(y - \beta_0) \mid x]$. Then $\rho(u) = u^2$ yields the conditional expectation that we analyzed mostly so far, and $\rho(u) = |u|$ yields the conditional median. This is just a special case for $\alpha = \frac{1}{2}$ of the loss function $\rho_\alpha(u) = |u| + (2\alpha - 1)u$, already mentioned in (12.17). The term ρ_α was introduced by Koenker and Bassett for parametric quantile estimation. The function $2\rho_\alpha(u)$ for various α values is exhibited in Figure 12.7.

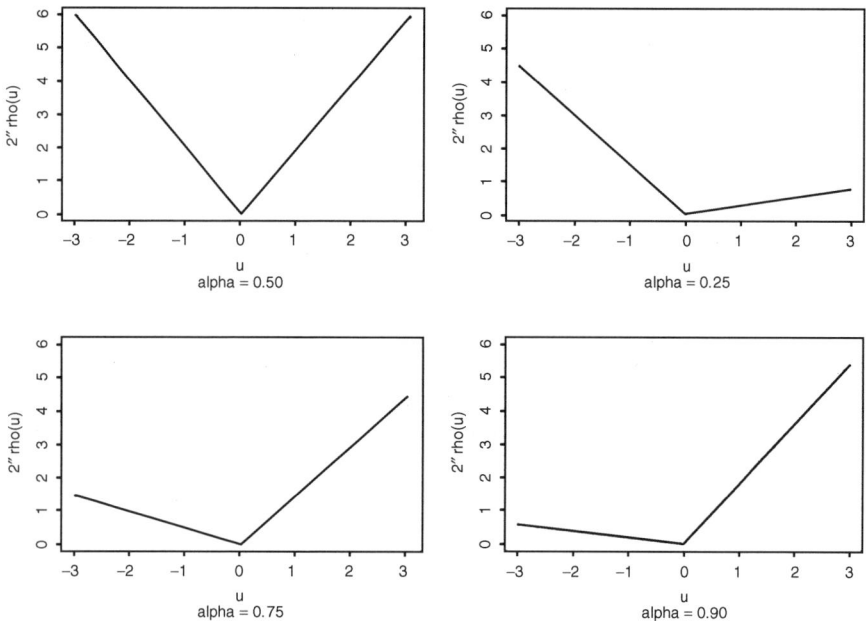

FIGURE 12.7 $2*\rho_\alpha(u)$ according to Koenker and Basset for several α.

In robustness considerations ρ functions were introduced, which increase less rapidly than the square function and for which ρ' is the so-called ψ function. (Huber 1981, Hampel et al. 1986).

A local constant estimator for m_ρ is

$$\hat{m}_\rho(x) = \operatorname{argmin}_{\beta_0} \sum_{i=1}^n \rho(y_i - \beta_0) K\left(\frac{x_i - x}{h}\right).$$

The known drawbacks of a local constant approach is that it cannot adapt to unbalanced design situations and that it has adverse boundary effects that require boundary corrections. This idea leads to the estimator

$$\hat{m}_\rho(x) = \hat{\beta}_0$$

where

$$(\hat{\beta}_0, \hat{\beta}) = \operatorname{argmin}_{\beta_0, \beta} \sum_{i=1}^n \rho(y_i - \beta_0 - \beta'(x - x_0)) K\left(\frac{x - x_0}{h}\right). \quad (12.41)$$

For a ρ function belonging to a robustness class, such as Huber's M-type estimators, known methods for robust estimation can be applied in order to solve the minimum problem (12.41). We would like to remark that the use of kernels automatically safeguards against large deviations in the design space. For nonparametric robust M, L, and R estimation in a time series setting, see Michels (1992).

For a local α-quantile regression with the ρ_α function (12.17), the local solution in (12.38) can be evaluated by solving a linear programming problem, as was shown in the paper of Koenker and Bassett (1978). An algorithm for evaluating this can be found in Koenker and Dorey (1987).

For the case of a general convex ρ function and iid observations, asymptotic normality is proved in Fan et al. (1994). The α-quantile estimation according to (12.41) is also considered by Yu and Jones (1998) and compared with the estimator (12.39). For reasons of practical performance, the authors prefer the double-smoothing approach (12.39). They also give an asymptotic expression for the mean-squared error for x scalar, which for the solution of (12.41) is given by

$$MSE(\hat{q}_\alpha(x)) = \text{Bias}^2(\hat{q}_\alpha(x)) + \text{Var}(\hat{q}_\alpha(x))$$
$$= \frac{1}{4}h^4 \mu_2^2 q_\alpha''(x) + \frac{v_0 \alpha(1-\alpha)}{nh f(x) f(q_\alpha(x)|x)^2}.$$

These expressions are used for suggestions of bandwidth choice.

The cases of robust locally linear regression and of quantile regression are also considered in Fan and Gijbels (1996).

12.6. APPLICATIONS OF LOCALLY WEIGHTED REGRESSION TO TIME SERIES

Local linear or higher-order polynomial regression, originally considered mainly for independent data, can be applied in the same way to stationary processes with certain memory restrictions. The reasons are the same as those mentioned at the beginning of Section 12.3. Given two (dependent) random variables x_s and x_t and a point x in the design space, the random variables $(1/h)K[(x_s - x)/h]$ and $(1/h)K[(x_t - x)/h]$ are nearly uncorrelated as $h \to 0$. This is the *whitening by windowing principle* and it is worthwhile mentionening that this property is not shared by parametric estimators. To handle memory restrictions in the proofs of consistency and asymptotic normality, mixing conditions (strong mixing, uniform mixing, or ϕ mixing) are used. They give a bound to the maximal dependence between events being at least k instants apart from each other. Short-term dependence does not have much effect on local regression. But local polynomial techniques are also applicable under weak dependence in medium or long term. If suitable mixing conditions are fulfilled, local polynomial estimators for dependent data have the same asymptotic properties as for independent data. Of course, the bias is not influenced by dependence, whereas the variance terms are affected. In proving asymptotic equivalence, then, the task consists in showing that the additional terms due to nonvanishing covariances between the variables are of smaller order asymptotically.

For a local linear estimation of $m(x) = m(x_1, \ldots, x_p)$ in the autoregressive model (12.13), the design matrix and the vector y have the form

$$X_x = \begin{pmatrix} z_p - x_1, & \ldots & z_1 - x_p \\ \vdots & & \\ z_{n-1} - x_1, & \ldots & z_{n-p} - x_p \end{pmatrix}, \quad y = \begin{pmatrix} z_{p+1} \\ \vdots \\ z_{t-1} \end{pmatrix}$$

and with $(x_t - x)' = (z_{t-1} - x_1, \ldots, z_{t-p} - x_p)'$ the esimator can be evaluated as in (12.27). For $x = x_{n+1} = (z_n, \ldots, z_{n-p+1})'$

$$\hat{m}(x_{n+1}) = \hat{\beta}_0$$

yields the one-step ahead predictor. A direct k-step-ahead predictor is given if $y = (z_{p+k}, \ldots, z_n)'$ and if the last row of the X_x matrix is $(z_{n-k} - z_n, \ldots, z_{n-k-p+1} - z_{n-p+1})$. But in this case a succession of one-step ahead predictions seems preferable, as mentioned in Section 12.3.

Asymptotic normality results for locally linear autoregression can be found in Härdle et al. (in press) and Fan and Gijbels (1996).

For the CHARN model $z_t = m(x_t) + \sigma(x_t)\xi_t$, the function $g(x_t)$ according to (12.20) can be estimated in a way similar to that shown above, where only in the vector y the time series values are replaced by the squares. Asymptotic normality for this case is shown in Härdle and Tsybakov (1997). For a residual-based estimator of $\sigma^2(x)$, see (12.51) or Feng and Heiler (1998a).

The local linear estimation of a conditional density in a time series setting with the abovementioned double-smoothing procedure as in (12.39) is considered in Fan et al. (1996) and in Fan and Gijbels (1996), where asymptotic results can also be found.

For the estimation of the conditional distribution function according to the proposal of Yu and Jones (1998) as in (12.39) and for a general solution of (12.41) asymptotic results are known for independent data. See the papers of Yu and Jones (1998), Härdle and Gasser (1984) and Tsybakov (1986). For dependent data, we have not found yet formally puplished proofs. But considering the *whitening by windowing* effect makes it clear that for these cases consistency results will hold under suitable mixing conditions.

12.7. PARAMETER SELECTION

One of the first questions to be answered in the application of kernel smoothing is which type of kernel to use for different choices of r and j. It is well known that for $r - j$ odd in the interior of the x space, the Epanechnikov kernel $K(u) = \frac{3}{4}(1-u^2)_+$ is the one that minimizes the mean-squared error in the class of all nonnegative, symmetric, and Lipschitz continuous functions and that for the endpoints x_* and x^*, the triangular kernels $(1-u)\mathbf{1}_{[0,1]}(u)$ resp. $(1+u)\mathbf{1}_{[-1,0]}$ are optimal. For other points

12.7. PARAMETER SELECTION

in the boundary area, optimal solutions are not known. It is easy to see that when looking at variance, only the uniform kernel $\frac{1}{2}\mathbf{1}_{[-1,1]}(u)$ is the one minimizing the variance.

It is well known that in practice the choice of the kernel is not very important compared to the choice of the bandwidth. The Epanechnikov kernel will therefore be a good choice in many cases. Nonetheless, in practice often higher-oder kernels such as the bisquare or the triweight are preferred. This has to do with the degree of smoothness, since the kernel estimates inherit the smoothness properties of the kernel. According to the degree of smoothness as introduced by Müller (1985), the uniform kernel has degree zero (not continuous), the triangle and the Epanecknikov kernel have degree 1 (continuous, but first derivate not continuous), the bisquare and the triweight kernels have degrees 2 and 3, respectively, and the Gaussion kernel has degree ∞.

The most crucial task in kernel smoothing is bandwidth selection. Much ink has been spoiled on papers concerning this problem. It is hence impossible to give a comprehensive survey here. Instead, we will discuss only a few basic ideas. The aim is to choose bandwidths such that the conditional mean-squared error, given by

$$MSE(\hat{m}^{(j)}(x)) = \text{Bias}^2(\hat{m}^{(j)}(x)) + \text{Var}(\hat{m}^{(j)}(x)) \tag{12.42}$$

becomes minimal. We have to distinguish between a locally optimal banwidth and a globally optimal, constant banwidth.

It is clear that a large bandwidth will lead to a low variance, but a high bias. Decreasing the bandwidth will increase the variance, but reduce the bias. An optimal bandwidth is achieved when the changes in bias and variance balance.

Using the asymptotic expressions (12.35) and (12.36) for the conditional variance and bias, then minimizing (12.42) with respect to h yields for the (asymptotically) optimal bandwidth at x for a scalar x

$$h_n^* = C_{r,j}(K) \left[\frac{\sigma^2(x)}{(m^{(r+1)}(x))^2 f(x)} \cdot \frac{1}{n} \right]^{1/(2r+3)} \tag{12.43}$$

where the constant

$$C_{r,j}(K) = \left[\frac{((r+1)!)^2 (2j+1) \int \tilde{K}^{(j)}(u)^2 du}{2(r+1-j)\{\int u^{r+1} \tilde{K}^{(j)}(u)du\}^2} \right]^{1/(2r+3)} \tag{12.44}$$

depends only on r, j and the used kernel and can be calculated beforehand.

In time series applications we are mainly interested in a constant, global bandwidth, for which the integrated mean squared error (IMSE)

$$\int \left[\text{Bias}(\hat{m}^{(j)}(x))^2 + \text{Var}(\hat{m}^{(j)}(x)) \right] w(x) dx$$

is chosen as a criterion, where w is a weight function going to zero at the boundaries to avoid boundary effects. Minimizing the IMSE with respect to h yields the optimal global bandwidth

$$h_n^* = C_{r,j}(K) \left[\frac{\int \frac{\sigma^2(x)}{f(x)} w(x) dx}{\int \{m^{(r+1)}(x)\}^2 w(x) dx} \cdot \frac{1}{n} \right]^{1/(2r+3)}. \qquad (12.45)$$

For local linear estimation of m when x is a p vector and the same bandwidth is chosen in each coordinate, a similar expression can be derived (see Feng and Heiler 1998a). Here

$$h_n^* = c_0 \left(\frac{p}{n} \right)^{1/(p+4)}$$

where

$$c_0 = \left[\frac{\nu_0}{\mu_2^2} \frac{\sigma^2(x)}{f(x) tr\{H_m(x)\}} \right]^{1/(p+4)}$$

and $H_m(x)$ is the matrix of second derivatives of m. All these expressions contain quantities that are unknown and are therefore not amenable in practice. "Plug-in techniques" substitute these quantities by pilot estimates. For more details, see Ruppert et al. (1995).

A simple procedure of bandwidth selection for independent data, first developed to find the smoothing parameter in spline smoothing, is *cross-validation*. Let $\hat{m}_{h,i}(x_i)$ be the so-called *leave-one-out* estimator of m at x_i, where the observation (y_i, x_i) is not used in the estimation procedure. Then the criterion is

$$CV(h) = n^{-1} \sum_{i=1}^{n} [y_i - \hat{m}_{h,i}(x_i)]^2 \qquad (12.46)$$

and $h_{CV} = \operatorname{argmin} CV(h)$ is the cross-validation bandwidth selector. The idea can also be used for $x \in \mathbb{R}^p$ and for estimating derivatives. See Härdle (1990) for details. It can be shown that it converges almost surely to the IMSE optimal bandwidth, but the convergence rate is with $n^{-1/10}$ very low. The cross-validation idea was developed for independent data. In a time series setting it is suggested to replace the leave-one-out estimator by a *leave-block-out* estimator, where for estimating at x_i not only the ith observation is omitted, but a whole block of data around (y_i, x_i). This idea was used by Abberger (1996, 1997) in smoothing the conditional α quantile, where the square function is replaced by the ρ_α function (12.17).

Let σ^2 be the variance of the residuals in an iid sample and in the time series case, the unconditional variance of the stationary process. Rice (1983, 1984) proposed a

12.7. PARAMETER SELECTION

criterion R that for a general linear smoother is given by

$$R(h) = RSS(h) - \hat{\sigma}^2 + 2\hat{\sigma}^2 n^{-1} \sum_{i=1}^{n} w_{ni}(x_i) \quad (12.47)$$

where the w_{ni} are the actual weights for estimating $m(x_i)$, $\hat{\sigma}^2$ is an estimate for σ^2 and

$$RSS(h) = n^{-1} \sum_{i=1}^{n} [y_i - \hat{m}_h(x_i)]^2 \quad (12.48)$$

is the mean *residual sum of squares*. Under the assumption that $\hat{\sigma}^2$ is a consistent estimator, Rice (1984) showed that the proposed estimator $h_R = argmin R(h)$ is asymptotically optimal in the sense that $(h_R - h_0)/h_0 \to 0$ in probability, where h_0 is the minimizer of the *mean averaged squared error*

$$MASE(h) = n^{-1} E \left\{ \sum_{i=1}^{n} [\hat{m}_h(x_i) - m(x_i)]^2 \right\}.$$

The rate of convergences of h_R is the same low rate $n^{-1/10}$ as for the cross-validation solution h_{CV}. The main differences between the two is that R involves an estimate of σ^2, whereas CV does not.

For $\hat{\sigma}^2$ Rice proposed an estimator based on first differences, whereas Gasser et al. (1986) suggested taking second differences (since they annihilate a local linear mean value function);

$$\hat{\sigma}_G^2 = \frac{2}{3(n-2)} \sum_{i=1}^{n-2} \left[y_{i+1} - \frac{1}{2}(y_i + y_{i+2}) \right]^2. \quad (12.49)$$

An estimator based on a general *difference sequence* $D_m = \{d_0, d_1, \ldots, d_m\}$ such that $\sum_0^m d_j = 0$ and $\sum_0^m d_j^2 = 1$ was considered by Hall et al. (1990). The variance estimator based on D_m is then

$$\hat{\sigma}_m^2 = (n-m)^{-1} \sum_{i=1}^{n-m} \left(\sum_{j=0}^{m} d_j y_{j+i} \right)^2. \quad (12.50)$$

Fan and Gijbels (1995) suggest the *residual sum-of-squares criterion* (RSC), which is based on a local estimator of the conditional variance derived under a local homogeneity assumption:

$$\hat{\sigma}^2(x) = \frac{\sum_{i=1}^{n}(y_i - \hat{y}_i)^2 K\left(\frac{x_i - x}{h}\right)}{tr[W_x - W_x(X_x' W_x X_x)^{-1} X_x' W_x]}. \quad (12.51)$$

With this the *RSC* is defined as

$$RSC(x;h) = \hat{\sigma}^2(x)[1+(r+1)V] \tag{12.52}$$

where V is the first diagonal element of the matrix $(X_x'W_xX_x)^{-1}(X_x'W_x^2X_x)$ $(X_x'W_xX_x)^{-1}$; V^{-1} reflects the effective number of local data points. *RSC* admits the following interpretation. If h is too large, then the bias is large and hence also $\hat{\sigma}^2(x)$. When the bandwidth is too small, then V will be large. Therefore *RSC* protects against extreme choices of h.

The minimizer of $E[RSC(x;h)]$ can be approximated by

$$h_{n0}(x) = \left[\frac{a_0\sigma^2(x)}{2C_r\beta_{r+1}^2 nf(x)}\right]^{1/(2r+3)} \tag{12.53}$$

where a_0 denotes the first diagonal element of the matrix $S^{-1}S^*S^{-1}$, that is, $a_0 = \int \tilde{K}^2(u)du$ and $C_r = \mu_{2r+2} - c_r'S^{-1}c_r$, with the definitions given in Section 12.5 and $\beta_{r+1} = m^{(r+1)}(x)/(r+1)!$. The value $h_{n0}(x)$ differs from the optimal bandwidth in (12.44) by an adjusting constant that depends only on r, j, and the kernel used. Hence the latter one can be evaluated,

$$h_n^*(x) = Adj_{j,r}h_{n0}(x), \tag{12.54}$$

where

$$Adj_{j,r} = \left[\frac{(2j+1)C_r\int(\tilde{K}^{(j)}(u))^2 du}{(r+1-j)\{\int u^{r+1}\tilde{K}^{(j)}(u)du\}^2\int\tilde{K}(u)^2 du}\right]^{1/(2r+3)}.$$

For the Epanechnikov and the Gaussian kernel these constants are tabulated for various r and j in Fan and Gijbels (1996).

For a global bandwidth the minimizer $\hat{h}$ of the integrated *RSC*

$$IRSC(h) = \int RSC(x;h)dx$$

is taken, which in practice breaks down to evaluating a mean over certain grid points $x_{i_1}, \ldots, x_{i_m}$. $\hat{h}$ is also selected from among a number of grid points in an interval $[h_{\min}, h_{\max}]$. The global bandwidth is then given by

$$\hat{h}_{j,r} = Adj_{j,r}\hat{h}. \tag{12.55}$$

The *RS* criterion suffers also from having a low convergence rate. Therefore the following refined bandwidth selection procedure is suggested. It is a *double-smoothing* (DS) procedure. The pilot smoothing consists in fitting a polynomial of

12.7. PARAMETER SELECTION

order $r + 2$ and selecting $\hat{h}_{j,r}$ as above. With the bandwidth $\hat{h}_{r+1,r+2}$ estimates of $\hat{\beta}_{r+1}, \hat{\beta}_{r+2}$ and $\hat{\sigma}^2(x)$ are evaluated. With these pilot estimates in a second stage the $\widehat{MSE}_{(j,r)}(x;h) = \widehat{\text{Bias}}^2_{j,r}(x) + \widehat{\text{Var}}_{j,r}(x)$ is evaluated, where $\widehat{\text{Bias}}_{j,r}(x)$ denotes the $(j+1)$th element of the estimated bias vector and $\widehat{\text{Var}}_{(j,r)}(x)$ is the $(j+1)$th diagonal element of the matrix $(X'_x W_x X_x)^{-1}(X'_x W_x^2 X_x)(X'_x W_x X_x)^{-1}\hat{\sigma}^2(x)$. With $S_{n,l} = \sum_{i=1}^{n} K[(x_i - x)/h](x_i - x)^l$, the bias vector is estimated by

$$\hat{b}_r(x) = (X'_x W_x X_x)^{-1} \begin{pmatrix} \hat{\beta}_{r+1} S_{n,r+1} + \hat{\beta}_{r+2} S_{n,r+2} \\ \vdots \\ \hat{\beta}_{r+1} S_{n,2r+1} + \hat{\beta}_{r+2} S_{n,2r+2} \end{pmatrix}.$$

In order to avoid collinearity effects, modification of the vector on the right side is suggested by putting $S_{n,r+3} = \cdots = S_{n,2r+2} = 0$, which yields

$$\hat{b}_r(x) = (X'_x W_x X_x)^{-1} \begin{pmatrix} \hat{\beta}_{r+1} S_{n,r+1} + \hat{\beta}_{r+2} S_{n,r+2} \\ \hat{\beta}_{r+1} S_{n,r+2} \\ 0 \\ \vdots \\ 0 \end{pmatrix}.$$

The global refined bandwidth selector is then given by the minimizer $\hat{h}^R_{j,r}$ of

$$\int \widehat{MSE}_{j,r}(x;h)\,dx. \qquad (12.56)$$

This refined technique leads to an important improvement over the *RSC* bandwidth selector.

For a balanced design, that is, for equally spaced x values, Heiler and Feng (1998) propose a simple double-smoothing procedure, where in the pilot estimation step the *R* criterion is used. In Feng and Heiler (1998b) a further improvement of this proposal can be found, where a variance estimator based on the bootstrap idea is used. Equally spaced x values are given in a time series setting, for instance, where the regressor is the time index or a function of the time index. This kind of smoothing will be discussed in the next section.

For order selection in a time series autoregression model with $x_t = (z_{t-1}, \ldots, z_{t-p})$ and $\hat{m}_t(x)$ as the leave-one-out estimator according to (12.27), Cheng and Tong (1992) use the cross-validation criterion

$$CV(p) = (n - r + 1)^{-1} \sum_t [z_t - \hat{m}_t(x_t)]^2 w(x_t). \qquad (12.57)$$

where w is a weight function to avoid boundary effects.

Because of the curse-of-dimensionality problem it may be advisable not to take all lagged values $z_{t-1}, \ldots, z_{t-p}$ into account but to look for a subset of lagged values that yields the best forecasts. For a lag constellation $x_t(i) = (z_{t-i_1}, \ldots, z_{t-i_p})'$ Tjøstheim and Auestad (1994b) propose using the final prediction error

$$FPE(x_t(i)) = n^{-1} \sum_t [z_t - \hat{m}(x_t(i))]^2 f(i) \quad (12.58)$$

where the factor

$$f(i) = \frac{1 + (nh^p)^{-1} v_0 b_p(i)}{1 - (nh^p)^{-1} [2K^p(o) - v_0^p] b_p(i)}$$

$$v_o = \int K^2(u) du, \quad b_p(i) = n^{-1} \sum \frac{w^2(x_t(i))}{\hat{f}(x_t(i))}$$

where $\hat{f}(x_t(i))$ is a multivariate kernel density estimator. *FPE* in (12.57) is essentially a sum of squares of one-step-ahead prediction errors multiplied by a factor that penalizes small bandwidths and a large-order p.

12.8. TIME SERIES DECOMPOSITION WITH LOCALLY WEIGHTED REGRESSION

As mentioned in Section 12.3, if x_t is the time index itself or a polynomial in t, then we arrive at trend smoothing. In a simple trend model

$$z_t = m(t) + a_t$$

the considerations at the beginning of Section 12.5 deliver an estimator of the smooth trend function or its derivatives. Now the matrix X_t has the rows $(1, s-t, \ldots, (s-t)^r)$ for $s = 1, \ldots, n$ and $W_t = \text{diag}(K[(s-t)/h])$. As an interesting fact, one can easily see that in the interior of the time series, that is, for $h \leq t \leq n - h$, the weights given in (12.28),

$$w_{nt}^j(s) = e'_{j+1}(X'_t W_t X_t)^{-1}(1, s-t, \ldots, (s-t)^r) K\left(\frac{s-t}{h}\right)$$

are shift-invariant in the sense $w_{n,t+1}^j(s+1) = w_{nt}^j(s)$. This means that in the interior of the time series the local polynomial fit works like a moving average. But the main advantage over other trend smoothing techniques lies in the automatic boundary adaptation of the procedure. This property makes the idea of extending the local regression approach to so-called unobserved components models very appealing.

12.8. TIME SERIES DECOMPOSITION WITH LOCALLY WEIGHTED REGRESSION

Nonparametric estimation of trend-cyclical movements and of seasonal variations and their separation by local regression represents an interesting alternative to procedures based on parametric models like X–12 or TRAMO–SEATS. (See Chapter 8.) These involve extrapolation methods on either end of the time series in order to be able to estimate the components also in the boundary parts of a time series. This can lead to serious problems if unusual observations in the end parts of time series yield grossly erroneous forecasts. The latter problem will not appear with a local regression approach. Note also that with a data-driven parameter selection the procedure works in a fully automatic way.

The decomposition of a time series into trend-cyclical and seasonal components by *lo*cally *w*eighted *s*catterplot *s*moothing (LOWESS) was suggested by Cleveland et al. (1990). The procedure discussed here is different from their procedure in essential features.

We consider the additive (unobserved) components model

$$z_t = T(t) + S(t) + a_t, \quad t = 1, 2, \ldots \tag{12.59}$$

For the sake of simplicity we assume that $\{a_t\}$ is a white-noise sequence with mean zero and constant variance σ^2. $T(t)$ represents the trend cyclical and $S(t)$, the seasonal component. The usual assumption with respect to T is that it has certain smoothness properties so that the considerations at the beginning of Section 12.5 apply, leading to a local polynomial representation of order r. With respect to the seasonal variations, the usual assumption is that they show a similar pattern from one seasonal period to the next, but they are allowed to vary slowly in the course of time. Hence a natural assumption is that they can be approximated locally by a Fourier series, containing the seasonal frequency and its harmonics,

$$S(s) = \sum_{j=1}^{q} [\alpha_j(t) \cos 2\pi \lambda j(s-t) + \gamma_j(t) \sin 2\pi \lambda j(s-t)] \tag{12.60}$$

where λ is the seasonal frequency, $\lambda = 1/P$, and P is the period of the season. Of course, $\lambda q \leq \frac{1}{2}$ (and for $\lambda q = \frac{1}{2}$ the last sine term has to be omitted).
Let

$$u_t(s) = (\cos 2\pi\lambda(s-t), \sin 2\pi\lambda(s-t), \ldots, \cos 2q\pi\lambda(s-t), \sin 2q\pi\lambda(s-t))'$$
$$\alpha(t) = (\alpha_1(t), \gamma_1(t), \ldots, \alpha_q(t), \gamma_q(t))'.$$

Then $S(s) = \alpha(t)'u_t(s)$.

With the local polynomial representation for the trend-cyclical part

$$T(s) = \sum_{j=0}^{r} \beta_j(t)(s-t)^j = \beta(t)'x_t(s)$$

where $\beta(t) = (\beta_0(t), \ldots, \beta_r(t))'$, $x_t(s) = (1, s-t, \ldots, (s-t)^r)'$, the local least-squares criterion is

$$\sum_{s=1}^{n} [z_t - \beta(t)' x_t(s) - \alpha(t)' u_t(s)]^2 K\left(\frac{s-t}{h}\right). \tag{12.61}$$

With the design matrices X_{1t} with rows $x_t(s)'$, X_{2t} with rows $u_t(s)'$, $X_t = (X_{1t} \vdots X_{2t})$, the composed vector $\gamma(t)' = (\beta(t)', \alpha(t)')$, and the weight matrix $W_t = \text{diag}(K[(s-t)/h]$, the solution is

$$\hat{\gamma}(t) = (X_t' W_t X_t)^{-1} X_t' W_t y \tag{12.62}$$

$$\hat{T}(t) = e_1'(X_t' W_t X_t)^{-1} X_t' W_t y \tag{12.63}$$

$$\hat{S}(t) = (o', \phi_s')(X_t' W_t X_t)^{-1} X_t' W_t y \tag{12.64}$$

where o' is a row of zeros of length $r+1$ and ϕ_s' is a row vector of length $2q$ with entries $\phi_s' = (1\ 0\ 1\ 0 \cdots 1\ 0)$. It picks out the $\hat{\alpha}_j(t)$, pertaining to the cosine terms in $\hat{S}(t)$. The estimator for the jth derivative $T^{(j)}$ of T is

$$\hat{T}^{(j)} = j! e_{j+1}'(X_t' W_t X_t)^{-1} X_t' W_t y. \tag{12.65}$$

All these estimators work as moving averages in the interior part of the time series and have for $r - j$ odd the simple boundary adaptation property discussed in Section 12.5. The decomposition $\hat{m}(t) = \hat{T}(t) + \hat{S}(t)$ is not unique, since the matrix $X_t' W_t X_t$ is not block diagonal. This could, of course, be achieved by an orthogonalization procedure but seems not to be compelling for practical purposes. We call such decomposition a *natural decomposition*.

For parameter selection first a decision has to be made about the degree of the trend polynomial T and the trigonometric polynomial S. Since the seasonal variations are involved in the local approach, the bandwidths should be such that at least three to five periods of the season are included. In order to achieve this, the modelization of T should be rather flexible. Hence, for the interior part of the time series, the polynomial degree $r = 3$ may be preferable to the choice $r = 1$. A data-driven choice for a joint selection of r and bandwidth h is a very difficult task since the two parameters are highly correlated. A higher r allows a larger bandwidth and vice versa. In our experience collected so far a data-driven procedure for the interior part always opted for the highest allowed degree $r_{\max}$ that was put beforehand even if the MSE criterion included a penalty term for overparameterization. As far as the trigonometric polynomial is concerned, all harmonic terms should be included, unless an inspection of the periodogram or the estimated spectrum reveals that one or even more of the seasonal frequences can be omitted.

After this preselection of parameters, a procedure for bandwidth selection is needed. Since for an equidistant time series the "design density" f is a constant,

12.8. TIME SERIES DECOMPOSITION WITH LOCALLY WEIGHTED REGRESSION

the procedure is somewhat simpler than in the general situation discussed in Section 12.7.

A variant of a double-smoothing procedure is recommended. In the pilot stage a polynomial of degree $r+2$ is fitted and the bandwidth is selected with the Rice criterion with respect to $\hat{m} = \hat{T} + \hat{S}$. But because of seasonal variations, the difference based variance estimator (12.49) has to be altered. Heiler and Feng (1996) and Feng (1998) propose a seasonal difference-based variance estimator of the form in (12.50), where not only a local linear function but also a local periodic function is allowed for. An example for monthly data ($P = 12$) is

$$D_{26,12} = c^{-1}\{-1, 2, -1, 0, 0, 0, 0, 0, 0, 0, 0, 0, 2,$$
$$-4, 2, 0, 0, 0, 0, 0, 0, 0, 0, 0, -1, 2, -1\},$$

where c is determined such that $\sum_{j=0}^{m} d_j^2 = 1$. The term $D_{26,12}$ annihilates a local linear trend and a local periodic function with periodicity $P = 12$. Similar sequences can easily be constructed.

Let $\hat{\sigma}_G^2$ be the resulting estimator and let g be the minimizer of the R criterion (12.47). With $\hat{m}_g = \hat{T}_g + \hat{S}_g$, the resulting estimator is denoted. For an arbitrary h, the weights $w_t^h(s)$ for estimating $\hat{T}_h(t) + \hat{S}_h(t)$ are the components of the vector $(1\ 0, \ldots, 0, \phi_s')(X_t'W_t X_t)^{-1} X_t' W_t$, where for W_t a kernel with bandwidth h is taken. Using the pilot estimates $\hat{m}_g(t)$, the bias part of the MSE at t for an estimator with bandwidth h is estimated by

$$\widehat{\text{Bias}}(\hat{m}_h(t)) = \sum_{s=1}^{n} w_t^h(s) \hat{m}_g(s) - \hat{m}_g(t)$$

which yields for the bias part of the *mean averaged squared error MASE (h)*:

$$B(h) = n^{-1} \sum_{t=1}^{n} \widehat{\text{Bias}}^2(\hat{m}_h(t))$$
$$= n^{-1} \sum_{t=1}^{n} \left\{ \sum_{s=1}^{n} w_t^h(s) \hat{m}_g(s) - \hat{m}_g(t) \right\}^2. \qquad (12.66)$$

The variance is estimated by

$$V(h) = n^{-1} \hat{\sigma}^2 \sum_{t=1}^{n} \sum_{s=1}^{n} w_t^h(s)^2 \qquad (12.67)$$

where $\hat{\sigma}^2$ should be a suitable root-n consistent estimator of $\hat{\sigma}^2$.

After the first pilot step a minimizer $\tilde{h}$ of the criterion

$$MASE(h) = B(h) + V(h) \qquad (12.68)$$

is evaluated over a grid, where, in the second step, the estimator $\hat{\sigma}_G^2$ is used in $V(h)$. This second step leads to a considerable improvement over the simple R criterion, but the estimator $\hat{\sigma}_G^2$ is still not very good. Hence an improved estimation with a lower polynomial degree and a bandwith g_v larger than g is proposed. For details, see Feng and Heiler (1998b). According to considerations therein an estimator for g_v can easily be found by multiplying the minimizer $\tilde{h}$ of (12.68) with a *correction factor*. This factor depends only on the used kernel and on the polynomial degree r, $\hat{g}_v = CF_r\tilde{h}$. For instance, we get for the Epanechnikov kernel $CF_1 = 1.431$, and $CF_3 = 1.291$, for the bi-square kernel $CF_1 = 1.451$, and $CF_3 = 1.300$, and for the Gaussian kernel $CF_1 = 1.489$ and $CF_3 = 1.305$. See Table 5.1 in Müller (1988) or Table 1 in Feng and Heiler (1998b).

Now let $\hat{m}_{g_v} = \hat{T}_{g_v} + \hat{S}_{g_v}$ be an estimator with bandwidth g_v. Then an improved variance estimator is obtained by taking the mean-squared residuals

$$\hat{\sigma}_B^2 = n^{-1} \sum_{t=1}^{n} [z_t - \hat{m}_{g_v}(t)]^2. \tag{12.69}$$

In a third step this variance estimator is plugged into (12.69) for $\hat{\sigma}^2$, and with this again a minimizer h^* of the MASE (12.68) is evaluated. In principle, this procedure can be iterated several times; in the next step, with a polynomial of degree $r + 2$, a new bias estimator is evaluated.

The abovementioned procedure yields a bandwidth h^* for the interior part of the time series, where, after the selection of h^*, the interior is given by $[h^*+1, n-h^*]$. As described in Section 12.5, the procedure automatically adapts toward the boundaries. But as also described there, because of increasing variance, the MSE will increase as well, particularly if $r = 3$ is chosen, as was recommended at the beginning of this section.

One possibility to at least partly compensate for this is to switch to a nearest-neighbor estimator in the boundary area, that is, to keep the total bandwidth $h_T = 2h^* + 1$ constant at both ends of the time series. This means that for estimating from $t = n - h^* + 1$ to $t = n$, the same local neighborhood is used (and similarly for the left boundary).

Instead of or in addition to that, a switch from a local polynomial of order 3 to a local linear approach (for T) may be recommended whenever the MSE for $r = 1$ becomes smaller than that for $r = 3$. In order to do that, for the given bandwidth and the asymmetric neighborhood situation at each timepoint in the boundary area with the corresponding active weighting systems, the MSEs for $r = 3$ and $r = 1$ have to be evaluated according to the procedure described above. As soon as $MSE_1 < MSE_3$, a local linear approach is chosen for T and maintained to the endpoint. According to practical experience collected so far, such a switch occurred close to the endpoints in almost all cases.

In Figures 12.8 and 12.9 we present two examples where the decomposition procedure discussed above is applied. The first time series is the quarterly series of the German GDP from 1968 to 1994. In the top panel in Figure 12.8 the time series

12.8. TIME SERIES DECOMPOSITION WITH LOCALLY WEIGHTED REGRESSION

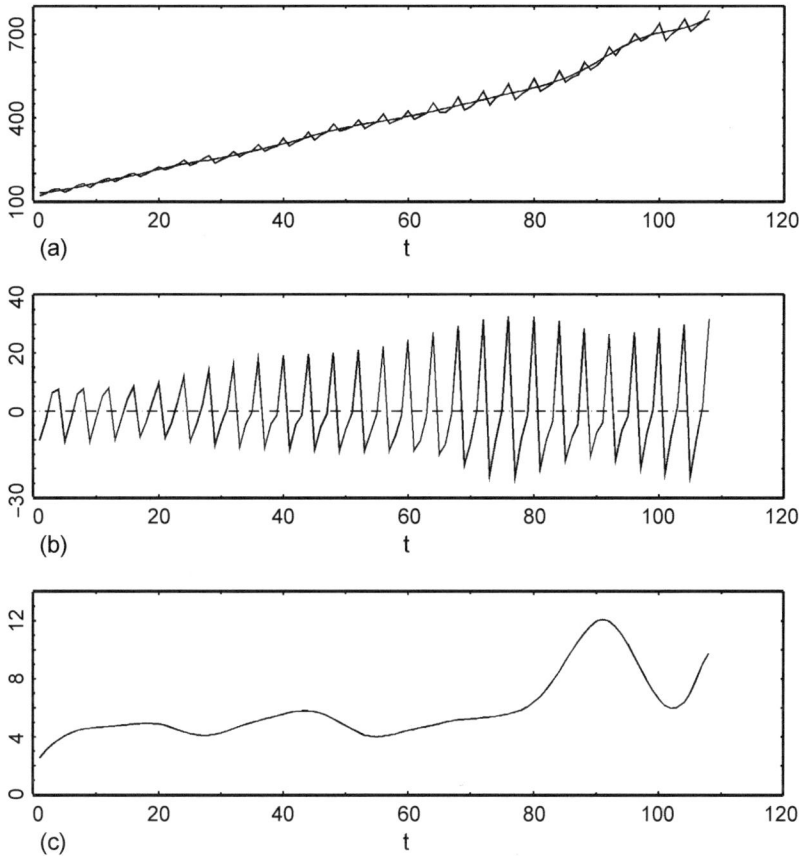

FIGURE 12.8 Decomposition results for the time series of the German GDP from 1968 to 1994: (a) the data and $\hat{T}$; (b) $\hat{S}$ and (c) $\hat{T}'$.

itself and the estimated trend-cyclical component are exhibited. The middle panel shows the estimated seasonal component, and in the bottom panel the first derivative of the trend-cyclical is exhibited. This latter picture shows clearly the temporary boom after German reunification. The double-smoothing procedure with bootstrap variance estimator selected $h = 11$ as bandwidth. The polynomial degree was 2 for estimating the first derivative and 3 for the other estimations. The second example presented in Fig. 12.9 shows corresponding results for the monthly series of the German unemployment rates (in percent) from January 1977 to April 1995. Here the selected bandwidth is $h = 21$. The polynomial degrees are the same as in the previous example.

Cleveland (1979) proposed an iterative robust locally weighted regression in a general regression context and later (Cleveland et al. 1990) this idea in time series decomposition. It can easily be adapted to the procedure discussed here, although in

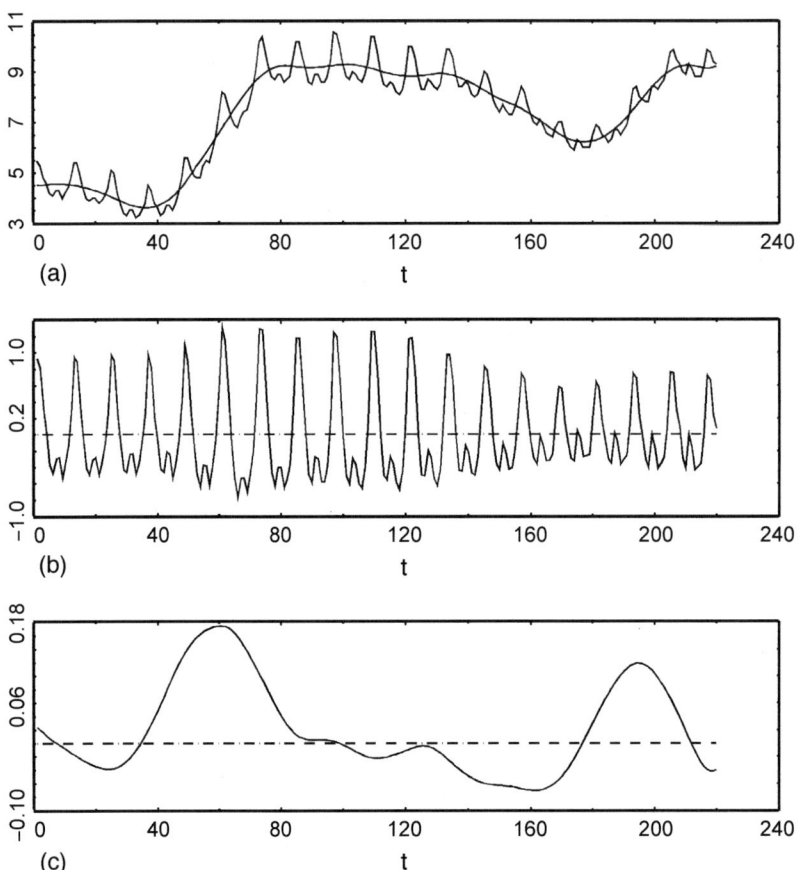

FIGURE 12.9 Decomposition results for the time series of the German unemployment rates (in percent) from January 1977 to April 1995: (a) the data and $\hat{T}$; (b) $\hat{S}$ and (c) $\hat{T}'$.

their proposal the subseries of equal weeks, month, quarters, and so on are treated separately. The idea consists in looking at the residuals $r_t = z_t - \hat{m}(t)$ of a first, nonrobust procedure and to evaluate a robust scale measure δ for the residuals. Cleveland suggests to take the median of the $|r_t|$. Since in many time series variability is different for different periods within the season depending on the size of the seasonal component, it seems reasonable to evaluate different scale measures for the different periods of the season.

For $t = 1, \ldots, n$ let $j = [(t-1)/P] + 1$ be the year index, $j = 1, \ldots, J = [(n-1)/P] + 1$, where [.] denotes the integer part and let $i = t - P(j-1)$ be the season index, that is, $z_t \longrightarrow z_{ij}$. Then for all $i = 1, \ldots, P$ a robust scale measure

$$\delta_i = median_j \left(|r_{ij}|\right)$$

is evaluated. From this so-called robustness weights are derived, which according to Cleveland's proposal are given by

$$\beta_{ij} = K\left(\frac{r_{ij}}{6\delta_i}\right)$$

where K is a kernel function (the bisquare kernel is suggested).

In a second step the local estimation procedure is repeated, where the neighborhood weights $k_{st} = K[(s-t)/h]$ in the diagonal weight matrices W_t are multiplied with the corresponding robustness weights β_{ij}, where i and j are the season and year index corresponding to s. Of course, with the time-dependent robustness weights, the procedure is no longer shift-invariant, so the least-squares solution must be evaluated for each t explicitely. Starting with the new residuals, the procedure can be iterated until the estimates stabilize. Since the robustness weights will change the active kernels, different bandwidths should be used in each iteration step. Cleveland (1979) claimed that two robust iterations should be adequate for almost all situations. Feng (1998) reported a stability criterion occuring in a higher number of iteration steps in most cases.

REFERENCES

Abberger, K. (1996). *Nichtparametrische Schätzung bedingter Quantile in Zeitreihen—Mit Anwendungen auf Finanzmarktdaten*. Hartung-Gorre Verlag, Konstanz.

Abberger, K. (1997). Quantile smoothing in financial time series. *Stat. Papers* **38**, 125–148.

Bongard, J. (1960). Some remarks on moving averages. In OECD (ed.), *Seasonal Adjustment on Electronic Computers. Proce. Int. conf., Paris*, pp. 361–387.

Breiman, L. and Friedman, J. H. (1985). Estimating optimal transformation for correlation and regression. *J. Am. Stat. Assoc.* **80**, 580–598.

Chen, R. (1996). A nonparametric multi-step prediction estimator in Markovian structures. *Statistica Sinica* **6**, 603–615.

Chen, R. and Tsay, R. S. (1993a). Functional-coefficient autoregressive models. *J. Am. Stat. Assoc.* **88**, 298–308.

Chen, R. and Tsay, R. S. (1993b). Nonlinear additive ARX models. *J. Am. Stat. Assoc.* **88**, 955–967.

Cheng, B. and Tong, H. (1992). On consistent non-parametric order determination and chaos (with discussion). *J. Roy. Stat. Soc. B* **54**, 427–474.

Cleveland, R. B., Cleveland, W. S., McRae, I. E., and Terpenning, I. (1990). STL: A seasonal-trend decomposition procedure based on LOWESS (with discussion). *J. Off. Stat.* **6**, 3–73.

Cleveland, W. S. (1979). Robust locally weighted regression and smoothing scatterplots. *J. Am. Stat. Assoc.* **74**, 829–836.

Collomb, G. (1980). Estimation nonparamétrique de probabilités conditionelles. *Co. Re. Acad. Sci. Paris A* **291**, 427–430.

Collomb, G. (1983). From nonparametric regression to nonparametric prediction: survey of the mean square error and original results on the predictogram. *Lect. Notes Stat.* **16**, 182–204.

Collomb, G. (1984). Propriétés de convergence presque complète du prédicteur à noyau. *Ze. Wahrschein. Verwandte Gebiete* **66**, 441–460.

Collomb, G. (1985). Nonparametric time series analysis and prediction: Uniform almost sure convergence of the k-NN autoregression estimates. *Statistics* **16**, 297–307.

Eubank, R. L. (1988). *Spline Smoothing and Nonparametric Regression*. Marcel Dekker, New York.

Fan, J. (1993). Local linear regression smoothers and their minimax efficiencies. *Ann. Stat.* **21**, 196–216.

Fan, J. and Gijbels, I. (1992). Variable bandwidth and local linear regression smoothers. *Ann. Stat.* **20**, 2008–2036.

Fan, J. and Gijbels, I. (1995). Data-driven bandwidth selection in local polynomial fitting: Variable bandwidth and spatial adaptation. *J. Roy. Stat. Soc. B* **57**, 371–394.

Fan, J. and Gijbels, I. (1996). *Local Polynomial Modelling and its Applications*. Chapman and Hall, London.

Fan, J., Hu, T-Ch., and Truong, Y. K. (1994). Robust Non-parametric function estimsation. *Scandi. J. Stat.* **21**, 433–446.

Fan, J., Yao, Q., and Tong, H. (1996). Estimation of conditional densities and sensitivity measures in nonlinear dynamic systems. *Biometrika* **83**, 189–216.

Feng, Y. (1998). *Kernel- and Locally Weighted Regression with Application to Time Series Decomposition*. Ph.D. thesis. Univ. Konstanz.

Feng, Y. and Heiler, S. (1998a). Locally weighted autoregression. In R. Galata and H. Küchenhoff (eds.), pp. 101–117. *Econometrics in Theory and Practice*. Festschrift for Hans Schneeweiss.

Feng, Y. and Heiler, S. (1998b). *Bandwidth Selection Based on Bootstrap*. Discussion paper, Univ. Konstanz.

Fisher, A. (1937). A brief note on seasonal variations. *J. Account.* **64**, 174.

Friedman, J. H. (1991). Multivariate adaptive regression splines (with discussion). *Ann. Stat.* **19**. 1–141.

Gasser, T., Kneip, A., and Köhler, W. (1991). A flexible and fast method for automatic smoothing. *J. Am. Stat. Assoc.* **86**, 643–652.

Gasser, T. and Müller, H. G. (1979). Kernel estimation of regression functions. In T. Gasser and M. Rosenblatt (eds.), *Smoothing Thecniques for Curve Estimation*, pp. 23–68. Springer-Verlag, Heidelberg.

Gasser, T. and Müller, H. G. (1984). Estimating regression functions and their derivatives by the kernel method. *Scandi. J. Stat.* **11**, 171–185.

Gasser, T., Müller, H. G., and Mammitzsch, V. (1985). Kernels for nonparametric curve estimation. *J. Roy. Stat. Soc. B* **47**, 238–252.

Gasser, T., Sroka, L., and Jennen-Steinmetz, C. (1986). Residual variance and residual pattern in nonlinear regression. *Biometrika* **73**, 625–633.

Gouriéroux, Ch. and Monfort, A. (1992). Qualitative threshold ARCH models. *J. Econ.* **52**, 159–199.

Hall, P., Kay, J. W., and Titterington, D. M. (1990). Asymptotically optimal difference-based estimation of variance in nonparametric regression. *Biometrika* **77**, 521–528.

Hampel, F. R., Ronchetti, E. M., Rousseeuw, P. J., and Stahel, W. A. (1986). *Robust Statistics: The Approach Based on the Influence Function*. Wiley, New York.

Härdle, W. (1990). *Applied Nonparametric Regression*. Cambridge Univ. Press, Cambridge, UK.

Härdle, W., Hall, P., and Marron, J. S. (1992). Regression smoothing parameters that are not far from their optimum. *J. Am. Stat. Assoc.* **87**, 227–233.

Härdle, W. and Gasser, T. (1984). Robust non-parametric function fitting. *J. Roy. Stat. Soc. B* **46**, 42–51.

Härdle, W., Lütkepohl, H., and Chen, R. (1997). A review of nonparametric time series analysis. *Int. Stat. Rev.* **65** 49–72.

Härdle, W. and Tsybakov, A. B. (1988). Robust nonparametric regression with simultaneous scale curve estimation. *Ann. Stat.* **16**, 120–135.

Härdle, W. and Tsybakov, A. B. (in press). Local polynomial estimators of the volatility function. *J. Econo.*

Härdle, W., Tsybakov, A. B. and Yang, L. (in press). Nonparametric vector autoregression. *J. Stat. Planning Inference.*

Härdle, W. and Yang, L. (1996). *Nonparametric Time Series Model Selection*. Discussion Paper, Humboldt-Univ. Berlin.

Hart, J. D. (1996). Some automated methods of smoothing time-dependent data. *J. Nonparametric Stat.* **6**, 115–142.

Hastie, T. J. and Tibshirani, R. J. (1990). *Generalized Additive Models*. Monographs on Statistics and Apllied Probability, Vol. 43, Chapman and Hall, London.

Heiler, S. (1995). Zur Glättung Saisonaler Zeitreihen. In H. Rinne, B. Rüger, and H. Strecker (eds.), *Grundlagen der Statistik und Ihre Anwendungen*, pp. 128–148. Festschrift für Kurt Weichselberger, Physika-Verlag, Heidelberg.

Heiler, S. and Feng Y. (1996). Datengesteuerte Zerlegung Saisonaler Zeitreihen. *ifo Studien*, 41–73.

Heiler, S. and Feng, Y. (1997). *A Bootstrap Bandwidth Selector for Local Polynomial Fitting*. Discussion Paper, SFB178, II–344, Univ. Konstanz.

Heiler, S. and Feng, Y. (1998). A simple root n bandwidth selector for nonparametric regression. *J. Nonparametric Stat.* **9**, 1–21.

Heiler, S. and Michels, P. (1994). *Deskriptive und Explorative Datenanalyse*. Oldenbourg-Verlag, Munich.

Horvath, L. and Yandell, B. S. (1988). Asymptotics of conditional empirical porcesses. *J. Multivar. Anal.* **26**, 184–206.

Huber, P. J. (1981). *Robust Statistics*. Wiley, New York.

Jones, H. L. (1943). Fitting of polynomial trends to seasonal data by the method of least squares. *J. Am. Stat. Assoc.* **38**, 453.

Jones, M. C. and Hall, P. (1990). Mean squared error properties of kernel estimates of regression quantiles. *Stat. Prob. Lett.* **10**, 283–289.

Koenker, R. and Bassett, G. (1978). Regression quantiles. *Econometrica* **46**, 33–50.

Koenker, R. and Dorey, V. (1987). Computing regression quantiles. *Appl. Stat.* **36**, 383–393.

Koenker, R., Portnoy, S., and Ng, P. (1992). Nonparametric estimation of conditional quantile functions. In Y. Dodge (ed.), L_1-*Statistical Analysis and Related Methods*. North-Holland, New York.

Macaulay, R. R. (1931). *The Smoothing of Time Series*. National Bureau of Economic Research, New York.

Messer, K. and Goldstein, L. (1993). A new class of kernels for nonparametric curve estimation. *Ann. Stat.* **21**, 179–195.

Michels, P. (1992). *Nichtparametrische Analyse und Prognose von Zeitreihen.* Physica-Verlag, Heidelberg.

Müller, H.-G. (1985). Empirical bandwidth choice for nonparametric kernel regression by means of pilot estimators. *Stat. Decisions* (Suppl. issue 2), 193–206.

Müller, H.-G. (1988). *Nonparametric Analysis of Longitudinal Data.* Springer-Verlag, Berlin.

Nadaraya, E. A. (1964). On estimating regression. *Theory Prob. Appl.* **9**, 141–142.

Priestley, M. B. and Chao, M. T. (1972). Nonparametric function fitting. *J. Roy. Stat. Soc. B* **34**, 385–392.

Rice, J. (1983). Methods for bandwidth choice in nonparametric kernel regression. In J. E. Gentle (ed.), *Computer Science and Statistics: The Interface*, pp. 186–190. North Holland, Amsterdam.

Rice, J. (1984). Bandwidth choice for nonparametric regression. *Ann. Stat.* **12**, 1215–1230.

Robinson, P. M. (1983). Nonparametric estimators for time series. *J. Time Ser. Anal.* **4**, 185–207.

Robinson, P. M. (1986). On the consistency and finite-sample properties of nonparametric kernel time series regression, autoregression and density estimators. *Ann. Inst. Stat. Math. A* **38**, 539–549.

Ruppert, D., Sheather, S. J., and Wand, M. P. (1995). An effective bandwidth selector for local least squares regression. *J. Am. Stat. Assoc.* **90**, 1257–1270.

Ruppert, D. and Wand, M. P. (1994). Multivariate locally weighted least squares regression, *Ann. Stat.* **22**, 1346–1370.

Silverman, B. W. (1984). Spline smoothing: The equivalent variable kernel method. *Ann. Stat.* **12**, 898–916.

Silverman, B. W. (1985). Some aspects of the spline smoothing approach to nonparametric regression curve fitting (with discussion). *J. Roy. Stat. Soc. B*, **47**, 1–52.

Stone, C. J. (1977). Consistent nonparametric regression (with discussion). *Ann. Stat.* **5**, 595–620.

Stute, W. (1984). Asymptotic normality of nearest neighbor regression function estimates. *Ann. Stat.* **12**, 917–926.

Stute, W. (1986). Conditional empirical processes. *Ann. Stat.* **14**, 638–647.

Tjøstheim, D. and Auestad, B. (1994a). Nonparametric identifixation of non-linear time series: projection. *J. Am. Stat. Assoc.* **89**, 1398–1409.

Tjøstheim, D. and Auestad, B. (1994b). Nonparametric identifixation of non-linear time series: selecting significant lags. *J. Am. Stat. Assoc.* **89**, 1410–1419.

Tsybakov, A. B. (1986). Robust reconstruction of function by the local approximation method. *Problems Inform. Transmission* **22**, 133–146.

Wahba, G. (1990). *Spline Models for Observational Data.* SIAM, Philadelphia.

Wand, M. P. and Jones, M. C. (1995). *Kernel Smoothing.* Chapman and Hall, London.

Watson, G. S. (1964). Smooth regression analysis. *Sankhyā, A* **26**, 359–372.

Yakowitz, S. (1979a). Nonparametric estimation of Markov transition functions. *Ann. Stat.* **7**, 671–679.

Yakowitz, S. (1979b). A nonparametric Markov model for daily river flow. *Water Resources Res.* **15**, 1035–1043.

Yakowitz, S. (1985). Markov flow models and the flood warning problem. *Water Resources Res.*, **21**, 81–88.

Yang, L. and Härdle, W. (in press). Nonparametric autoregression with multiplicative volatility and additive mean. *J. Time Ser. Anal.*

Yang, S. (1981). Linear functions of concomitants of order statistics with application to nonparametric estimation of a regression function. *J. Am. Stat. Assoc.* **76**, 658–662.

Yu, K. and Jones, M. C. (1998). Local linear quantile regression. *J. Am. Stat. Assoc.* **93**, 228–237.

CHAPTER 13

Neural Network Models

Kurt Hornik and Friedrich Leisch
Technische Universitat Wien

Certainly, everyone has heard success stories about the use of "neural networks" in a variety of tasks, including the recognition of speech, speakers, fingerprints, and handwritten characters (zip codes), the classification of medical images, decision support for loan applications, and in particular the forecasting of electric load and financial time series. In fact, there are now two conferences (Neural networks in the capital markets and Computational intelligence in financial engineering) dealing mostly with the use of neural networks for the processing of financial data.

In this chapter, we shall provide a brief introduction to neural networks and indicate how they can be used for temporal processing. We shall only outline the basic ideas, which will definitely not suffice for building moneymaking machines right away.

13.1. INTRODUCTION

To start with, there is no generally applicable definition of the term *neural network* (NN). Here, we shall not be concerned with biological neural networks (such as the human brain or nervous system), but artificial neural networks (ANNs), although the term *computational neural networks* would be more appropriate, (Bezdek 1992). The following two characteristics are of key importance:

- NNs are built from simple processing elements (PEs) called *nodes*, *units*, or *neurons*, and contain certain adjustable parameters ("weights").
- NNs use rules to modify these parameters upon the presentation of data, specifically to "learn from the environment."

A Course in Time Series Analysis, Edited by Daniel Peña, George C. Tiao, and Ruey S. Tsay.
ISBN 0-471-36164-X. © 2001 John Wiley & Sons, Inc.

13.2. THE MULTILAYER PERCEPTRON

This characterization is rather vague, and clearly applies also to many traditional learning systems, such as a linear regression model. On the other hand, it emphasizes the fact that an NN paradigm (model) is always a combination of rules for doing computations and rules for adapting (learning). In other words, neural networks are adaptive computational models.

The terminology used in the field of neurocomputing is oriented toward machine learning and biology, and may at first be rather confusing to statisticians and econometricians. For example, variables are called "features"; independent and dependent variables are referred to as "inputs" and "targets"; respectively, and the process of parameter estimation is called "learning" or "training." For that reason, publications that explain neural networks from a statistician's point of view have enjoyed great popularity (see, e.g., Ripley 1993, Cheng and Titterington 1994, Sarle 1994). In fact, there is even a dictionary for translating between neurocomputing and statistics (URL: *ftp://ftp.sas.com/pub/neural/jargon*). As this chapter is meant as a starting point for further reading, we have chosen to use the standard NN terminology.

13.2. THE MULTILAYER PERCEPTRON

The most popular processing elements used in neurocomputing are the *McCulloch–Pitts units*, which compute functions of the form

$$x \mapsto g(\alpha'x + \delta)$$

where x and α are vectors, $'$ denotes *transpose* so that $\alpha'x$ is the inner product of α and x, δ is a scalar, and g is some (typically nonlinear) function.

The original McCulloch–Pitts neuron as introduced in McCulloch and Pitts (1943) had integer weights α and δ and the Heaviside function as its activation function g (see Fig. 13.1):

$$g(t) = \begin{cases} 0, & t < 0 \\ 1, & t \geq 0 \end{cases}.$$

This special case is also called *threshold units*. It is an oversimplified model for biological neurons; the components of x can be interpreted as the outputs of other neurons that provide information to the neuron, the components of α as the synaptic coupling strengths, so that $\alpha'x$ is the actual input into the neuron, and $-\delta$ as the threshold above which the neuron "fires." McCulloch and Pitts showed that by suitable combination of such units with suitably chosen α's and δ's one can implement all Boolean functions, but failed to provide a rule for constructing such NN implementations.

A first learning rule was introduced by Rosenblatt(1958). His *perceptron* is a threshold neuron for binary classification (similar to Fisher's discriminant analysis), where the parameters are fitted from a collection of training patterns with the aid of the so-called perceptron algorithm. In some sense, the perceptron was the first learning machine.

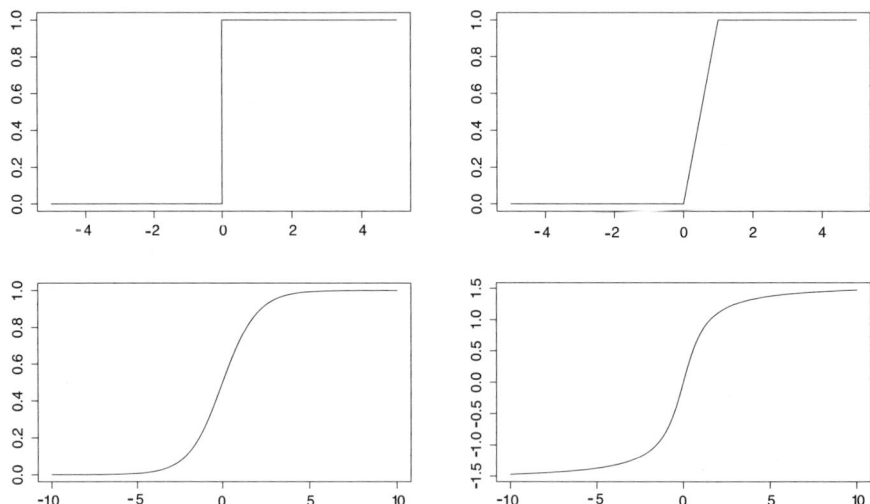

FIGURE 13.1 Activation functions: Heaviside (top left), ramp (top right), logistic (bottom left), and arctangent (bottom right).

The perceptron can solve problems only where the discriminant function is a hyperplane. For example, it cannot implement the XOR (exclusive OR) function where

x			y
0	0	↦	0
0	1	↦	1
1	0	↦	1
1	1	↦	0

Although it was clear that combinations of threshold units should be used because they offer more computational power (Rosenblatt 1962), no appropriate rule for adjusting the parameters of such a model was available for a long time (credit assignment problem).

The real breakthrough was the introduction of the MLP/BP paradigm in Rumelhart et al. (1986) based on two innovations:

- Use of the *multilayer perceptron* (MLP) with *sigmoidal* and typically differentiable activation functions g. MLPs are obtained by combining processing units in a layered feedforward manner. Sigmoidal functions are nondecreasing with finite limits at infinity, and hence are very similar in appearance to the Heaviside function in high-gain situations. Popular examples include the "ramp" function [the cdf (cumulative distribution function) of the uniform distribution on [0, 1]], the logistic function $g(t) = 1/(1 + e^{-t})$, or the arctangent (see Fig. 13.1).

13.2. THE MULTILAYER PERCEPTRON

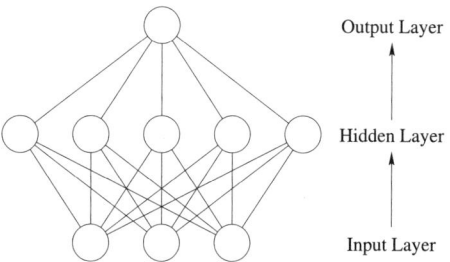

FIGURE 13.2 A single (hidden) layer perceptron with three inputs, five hidden units, and one output.

- (Re)introduction of the backpropagation (BP) learning algorithm [formerly introduced by Bryson and Ho (1969) and Werbos (1974)], which, given a new input pattern x and desired output pattern y, modifies the adjustable weights w of the MLP according to the rule

$$\Delta w \propto \text{grad}_w \|y - f(x, w)\|^2$$

where $f(x, w)$ is the output of the MLP with input x and weights w.

The combination of generalized McCulloch–Pitts units in layers yields a rather powerful computational model. For example, perceptrons with one intermediate ("hidden") layer and one linear output unit (Fig. 13.2) implement functions of the form

$$f(x, w) = \gamma + \sum_{j=1}^{h} \beta_j g(\alpha'_j x + \delta_j)$$

where h is the number of hidden units, g their common activation function and w the vector of all parameters of the model $(\alpha_j, \beta_j, \gamma, \ldots)$. The α_j's are vectors and define the weights corresponding to the links between the input layer and the jth hidden unit. The β_j's are scalars corresponding to the links between the hidden layer and the output unit.

Let $\mathcal{G}(g; A, \Theta)$ be the set of all such function with the α_j and δ_j in A and Δ, respectively. The following result is Theorem 2 in Hornik (1993).

Theorem 13.1. *Let g be essentially bounded and nonpolynomial on some nondegenerate compact interval Δ, and let A contain a neighborhood of the origin. Then for all compactly supported finite measures μ on the input space and $1 \le p < \infty$, $\mathcal{G}(g; A, \Theta)$ contains a subset that is dense in $L^p(\mu)$.*

Similar results can be given for uniform approximation of continuous functions on compacta. This is the famous universal approximation property of the MLP model, which states that almost "arbitrary" functions can be approximated using an MLP. Other classes of universal approximators that are commonly used include polynomials (Taylor series) and sine/cosine (Fourier series).

We now know that it makes sense to use an MLP for approximating an unknown function; however, we need rules to find the correct parameters of the model (or at least an estimate). As mentioned above, in NN jargon parameter estimation is called "learning" and backpropagation is such a learning algorithm.

Assume that we are given a sample $\{(x^1, y^1), \ldots, (x^N, y^N)\}$ of examples and let $L(x, y, w) = (y - f(x, w))^2$ denote the usual square loss. Backpropagation is an on-line gradient rule for the minimization of

$$\sum_{i=1}^{N} L(x^i, y^i, w) = \sum_{i=1}^{N} (y^i - f(x^i, w))^2$$

with respect to the network weights w. Its name stems from the fact that (by a simple application of the chain rule), the gradient can be efficiently computed by first propagating the input forward and then the error signal backwards through the net.

If we use the logistic activation function, we get

$$\frac{d}{dt} g(t) = \frac{d}{dt} \frac{1}{1+e^{-t}} = \frac{e^{-t}}{(1+e^{-t})^2} = g(t)[1 - g(t)].$$

Gradient descent amounts to iteratively adapting the weight vector w proportional to the gradient of loss L. The gradient is given by the partial derivatives of L with respect to the components of w, see a standard textbook on numerical optimization for details.

Let $u_j = g(\delta_j + \sum_i \alpha_{ij} x_i)$ denote the output of the jth hidden unit such that

$$f(x, w) = \gamma + \sum_j \beta_j u_j$$

and by simple application of the chain rule, we get

$$\frac{\partial L}{\partial \beta_j} = -2(y - f(x, w)) u_j$$

$$\frac{\partial L}{\partial \delta_j} = -2(y - f(x, w)) \beta_j u_j (1 - u_j)$$

$$\frac{\partial L}{\partial \alpha_{ij}} = -2(y - f(x, w)) \beta_j u_j (1 - u_j) x_i$$

The gradient contains only inputs x, targets y, the network output $f(x, w)$, and activations u_j of the hidden units and can therefore be computed very efficiently.

13.2. THE MULTILAYER PERCEPTRON

Backpropagation starts with some initial random weight vector w^0, and iteratively updates the weights according to

$$w^{t+1} = w^t - \eta_t \text{grad}_w L(x^t, y^t, w^t)$$

where η_t is a decreasing learning rate and (x^t, y^t) is a pattern randomly drawn from the training patterns.

At least if the training patterns are iid, the g thus obtained should be a good approximation to the conditional expectation $\mathbb{E}(y|x)$. For classification tasks, one can choose the targets as Cartesian unit vectors indicating the class numbers (one-in-k coding); the MLP should then approximate the posterior class probabilities and hence the Bayes decision given uniform loss.

Of course, nothing prevents us from restricting the above approach to situations with iid training patterns. In time series applications, the "input" at time t would contain lagged targets $y(t-1), \ldots, y(t-p)$, such that in the simplest case "using an NN" amounts to fitting a model

$$y_t = f(y_{t-1}, \ldots, y_{t-p}) + a_t$$

where the unknown f is implemented by an MLP with a certain architecture; see Section 13.3.

The actual use of the above general recipe is hampered by at least the following two facts:

- Backpropagation is a gradient descent algorithm that typically converges rather slowly and in general only finds local minima of the error function. As initial values for parameters are typically chosen at random, the training process seldom results in unique solutions for different initializations.
- The choice of a "suitable MLP architecture" (i.e., the specification of the connection patterns of the units, and in particular of the numbers of units and layers employed) is extremely difficult. Intuitively speaking, if the MLPs are too small, then it may not be possible to approximate the unknown $\mathbb{E}(y|x)$ to a satisfactory degree of accuracy; on the other hand, if they are too large, the out-of-sample performance may be bad ("bias-variance dilemma").

Therefore, there is a huge number of papers dealing with improvements of the basic MLP/BP model, ranging from the introduction of ad hoc strategies (pruning, "quickprop," "weight decay," "optimal brain damage," etc.) to the systematic use of superior numerical methods for parameter estimation (conjugate gradient, Newton-type methods, etc.) and the application of model selection and regularization methods (effective number of units, network information criterion, Bayesian approaches, cross-validation, bootstrapping, etc.). The "frequently asked questions" (FAQs) of the newsgroup comp.ai.neural-nets (URL: *ftp://ftp.sas.com/pub/neural/FAQ.html*) contains a large number of corresponding references.

13.3. AUTOREGRESSIVE NEURAL NETWORK MODELS

Probably the most popular way of using ANNs for time series analysis is to generalize the standard AR(p) model

$$y_t = \phi_1 y_{t-1} + \cdots + \phi_p y_{t-p} + a_t \tag{13.1}$$

to nonlinear models of form

$$y_t = f(y_{t-1}, \ldots, y_{t-p}; w) + a_t \tag{13.2}$$

where a_t is an iid noise process. If $f(\cdots; w)$ is a feedforward neural network with parameter ("weight") vector w, we call equation (13.2) an autoregressive neural network process of order p, in short, AR-NN(p), in the following. If $\mathbb{E}a_t = 0$, then f equals the conditional expectation $\mathbb{E}(y_t | y_{t-1}, \ldots, y_{t-p})$ and $f(y_{t-1}, \ldots, y_{t-p}; w)$ is the natural predictor for y_t in the mean square sense.

The following two (closely related) MLP architectures are provided with most NN software packages:

Single-hidden-layer perceptrons:

$$f(y_{t-1}, \ldots, y_{t-p}; w) = \gamma_0 + \sum_{i=1}^{h} \beta_i g(\delta_i + \alpha_{i1} y_{t-1} + \cdots \alpha_{ip} y_{t-p}) \tag{13.3}$$

where h is the number of hidden units, δ_i, β_i and γ_0 are scalar weights, α_i. are p-dimensional weight vectors, and $\sigma(\cdot)$ is a bounded sigmoid function such as $\tanh(\cdot)$ (see Fig. 13.3).

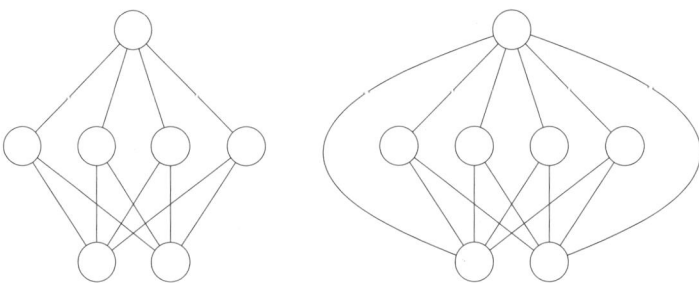

FIGURE 13.3 Single hidden layer perceptron with $p =$ two inputs, $h =$ four hidden units, and one output. The left figure shows a network without shortcut connections [equation (13.3)]; the right figure, a network with shortcut connections [equation (13.4)].

13.3. AUTOREGRESSIVE NEURAL NETWORK MODELS

Single-hidden-layer perceptrons with shortcut connections:

$$f(y_{t-1}, \ldots, y_{t-p}; w) = \gamma_0 + \phi_1 y_{t-1} + \cdots \phi_p y_{t-p} \quad (13.4)$$

$$+ \sum_{i=1}^{h} \beta_i g(\delta_i + \alpha_{i1} y_{t-1} + \cdots \alpha_{ip} y_{t-p})$$

where ϕ is an additional weight vector for shortcut connections between inputs and output. In this case we define the characteristic polynomial $\Phi(z)$ associated with the linear shortcuts (see Fig. 13.3) as

$$\Phi(z) = 1 - \phi_1 z - \phi_2 z^2 - \cdots - \phi^p z^p, \quad z \in \mathbb{C}.$$

To check the stationarity of such models, we can use the following theorem (Leisch et al. 1999).

Theorem 13.2. *Let $\{y_t\}$ be defined by (13.2); further let $\mathbb{E}|a_t| < \infty$ and the density of a_t be positive everywhere in $\mathbb{R}$. Then*

1. *If f is a network without linear shortcuts as defined in (13.3), then $\{y_t\}$ is asymptotically stationary.*
2. *If f is a network with linear shortcuts as defined in (13.4) and additionally $c(z) \neq 0, \forall z \in \mathbb{C} : |z| \leq 1$, then $\{y_t\}$ is asymptotically stationary.*

The time series $\{y_t\}$ remains stationary if we allow for more than one hidden layer [→ multilayer perceptron (MLP)] or nonlinear output units, as long as the overall mapping has bounded range. An MLP with shortcut connections combines a (possibly nonstationary) linear AR(p) process with a nonlinear stationary NN part. Thus, the NN part can be used to model nonlinear fluctuations around a linear process like a random walk.

The only part of the network that controls whether the overall process is stationary is the linear shortcut connections (if present). If there are no shortcuts, then the process is always stationary. With shortcuts, the usual test for stability of a linear system applies.

13.3.1. Example: Sunspot series

As an example, we demonstrate how to fit an AR-NN(p) model to the well known sunspot series. The series gives the mean number of sunspots in the years 1700–1988; hence we have 289 obervations. As mentioned above, there currently exist no analytical model selection methods for ANN models such as the AIC for linear models. We will therefore rely on cross-validation techniques.

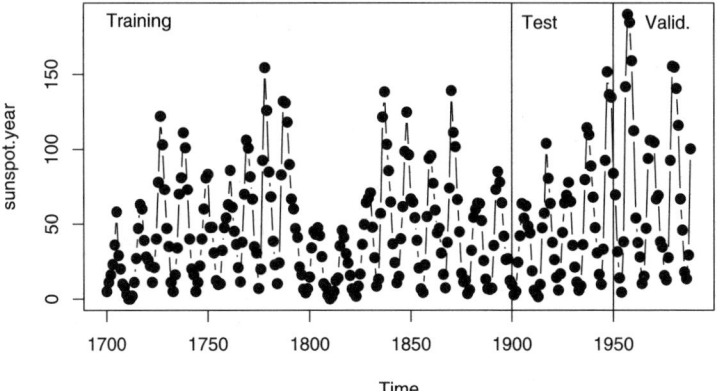

FIGURE 13.4 The sunspot series.

The sunspot series is shown in Figure 13.4. Suppose that we are interested in one-step predictions; for instance, at time t we want to predict value y_{t+1} using $y_1, \ldots, y_t$. To get a valid estimate for the performance of our model, we can split the sample into three independent subsamples, use the first one to get parameter estimates, use the second subsample to choose the order of the model, and finally use the third subsample to get an independent estimate for the mean prediction error. Of course, this approach is valid only if the data generating process are unchanged between the three periods of time—an assumption that should be reasonable for the sunspot series.

In neural network language this is called splitting the available data into a *training set*, a *test set*, and a *validation set*. Only the training set is used for training the network; the test set is used for model selection, and the validation set is reserved to get an independent estimate for the performance. In our case we use observations 1–200 (corresponding to years 1700 to 1899) as training set, observations 201–250 (years 1900–1949) as test set and the remaining observations (years 1950–1988) as test set.

First we estimate linear AR(p) models with $p = 1, \ldots, 15$ as benchmark, using only the training set for parameter estimation. Figure 13.5 shows the error on the test set, indicating that an AR(9) model is best with a mean square error (MSE) of 226. This result is consistent with previous findings in the literature (Tong 1990).

Then we train neural networks with $h = 2$ and $h = 4$ hidden units using also lags of $p = 1, \ldots, 15$. Figure 13.5 shows that the NNs do better on the test set than does the standard AR model for smaller lags (including the best linear model). The increase in error for larger lags is due to overfitting; an MLP with 15 inputs and 4 hidden units has almost 100 free parameters, which cannot be estimated from only 200 training samples. The best network with $h = 2$ hidden nodes uses $p = 6$ inputs and has a test set error of 213; the best network with $h = 4$ hidden nodes uses $p = 7$ inputs and has a test set error of 207.

The validation set errors are 653 for the AR(9) compared to 387 of the 6–2–1 MLP and 454 for the 7–4–1 MLP. Note that neural networks must be seen as "black-

13.4. THE RECURRENT PERCEPTRON

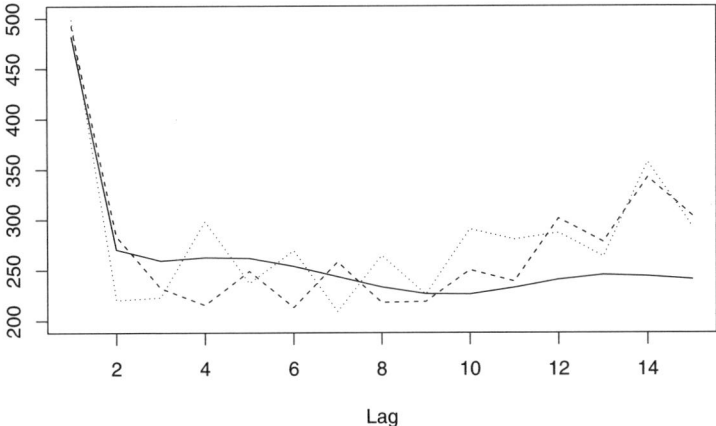

FIGURE 13.5 Model selection for the sunspot series. Mean square error on a test set versus lag p of the model for a standard AR model (solid line), NN with two hidden nodes (dashed line), and four hidden nodes (dotted line).

boxes" for function approximation (but very powerful ones), and any "interpretation" of the models chosen above other than the number of lags used is only of limited value.

13.4. THE RECURRENT PERCEPTRON

Recurrent neural networks differ from feedforward nets in that feedback connections among units are permitted. This provides them with dynamic properties that allow for dealing with *time*, which is important in many tasks where time plays a role (such as the analysis of economic or financial time series, but also for speech, vision, and control). For these and many other reasons, recurrent nets were the subject of broad research interest in the late 1990s.

In this section we shortly present three popular recurrent network models as an introduction to this field. Parameter estimation is essentially harder in the recurrent case such that a detailed treatment is beyond the scope of this introduction to NNs for time series analysis. For details, especially on applications and network training, we refer to the neural network literature.

13.4.1. Examples of recurrent neural network models

Elman and Jordan networks
The so-called Elman network (Elman 1990) and Jordan network (Jordan 1986, 1992), enhance the standard two-layer (i.e., one hidden layer) perceptron architecture by providing previous hidden unit activations and outputs, respectively, as additional inputs.

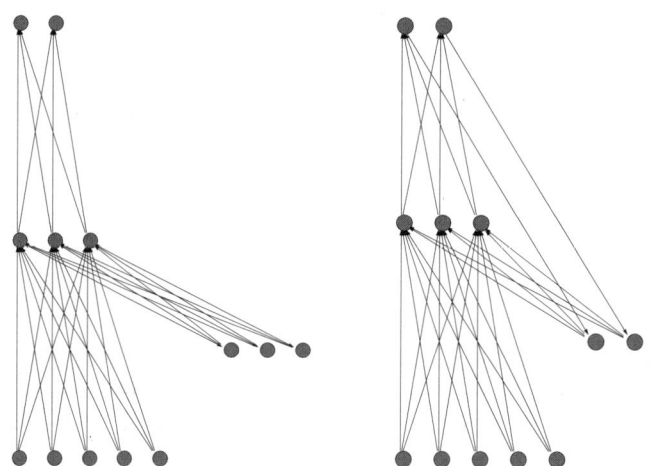

FIGURE 13.6 Elman network (left) and Jordan network (right).

Elman (1990) was interested in learning sequential patterns in linguistics such as the syntactic or semantic features of words. In such applications, it is useful to "memorize" internal states, namely, hidden unit activations, of the net, thereby encoding the temporal properties of sequential inputs. The structure of an Elman network is shown in Figure 13.6. The three units on the righthand side (RHS) depict a time delay, namely, the activations of the hidden units at time $t-1$ are used as additional inputs at time t. Let

$$u_j(t) = g\left(\delta_j + \sum_{i=1}^{p} \alpha_{ij} y_{t-i}\right), \quad j = 1, \ldots, h$$

denote the activation of the jth hidden unit at time t. Then an Elman network amounts to using a model of form

$$y_t = f(y_{t-1}, \ldots, y_{t-p}; u_1(t-1), \ldots, u_h(t-1); w) + a_t$$

On the other hand, Jordan (1986, 1992) was concerned with the control and learning of robot movements. There, it is desirable to have the neural net "memorize" previous positions of the robot, suggesting to use them as additional inputs. More generally, the Jordan network is particularly suitable when the serial order of the outputs is important. Using a Jordan network amounts to fitting models of form

$$y_t = f(y_{t-1}, \ldots, y_{t-p}; \hat{y}_{t-1}; w) + a_t$$

where $\hat{y}_t = y_t - a_t$ denotes the output of the network at time t. This is a nonlinear generalization of the ARMA(p,1) model, because using both y_{t-1} and $\hat{y}_{t-1}$ is equivalent to using y_{t-1} and a_{t-1} as inputs (remember that the hidden nodes compute linear combinations of the inputs).

13.4. THE RECURRENT PERCEPTRON

Both Elman and Jordan networks have in common (as do all recurrent networks) that the parameters cannot be estimated using standard backpropagation. The gradient of the error function at time t depends on the gradient at time $t-1$ because of the recursive definition of the model, such that at time t all gradients back to $t = 0$ must be used.

Language recognition

Language recognition is a typical task where the correct classification of temporal patterns of arbitrary (finite) length is required. The following model for recognizing formal languages L, that is, families of 0–1 strings, is considered and rigorously analyzed in Siegelmann and Sontag (1992, 1994). The net evolves according to

$$u(t+1) = g(Au(t) + By(t) + c),$$

where $u(t)$ and $y(t)$ are the network activations and inputs at time t, respectively, and g performs coordinatewise application of the saturated linear function (ramp function)

$$g(\alpha) = \begin{cases} 0 & \text{if } \alpha < 0 \\ \alpha & \text{if } 0 \leq \alpha \leq 1 \\ 1 & \text{if } \alpha > 1. \end{cases}$$

There are two binary inputs, one representing the data line over which the patterns are presented, and the other the validation line, which takes the value "1" when an input is applied, and "0" otherwise. Similarly, two units are taken as output units for data and validation. The initial state of the net is always taken to be a zero equilibrium. The network classifies an input word $\omega = \omega_1 \cdots \omega_k$ in time τ if the output sequences are of the form

$$u_{\text{data}} = \underbrace{0 \cdots 0}_{\tau-1} \eta(\omega) 0 \cdots, \qquad u_{\text{validation}} = \underbrace{0 \cdots 0}_{\tau-1} 10 \cdots$$

where $\eta(\omega)$, the network's "final response" to the input pattern ω, is either 0 or 1.

13.4.2. A unifying view

Recurrent neural networks can be described as dynamical systems of the form

$$u(t+1) = f(u(t), x(t))$$

perhaps with an additional observation equation of the form $o(t) = g(u(t))$, or in continuous time by a differential equation of the form $\dot{u} = f(u, x)$ (the dot denotes the derivative with respect to time). In the simplest recurrent perceptron case, this gives

$$u(t+1) = \sigma(Au(t) + Bx(t) + c)$$

which is obviously closely related to nonlinear state-space models.

Given the familiar universality of approximation results for two-layer perceptrons (e.g., Hornik 1993), it is rather straightforward to show that, for instance, recurrent two-layer perceptrons with output feedback (i.e., the Jordan networks of Example 13.1) can uniformly approximate arbitrary dynamical systems on compact time intervals (of course, under suitable regularity conditions). Rigorous theorem are given, for example, in Chen and Chen (1993, 1995).

The idea of adjusting the parameters of a (general nonlinear) dynamical system by gradient descent on some performance functional is neither new nor specific to the neural network field. (Nevertheless, similar to the derivation of the backpropagation learning algorithm for feedforward perceptrons, the development of gradient-based learning algorithms for recurrent neural networks has been rather ad hoc and and lacking a unifying general perspective for quite a while.) Such methods have been successfully employed in systems theory and control for a considerable amount of time (although the emphasis in these areas has clearly been on linear systems).

For sake of notational simplicity, we shall base our derivations on continuous-time systems; discrete-time systems can be dealt with analogously.

Hence, consider a dynamical system of the form

$$\dot{u} = f(t, u; w) \tag{13.5}$$

with initial condition $u(t_0) = u_0$. Here, u is the N_u-dimensional vector of state variables, and w is a N_w-dimensional vector of parameters ("weights") that is to be adjusted to achieve a certain "suitable" behavior of the system. The exact form of f is irrelevant for the derivation of the basic equations, but may heavily influence their actual implementation. Note that the "t" in f can also represent external inputs $x(t)$ applied to the system.

We have already encountered different possible tasks that such a system could be used for. In Example 13.2, the object of interest was the equilibrium state $u(\infty)$. In language recognition (Example 13.3), the network's "final" classification response $u_{\text{data}}(t_f)$ to the presented input should be correct. In trajectory learning (Example 13.4), the system should approximate a target trajectory as well as possible on some time interval $[t_0, t_f]$. Note that in the NN literature, "trajectory learning" is typically contrasted to "fixed-point learning." But clearly, the latter concept only adequately characterizes the situations where performance is measured in terms of a static equilibrium reached by the system for $t \to \infty$, as is appropriate when training a network to be an associative memory, for instance, but not the state-based cases where such an equilibrium is not reached or finite time horizons are of interest (e.g., in language recognition).

Let $\ell(t) = \ell(t, u(t))$ be the instantaneous performance of the system at time t. In typical cases, $\ell(t)$ is proportional to the (prediction) error

$$\ell(t) = \frac{1}{2} \sum_{i \in O(t)} (u_i(t) - y_i(t))^2,$$

where $O(t)$ and $y(t)$ are the set of output units and the target, respectively, employed at time t. Note that u and hence also ℓ are, of course, functions of the adjustable parameters w as well, although this dependence is not made notationally explicit.

Then in all examples above, the overall performance of the system is of the form

$$\Lambda = \int_{t_0}^{t_f} \ell(t)\,dt + \ell(t_f).$$

Trajectory learning is obtained with $\ell(t_f) = 0$, and conversely the final state-based cases correspond to $\ell(t) = 0$ for $t_0 \leq t < t_f$. In fact, the general form of Λ might, for instance, be used in a control application where a desired final state is to be approximated with as little cost as possible.

Clearly, gradient descent learning simply amounts to modifying the parameters w according to

$$\Delta w = -\eta \nabla_w \Lambda$$

which, due to the linear structure of Λ, is equivalent to the rule

$$\Delta w = -\eta \left(\int_{t_0}^{t_f} \nabla_w \ell(t)\,dt + \nabla_w \ell(t_f) \right)$$

where $\nabla_w \ell$ denotes the gradient of ℓ with respect to w. For efficient methods to compute this gradient such as "backpropagation through time," we refer to Haykin(1994), for example.

REFERENCES

Bezdek, J. C. (1992). On the relationship between neural networks, pattern recognition and intelligence. *Int. J. Approx. Reason.* **6**, 95–107.

Bryson, A. E. and Ho, Y.-C. (1969). *Applied Optimal Control*. Blaisdell, New York.

Chen, T. and Chen, H. (1993). Approximations of continuous functionals by neural networks with application to dynamic systems. *IEEE Trans. Neural Networks* **4**(6), 910–918.

Chen, T. and Chen, H. (1995). Universal approximation to nonlinear operators by neural networks with arbitrary activation functions and its application to dynamical systems. *IEEE Trans. Neural Networks* **6**(4), 911–917.

Cheng, B. and Titterington, D. M. (1994). Neural networks: A review from a statistical perspective. *Stati. Sci.* **9**, 2–54.

Elman, J. L. (1990). Finding structure in time. *Cogn. Sci.* **14**, 179–211.

Haykin, S. (1994). *Neural Networks. A Comprehensive Foundation*. Macmillan College Publishing, New York.

Hornik, K. (1993). Some new results on neural network approximation. *Neural Networks* **6**, 1069–1072.

Jordan, M. (1986). *Serial Order: A Parallel Distributed Processing Approach*. Technical Report ICS 8604, Institute of Cognitive Science, Univ. California at San Diego.

Jordan, M. (1992). Constrained supervised learning. *J. Math. Psychol.* **36**, 396–425.

Leisch, F., Trapletti, A., and Hornik, K. (1999). Stationarity and stability of autoregressive neural network processes. In *Advances in Neural Information Processing Systems*, Vol. 11.

McCulloch, W. S. and Pitts, W. (1943). A logical calculus of the ideas immanent in nervous activity. *Bulle. Math. Biophys.* **5**, 115–133.

Murata, N., Yoshizawa, S., and Amari, S.-I. (1994). Network information criterion—determining the number of hidden units for an artificial neural network model. *Neural Networks* **5**(6), 865–872.

Pineda, F. J. (1987). Generalization of back-propagation to recurrent neural networks. *Phys. Rev. Lett.* **59**, 2229–2232.

Ripley, B. (1993). Statistical aspects of neural networks. In O. E. Barndorff-Nielsen, J. L. Jensen, and W. S. Kendall (eds.), *Networks and Chaos—Statistical and Probabilistic Aspects*, Vol. 50 of *Monographs on Statistics and Applied Probability*, pp. 40–123. Chapman and Hall, London.

Rosenblatt, F. (1958). The perceptron: A probabilistic model for information storage and organization in the brain. *Psycholo. Rev.* **65**, 386–408.

Rosenblatt, F. (1962). *Principles of Neurodynamics*. Spartan, New York.

Rumelhart, D. E., Hinton, G. E., and Williams, R. J. (1986). Learning internal representations by error propagation. In D. E. Rumelhart, and J. L. McClelland (eds.), *Parallel Distributed Processing*. MIT Press, Cambridge, MA.

Sarle, W. S. (1994). Neural networks and statistical models. In *Proc. 19th of Annual SAS Users Group International Conference*, pp. 1538–1550, Cary, NC. SAS Institute. ftp://ftp.sas.com/pub/neural/neural1.ps.

Siegelmann, H. T. and Sontag, E. D. (1992). On the computational power of neural nets. In *Proc. 5th ACM Workshop on Computational Learning Theory*, pp. 440–449.

Siegelmann, H. T. and Sontag, E. D. (1994). Analog computation, neural networks, and circuits. *Theor. Comp. Sci.* **131**, 331–360.

Sontag, E. (1993). *Some Topics in Neural Networks and Control*. Technical Report LS93-02, Siemens Corporate Research, Princeton, NJ.

Tong, H. (1990). *Non-linear Time Series: A Dynamical System Approach*. Oxford Univ. Press, New York.

Werbos, P. J. (1974). *Beyond Regression: New Tools for Prediction and Analysis in the Behavioural Sciences*. Ph.D. thesis, Harvard Univ.

Widrow, B. and Hoff, Jr., M. E. (1960). Adaptive switching circuits. *IRE WESCON Convention Rec.* **4**, 96–104.

PART III

Multivariate Time Series

CHAPTER 14

Vector ARMA Models

George C. Tiao
University of Chicago

14.1. INTRODUCTION

Business, economic, engineering and environmental data are often collected in roughly equally spaced time intervals, such as hour, week, month, or quarter. In many problems, such time series data may be available on several related variables of interest. Two of the reasons for analyzing and modeling such series jointly are:

1. To understand the dynamic relationships among them. They may be contemporaneously related, one series may lead the others or there may be feedback relationships.
2. To improve accuracy of forecasts. When there is information on one series contained in the historical data of another, better forecasts can result when the series are modeled jointly.

In addition, one may be interested in the structure of the relationship among the series. In particular, there may be hidden factors responsible for the dynamic movement of the component series. For examples, there may be combinations that underline the growth of all the components, and there may also be combinations which are more stable than each individual series.
   Let

$$\{z_{1t}\}, \ldots, \{z_{kt}\}, \quad t = 0, \pm 1, \pm 2, \ldots \qquad (14.1)$$

be k series taken in equally spaced time intervals. Writing

$$\mathbf{z}_t = (z_{1t}, \ldots, z_{kt})' \qquad (14.2)$$

A Course in Time Series Analysis, Edited by Daniel Peña, George C. Tiao, and Ruey S. Tsay.
ISBN 0-471-36164-X. © 2001 John Wiley & Sons, Inc.

we shall refer to the k series as a k-dimensional vector time series. Models that are of possible use in representing such multiple time series, considerations of their properties and methods for relating them to actual data have been extensively discussed in the literature. See in particular, Quenouille (1957), Whittle (1963), Hannan (1970), Zellner and Palm (1974), Brillinger (1975), Dunsmuir and Hannan (1976), Box and Haugh (1977), Parzen (1977), Wallis (1977), Chan and Wallis (1978), Deistler et al. (1978), Hallin (1978), Jenkins (1979), Hsiao (1979), Akaike (1980), Hannan (1980), Hannan et al. (1980), Quinn (1980), and Granger and Newbold (1986). The issue of multivariate linear structure has been given by Hannan and Kavalieris (1984), Hannan and Deistler (1988), Lutkepohl (1991), Reinsel (1993), Reinsel and Velu (1998), and many others.

The principal objective of this chapter is to describe the approach to analyzing multiple time series that have been initiated by Tiao and Box (1981). Our main goal will be on motivating, describing and illustrating the methods used in an iterative model building process with special emphasis on tentative model specification. Much, if not all, of the underlying theory can be found in the references given and, therefore, will not be repeated. Section 14.2 presents a short review of the widely used transfer function models as developed in Box et al. (1994). Section 14.3 discusses a class of vector autoregressive moving-average models. Model building procedures are discussed in Section 14.4 and applied to three actual examples in Section 14.5. We then move on to provide a discussion of issues involved in structural analysis of multivariate time series. The canonical analysis method proposed by Box and Tiao (1977) is discussed in Section 14.6, and finally an introduction to the scalar component model (SCM) approach proposed by Tiao and Tsay (1989) for structural analysis is given in Section 14.7.

14.2. TRANSFER FUNCTION OR UNIDIRECTIONAL MODELS

When $k = 1$ we shall write $\mathbf{z}_t = z_t$. First, recall from equation (3.3) (of Chapter 3) that a widely used linear model for discrete univariate time series is the ARMA(p,q) model

$$\varphi(B)z_t = c + \theta(B)a_t. \tag{14.3}$$

This model can be alternatively written as

$$z_t = c^* + \psi(B)a_t, \quad \psi(B) = \theta(B)/\varphi(B). \tag{14.4}$$

When k series $\{z_{1t}\}, \ldots, \{z_{kt}\}$ are of interest, relationships sometime exist that can be represented by a linear *transfer function* or *unidirectional* model of the form

$$\begin{aligned}
z_{1t} &= L_1(B)\alpha_{1t} \\
z_{2t} &= v_{21}(B)z_{1t} + L_2(B)\alpha_{2t} \\
&\vdots \\
z_{1t} &= v_{k1}(B)z_{1t} + \cdots + v_{k(k-1)}(B)z_{(k-1)t} + L_k(B)\alpha_{kt}
\end{aligned} \tag{14.5}$$

14.2. TRANSFER FUNCTION OR UNIDIRECTIONAL MODELS

where, for convenience the constant terms are suppressed, the $v(B)$'s and $L(B)$'s are polynomials in B, and $\{\alpha_{1t}\}, \ldots, \{\alpha_{kt}\}$ are k independent white noise processes with zero means and variances $\sigma_1^2, \ldots, \sigma_k^2$. Often the polynomials $v(B)$'s and $L(B)$'s can be parsimoneously represented as ratios of polynomials of the form

$$v(B) = \frac{\omega(B)}{\delta(B)} B^b, \qquad L(B) = \frac{\theta(B)}{\varphi(B)} \qquad (14.6)$$

where b is a nonnegative integer, and

$$\omega(B) = \omega_0 - \omega_1 B - \cdots - \omega_s B^s, \qquad \delta(B) = 1 - \delta_1 B - \cdots - \delta_r B^r$$
$$\theta(B) = 1 - \theta_1 B - \cdots - \theta_q B^q, \qquad \varphi(B) = 1 - \varphi_1 B - \cdots - \varphi_p B^p.$$

It will be understood that the polynomials $\omega(B)$, $\delta(B)$, $\theta(B)$, $\varphi(B)$ and the exponent b will in general be different for the different $v(B)$'s and $L(B)$'s.

For $k = 2$, the model in (14.5) reduces to the transfer function model for the output variable $y_t = z_{2t}$ with one stochastic input variable $x_t = z_{1t}$, as discussed in detail in Box et al. (1994). Their model building procedure has been widely adopted, but it becomes cumbersome to apply to situations when there are more than one input variable. Their procedure is compared with an alternative method using the vector ARMA model later in the chapter. It is also worth noting that if the input series is deterministic, the model for the output series is of the same form used in intervention analysis (Box and Tiao, 1975).

Transfer function models of the form (14.5) assume that the k series, when suitably arranged, possess a triangular unidirectional relationship; that is to say, for example, that z_1 depends only on its own past, z_2 depends on its own past and on the present and past of z_1, z_3 on its own past and on the present and past of z_2 and z_1, and so on. A graphical illustration of the model (14.5) for $k = 3$ is given in Figure 14.1. On the other hand, if z_1 depends on the past of z_2, and also z_2 depends on the past of z_1, then we must have a model which allows for this *feedback*.

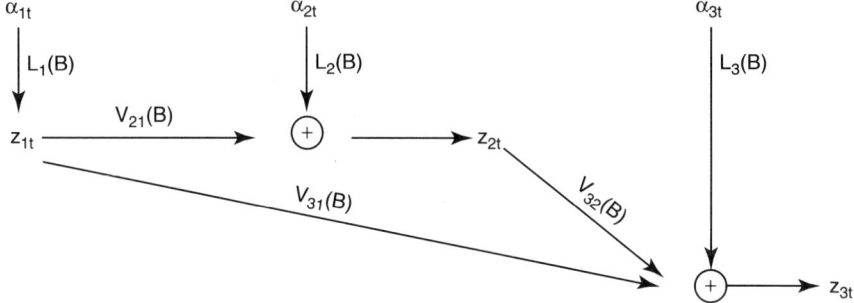

FIGURE 14.1 A trivariate transfer function model.

14.3. THE VECTOR ARMA MODEL

For k series $\{\mathbf{z}_t\}$, the vector autoregressive moving average model takes the form

$$\Phi(B)\mathbf{z}_t = \mathbf{c} + \Theta(B)\mathbf{a}_t \tag{14.7}$$

where $\Phi(B) = \mathbf{I} - \Phi_1 B - \cdots - \Phi_p B^p$, $\Theta(B) = \mathbf{I} - \Theta_1 B - \cdots - \Theta_q B^q$ are matrix polynomials in B, the Φ's and Θ's are $k \times k$ parameter matrices, $\mathbf{c}$ is a $k \times 1$ vector of constants, and $\{\mathbf{a}_t\}$ with $\mathbf{a}_t = (a_{1t}, \ldots, a_{kt})'$ is a series of random shock vectors normally, identically and independently distributed with zero mean vectors and covariance matrix Ω. We shall suppose that the zeros of the determinantal polynomials $|\Phi(B)|$ and $|\Theta(B)|$ are on or outside the unit circle. The series $\{\mathbf{z}_t\}$ will be stationary when the zeros of $|\Phi(B)|$ are all outside the unit circle, and will be invertible when those of $|\Theta(B)|$ are all outside the unit circle. The model in (14.7) will be denoted as ARMA(p,q). The $k \times k$ covariance matrix Ω may either be positive definite or positive semidefinite. In what follows we shall use the notation $N_k(\mathbf{0}, \Omega)$ to denote a k-dimensional multivariate normal distribution with mean vector $\mathbf{0}$ and covariance matrix Ω.

In the literature, the vector ARMA model in (14.7) has been extensively studied by a large number of researchers. Properties of the model have been discussed, for example, in Hannan (1970), Anderson (1971), and Granger and Newbold (1986). The parameter estimation for a given model has been investigated by Tunnicliffe Wilson (1973), Phadke and kedem (1978), Nicholls and Hall (1979), Hillmer and Tiao (1979), and Solo (1984). The problem of model building has been discussed by Akaike (1976), Chan and Wallis (1978), Parzen (1977), Jenkins and Alavi (1981), Cooper and Wood (1982), Tiao and Tsay (1983), and others. Various applications of model (14.7) can be found in Zellner and Palm (1974), Anderson and Moore (1979), and many others.

14.3.1. Some simple examples

To illustrate the behavior of observations from these models, Figure 14.2 shows two series with 250 observations generated from the bivariate $k = 2$ first-order moving-average MA(1) or ARMA(0,1) model

$$\mathbf{z}_t = \mathbf{c} + (\mathbf{I} - \Theta B)\mathbf{a}_t \tag{14.8}$$

with

$$\mathbf{c} = \begin{bmatrix} 17 \\ 25 \end{bmatrix}, \quad \Theta = \begin{bmatrix} .2 & .3 \\ -.6 & 1.1 \end{bmatrix} \quad \text{and} \quad \Omega = \begin{bmatrix} 4 & 1 \\ 1 & 1 \end{bmatrix}.$$

Figure 14.3 shows two series with 150 observations generated from the bivariate first order autoregressive AR(1) or ARMA(1,0) model,

$$(\mathbf{I} - \Phi B)\mathbf{z}_t = \mathbf{c} + \mathbf{a}_t \tag{14.9}$$

14.3. THE VECTOR ARMA MODEL

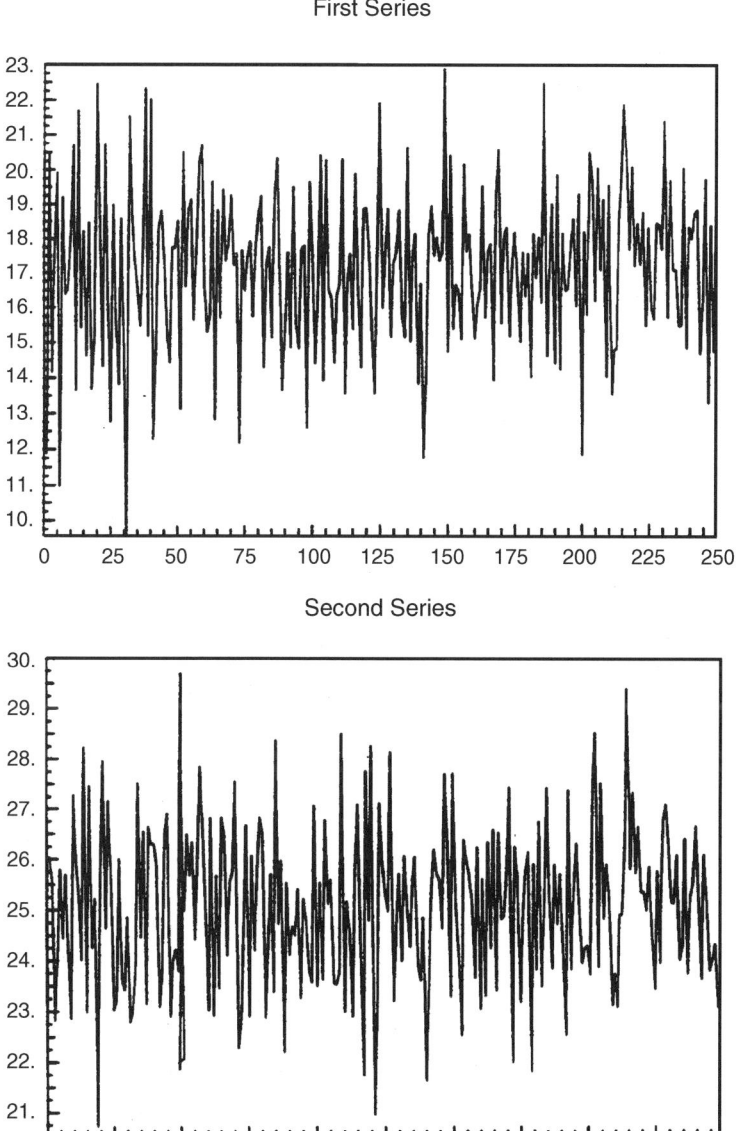

FIGURE 14.2 Data generated from a bivariate MA(1) model with parameter values as in (14.8).

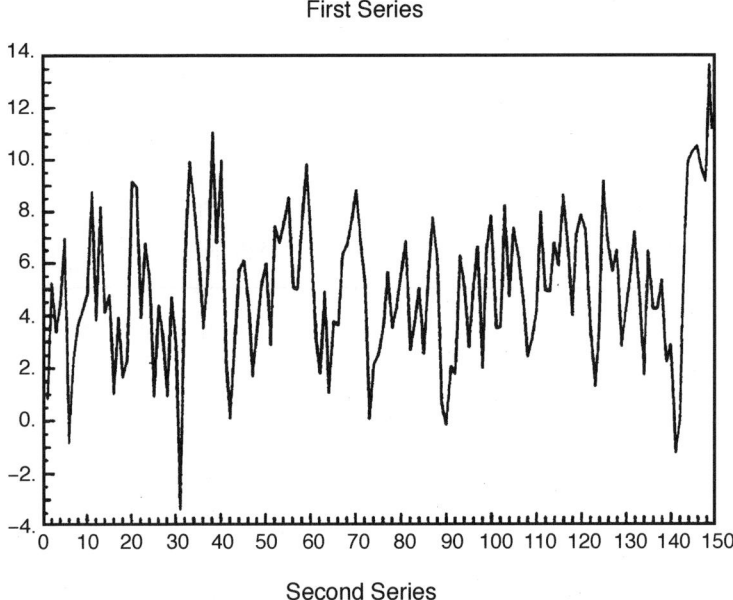
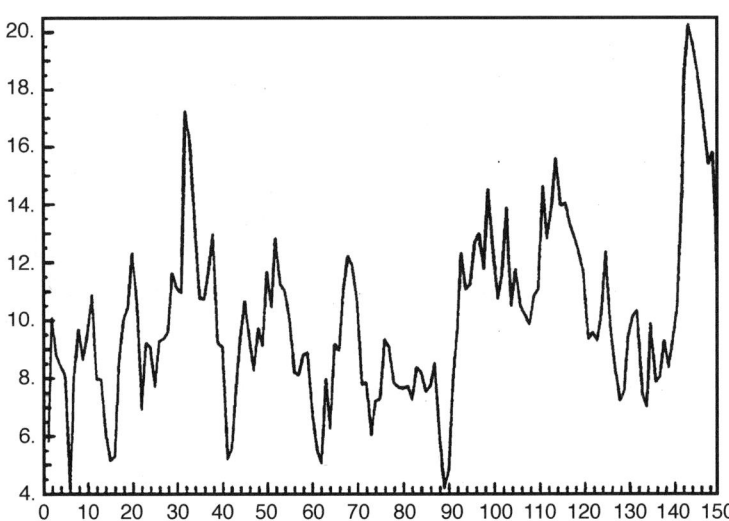

FIGURE 14.3 Data generated from a bivariate AR(1) model with parameter values in (14.9).

with

$$c = \begin{bmatrix} 5 \\ 10 \end{bmatrix}, \quad \Phi = \begin{bmatrix} .2 & .3 \\ -.6 & 1.1 \end{bmatrix} \quad \text{and} \quad \Omega = \begin{bmatrix} 4 & 1 \\ 1 & 1 \end{bmatrix}.$$

Although in both cases the series are seen to be stationary, observations from the autoregressive model are seen to have more "momentum" than those from the moving average model.

In practice, time series often exhibit nonstationary behavior. When several such series are considered jointly, nonstationarity may be modeled by allowing the zeros of $|\Phi(B)|$ in (14.7) to lie on the unit circle. A particular example is the model $(1 - B)\mathbf{z}_t = (\mathbf{I} - \Theta B)\mathbf{a}_t$; after differencing each series we obtain a vector MA(1) model. This is a vector analog of the commonly used univariate nonstationary model $(1 - B)z_t = (1 - \theta B)a_t$. However, it should be noted here that for vector time series, linear combinations of the elements of $\mathbf{z}_t$ may often be stationary and simultaneous differencing of all series can lead to unnecessary complications in model fitting, see, for instance the discussions in Box and Tiao (1977) and Hillmer and Tiao (1979).

14.3.2. Relationship to transfer function model

For the vector model in (14.7), in general, all elements of $\mathbf{z}_t$ are related to all elements of $\mathbf{z}_{t-j}$ ($j = 1, 2, \ldots$) and there can be feedback relationships between all the series. However, if the z_{kt}'s can be arranged so that the coefficient matrices Φ's and Θ's are all lower triangular, then (14.7) can be written as a transfer function model of the form (14.5).

More generally, if the Φ's and Θ's are all lower block triangular, then we obtain a generalization of the transfer function form of (14.5) in which both the input vector series and the output vector series are allowed to have feedback relationships. We note here that relationships between the vector transfer function model and the econometric linear simultaneous equation model have been discussed in Zellner and Palm (1974), and Wallis (1977).

14.3.3. Cross-covariance and correlation matrices

For a stationary vector time series $\{\mathbf{z}_t\}$ with mean vector $\mathbf{0}$, let

$$\Gamma(\ell) = E(\mathbf{z}_{t-\ell}\mathbf{z}'_t) = \{\gamma_{ij}(\ell)\}, \quad i, j = 1, \ldots, k; \quad \ell = 0, \pm 1, \pm 2, \ldots \quad (14.10)$$

be the lag ℓ cross-covariance matrix and $\rho(\ell) = \{\rho_{ij}(\ell)\}$ be the corresponding cross-correlation matrix. When the vector ARMA model in (14.7) is stationary, explicit expressions for the $\Gamma(\ell)$'s can be obtained as follows. First, assuming without loss of generality that $E(\mathbf{z}_t) = \mathbf{0}$, we can write the model for $\mathbf{z}_t$ in the Ψ form

$$\mathbf{z}_t = \Psi(B)\mathbf{a}_t, \quad \Psi(B) = \mathbf{I} + \Psi_1 B + \Psi_2 B^2 + \cdots \quad (14.11)$$

where the Ψ's are obtained by equating coefficients from the relation $\Phi(B)\Psi(B) = \Theta(B)$. Now, from (14.7)

$$\mathbf{z}_{t-\ell}\{(\mathbf{I} - \Phi_1 B - \cdots - \Phi_p B^p)\mathbf{z}_t\}' = \mathbf{z}_{t-\ell}\{(\mathbf{I} - \Theta_1 B - \cdots - \Theta_q B^p)\mathbf{a}_t\}'. \quad (14.12)$$

Taking expectation on both sides of (14.12), and making use of the fact from (14.11) that $E(\mathbf{z}_t \mathbf{a}'_{t-j}) = \Psi_j \Omega$, where $\Psi_0 = \mathbf{I}$, it is straightforward to verified that

$$\Gamma(\ell) = \begin{cases} \sum_{j=\ell-r}^{\ell-1} \Gamma(j)\Phi'_{\ell-j} - \sum_{j=0}^{r-\ell} \Psi_j \Omega \Theta'_{j+\ell}, & \ell = 0, \ldots, r \\ \sum_{j=1}^{r} \Gamma(\ell-j)\Phi'_j, & \ell > r \end{cases} \quad (14.13)$$

where $\Theta_0 = -\mathbf{I}$, $r = \max(p,q)$ and it is understood that (i) if $p < q$, $\Phi_{p+1} = \cdots = \Phi_r = \mathbf{0}$, and (ii) if $q < p$, $\Theta_{q+1} = \cdots = \Theta_r = \mathbf{0}$. In particular, when $p = 0$, i.e. we have a vector MA(q) model, then

$$\Gamma(\ell) = \begin{cases} \sum_{j=0}^{q-\ell} \Theta_j \Omega \Theta'_{j+\ell}, & \ell = 0, \ldots, q \\ \mathbf{0}, & \ell > q. \end{cases} \quad (14.14)$$

Thus, $\Gamma(\ell) \neq 0$ for $\ell = q$ but all auto- and cross-correlations are zero when $\ell > q$. On the other hand, for a vector autoregressive model, the auto- and cross-correlations in general will decay gradually to zero as $|\ell|$ increases.

14.3.4. The partial autoregression matrices

From the moment equations in (14.13) for a stationary ARMA(p,q) model, we see that, for $m \geq q$, the autocovariance matrices $\Gamma(\ell)$'s and the autoregressive coefficient matrices Φ's are related as follows:

$$\begin{bmatrix} \mathbf{A}(p, m) & \mathbf{b}(p, m) \\ \mathbf{g}'(p, m) & \Gamma(m) \end{bmatrix} \begin{bmatrix} \Lambda(p-1) \\ \Phi'_p \end{bmatrix} = \begin{bmatrix} \mathbf{c}(p, m) \\ \Gamma(p \mid m) \end{bmatrix}, \quad m = q, q+1, \ldots$$

(14.15)

where

$$\mathbf{A}(p, m) = \begin{bmatrix} \Gamma(m) & \Gamma(m-1) & \cdots & \Gamma(m-p+2) \\ \Gamma(m+1) & \Gamma(m) & & \\ \vdots & & \ddots & \vdots \\ & & \Gamma(m) & \Gamma(m-1) \\ \Gamma(m+p-2) & \cdots & \Gamma(m+1) & \Gamma(m) \end{bmatrix},$$

14.4. MODEL BUILDING STRATEGY FOR MULTIPLE TIME SERIES

$$\mathbf{b}(p,m) = \begin{bmatrix} \Gamma(m-p+1) \\ \vdots \\ \Gamma(m-1) \end{bmatrix}, \quad \mathbf{c}(p,m) = \begin{bmatrix} \Gamma(m+1) \\ \vdots \\ \Gamma(m+p-1) \end{bmatrix},$$

$$\mathbf{g}(p,m) = \begin{bmatrix} \Gamma'(m+p-1) \\ \vdots \\ \Gamma'(m+1) \end{bmatrix}$$

and $\Lambda'(p-1) = [\Phi_1, \ldots, \ldots, \Phi_{p-1}]$.

In the special case $m = q = 0$, (14.15) is a multivariate generalization of the Yule–Walker equations for autoregressive models in univariate time series. In particular, by partitioning inversion, and on writing $\mathbf{A}(p) = \mathbf{A}(p,0)$, $\mathbf{b}(p) = \mathbf{b}(p,0)$, $\mathbf{c}(p) = \mathbf{c}(p,0)$, $\mathbf{g}(p) = \mathbf{g}(p,0)$, we have that

$$\Phi'_p = [\Gamma(0) - \mathbf{g}'(p)\mathbf{A}^{-1}(p)\mathbf{b}(p)]^{-1}[\Gamma(p) - \mathbf{g}'(p)\mathbf{A}^{-1}(p)\mathbf{c}(p)] \quad (14.16)$$

Motivated by this result, we may define a *partial autoregression matrix function* $\wp(\ell)$, which is analogous to the partial autocorrelation function for the univariate case, such that

$$\wp'(\ell) = \begin{cases} \Gamma^{-1}(0)\Gamma(1), & \ell = 1 \\ [\Gamma(0) - \mathbf{g}'(\ell)\mathbf{A}^{-1}(\ell)\mathbf{b}(\ell)]^{-1}[\Gamma(\ell) - \mathbf{g}'(\ell)\mathbf{A}^{-1}(\ell)\mathbf{c}(\ell)], & \ell > 1. \end{cases} \quad (14.17)$$

This function has the property that for a vector AR(p) model

$$\wp'(\ell) = \begin{cases} \Phi_\ell, & \ell = p \\ 0, & \ell > p. \end{cases} \quad (14.18)$$

14.4. MODEL BUILDING STRATEGY FOR MULTIPLE TIME SERIES

In this section we sketch an iterative approach consisting of (1) tentative specification (identification), (2) estimation, and (3) diagnostic checking for the vector ARMA models in (14.7).

14.4.1. Tentative specification

The aim here is to employ statistics that (1) can be readily calculated from the data and (2) facilitate the choice of a subclass of models worthy of further examination.

Sample cross-correlations. The sample cross-correlations

$$\hat{\rho}_{ij}(\ell) = \frac{\sum(z_{it} - \bar{z}_i)(z_{j(t+\ell)} - \bar{z}_j)}{\{\sum(z_{it} - \bar{z}_i)^2 \sum(z_{jt} - \bar{z}_j)^2\}^{1/2}} \quad (14.19)$$

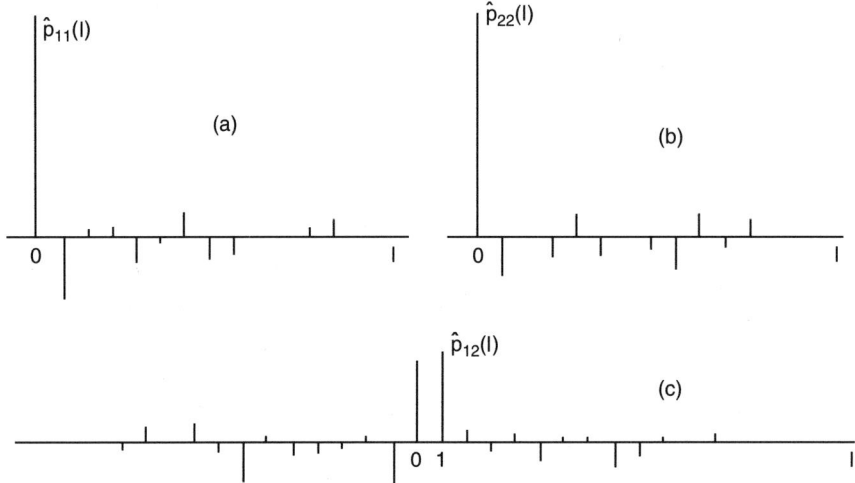

FIGURE 14.4 Sample auto- and cross-correlations for the data in Figure 14.2.

where $\bar{z}_i$ is the sample mean of the ith component series of $\mathbf{z}_t$, are particularly useful in spotting low-order vector moving average models since from (14.14) $\rho_{ij}(\ell) = 0$ for $\ell > q$.

For the data shown in Figure 14.2 generated from a bivariate MA(1) model, Figures 14.4a–14.4c show, respectively, the sample autocorrelations $\hat{\rho}_{11}(\ell)$ and $\hat{\rho}_{22}(\ell)$, and the sample cross-correlations $\hat{\rho}_{12}(\ell)$. The large values occurring at $|\ell| = 1$ would lead to tentative specification of the model as an MA(1). However, graphs of this kind become increasingly cumbersome as the number of series is increased. Furthermore, identification is also not easy from a listing of sample cross-correlation matrices $\hat{\rho}(\ell)$ like that in Table 14.1 (a), particularly when k is greater than 4 or 5.

In this circumstance, we have found the following simple device of great practical value. Instead of the numerical values, a plus sign "+" is used to indicate a value greater than $2n^{-1/2}$, a minus sign "−" for a value less than $-2n^{-1/2}$ and a dot "·" to indicate a value in between $-2n^{-1/2}$ and $2n^{-1/2}$. The motivation is that if the series were white noise, the $\hat{\rho}_{ij}(\ell)$'s would, for large n, be normally distributed with mean 0 and variance n^{-1}. The symbols can be arranged either as in Table 14.1 (b) or as in Table 14.1 (c), both of which clearly suggest a vector MA(1) model for the data.

We realize that the variances of the $\hat{\rho}_{ij}(\ell)$'s can be considerably greater than n^{-1} when the series are highly autocorrelated, so that these "indicator symbols," if taken literally, can lead to over parameterization. However, we do not interpret these "indicator symbols" in the sense of formal significance tests but as a rather crude "signal-to-noise ratio" guide. Taken together they can give useful and assimilable indicators of the general correlation pattern.

Table 14.2 shows sample cross correlation matrices in terms of these "indicator symbols" for the series (Fig. 14.3) generated from a bivariate AR(1) model. The

14.4. MODEL BUILDING STRATEGY FOR MULTIPLE TIME SERIES

TABLE 14.1. Sample Cross-Correlation Matrices for Data in Figure 14.2

(a) Sample Cross-Correlation Matrices $\hat{\rho}(\ell)$

Lags 1–6

$$\begin{bmatrix} -.28 & .37 \\ -.21 & -.19 \end{bmatrix} \begin{bmatrix} .03 & .08 \\ .02 & .01 \end{bmatrix} \begin{bmatrix} .04 & -.03 \\ -.01 & -.08 \end{bmatrix} \begin{bmatrix} -.11 & .04 \\ -.03 & -.09 \end{bmatrix} \begin{bmatrix} -.02 & -.09 \\ -.02 & -.08 \end{bmatrix} \begin{bmatrix} .10 & .01 \\ .01 & -.00 \end{bmatrix}$$

Lags 7–12

$$\begin{bmatrix} -.11 & .01 \\ -.17 & -.06 \end{bmatrix} \begin{bmatrix} -.09 & -.12 \\ -.03 & -.16 \end{bmatrix} \begin{bmatrix} .01 & -.06 \\ .08 & .10 \end{bmatrix} \begin{bmatrix} -.00 & .02 \\ .01 & -.04 \end{bmatrix} \begin{bmatrix} .03 & .00 \\ .08 & .08 \end{bmatrix} \begin{bmatrix} .06 & .04 \\ -.01 & .01 \end{bmatrix}$$

(b) $\hat{\rho}(\ell)$ in Terms of Indicator Symbols

Lags 1–6

$$\begin{bmatrix} - & + \\ - & - \end{bmatrix} \begin{bmatrix} . & . \\ . & . \end{bmatrix} \begin{bmatrix} . & . \\ . & . \end{bmatrix} \begin{bmatrix} . & . \\ . & . \end{bmatrix} \begin{bmatrix} . & . \\ . & . \end{bmatrix} \begin{bmatrix} . & . \\ . & . \end{bmatrix}$$

Lags 7–12

$$\begin{bmatrix} . & . \\ - & . \end{bmatrix} \begin{bmatrix} . & . \\ . & - \end{bmatrix} \begin{bmatrix} . & . \\ . & . \end{bmatrix} \begin{bmatrix} . & . \\ . & . \end{bmatrix} \begin{bmatrix} . & . \\ . & . \end{bmatrix} \begin{bmatrix} . & . \\ . & . \end{bmatrix}$$

(c) Pattern of Correlations for Each Element in Matrix Over All Lags

	z_1	z_2
z_1	....	+...........
z_2	....	....

persistence of large sample auto- and cross-correlations indicates that the data are not likely to have come from a low-order MA model, and suggests the possibility of autoregressive behavior. In general, the pattern of indicator symbols for the cross correlation matrices makes it very easy to identify a low-order moving-average model.

TABLE 14.2. Sample Cross-Correlation Matrices $\hat{\rho}(\ell)$ for Data in Figure 14.3 in Terms of Indicator Symbols

Lags 1–6

$$\begin{bmatrix} + & . \\ + & + \end{bmatrix} \begin{bmatrix} . & . \\ + & + \end{bmatrix} \begin{bmatrix} . & . \\ + & + \end{bmatrix} \begin{bmatrix} . & - \\ + & + \end{bmatrix} \begin{bmatrix} . & . \\ + & + \end{bmatrix} \begin{bmatrix} . & . \\ + & . \end{bmatrix}$$

Lags 7–12

$$\begin{bmatrix} - & - \\ . & . \end{bmatrix} \begin{bmatrix} . & - \\ . & . \end{bmatrix} \begin{bmatrix} . & . \\ . & . \end{bmatrix} \begin{bmatrix} . & . \\ . & . \end{bmatrix} \begin{bmatrix} . & . \\ . & . \end{bmatrix} \begin{bmatrix} . & . \\ . & . \end{bmatrix}$$

Sample partial autoregression and related summary statistics

For an AR(p) process the partial autoregression matrices $\wp(\ell)$ in (14.17) are zero for $\ell > p$. They are particularly useful, therefore, for identifying an autoregressive model. Estimates of $\wp(\ell)$ and their standard errors can be obtained by fitting autoregressive models of successively high-order 1,2... by standard multivariate least squares.

It is well known (see, e.g., Anderson 1971) that for a stationary AR(p) model, asymptotically the estimates $\hat{\Phi}_1, \ldots, \hat{\Phi}_p$ are jointly normally distributed. A useful summary of the pattern of the partials is obtained by listing indicator symbols, assigning a plus (minus) sign when a coefficient in $\hat{\wp}(\ell)$ is greater (resp. less) than 2 (resp. −2) times its estimated standard errors, and a dot for in between values.

To help determine tentatively the order of an autoregressive model, we may also employ the likelihood ratio statistics corresponding to testing the null hypothesis $\Phi_\ell = \mathbf{0}$ against the alternative $\Phi_\ell \neq \mathbf{0}$ when an AR(ℓ) model is fitted. Let

$$\mathbf{S}(\ell) = \sum_{t=\ell+1}^{n} (\mathbf{z}_t - \hat{\Phi}_1 \mathbf{z}_{t-1} - \cdots - \hat{\Phi}_\ell \mathbf{z}_{t-\ell})(\mathbf{z}_t - \hat{\Phi}_1 \mathbf{z}_{t-1} - \cdots - \hat{\Phi}_\ell \mathbf{z}_{t-\ell})' \quad (14.20)$$

be the matrix of residual sum of squares and cross-products after fitting an AR(ℓ). The likelihood ratio statistic is the ratio of the determinants

$$U = |\mathbf{S}(\ell)|/|\mathbf{S}(\ell-1)|. \quad (14.21)$$

Using Bartlett's (1938) approximation, the statistic

$$M(\ell) = -(N - .5 - \ell k) \log_e U \quad (14.22)$$

is, on the null hypothesis, asymptotically distribute.d as χ^2 with k^2 degrees of freedom where $N = n - p - 1$ is the effective number of observations, assuming that a constant term is included in the model.

Finally, a measure of the extent to which the fit is improved as the order is increased is provided by the diagonal elements of the estimated residual covariance matrices $\hat{\Omega}$ corresponding to the successive AR models.

For illustration, for the series in Figure 14.3 generated from a bivariate AR(1) model, the matrices of summary symbols, the $M(\ell)$ statistics and the diagonal elements of the residual covariance matrices are shown in Table 14.3 for $\ell = 1, \ldots, 5$. They indicate that a bivariate AR(l) or at most an AR(2) would be adequate for the data.

For the series shown in Figure 14.2, the pattern of the partials and related statistics are given in Table 14.4. Notice here that if we had confined attention to autoregressive models as is advocated in many econometric textbooks, we would have needed p to be as high as 7. This is not surprising since with the bivariate MA(l) model of (14.8) written in the autoregressive form (ignoring the constant term $\mathbf{c}$)

$$(\mathbf{I} - \Theta B)^{-1} \mathbf{z}_t = \mathbf{a}_t, \quad \text{or} \quad \mathbf{z}_t = \Pi_1 \mathbf{z}_{t-1} + \Pi_2 \mathbf{z}_{t-2} + \cdots + \mathbf{a}_t$$

14.4. MODEL BUILDING STRATEGY FOR MULTIPLE TIME SERIES

TABLE 14.3. Indicator Symbols for Partial Autoregression and Related Statistics for Data in Figure 14.3

Lag ℓ	Indicator Symbols	$M(\ell)^a \simeq \chi_4^2$	Diagonal Elements of $\hat{\Omega}$
1	$\begin{bmatrix} \cdot & + \\ - & + \end{bmatrix}$	356.96	5.30 1.08
2	$\begin{bmatrix} \cdot & \cdot \\ \cdot & + \end{bmatrix}$	7.04	5.16 1.03
3	$\begin{bmatrix} \cdot & \cdot \\ \cdot & \cdot \end{bmatrix}$	2.63	5.07 1.03
4	$\begin{bmatrix} \cdot & \cdot \\ \cdot & \cdot \end{bmatrix}$	4.38	5.01 1.02
5	$\begin{bmatrix} \cdot & \cdot \\ \cdot & \cdot \end{bmatrix}$	2.42	4.95 1.01

$^a \simeq$ means approximately distributed as.

TABLE 14.4. Pattern of Partial Autoregression and Related Statistics for Data in Figure 14.2

Lag	Pattern of $\hat{\rho}(\ell)$	$M(\ell) \simeq \chi_4^2$	Diagonal Elements of $\hat{\Omega}$
1	$\begin{bmatrix} - & - \\ + & - \end{bmatrix}$	123.2	4.78 1.88
2	$\begin{bmatrix} \cdot & \cdot \\ + & - \end{bmatrix}$	75.9	4.75 1.43
3	$\begin{bmatrix} + & \cdot \\ + & - \end{bmatrix}$	35.2	4.63 1.23
4	$\begin{bmatrix} \cdot & \cdot \\ + & \cdot \end{bmatrix}$	27.5	4.63 1.08
5	$\begin{bmatrix} \cdot & \cdot \\ + & \cdot \end{bmatrix}$	16.6	4.61 1.04
6	$\begin{bmatrix} \cdot & \cdot \\ + & \cdot \end{bmatrix}$	13.5	4.53 .98
7	$\begin{bmatrix} \cdot & - \\ + & \cdot \end{bmatrix}$	16.5	4.38 .94
8	$\begin{bmatrix} \cdot & \cdot \\ \cdot & - \end{bmatrix}$	8.1	4.31 .91

we find

$$\Pi_1 = \begin{bmatrix} -.2 & -.3 \\ .6 & -1.1 \end{bmatrix}, \Pi_2 = \begin{bmatrix} .14 & -.39 \\ .78 & -1.03 \end{bmatrix}, \ldots, \Pi_6 = \begin{bmatrix} .23 & -.25 \\ .49 & -.51 \end{bmatrix} \quad (14.23)$$

$$|\Pi_1| = .4, |\Pi_2| = .16, \ldots, |\Pi_6| = .0041.$$

Thus, although the determinants $|\Pi_j|$ decrease rapidly towards zero as j increases, the elements of Π_j converge to zero very slowly so that many autoregressive terms would be needed to provide an adequate approximation.

In general, the pattern of the partial autoregression matrices, the $M(\ell)$ statistic, and the diagonal elements of the residual covariance matrix are useful to distinguish between moving-average and low-order autoregressive models and, for the latter to tentatively select the appropriate order.

14.4.2. Estimation

Once the order of the model in (14.7) has been tentatively selected, efficient estimates of the associated parameter matrices $\Phi = (\Phi_1, \ldots, \Phi_p)$ and $\Theta = (\Theta_1, \ldots, \Theta_q)$ are then determined by maximizing the likelihood function. Approximate standard errors and correlation matrix of the estimates of elements of the Φ_j's and Θ_j's can also be obtained.

Conditional likelihood
For the ARMA(p,q) model in (14.7), we can write

$$\mathbf{a}_t = \mathbf{z}_t - \mathbf{c} - \Phi_1 \mathbf{z}_{t-1} - \cdots - \Phi_p \mathbf{z}_{t-p} + \Theta_1 \mathbf{a}_{t-1} + \cdots + \Theta_q \mathbf{a}_{t-q}. \quad (14.24)$$

As in the univariate case discussed in Box et al. (1994), the likelihood function can be approximated by a "conditional" likelihood function as follows. The series is regarded as consisting of the n-p vector observations $\mathbf{z}_{p+1}, \ldots, \mathbf{z}_n$. The likelihood function is then determined from $\mathbf{a}_{p+1}, \ldots, \mathbf{a}_n$ using $\mathbf{z}_1, \ldots, \mathbf{z}_p$ as preliminary values and conditional on zero values for $\mathbf{a}_p, \ldots, \mathbf{a}_{p-q+1}$. Thus, as shown in Tunnicliffe-Wilson (1973)

$$\ell_{\text{con}}(\mathbf{c}, \Phi, \Theta, \Omega \mid \mathbf{z}) \propto |\Omega|^{-\frac{n-p}{2}} \exp\left\{ -\frac{1}{2} tr[\Omega^{-1} S(\mathbf{c}, \Phi, \Theta)] \right\} \quad (14.25)$$

where

$$S(\mathbf{c}, \Phi, \Theta) = \sum_{t=p+1}^{n} \mathbf{a}_t \mathbf{a}_t'.$$

14.4. MODEL BUILDING STRATEGY FOR MULTIPLE TIME SERIES

Properties of the maximum likelihood estimates obtained from (14.25) have been discussed in Nicholls (1976, 1977) and Anderson (1980).

It has been shown in Hillmer and Tiao (1979) that this approximation can be seriously inadequate if n is not sufficiently large and if one or more zeros of $|\Theta(B)|$ lie on or close to the unit circle. Specifically, this would lead to estimates of the moving-average parameters with large bias.

Exact likelihood function

For univariate ARMA models, the exact likelihood function has been considered by Tiao and Ali (1971), Newbold (1974), Dent (1977), Ansley (1979), and others. For vector models, this function has been studied by Osborn (1977) for the pure moving-average case, and by Phadke and Kedem (1978), Nicholls and Hall (1979), and Hillmer and Tiao (1979). It takes the form

$$\ell(\mathbf{c}, \Phi, \Theta, \Omega \mid \mathbf{z}) \propto \ell_{\text{con}}(\mathbf{c}, \Phi, \Theta, \Omega \mid \mathbf{z}) \ell_1(\mathbf{c}, \Phi, \Theta, \Omega \mid \mathbf{z}) \qquad (14.26)$$

where $\ell_1(\cdot)$ depends (1) only on. $\mathbf{z}_1, \ldots, \mathbf{z}_p$ if $q = 0$, and (2) on all the data vectors $\mathbf{z}_1, \ldots, \mathbf{z}_n$ if $q \neq 0$. For the general vector ARMA(p,q) model, it has been shown that a close approximation to the exact likelihood can be obtained by considering the transformation $\mathbf{w}_t = \Phi(B)\mathbf{z}_t$ so that

$$\mathbf{w}_t = \Theta(B)\mathbf{a}_t \qquad (14.27)$$

and then apply the exact likelihood results for vector MA(q) model to $\mathbf{w}_t, t = p + 1, \ldots n$.

Because estimation of moving-average parameters using the exact likelihood is computationally rather complex, we propose to employ the conditional method in the preliminary stages of iterative model building and switch to the exact method toward the end.

14.4.3. Diagnostic checking

To guard against model misspecification and to search for directions of improvement, a detailed diagnostic analysis of the residual series $\{\hat{\mathbf{a}}_t\}$ where

$$\hat{\mathbf{a}}_t = \mathbf{z}_t - \hat{\mathbf{c}} - \hat{\Phi}_1 \mathbf{z}_{t-1} - \cdots - \hat{\Phi}_p \mathbf{z}_{t-p} + \hat{\Theta}_1 \hat{\mathbf{a}}_{t-1} + \cdots + \hat{\Theta}_q \hat{\mathbf{a}}_{t-q}. \qquad (14.28)$$

is performed. Useful diagnostic checks include (1) plots of standardized residual series against time and/or other variables and (2) cross-correlation matrices of the residuals $\hat{\mathbf{a}}_t$'s. As before, the structure of the correlations is summarized by indicator symbols. Hosking (1980) and Li and McLeod (1981) have proposed overall χ^2 tests based on the sample cross-correlations of the residuals. However, as noted in Box et al. (1994), such overall tests are not substitutes for more detailed study of the correlation structure.

14.5. ANALYSES OF THREE EXAMPLES

We now apply the model building approach introduced in the preceding section to the following three sets of data:

1. The *Financial Times* ordinary share index, U.K. car production and the *Financial Times* commodity price index: quarterly data 3/1952–4/1967, obtained from Coen et al. (1969). This will be referred to as the SCC data. (See Fig. 1.11.)
2. The gas furnace data given in Box et al. (1994). (See Fig. 1.10.)
3. The census housing data (Hillmer and Tiao 1979).

14.5.1. The SCC data

Let

z_{1t} = *Financial Times* ordinary share index
z_{2t} = U.K. car production index
z_{3t} = *Financial Times* commodity price index

The authors of the original study were interested in the possibility of predicting z_{1t} from lagged values of z_{2t} and z_{3t} using a standard regression analysis in which z_{1t} was treated as a dependent variable and $z_{2(t-6)}$ and $z_{3(t-7)}$ as regressors or independent variables. For a critical evaluation of this approach, see Box and Newbold (1970). Here we consider what structure is revealed by the present multiple time series analysis, in which the three series are jointly modeled.

Tentative specification
We see in Table 14.5 that the original series show high and persistent auto- and cross-correlations. Examination of the partials and related statistics in Table 14.6 shows that for $\ell > 2$, most of the elements of $\hat{\wp}(\ell)$ are small compared with their estimated standard errors, and the $M(\ell)$ statistic fails to show significant improvement. Table 14.7 shows that the pattern of the cross correlations of the residuals after AR(2) is consonant with estimated white noise. However, note that there is one large residual correlation at lag 1 after the AR(1) fit, suggesting also the possibility of an ARMA(1,1) model.

Estimation
Both an AR(2) model and an ARMA(1,1) model were fitted using the exact likelihood method but results are given only for the ARMA(1,1) model which produced a marginally better representation. For this model,

$$(\mathbf{I} - \Phi)\mathbf{z}_t = \mathbf{c} + (\mathbf{I} - \Theta B)\mathbf{a}_t \qquad (14.29)$$

14.5. ANALYSES OF THREE EXAMPLES

TABLE 14.5. Pattern of Sample Cross-Correlations for the SCC Data

	Z_1 Stocks	Z_2 Cars	Z_3 Commodities
Z_1 Stocks	+++++++ ···· ······	------------ --------	------------ --------
Z_2 Cars	············ ····· +++	+++++++++++ +++ ·····	+++++++++++ ++ ·····
Z_3 Commodities	---------···· ··· +++++	+++++++++++ +++++ ··	+++++++++++ +++++ ···

TABLE 14.6. Partial Autoregression and Related Statistics: SCC Data

Lag	Indicator Symbols for Partials	$M(\ell)$ Statistic $\simeq \chi_9^2$	Diagonal Elements of $\hat{\Omega} \times 10$
1	$\begin{bmatrix} + & \cdot & \cdot \\ \cdot & + & \cdot \\ \cdot & \cdot & + \end{bmatrix}$	301.3	.44 .89 1.62
2	$\begin{bmatrix} - & \cdot & \cdot \\ \cdot & \cdot & \cdot \\ - & + & - \end{bmatrix}$	18.6	.40 .84 1.23
3	$\begin{bmatrix} \cdot & \cdot & \cdot \\ \cdot & \cdot & \cdot \\ \cdot & \cdot & \cdot \end{bmatrix}$	9.6	.37 .81 1.21
4	$\begin{bmatrix} \cdot & \cdot & \cdot \\ \cdot & \cdot & \cdot \\ \cdot & \cdot & \cdot \end{bmatrix}$	3.6	.36 .79 1.19
5	$\begin{bmatrix} \cdot & + & \cdot \\ \cdot & + & \cdot \\ \cdot & \cdot & \cdot \end{bmatrix}$	11.9	.32 .70 1.11

TABLE 14.7. Pattern of Cross-Correlation Matrices of Residuals: SCC Data

Lag	1	2	3	4	5	6	7	8
(a) AR(1) Model								
	··+	···	···	···	···	···	···	···
	···	···	···	···	···	···	···	···
	···	···	···	···	···	···	··+	···
(b) AR(2) Model								
	···	···	···	···	···	···	···	···
	···	···	···	−··	···	···	···	···
	···	···	···	···	···	···	··+	···

TABLE 14.8. Estimation Results for the Model (14.29): SCC Data (Exact Likelihood)[a]

$\hat{c}$	$\hat{\Phi}$	$\hat{\Theta}$	$\hat{\Omega}$

(a) Full Model

$$\begin{bmatrix} 1.11 \\ (.64) \\ 1.74 \\ (.82) \\ 4.08 \\ (1.47) \end{bmatrix} \quad \begin{bmatrix} .81 & .15 & -.06 \\ (.08) & (.07) & (.04) \\ -.07 & .98 & -.09 \\ (.10) & (.10) & (.05) \\ -.32 & .30 & .76 \\ (.18) & (.17) & (.08) \end{bmatrix} \quad \begin{bmatrix} -.29 & .23 & .06 \\ (.15) & (.11) & (.07) \\ -.45 & .20 & -.15 \\ (.22) & (.17) & (.11) \\ -.79 & .57 & -.44 \\ (.28) & (.21) & (.13) \end{bmatrix} \quad \begin{bmatrix} .037 & & \\ .022 & .078 & \\ .013 & .022 & .129 \end{bmatrix}$$

(b) Restricted Model (Intermediate)

$$\begin{bmatrix} .13 \\ (.09) \\ .59 \\ (.05) \\ 2.48 \\ (1.10) \end{bmatrix} \quad \begin{bmatrix} .90 & .08 & \cdot \\ (.06) & (.06) & \cdot \\ \cdot & .92 & -.02 \\ \cdot & (.04) & (.04) \\ \cdot & \cdot & .85 \\ \cdot & \cdot & (.07) \end{bmatrix} \quad \begin{bmatrix} -.22 & .15 & \cdot \\ (.14) & (.10) & \cdot \\ -.31 & \cdot & \cdot \\ (.17) & \cdot & \cdot \\ -.55 & .22 & -.44 \\ (.23) & (.15) & (.12) \end{bmatrix} \quad \begin{bmatrix} .042 & & \\ .022 & .079 & \\ .017 & .021 & .131 \end{bmatrix}$$

(c) Restricted Model (Final)

$$\begin{bmatrix} .12 \\ (.08) \\ .24 \\ (.10) \\ 2.76 \\ (1.07) \end{bmatrix} \quad \begin{bmatrix} .98 & \cdot & \cdot \\ (.03) & \cdot & \cdot \\ \cdot & .93 & \cdot \\ \cdot & (.04) & \cdot \\ \cdot & \cdot & .83 \\ \cdot & \cdot & (.06) \end{bmatrix} \quad \begin{bmatrix} \cdot & \cdot & \cdot \\ \cdot & \cdot & \cdot \\ -.40 & \cdot & -.41 \\ (.23) & \cdot & (.12) \end{bmatrix} \quad \begin{bmatrix} .045 & & \\ .024 & .085 & \\ .019 & .023 & .134 \end{bmatrix}$$

[a] For this example, estimates from the conditional likelihood for the ARMA(1,1) case are very close to the exact results.

Table 14.8 shows the initial unrestricted fit and also the fits for two simpler versions obtained by setting to zero those coefficients whose estimates were small compared to their standard errors.

Diagnostic checking
Table 14.9 suggests that the restricted ARMA(1,1) model provides an adequate representation of the data.

Implication of the model
The final model implies that the system is approximately

$$(1 - .98B)z_{1t} = a_{1t}$$
$$(1 - .93B)z_{2t} = .2 + a_{2t} \tag{14.30}$$
$$(1 - .83B)z_{3t} = 2.8 + .40a_{1(t-1)} + (1 + .41B)a_{3t}.$$

14.5. ANALYSES OF THREE EXAMPLES

TABLE 14.9. Pattern of Residual Cross-Correlations after Final Restricted ARMA(1,1) Model Fit: SCC Data

	$\hat{a}_1$	$\hat{a}_2$	$\hat{a}_3$
$\hat{a}_1$	....	....	....
$\hat{a}_2$	....	....	....
$\hat{a}_3$	....	....	....

Thus, the ordinary share $\{z_{1t}\}$ series behaves like a random walk and does not depend on the lagged values of the other two series. The same can roughly be said about the car production index $\{z_{2t}\}$ series. From (14.30) we can make use of the first equation to write the model for z_{3t} as

$$(1 - .83B)z_{3t} = 2.8 + .40(1 - .98B)z_{1(t-1)} + (1 + .41B)a_{3t} \quad (14.31)$$

which implies that the ordinary share z_{1t} is a *leading indicator* at lag 1 of the commodity index z_{3t}. The impact of z_{1t} is, however, small, as can be seen, for example, by the improvement achieved over the corresponding best fitting univariate model for z_{3t}, which was

$$(1 - .78B)z_{3t} = 3.6 + (1 + .53B)a_t. \quad (14.32)$$

The estimated residual variance $\hat{\sigma}_a^2 = .151$ from the univariate model is not much larger than the value .134 for the estimated variance of a_{3t} obtained from the final vector model.

In conclusion, we see that the three series z_{1t}, z_{2t}, and z_{3t} are essentially *not* dynamically related. Although the multiple time series analysis fails to reveal anything very surprising for this example, it shows what is there and does not mislead.

14.5.2. The gas furnace data

The two series shown in Figure 1.10 consist of (1) input gas rate and (2) output CO_2 concentration at 9-s intervals from a gas furnace. We shall let z_{1t} = gas rate + .057 and $z_{2t} = CO_2 - 5.35$. This set of data was employed in Box et al. (1994) to illustrate a procedure of tentative specification, fitting, and checking transfer function model of the form (14.5) for $k = 2$ relating two time series, one of which is *known* to be input for the other. Using this approach, the following models were found for the input z_{1t} and the output z_{2t}

$$(1 - 1.97B + 1.37B^2 - .34B^3)z_{1t} = a_{1t}, \quad \hat{\sigma}_{a_1}^2 = .0353$$
$$z_{2t} = \frac{\omega(B)}{\delta(B)} B^3 z_{1t} + \varphi^{-1}(B)a_{2t}, \quad \hat{\sigma}_{a_2}^2 = .0561 \quad (14.33)$$

TABLE 14.10. Tentative Identification for the Gas Furnace Data

(a) Pattern of Cross-Correlations of Original Data

	Z_{1t}	Z_{2t}
Z_{1t}	+++++++++++	- - - - - - - - -
Z_{2t}	- - - - - - - - -	+++++++++++

(b) M Statistic for Partial Autoregression

Lag ℓ	1	2	3	4	5	6	7	8	9	10	11
$M(\ell)$	1650	665	31.7	22.5	5.6	12.9	1.8	8.0	3.5	0	2.0

(c) Pattern of Cross-Correlations of Residuals after AR(6) Fit

	$\hat{a}_{1t}$	$\hat{a}_{2t}$
$\hat{a}_{1t}$	. . . .	. . . .
$\hat{a}_{2t}$	. . . .	. . . .

where $\omega(B) = -(.53 + .37B + .51B^2)$, $\delta(B) = 1 - .57B$, $\varphi(B) = 1 - 1.53B + .63B^2$, and the $\{a_{1t}\}$ and $\{a_{2t}\}$ series are assumed independent.

Particularly when we are dealing with econometric rather than engineering models, feedback relationships may not be known a priori, and it is of interest, therefore, to analyze the data using the present multivariate approach where no distinction is made between an input and output variable, and the fact that no feedback could occur in the system is not used.

Tentative specification

In Table 14.10, we see that the auto- and cross-correlations of the original data in part (a) are persistently large in magnitude, ruling out low-order moving-average models, the $M(\ell)$ statistic in part (b) suggests that an AR(6) model might be appropriate, and the residual cross-correlation pattern after an AR(6) fit in part (c) seems to verify the appropriateness of this model.

Estimation results

Estimation results corresponding to an unrestricted AR(6) model

$$(I - \Phi_1 B - \cdots - \Phi_6 B^6)z_t = a_t \tag{14.34}$$

are as follows:

$$\hat{\Phi}_1 = \begin{bmatrix} 1.93 & -.05 \\ (.06) & (.05) \\ .06 & 1.55 \\ (.08) & (.06) \end{bmatrix}, \quad \hat{\Phi}_2 = \begin{bmatrix} -1.20 & .10 \\ (.13) & (.08) \\ -.14 & -.59 \\ (.16) & (.11) \end{bmatrix}, \quad \hat{\Phi}_3 = \begin{bmatrix} .17 & -.08 \\ (.15) & (.09) \\ -.44 & -.17 \\ (.19) & (.11) \end{bmatrix},$$

14.5. ANALYSES OF THREE EXAMPLES

$$\hat{\Phi}_4 = \begin{bmatrix} -.16 & .03 \\ (.15) & (.09) \\ .15 & .13 \\ (.19) & (.11) \end{bmatrix}, \quad \hat{\Phi}_5 = \begin{bmatrix} .38 & -.04 \\ (.14) & (.08) \\ -.12 & .06 \\ (.18) & (.10) \end{bmatrix}, \quad \hat{\Phi}_6 = \begin{bmatrix} -.22 & .03 \\ (.08) & (.03) \\ .25 & -.04 \\ (.11) & (.04) \end{bmatrix},$$

$$\hat{\Omega} = \begin{bmatrix} .0345 & \\ -.0023 & .0566 \end{bmatrix}, \quad \text{and} \quad \hat{\rho}_{a_1,a_2}(0) = .045.$$

If we let

$$\hat{\Phi}_\ell = \begin{bmatrix} \hat{\phi}_{11}^{(\ell)} & \hat{\phi}_{12}^{(\ell)} \\ \hat{\phi}_{21}^{(\ell)} & \hat{\phi}_{22}^{(\ell)} \end{bmatrix} \tag{14.35}$$

then we see that $\hat{\phi}_{12}^{(\ell)}$ are small compared with their standard errors over all lags ℓ, confirming (as in this case is known from the physical nature of the apparatus generating the data) that there is a unidirectional relationship between z_{1t} and z_{2t} involving no feedback. Also, $\hat{\phi}_{21}^{(\ell)}$ is small for $\ell = 1, 2$, and the residuals $\hat{a}_{1t}$ and $\hat{a}_{2t}$ are essentially uncorrelated, implying a delay of three periods. It should be noted in addition, that the estimated variances for a_{1t} and a_{2t} are very close to those for a_{1t} and a_{2t} in (14.33), and their correlation is negligible.

To facilitate comparison with (14.33), we set

$$\phi_{11}^{(\ell)} = 0, \ell > 3; \quad \phi_{12}^{(\ell)} = 0, \text{ all } \ell; \quad \phi_{21}^{(\ell)} = 0, \ell = 1, 2; \quad \phi_{22}^{(\ell)} = 0, \ell = 5, 6.$$

Estimation results for this restricted AR(6) model are then

$$\hat{\phi}_1 = \begin{bmatrix} 1.98 \\ (.06) \\ & 1.53 \\ & (.06) \end{bmatrix}, \quad \hat{\phi}_2 = \begin{bmatrix} -1.38 \\ (.10) \\ & -.58 \\ & (.11) \end{bmatrix}, \quad \hat{\phi}_3 = \begin{bmatrix} .35 \\ (.06) \\ -.53 & -.14 \\ (.07) & (.10) \end{bmatrix},$$

(14.36)

$$\hat{\phi}_4 = \begin{bmatrix} & \\ .11 & .12 \\ (.16) & (.04) \end{bmatrix}, \quad \hat{\phi}_5 = \begin{bmatrix} & \\ -.04 \\ (.17) \end{bmatrix}, \quad \hat{\phi}_6 = \begin{bmatrix} & \\ .21 \\ (.11) \end{bmatrix},$$

$$\hat{\Omega} = \begin{bmatrix} .0359 & \\ -.0029 & .0561 \end{bmatrix}, \quad \hat{\rho}_{a_1,a_2}(0) \cong 0.$$

Examination of the pattern of the cross-correlations of the residuals suggests that the model is adequate.

Implication of the bivariate model

The final AR(6) model (14.36) can be written

$$\begin{bmatrix} \phi_{11}(B) & \\ \phi_{21}(B) & \phi_{22}(B) \end{bmatrix} \begin{bmatrix} z_{1t} \\ z_{2t} \end{bmatrix} = \begin{bmatrix} a_{1t} \\ a_{2t} \end{bmatrix} \qquad (14.37)$$

where $\phi_{11}(B) = (1 - 1.98B + 1.38B^2 - .35B^3)$, $\phi_{21}(B) = (.53 - .11B - .21B^3)B^3$, and $\phi_{22}(B) = (1 - 1.53B + .58B^2 + .14B^3 - .12B^4)$. Assuming that a_{1t} and a_{2t} are uncorrelated, the input model $\phi_{11}(B)z_{1t} = a_{1t}$ with Var$(a_{1t}) = .0359$ is essentially the same as that in (14.33). Now the model relating the output z_{2t} to the input z_{1t} is

$$z_{2t} = -\frac{\phi_{21}(B)}{\phi_{22}(B)} z_{1t} + \frac{1}{\phi_{22}(B)} a_{2t} \qquad (14.38)$$

with Var$(a_{2t}) = .0561$. The noise model $\phi_{22}^{-1}(B)a_{2t}$ is not very different from the corresponding one $\varphi^{-1}(B)a_{2t}$ in (14.33), but the dynamic model $-\phi_{22}^{-1}(B)\phi_{21}(B)z_{1t}$ at first sight appears markedly different from the first term on the RHS of the output model in (14.33). The reason is that in the form (14.38) the denominators of the dynamic model and that of the noise model are constrained to be identical. This restriction is not present in the transfer function model (14.33). This less restrictive form can however be written in the form of (14.38) if we set

$$\phi_{22}(B) = \varphi(B); \qquad -\phi_{21}(B) = \omega(B)B^3\{\varphi(B)\delta^{-1}(B)\}.$$

For this example, the factor $\varphi(B)\delta^{-1}(B) \cong 1 - .96B$ and it is then seen that the models are in fact very similar. This may be confirmed by comparing the impulse response weights in Table 14.11, where

$$\omega(B)B^3\delta^{-1}(B) = \sum_{j=0}^{\infty} v_j B^j; \qquad -\phi_{21}(B)\phi_{22}^{-1}(B) = \sum_{j=0}^{\infty} v_j^* B^j.$$

Implications on general time series model building

The relative merit of the vector ARMA procedure and the transfer function modeling of the system will depend on how much is known or we are prepared to assume. In some applications, particularly in engineering and most examples of intervention analysis,

TABLE 14.11. Impulse Response Weights for the Gas Furnace Data

j	0	1	2	3	4	5	6	7	8	9	10	11	12
v_j				$-.53$	$-.67$	$-.89$	$-.51$	$-.29$	$-.17$	$-.09$	$-.05$	$-.03$	$-.02$
v_j^*				$-.53$	$-.70$	$-.77$	$-.48$	$-.26$	$-.09$	$-.01$	$.01$	$.00$	$-.01$

14.5. ANALYSES OF THREE EXAMPLES

an adequate initial specification may be possible from knowledge of the nature of the problem. This may allow a flow diagram showing the feedback structure to be drawn and likely orders to be guessed for the various dynamic components. The resulting models can then be directly fitted in the manner described and illustrated in Box and MacGregor (1974, 1976) and Box and Tiao (1975). For a single input with feedback known to be absent, a prewhitening method is given in Box et al. (1994) for *identifying* an unknown dynamic system but extension of this identification method to multiple inputs is rather complex.

Particularly for economic and business examples, however, the feedback structure and orders of the multiple system are seldom known. The present multiple time series procedure has the great advantage that it allows *identification* of the feedback and dynamic structure. Furthermore

1. A one-sided causal relationship, if it exists, will emerge in the identification process, and the stochastic structures of the input as well as the transfer function relationship between input and output will be modeled simultaneously.
2. Stochastic multiple input and multiple output situations are readily handled.
3. A useful method is provided for seeking leading indicators in economic and business applications. In this context it should be noted that a unidirectional dynamic relationship may not exist between two time series even when one variable is known to be the input for the other. One reason for this phenomenon is the effect of temporal aggregation. As shown in Tiao and Wei (1976), pseudofeedback relationships could occur because of this temporal aggregation effect, and it would be a mistake to impose a transfer function model in such a situation;
4. However, when a simple transfer function structure of the form (14.5) *is* appropriate, the present multiple time series approach could rarely reproduce it directly [see, e.g., (14.33) and (14.38)], and some analysis of the fitted form might be necessary to reveal a more parsimonious and more easily understood structure.

14.5.3. The census housing data

As an further example, we illustrate the modeling and analysis of seasonal time series by considering the monthly single-family housing starts z_{1t} and houses sold z_{2t} for the period January 1965 through May 1975. The data were obtained from the "survey of current business," and are plotted in Figure 14.5. Because of the strong seasonal behavior in these series, one might well be led to consider the seasonally difference data $u_{it} = (1 - B^{12})z_{it}$, $i = 1,2$. The seasonally differenced series are shown in Figure 14.6.

Models for individual series
In Hillmer and Tiao (1979), it is found that each series individually can be adequately represented by a univariate multiplicative model of the form

$$(1 - B)(1 - B^{12})z_t = (1 - \theta_1 B)(1 - \theta_{12} B^{12})a_t \qquad (14.39)$$

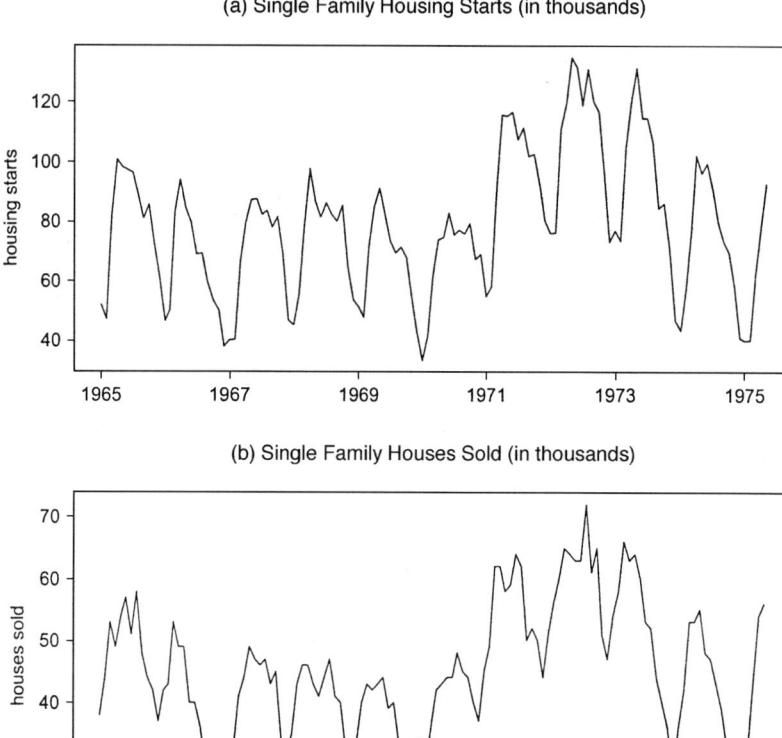

FIGURE 14.5 The census housing data for January 1965–May 1975.

Table 14.12 compares the parameter estimates obtained by employing the exact (E) likelihood (14.26) with those from the conditional (C) likelihood (14.25), where the numbers in the parentheses are the associated estimated standard errors.

We observe that for both series the estimates $\hat{\theta}_{12}$ of the seasonal moving average parameter are appreciably smaller and the corresponding estimates $\hat{\sigma}_a^2$ larger for C

TABLE 14.12. Parameter Estimates for Individual Models: Census Housing Series

		$\hat{\theta}_1$	$\hat{\theta}_{12}$	$\hat{\sigma}_a^2$
Housing starts z_{1t}	E	.28 (.09)	.91 (.06)	41.61
	C	.30 (.09)	.75 (.07)	50.49
Houses sold z_{2t}	E	.16 (.10)	1.00 (.06)	11.93
	C	.24 (.10)	.72 (.08)	16.46

14.5. ANALYSES OF THREE EXAMPLES

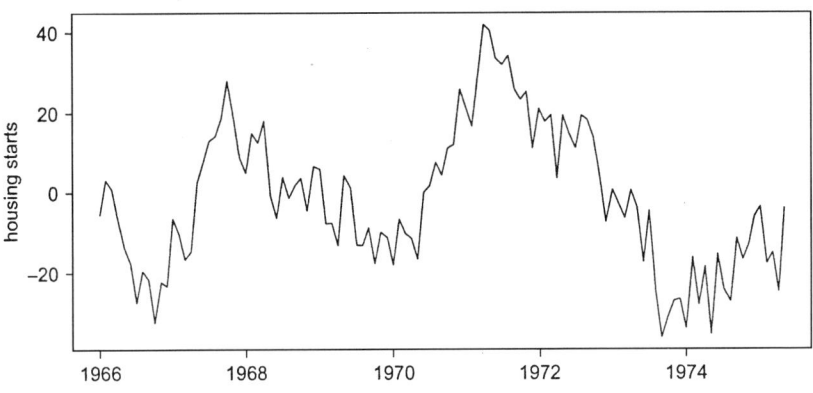

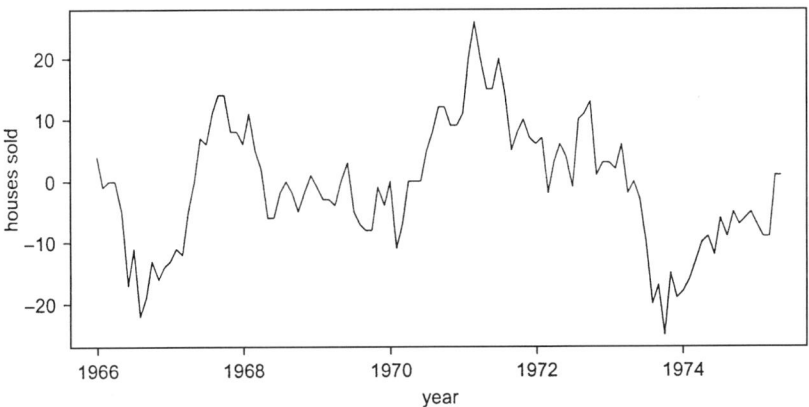

FIGURE 14.6 The seasonally differenced series: January 1966–May 1975.

than for E. It appears that θ_{12} is close to unity, especially for the "houses sold" series z_{2t}, implying a possibly deterministic seasonal structure:

$$(1 - B)z_t = S_t + (1 - \theta B)a_t \qquad (14.40)$$

where $(1 - B^{12})S_t = 0$. Such a structure, however, would not be detected if the conditional likelihood method were employed.

Bivariate model

Part (a) of Table 14.13 shows the pattern of the sample cross correlations of $\mathbf{u}_t = (u_{1t}, u_{2t})'$ indicating that low order vector MA model would not be appropriate. Part (b) of the same table gives the $M(\ell)$ statistics for $\ell = 1, \ldots, 5$, and part (c) shows the pattern of the cross correlations of the residuals after a bivariate AR(1) fit. These

TABLE 14.13. Tentative Identification for the Seasonally Differenced Housing Data U_t

(a) Pattern of Cross-Correlations

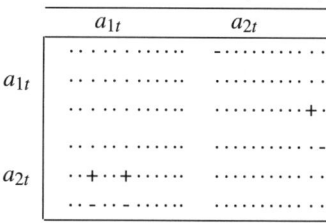

(b) $M(\ell)$ Statistics Corresponding to Partial Autoregression Matrices

Lag ℓ	1	2	3	4	5
$M(\ell)$	218.6	3.5	2.3	4.7	5.4

(c) Pattern of Residual Cross-Correlations after AR(1) Fit

	a_{1t}	a_{2t}
a_{1t}	....	....+. ....
a_{2t}	..+..+...... ...-........	....

summaries suggest the tentative model

$$(\mathbf{I} - \Phi B)(1 - B^{12})\mathbf{z}_t = (\mathbf{I} - \Theta B^{12})\mathbf{a}_t \qquad (14.41)$$

Estimation and checking

Table 14.14 summaries the estimation results corresponding to

1. The full bivariate model in (14.41) using the conditional likelihood method
2. The full bivariate model using the exact likelihood method
3. The restricted model by setting "small" parameter estimates to zero

Table 14.15 shows the pattern of the sample cross-correlations of the residuals corresponding to the restricted case, showing that the model gives an adequate representation of the series.

Interpretation

For the full bivariate model, comparing the results of the conditional likelihood with those from the exact likelihood, we see that

14.5. ANALYSES OF THREE EXAMPLES

TABLE 14.14. Estimation Results for the Model (16.41): Census Housing Data

(1) Full Model Conditional Likelihood	(2) Full Model Exact Likelihood	(3) Restricted Model Exact Likelihood
$\hat{\Phi} = \begin{bmatrix} .47 & .89 \\ (.07) & (.13) \\ .14 & .69 \\ (.05) & (.08) \end{bmatrix}$	$\begin{bmatrix} .46 & .95 \\ (.07) & (.14) \\ .10 & .76 \\ (.05) & (.09) \end{bmatrix}$	$\begin{bmatrix} .42 & 1.03 \\ (.07) & (.13) \\ & .93 \\ & (.04) \end{bmatrix}$
$\hat{\Theta} = \begin{bmatrix} .75 & .06 \\ (.07) & (.11) \\ .07 & .69 \\ (.05) & (.08) \end{bmatrix}$	$\begin{bmatrix} 1.01 & -.04 \\ (.07) & (.12) \\ .05 & .99 \\ (.05) & (.07) \end{bmatrix}$	$\begin{bmatrix} .94 & \\ (.06) & \\ & 1.00 \\ & (.06) \end{bmatrix}$
$\hat{\Omega} = \begin{bmatrix} 37.51 & \\ 6.29 & 15.15 \end{bmatrix}$	$\begin{bmatrix} 28.09 & \\ 4.98 & 11.13 \end{bmatrix}$	$\begin{bmatrix} 29.75 & \\ 5.89 & 11.83 \end{bmatrix}$
		$\hat{\rho}(a_1, a_2) \doteq .31$

1. There is a substantial increase in the estimated values of the diagonal elements of Θ to near unity, and a corresponding decrease in the estimated variances of the residuals, when the exact method is used
2. In contrast, little change occurs in the estimates of the elements of Φ

Now from the restricted model, we can write, for the "houses sold" series z_{2t}

$$(1 - .93B)(1 - B^{12})z_{2t} = (1 - B^{12})a_{2t} \tag{14.42}$$

which means that

$$(1 - .93B)z_{2t} = S_{2t} + a_{2t}, \quad \hat{\sigma}_{22} = 11.8 \tag{14.43}$$

where S_{2t} satisfies the relation $S_{2t} = S_{2(t-12)}$. Thus, z_{2t} behaves nearly like a random walk with a deterministic seasonal component and does not depend on the past of z_{1t}. We note that this result is essentially the same as the individual model, shown in Table 14.12, for the "house sold" series fitted using the exact likelihood.

TABLE 14.15. Residual Cross-Correlations for the Restricted Model: Census Housing Data

	a_{1t}	a_{2t}
a_{1t}	....	......+..... ....
a_{2t}	....+......	....+-----

On the other hand, for the "housing starts" series z_{1t}, we have that

$$(1 - .42B)(1 - B^{12})z_{1t} = 1.03(1 - B^{12})z_{2(t-1)} + (1 - .94B^{12})a_{1t} \quad (14.44)$$

so that the seasonal differencing operator $(1 - B^{12})$ again nearly cancels yielding approximately

$$(1 - .42B)z_{1t} = S_{1t} + 1.03z_{2(t-1)} + a_{1t}, \quad \hat{\sigma}_{11} = 29.8 \quad (14.45)$$

where $S_{1t} = S_{1(t-12)}$. Thus, housing starts z_{1t} depends not only on its own past, $z_{1(t-1)}$ but also on the past of houses sold, $z_{2(t-1)}$. From Table 14.12 the appropriate individual model for z_{1t} is

$$(1 - B)(1 - B^{12})z_{1t} = (1 - .28B)(1 - .91B^{12})a_t \quad (14.46)$$

or approximately

$$(1 - B)z_{1t} = S_{1t} + (1 - .28B)a_t, \quad \hat{\sigma}_a^2 = 41.61 \quad (14.47)$$

We see from (14.45) that the difference operator $(1 - B)$ in (14.47), indicating that z_{1t} is nonstationary, arises because of the dependence of z_{1t} on $z_{2(t-1)}$. Also, by comparing the estimated variance $\hat{\sigma}_a^2$ in (14.47) with the corresponding estimated variance $\hat{\sigma}_{11}$ in (14.45), a substantial reduction occurs when the information $z_{2(t-1)}$ is utilized.

In summary, this example shows that (1) the existence of a deterministic seasonal component can be detected when the exact likelihood method is employed, (2) an appreciable reduction in the one step ahead forecast variance can occur by modeling several series jointly, and (3) although individually both z_{1t} and z_{2t} are nonstationary, there is really only one nonstationary component z_{2t} since nonstationarity of z_{1t} stems from its dependence on $z_{2(t-1)}$. The last point is closely related to the concept of cointegration discussed below.

14.6. STRUCTURAL ANALYSIS OF MULTIVARIATE TIME SERIES

In modeling and analysis of multivariate time series, it is of interest, and often critical, to examine the structure of the relations among the component series. The vector ARMA model (14.7) is a straightforward generalization of the univariate ARMA model (14.3). This direct generalization, however, creates a number of difficulties. In particular, each AR or MA term in the vector model contains k^2 number of parameters so that even for a moderate k, there will be an overflow of parameters whose estimates are often highly correlated. This will make it difficult, if not impossible, to comprehend the fitting result. For univariate ARIMA models, parsimony is achieved by making the orders p and q as small as possible, but this is not sufficient for the multivariate case. Here we need to simplify the structure of the parameter matrices Φ's and Θ's, and a possible way to achieve this is by considering *linear transformations* of the original component series. In addition, in many problems it is also of interest to explore the

14.6. STRUCTURAL ANALYSIS OF MULTIVARIATE TIME SERIES

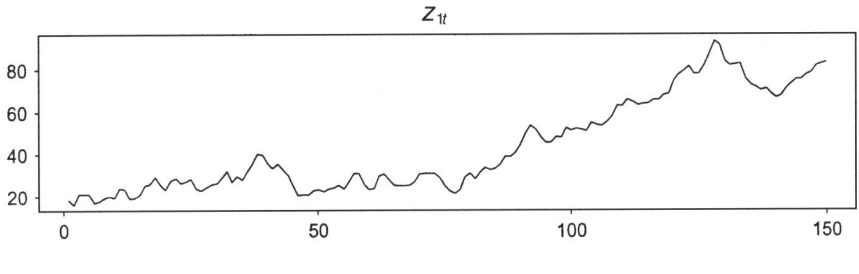

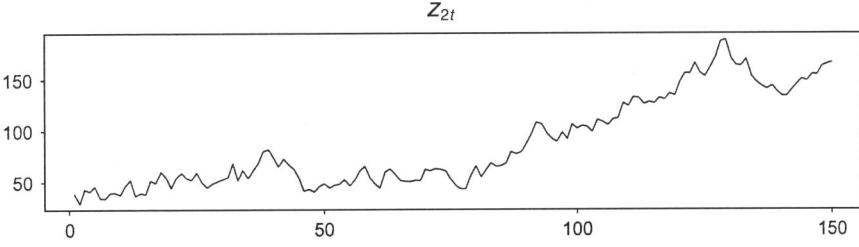

FIGURE 14.7 Generated bivariate ARMA(1,1) series.

existence of possibly simpler underlying structures which explain important features of the observed series. For example, there may exist a linear combination of the component series which is responsible for most of the variation of the data, or another combination responsible for the dynamic growth of a number of the components. In this and the next sections, we discuss some useful types of linear transformations for structural analysis.

As an illustration of the usefulness of linear transformations, consider the bivariate time series $\mathbf{z}_t = (z_{1t}, z_{2t})'$ of 150 observations shown in Figure 14.7, generated from the following ARMA(1,1) model

$$(\mathbf{I} - \Phi B)\mathbf{z}_t = (\mathbf{I} - \Theta B)\mathbf{a}_t \qquad (14.48)$$

where

$$\Phi = \begin{bmatrix} 3 & -1 \\ 6 & -2 \end{bmatrix}, \quad \Theta = \begin{bmatrix} -.5 & .5 \\ -1 & 1 \end{bmatrix} \quad \text{and} \quad \Omega = \begin{bmatrix} 1 & .3 \\ .3 & 1 \end{bmatrix}.$$

The matrices Φ and Θ contain a total of eight non-zero parameters. Both series appear to be nonstationary and move in tandem. It is easily shown that individually each component in $\mathbf{z}_t = (z_{1t}, z_{2t})'$ follows a nonstationary ARIMA(0,1,1) model. Consider, however, the linear transformation

$$\mathbf{y}_t = \mathbf{T}\mathbf{z}_t \quad \text{where} \quad \mathbf{T}' = \begin{bmatrix} \mathbf{v}_0^{(1)}, & \mathbf{v}_0^{(2)} \end{bmatrix} = \begin{bmatrix} 2 & -1 \\ -1 & 1 \end{bmatrix}. \qquad (14.49)$$

The transformed bivariate series $\mathbf{y}_t = (y_{1t}, y_{2t})'$ is shown in Figure 14.8, where y_{1t}

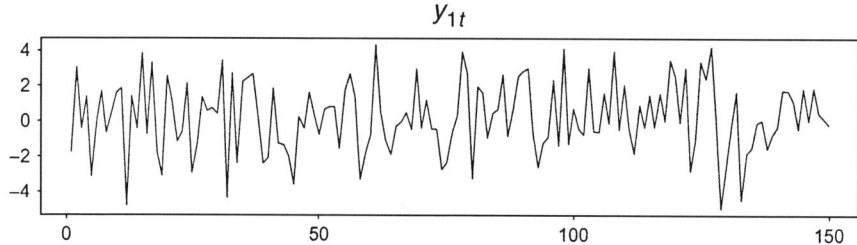

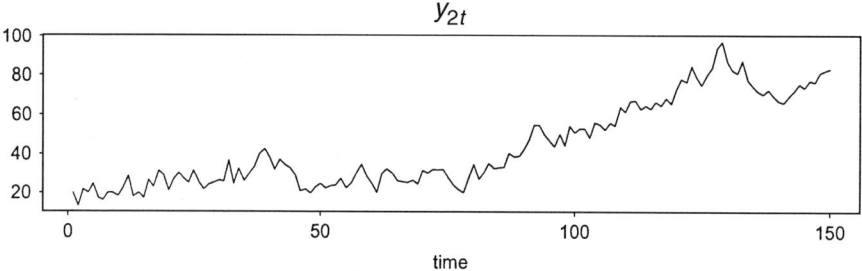

FIGURE 14.8 Transformed bivariate ARMA(1,1) series.

is clearly seen to be stationary and y_{2t}, nonstationary. It is easily shown that the transformed process $\mathbf{y}_t$ has the bivariate ARMA(1,1) model

$$(\mathbf{I} - \Phi^* B)\mathbf{y}_t = (\mathbf{I} - \Theta^* B)\mathbf{b}_t \qquad (14.50)$$

where

$$\Phi^* = \mathbf{T}\Phi\mathbf{T}^{-1} = \begin{bmatrix} 0 & 0 \\ 2 & 1 \end{bmatrix}, \quad \Theta^* = \mathbf{T}\Theta\mathbf{T}^{-1} = \begin{bmatrix} 0 & 0 \\ 0 & .5 \end{bmatrix}, \quad \text{and}$$

$$\mathbf{b}_t = \mathbf{T}\mathbf{a}_t = \begin{bmatrix} b_{1t} \\ b_{2t} \end{bmatrix}.$$

From the model for $\mathbf{y}_t$ in (14.50), we see that

1. Φ^* and Θ^* together contain only three nonzero parameters.
2. The first component $y_{1t} = b_{1t}$ is simply a white-noise process.
3. There is only one nonstationary component y_{2t}.
4. The nonstationarity of the original components z_{1t} and z_{2t}, is due to this common component y_{2t}.

For this example, the linear transformation $\mathbf{T}\mathbf{z}_t$ produces not only parsimony in parameterization but also components whose structure can be substantively meaningful. In the analysis of economic time series, it can happen that a linear combination of

14.6. STRUCTURAL ANALYSIS OF MULTIVARIATE TIME SERIES

several individually nonstationary time series becomes stationary, implying a stable relationship among the component series. The basic idea is discussed in Box and Tiao (1977) and was later popularized by Engle and Granger (1987) in the econometric literature in what has since been called *cointegration*.

14.6.1. A canonical analysis of multiple time series

Here we discuss the linear transformation method proposed by Box and Tiao in their 1977 paper. The basic objective is to assess linear combinations of the observed series $\mathbf{z}_t$ according to their dynamic dependence on the past history of the series. The vector ARMA(p,q) model in (14.7) can be written as

$$\mathbf{z}_t = \hat{\mathbf{z}}_{t-1}(1) + \mathbf{a}_t \qquad (14.51)$$

where $\hat{\mathbf{z}}_{t-1}(1)$ is the one-step-ahead forecast of $\mathbf{z}_t$ made at time $t-1$. Now, $\hat{\mathbf{z}}_{t-1}(1)$ is a linear function of the past observations $(\mathbf{z}_{t-1}, \mathbf{z}_{t-2}, \ldots)$ and is independent of $\mathbf{a}_t$. When the model is invertible, we have that

$$\hat{\mathbf{z}}_{t-1}(1) = [\mathbf{I} - \Pi(B)]\mathbf{z}_t. \qquad (14.52)$$

where $\Pi(B) = \mathbf{I} - \Pi_1 B - \Pi_2 B^2 - \cdots$ and $\Pi(B)$ satisfies the relation $\Theta(B)\Pi(B) = \Phi(B)$.

For stationary series with zero mean vector, let

$$\Gamma_\mathbf{z}(0) = E(\mathbf{z}_t \mathbf{z}_t') \quad \text{and} \quad \Gamma_{\hat{\mathbf{z}}}(0) = E(\hat{\mathbf{z}}_{t-1}(1)\hat{\mathbf{z}}_{t-1}(1)'), \qquad (14.53)$$

Consider the linear transformation $y_t = \mathbf{h}'\mathbf{z}_t$ where $\mathbf{h}$ is a $k \times 1$ vector of constants. The variance of y_t is $\mathbf{h}'\Gamma_\mathbf{z}(0)\mathbf{h}$ where

$$\mathbf{h}'\Gamma_\mathbf{z}(0)\mathbf{h} = \mathbf{h}'\Gamma_{\hat{\mathbf{z}}}(0)\mathbf{h} + \mathbf{h}'\Omega\mathbf{h} \qquad (14.54)$$

Thus, an appropriate measure of the dynamic dependence of y_t on the past history $(\mathbf{z}_{t-1}, \mathbf{z}_{t-2}, \ldots)$ is the variance ratio

$$\lambda = \frac{\mathbf{h}'\Gamma_{\hat{\mathbf{z}}}(0)\mathbf{h}}{\mathbf{h}'\Gamma_\mathbf{z}(0)\mathbf{h}}. \qquad (14.55)$$

It follows from standard canonical correlation analysis that the combination y_t which maximizes this ratio is such that λ is the largest eigenvalue and $\mathbf{h}$ the corresponding eigenvector of

$$\Gamma_\mathbf{z}^{-1}(0)\Gamma_{\hat{\mathbf{z}}}(0) \qquad (14.56)$$

After the largest eigenvalue and the corresponding eigenvector, it may be of interest to consider the next largest, and so on. In general, let

$$0 \le \lambda_1 \le \cdots \le \lambda_k \le 1 \quad \text{and} \quad \mathbf{M}' = [\mathbf{m}_1, \ldots, \mathbf{m}_k] \qquad (14.57)$$

be the k (ordered) eigenvalues and the corresponding matrix of eigenvectors. Consider the linear transformation $\mathbf{y}_t = \mathbf{M}\mathbf{z}_t$. Then

$$\mathbf{y}_t = \hat{\mathbf{y}}_{t-1}(1) + \mathbf{b}_t \tag{14.58}$$

where $\hat{\mathbf{y}}_{t-1}(1) = \mathbf{M}\hat{\mathbf{z}}_{t-1}(1)$, and $\mathbf{b}_t = \mathbf{M}\mathbf{a}_t$, $\hat{\mathbf{y}}_{t-1}(1)$, and $\mathbf{b}_t$ are independent, and it can be readily shown that the covariance matrices of $\mathbf{y}_t$, $\hat{\mathbf{y}}_{t-1}(1)$ and $\mathbf{b}_t$ are all diagonal. In particular, if the eigenvectors are nomalized so that all the components of $\mathbf{y}_t = (y_{1t}, \ldots, y_{kt})'$ have unit variance, then the covariance matrix of $\hat{\mathbf{y}}_{t-1}(1)$ will be $\Lambda = \text{diag}\{\lambda_1, \ldots, \lambda_k\}$ and that of $\mathbf{b}_t$, $\mathbf{I} - \Lambda$.

These results highlight the characteristics of the canonical transformation (14.58). First, the transformed component y_{1t} is the least dynamically dependent linear combination of $\mathbf{z}_t$ and y_{kt} is the most dynamically dependent linear combination. The dynamical dependency of the y_{it} series ranges from λ_1 to λ_k. If $\lambda_1 = 0$, $y_{1t} = b_{1t}$ so that y_{1t} is a white noise process with a constant variance. Since y_{1t} is a linear combination of $\mathbf{z}_t$, this implies that there exists a very stationary contemporaneous relationship among the k original series. On the other extreme, when $\lambda_k \to 1$, it can be shown that y_{kt} will approach a series with a root on the unit circle in its model so that $\mathbf{z}_t$ will contain a nonstationary component. Thus, one may also regard the λ's as measures of the stationarity of the transformed components, with small values signifying the existence of very stationary components and values close to unity, nonstationart components. The idea here is, of course, closely related to cointegration.

For illustration, Box and Tiao have applied the canonical transformation method to the hog data given in Quenouille (1957) consisting of annual observations of hog supply, hog prices, corn supply, corn prices, and farm wages for the period 1867–1948. Individually, all five series are apparently nonstationary. (See Fig. 1.12 of Chapter 1.) By fitting a vector AR(1) model to the data, they have found that

$$\hat{\lambda}_1 = .0232, \quad \hat{\lambda}_2 = .1421, \quad \hat{\lambda}_3 = .5061, \quad \hat{\lambda}_4 = .6901, \quad \hat{\lambda}_5 = .8868.$$

The five transformed series are shown in Figure 14.9. The first two transformed series corresponding to the two smallest eignvalues are seen to be very stable over time, and the last series corresponding to the largest eigenvalue underlines much of the growth in the data. Out of the originally five nearly nonstationary series, there seem to exist two linear combinations that are very stationary. For further details of the analysis and substantive interpretation, see the 1977 paper.

14.7. SCALAR COMPONENT MODELS IN MULTIPLE TIME SERIES

This section introduces an approach for modeling vector time series proposed by Tiao and Tsay (1989). The goal is to find

1. An overall parsimonious model
2. Possibly simplifying structures that may not be obvious by direct modeling of the observed data

14.7. SCALAR COMPONENT MODELS IN MULTIPLE TIME SERIES

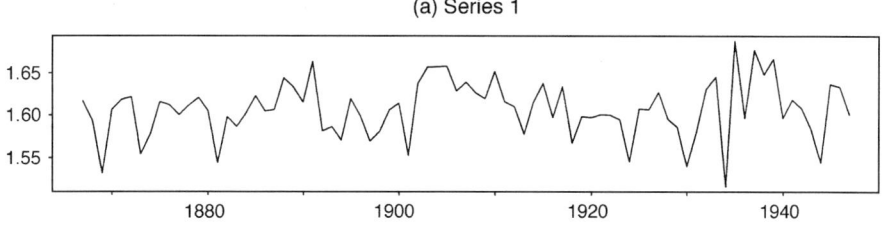

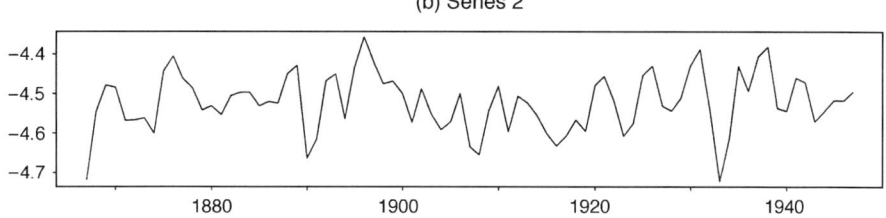

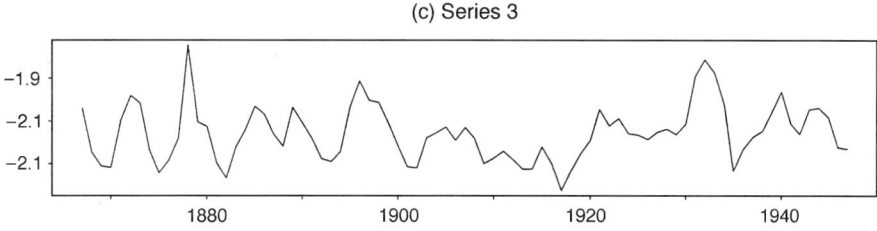

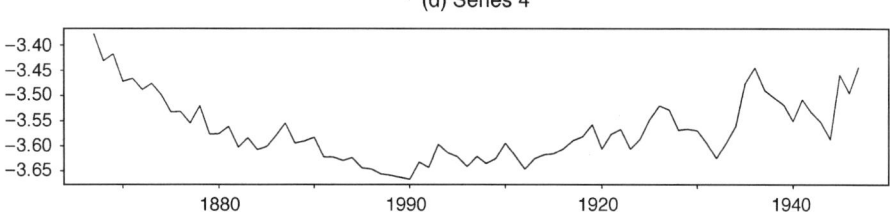

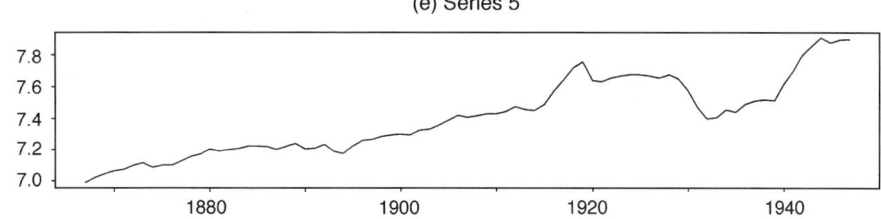

FIGURE 14.9 Transformed hog data.

This is achieved at the model specification stage by investigating linear combinations of the observed series.

As mentioned earlier, while the vector ARMA(p,q) model (14.7) is a straight forward generalization of the univariate ARIMA model (14.3), this direct generalization creates a major difficulty that, for moderate or large k, there will be an overflow of parameters whose estimates are often highly correlated. Another major difficulty in this generalization is the lack of identifiable models. To illustrate, consider again the bivariate ARMA(1,1) model in (14.48) and its transform in (14.50). Since $y_{1(t-1)} = b_{1(t-1)}$, the model for $\mathbf{y}_t$ in (14.50) can alternatively be written as

$$(\mathbf{I} - \Phi^{**}B)\mathbf{y}_t = (\mathbf{I} - \Theta^{**}B)\mathbf{b}_t \tag{14.59}$$

where

$$\Phi^{**} = \begin{bmatrix} 0 & 0 \\ 0 & 1 \end{bmatrix}, \quad \text{and} \quad \Theta^{**} = \begin{bmatrix} 0 & 0 \\ -2 & .5 \end{bmatrix}.$$

The two bivariate ARMA(1,1) models, (14.50) and (14.59), are exchangeable in the sense that they have different parameter values for the autoregressive (AR) and moving-average (MA) matrices but yield the same probabilistic structure of the process $\mathbf{y}_t$. In general, the vector ARMA model representation allows for exchangeable models because it opens the possibility of parameterization in both the AR and the MA parts. The possibility of multiple-model representations gives rise to the problem of identifiability of the vector ARMA model. This identifiability problem has been discussed extensively in the literature. See Hannan (1969), Akaike (1976), Solo (1986), and the references cited therein. Here, we shall not discuss the theory of identifiability or the sufficient conditions for a unique model representation, but consider methods that can recognize exchangeable models when they exist.

In what follows we first introduce the idea of *scalar component models* (SCMs) to describe a components structure in a multivariate framework. Some issues concerned with exchangeable models are then discussed. Finally we present an example illustrating the use of canonical correlation analysis to specify scalar component models. For details, see the paper by Tiao and Tsay.

14.7.1. Scalar component models

To introduce the concept of SCMs, we begin with another simple example. Suppose that $\mathbf{z}_t$ follows the bivariate ARMA(1,1) model

$$(\mathbf{I} - \Phi B)\mathbf{z}_t = (\mathbf{I} - \Theta B)\mathbf{a}_t \tag{14.60}$$

where

$$\Phi = \begin{bmatrix} .3 & .2 \\ .9 & .6 \end{bmatrix}, \quad \Theta = \begin{bmatrix} .25 & .175 \\ .5 & .35 \end{bmatrix}$$

14.7. SCALAR COMPONENT MODELS IN MULTIPLE TIME SERIES

containing eight nonzero parameters. Consider the linear transformation

$$\mathbf{y}_t = \mathbf{T}\mathbf{z}_t \quad \text{where} \quad \mathbf{T}' = [\mathbf{v}_0^{(1)}, \ \mathbf{v}_0^{(2)}] = \begin{bmatrix} 2 & -3 \\ -1 & 1 \end{bmatrix}.$$

Premultiplying expression (14.60) by $\mathbf{T}$ we have that

$$\mathbf{T}\mathbf{z}_t + \mathbf{G}\mathbf{z}_{t-1} = \mathbf{T}\mathbf{a}_t + \mathbf{H}\mathbf{a}_{t-1} \tag{14.61}$$

where

$$\mathbf{G} = \begin{bmatrix} .3 & .2 \\ 0 & 0 \end{bmatrix}, \quad \mathbf{H} = \begin{bmatrix} 0 & 0 \\ .25 & .175 \end{bmatrix}$$

and it is seen that $\mathbf{G}$ and $\mathbf{H}$ each contains a row of zero values. By inserting $\mathbf{T}^{-1}\mathbf{T}$ in front of $\mathbf{z}_{t-1}$ and $\mathbf{a}_{t-1}$, the model for $\mathbf{y}_t$ is

$$(\mathbf{I} - \mathbf{\Phi}^*B)\mathbf{y}_t = (\mathbf{I} - \mathbf{\Theta}^*B)\mathbf{b}_t \tag{14.62}$$

where

$$\mathbf{\Phi}^* = \begin{bmatrix} .9 & .7 \\ 0 & 0 \end{bmatrix} \quad \text{and} \quad \mathbf{\Theta}^* = \begin{bmatrix} 0 & 0 \\ .775 & .6 \end{bmatrix}.$$

Now for the transformed $\mathbf{y}_t$ process, the component y_{1t} does not need any non-zero MA coefficient and the component y_{2t} has no AR coefficient. In other words, corresponding to the zero rows of $\mathbf{G}$ and $\mathbf{H}$, $\mathbf{\Phi}^*$ and $\mathbf{\Theta}^*$ each contains a row of zero values. This is the type of simplification that we intend to capture in this approach.

To describe the structure of y_{1t}, we say that it follows an SCM of order $(1,0)$. The order $(1,0)$ signifies that, within the bivariate ARMA(1,1) framework, y_{1t} needs AR parameters at lag 1, but it does not require any nonzero MA coefficient. This is because the vector $\mathbf{v}_0^{(1)}$ has the property that $\mathbf{v}_0^{(1)'}\mathbf{\Phi} = -(.3, .2) \neq \mathbf{0}'$ but $\mathbf{v}_0^{(1)'}\mathbf{\Theta} = \mathbf{0}'$. More generally, given the vector ARMA(p,q) model (14.7) we say that a non-zero linear combination $y_t = \mathbf{v}_0'\mathbf{z}_t$ follows an SCM of order (p_1, q_1) if $\mathbf{v}_0$ has the properties

$$\begin{aligned} \mathbf{v}_0'\mathbf{\Phi}_{p_1} &\neq \mathbf{0}', & 0 &\leq p_1 \leq p \\ \mathbf{v}_0'\mathbf{\Phi}_\ell &= \mathbf{0}', & \ell &= p_1 + 1, \ldots, p \\ \mathbf{v}_0'\mathbf{\Theta}_{q_1} &\neq \mathbf{0}', & 0 &\leq q_1 \leq q \\ \mathbf{v}_0'\mathbf{\Theta}_\ell &= \mathbf{0}', & \ell &= q_1 + 1, \ldots, q. \end{aligned} \tag{14.63}$$

Since $\mathbf{v}_0'\mathbf{\Phi}(B)\mathbf{z}_t = \mathbf{v}_0'\mathbf{\Theta}(B)\mathbf{a}_t$, the structure of y_t can be written as

$$y_t + \sum_{\ell=1}^{p_1} \mathbf{g}_\ell'\mathbf{z}_{t-\ell} = \mathbf{v}_0'\mathbf{a}_t + \sum_{\ell=1}^{q_1} \mathbf{h}_\ell'\mathbf{a}_{t-\ell} \tag{14.64}$$

where $\mathbf{g}_\ell' = -\mathbf{v}_0'\mathbf{\Phi}_\ell$ and $\mathbf{h}_\ell' = -\mathbf{v}_0'\mathbf{\Theta}_\ell$. We shall denote the model structure of y_t

as SCM(p_1, q_1). Thus, by allowing $\boldsymbol{v}_0$ to be an arbitrary nonzero vector, the SCM is a direct generalization of the model of each component z_{jt} in the vector ARMA framework so that the model structure can be simplified.

The SCM is a device designed to capture the structure of a component within a vector model. It is *not* a univariate ARMA model. Thus, the fact that y_t has an SCM(p_1, q_1) structure in a vector model does not necessarily imply that y_t follows an univariate ARMA(p_1, q_1) process. For example, y_{1t} in model (14.48) follows an SCM(1, 0) structure but it is not a univariate ARMA(1, 0) process, because the SCM structure of y_{1t} involves nonzero coefficients of $y_{2(1-t)}$.

Given a k-dimensional vector ARMA process $\mathbf{z}_t$, let $\mathbf{y}_t = \mathbf{T}\mathbf{z}_t$, where $\mathbf{T} = [\boldsymbol{v}_0^{(1)}, \ldots, \boldsymbol{v}_0^{(k)}]'$ is a $k \times k$ nonsingular matrix, be the transformed process associated with k SCMs of orders (p_i, q_i), $i = 1, \ldots, k$. Such a transformation can lead to considerable parsimony in parameterization of the transformed model. As a further illustration, suppose that the observed process $\mathbf{z}_t$ follows a four dimensional ARMA(2,1) model and that the transformation $\mathbf{T}\mathbf{z}_t$ produces four scalar component models of orders (0,0), (0,1), (1,0), and (2,1). We then have, for the transformed process $\mathbf{y}_t$, the ARMA(2,1) model

$$(\mathbf{I} - \Phi_1 B - \Phi_2 B^2)\mathbf{y}_t = (\mathbf{I} - \Theta_1 B)\mathbf{b}_t \qquad (14.65)$$

where (suppressing the superscript *)

$$\Phi_1 = \begin{bmatrix} 0 & 0 & 0 & 0 \\ 0 & 0 & 0 & 0 \\ \times & \times & \times & \times \\ \times & \times & \times & \times \end{bmatrix}, \quad \Phi_2 = \begin{bmatrix} 0 & 0 & 0 & 0 \\ 0 & 0 & 0 & 0 \\ 0 & 0 & 0 & 0 \\ \times & \times & \times & \times \end{bmatrix}, \quad \Theta_1 = \begin{bmatrix} 0 & 0 & 0 & 0 \\ \times & \times & \times & \times \\ 0 & 0 & 0 & 0 \\ \times & \times & \times & \times \end{bmatrix},$$

with 0 and × denoting zero and nonzero parameters, respectively. In this particular instance then, modeling the transformed series $\mathbf{y}_t$ would involve 20 parameters in the coefficient matrices instead of 48, a saving of 28 parameters in estimation. As will be shown in the next section, because of the structure of the first and the third SCMs, we may further set the (4,1)th and the (4,3)th elements of Θ_1 to zero, achieving a total reduction of 30 parameters. The reduction could be even more substantial when the dimension k is relatively large. In view of the possibility of high correlations among the unconstrained 48 parameter estimates, the reduction could drastically simplify the complexity in estimation.

14.7.2. Exchangeable models and overparameterization

Exchangeable models are a special feature of vector time series that does not occur in the univariate case. In what follows, we discuss two issues:

1. Models with exchangeable structures
2. Elimination of redundant parameters

14.7. SCALAR COMPONENT MODELS IN MULTIPLE TIME SERIES

Models with exchangeable structures

Two vector ARMA models are exchangeable if they are of finite order and give the same probability distribution of $\mathbf{z}_t$. This can happen, for instance, when $|\Phi(B)|$ or $|\Theta(B)|$ is a constant. A simple example is that a bivariate AR(1) model $\mathbf{z}_t = \Phi \mathbf{z}_{t-1} + \mathbf{a}_t$ can be written exactly as a bivariate MA(1) model $\mathbf{z}_t = -\Theta \mathbf{a}_{t-1} + \mathbf{a}_t$ if

$$\Phi = \begin{bmatrix} 0 & 2 \\ 0 & 0 \end{bmatrix}, \quad \Theta = \begin{bmatrix} 0 & -2 \\ 0 & 0 \end{bmatrix}. \tag{14.66}$$

Here the lack of identifiability exists because $z_{2t} = a_{2t}$ so that $z_{2(t-1)}$ and $a_{2(t-1)}$ can appear as alternatives in the model structure of z_{1t}: $z_{1t} = 2z_{2(t-1)} + a_{1t} = 2a_{2(t-1)} + a_{1t}$ and generally $z_{1t} = (2-\delta)z_{2(t-1)} + \delta a_{2(t-1)} + a_{1t}$ so that we can also write the model structure for $\mathbf{z}_t$ in the form of a bivariate ARMA(1,1) model. In general, when the AR matrix polynomial $\Phi(B)$ and the MA matrix polynomial $\Theta(B)$ of model (4.7) are left coprime and if $\Phi(B)$ can be written as the product $\Phi(B) = \Xi(B)\Lambda(B)$ where $\Xi(B)$ and $\Lambda(B)$ are two nontrivial matrix polynomials such that $|\Xi(B)|$ is a non-zero constant, then $\Phi(B)\mathbf{z}_t = \Theta(B)\mathbf{a}_t$ is equivalent to $\Lambda(B)\mathbf{z}_t = \Xi^{-1}(B)\Theta(B)\mathbf{a}_t$, which is also of finite order. A similar condition applies to $\Theta(B)$.

Existence of exchangeable vector ARMA models leads to alternative specifications of the structure of some associated SCMs. Consider again the simple exchangeable bivariate AR(1) and MA(1) model corresponding to (14.66). For the scalar component $z_{1t} = \mathbf{v}_0' \mathbf{z}_t$ where $\mathbf{v}_0 = (1,0)'$, the structure will be SCM(1, 0) with $\mathbf{g}_1 = (0, -2)'$ for the bivariate AR(1) representation, and it becomes SCM(0, 1) with $\mathbf{h}_1 = (0, 2)'$ for the bivariate MA(1) model. In this case, we shall say that $\mathbf{v}_0' \mathbf{z}_t$ follows an SCM of exchangeable orders (1,0) and (0,1). As a further illustration, consider the two exchangeable models (14.50) and (14.59). For the second scalar component y_{2t}, the structure is SCM(1,1) for both model (14.50) and model (14.59) but the coefficients are different.

Since exchangeable models have the same probability distribution, they give the same covariance structure and provide the same inference. For this reason, we shall not discuss the conditions needed for unique model representation. Since exchangeable models are synonymous with exchangeable SCMs, we focus instead on methods that can point out the existence of exchangeable SCMs.

Redundant parameters

Redundant parameters are one of the most troublesome features in fitting vector ARMA models. One type of redundant (or unidentifiable) parameters may be eliminated by a careful study of the SCM structure. Consider, for example, the transformed ARMA(2,1) model (14.65). Let $\mathbf{y}_t = (y_{1t}, \ldots, y_{4t})'$, $\mathbf{b}_t = (b_{1t}, \ldots, b_{4t})'$, $\theta_{ij}^{(\ell)}$ and $\phi_{ij}^{(\ell)}$ be the (i, j)th elements of Θ_ℓ and Φ_ℓ, respectively. For the component y_{4t}, we have that

$$y_{4t} = \sum_{j=1}^{4} \phi_{4j}^{(1)} y_{j(t-1)} + \sum_{i=1}^{4} \phi_{4i}^{(2)} y_{i(t-2)} + b_{4t} - \sum_{j=1}^{4} \theta_{4j}^{(1)} b_{j(t-1)}. \tag{14.67}$$

Now, since y_{1t} is white noise, $y_{1(t-1)} = b_{1(t-1)}$. It is clear that in (14.67) $\theta^{(1)}_{41} - \phi^{(1)}_{41}$ is a constant, and we may set either $\theta^{(1)}_{41}$ or $\phi^{(1)}_{41}$ to zero. Next, from the structure of y_{3t} we have

$$y_{3(t-1)} - b_{3(t-1)} = \sum_{i=1}^{4} \phi^{(1)}_{3i} y_{i(t-2)}.$$

Since $y_{i(t-2)}$, for $i = 1, \ldots, 4$, are already in (14.67), it is clear that either $y_{3(t-1)}$ or $b_{3(t-1)}$ can be eliminated and we may set $\theta^{(1)}_{43}$ to zero.

In general, the structure of SCMs for a vector ARMA model makes it particularly convenient to spot redundant parameters. More specifically, consider a transformed vector ARMA(p,q) model for $\mathbf{y}_t$ such that y_{1t} has an SCM(p_1,q_1) structure and y_{2t} has an SCM(p_2,q_2) structure where $p_2 > p_1$ and $q_2 > q_1$. In this case, we can write the SCMs for y_{1t} and y_{2t} as

$$y_{it} - \{\phi^{(1)}_i B + \cdots + \phi^{(p_i)}_i B^{p_i}\}\mathbf{y}_t = b_{it} - \{\theta^{(1)}_i B + \cdots + \theta^{(q_i)}_i B^{q_i}\}\mathbf{b}_t \quad (14.68)$$

where $i = 1, 2$ and $\phi^{(\ell)}_i$ and $\theta^{(\ell)}_i$ are the ith rows of the matrices Φ_ℓ and Θ_ℓ, respectively. Now for $i = 2$ we see from (14.68) that y_{2t} is related to $y_{1(t-1)}, \ldots, y_{1(t-p_2)}$ and $b_{1(t-1)}, \ldots, b_{1(t-q_2)}$ via

$$\left(\phi^{(1)}_{21} B + \cdots + \phi^{(p_2)}_{21} B^{p_2}\right) y_{1t} - \left(\theta^{(1)}_{21} B + \cdots + \theta^{(q_2)}_{21} B^{q_2}\right) b_{1t}. \quad (14.69)$$

Since

$$B^s(y_{1t} - b_{1t}) = \{\phi^{(1)}_1 B + \cdots + \phi^{(p_1)}_1 B^{p_1}\}\mathbf{y}_{t-s} - \{\theta^{(1)}_1 B + \cdots + \theta^{(q_1)}_1 B^{q_1}\}\mathbf{b}_{t-s}, \quad (14.70)$$

it is clear that, if all the $\mathbf{y}$'s and the $\mathbf{b}$'s on the RHS of (14.70) are in the component model for y_{2t}, then either the coefficient of $y_{1(t-s)}$ or that of $b_{1(t-s)}$ is redundant given that the other is in the model. Therefore, if $p_2 > p_1$ and $q_2 > q_1$, then for each pair of parameters ($\phi^{(s)}_{21}, \theta^{(s)}_{21}$) in (14.69) $s = 1, \ldots, \min(p_2 - p_1, q_2 - q_1)$, only one parameter is needed. For convenience, we refer to this rule of spotting redundant parameters as *the rule of elimination*.

In general, by considering a vector ARMA model in its SCM form, and applying the rule of elimination in a pairwise fashion, all such redundant AR or MA parameters can be eliminated.

14.7.3. Model specification via canonical correlation analysis

Tiao and Tsay use canonical correlation analysis to search for SCMs of minimal orders at the model specification stage. The basic idea of the procedure is as follows.

14.7. SCALAR COMPONENT MODELS IN MULTIPLE TIME SERIES

First, for $m \geq 0$ let

$$\mathbf{z}_{m,t} = (\mathbf{z}'_t, \ldots, \mathbf{z}'_{t-m})' \tag{14.71}$$

From (14.63), a SCM of order (p_1, q_1) implies that there exists a linear combination of $\mathbf{z}_{p_1,t}$ which is independent of $\mathbf{a}_{t-j-1}$, and hence of $\mathbf{z}_{t-j-1}$, for $j \geq q_1$. Thus, corresponding to a SCM(p_1, q_1), there will be a zero canonical correlation between the two processes $\mathbf{z}_{p_1,t}$ and $\mathbf{z}_{h,t-j-1}$ for $j \geq q_1$ and $h \geq 0$. In general, for $m = 0, 1, \ldots$; $j = 0, 1, \ldots$; and fixed h, the procedure searches for k linearly independent SCMs of minimal orders by examining the number of zero canonical correlations (and corresponding eigenvectors) between $\mathbf{z}_{m,t}$ and $\mathbf{z}_{h,t-j-1}$. The discriminating criteria involve the choice of lags of sample autocovariance matrices and associated χ^2 test statistics. For further details, the reader is referred to the Tiao and Tsay (1989) paper.

14.7.4. An illustrative example

To illustrate, consider again the 5-series U.S. hog data discussed earlier in Section 14.6. Following the SCM search procedure, an overall ARMA(2,1) model for the transformed process with five SCMs of orders (0,0), (0,1), (1,0), (1,1) and (2,0) is tentatively specified:

$$(\mathbf{I} - \Phi_1 B - \Phi_2 B^2)\mathbf{y}_t = \mathbf{c} + (\mathbf{I} - \Theta_1 B)\mathbf{b}_t \tag{14.72}$$

where the detailed structure of Φ_1, Φ_2, and Θ_1 after applying the rule of elimination is

$$\Phi_1 = \begin{bmatrix} 0 & 0 & 0 & 0 & 0 \\ 0 & 0 & 0 & 0 & 0 \\ \times & \times & \times & \times & \times \\ \times & \times & \times & \times & \times \\ \times & \times & \times & \times & \times \end{bmatrix}, \quad \Phi_2 = \begin{bmatrix} 0 & 0 & 0 & 0 & 0 \\ 0 & 0 & 0 & 0 & 0 \\ 0 & 0 & 0 & 0 & 0 \\ 0 & 0 & 0 & 0 & 0 \\ \times & \times & \times & \times & \times \end{bmatrix},$$

$$\Theta_1 = \begin{bmatrix} 0 & 0 & 0 & 0 & 0 \\ \times & \times & \times & \times & \times \\ 0 & 0 & 0 & 0 & 0 \\ 0 & \times & \times & \times & \times \\ 0 & 0 & 0 & 0 & 0 \end{bmatrix}.$$

There are 29 parameters in the coefficient matrices. In contrast, an unconstrained 5D ARMA(2,1) model would have 75 parameters in these matrices. This illustrates the usefulness of structural specification in multiple time series analysis. In fact, after further simplification in the estimation stage of the transformed model, there are only 16 nonzero coefficients in the final model. For further details; see the 1989 paper.

The five transformed series are shown in Figure 1.13 (in Chapter 1). The first two series are quite stable, while the other three show certain nonstationary behavior. In particular, the third series capture much of the growth of the original data. These features are broadly similar to the first two and the last transformed series in Figure 14.9 using the canonical analysis method of Box and Tiao (1977). Note that the last transformed component in Figure 1.13 seems to capture the periodic behavior of the observed processes. This feature was discussed in Quenouille (1957) but seems less apparent in Figure 14.9.

14.7.5. Some further remarks

In the structural analysis discussed in this and the preceding sections, the method of canonical correlation analysis has been employed. This method is commonly used in statistics for extracting information in a multivariate problem (Hotelling 1936, Anderson 1984). Its potential for time series analysis has long been recognized. See, for instance, Robinson (1973), Akaike (1976), Brillinger (1975, 1981), Cooper and Wood (1982), Jewell and Bloomfield (1983), Jewell et al. (1983), Tsay and Tiao (1985), Velu et al. (1986) and Peña and Box (1987).

Finally, it is worth noting that apart from the two methods presented in this chapter, the principal component analysis is also of value in structural simplification. In particular, it should be applied to the sample covariance of the original series to eliminate redundant series that are exact linear combinations of the other components. It should also be applied to the sample covariance matrics of the residuals after vector autoregressive fitting to uncover exact lagged linear relationship among the series. Apart from finding exact linear relations, principal component analysis can also help achieve structural simplification in multiple time series, as illustrated in Tiao et al. (1993).

REFERENCES

Abraham, B. (1980). Intervention analysis and multiple time series. *Biometrika* **67**, 73–78.

Akaike, H. (1976). A new look at the statistical model identification. *JEEE Trans. Aut. Contr.* **19**, 716–723.

Akaike, H. (1980). On the identification of state space models and their use in control. In D. R. Brillinger and G. C. Tiao (eds.), *Directions in Time Series*, pp. 175–187. Institute of Mathematical Statistics.

Anderson, B. D. O. and Moore, J. B. (1979). *Optimal Filtering*, Prentice-Hall, Englewood Cliffs, NJ.

Anderson, T. W. (1971). *The Statistical Analysis of Time Series*. Wiley, New York.

Anderson. T. W. (1980). Maximum likelihood estimation for vector autoregressive moving average models. In D. R. Brillinger and G. C. Tiao (eds.), *Directions in Time Series*, pp. 49–59. Institute of Mathematical Statistics.

Anderson, T. W. (1984). *An Introduction to Multivariate Statistical Analysis*, 2nd ed. Wiley, New York.

REFERENCES

Ansley, C. F. (1979). An algorithm for the exact likelihood of a mixed autoregressive moving average process. *Biometrika* **66**, 59–65.

Bartlett, M. S. (1938). Further aspects of the theory of multiple regression. *Proc. Cambridge Phil. Soc.* **34**, 33–40.

Box, G. E. P. and Haugh, L. (1977). Identification of dynamic regression models connecting two time series. *J. Am. Stat. Assoc.* **72**, 121–130.

Box, G. E. P., Jenkins, G. M., and Reinsel, G. C. (1994). *Time Series Analysis: Forecasting and Control*, Holden-Day, San Francisco.

Box, G. E. P. and MacGregor, J. F. (1974). The analysis of closed loop dynamic stochastic system. *Technometrics* **16**, 391–398.

Box, G. E. P. and MacGregor, J. F. (1976). Parameter estimation with closed loop operating data. *Technometrics* **18**, 371–380.

Box, G. E. P. and Newbold, P. (1970). Some comments on a paper by Coen, Gomme and Kendall. *J. Roy. Stat. Soc. A.* **134**, 229–240.

Box, G. E. P. and Tiao, G. C. (1975). Intervention analysis with applications to environmental and economic problems. *J. Am. Stat. Assoc.* **70**, 70–79.

Box, G. E. P. and Tiao, G. C. (1977). A canonical analysis of multiple time series. *Biometrika* **64**, 355–365.

Brillinger, D. R. (1975). *Time Series: Data Analysis and Theory*, Holt, Rinehart and Winston, New York.

Brillinger, D. R. (1981). *Time Series: Data Analysis and Theory* (expanded ed.), Chapters 9 and 10. Holden-Day, San Francisco.

Chan, W. Y. T. and Wallis, K. F. (1978). Multiple time series modelling: Another look at the Mink-Muskrat Interaction. *Appl. Stat.* **27**, 168–175.

Coen, P. G., Gomme, E. D., and Kendall, M. G. (1969). Lagged relationships in economic forecasting. *J. Roy. Star. Soc. A.* **132**, 133–163.

Cooper, D. M. and Wood, E. F. (1982). Identifying multivariate time series models. *J. Time Ser. Anal.* **3**, 153–164.

Deistler, M., Dunsmuir, W., and Hannan, E. J. (1978). Vector linear time series models: Corrections and extensions. *Adv. Appl. Prob.* **10**, 360–372.

Dent, W. (1977). Computation of the exact likelihood function of an ARIMA process. *J. Stat. Comp. Simul.* **5**, 193–206.

Dunsmuir, W. and Hannan, E. J. (1976). Vector linear time series models. *Adv. Appl. Prob.* **8**, 339–364.

Engle, R. F. and Granger, C. W. J. (1987). Co-integration and error correction: Representation, estimation and testing. *Econometrica* **55**, 251–276.

Granger, C. W. J. and Newbold, P. (1986). *Forecasting Economic Time Series*, 2nd ed. Academic Press, Orlando, FL.

Gray, H. L. Kelley, G. D., and McIntire, D. D. (1978). A new approach to ARMA modelling. *Commun. Stat.* **B7**, 1–77.

Hallin, M. (1978). Mixed autoregressive moving average multivariate processes with time-dependent coefficients. *J. Multivar. Anal.* **8**, 567–572.

Hannan, E. J. (1969). The identification of vector mixed autoregressive moving average systems. *Biometrika* **57**, 223–225.

Hannan, E. J. (1970). *Multiple Time Series*. Wiley, New York.

Hannan, E. J. (1980). The estimation of the order of an ARMA process. *Ann. Stat.* **8**, 1071–1081.

Hannan, E. J., and Deistler, M. (1988). *The Statistical Theory of Linear Systems.* Wiley, New York.

Hannan, E. J., Dunsmuir, W. T. M., and Deistler, M. (1980). Estimation of vector ARMAX models. *J. Multivar. Anal.* **10**, 275–295.

Hannan, E. J. and Kavalieris, L. (1984). Multivariate linear time series models. *Adv. Appl. Prob.* **16**, 492–561.

Hillmer, S. C. and Tiao, G. C. (1979). Likelihood function of stationary multiple autoregressive moving average models. *J. Am. Stat. Assoc.* **74**, 652–660.

Hosking, J. R. M. (1980). The multivariate portmanteau statistic. *J. Am. Stat. Assoc.* **75**, 602–607.

Hotelling, H. (1936). Relations between two sets of variates. *Biometrika* **28**, 321–327.

Hsiao, C. (1979). Autoregressive modeling of Canadian money and income data. *J. Am. Stat. Assoc.* **74**, 553–560.

Jenkins, G. J. (1979). *Practical Experiences with Modelling and Forecasting Time Series*, GJP Ltd., Channel Islands, UK.

Jenkins, G. J. and Alavi, A. S. (1981). Some aspects of modelling and forecasting multivariate time series. *J. Time Ser. Anal.* **2**, 1–47.

Jewell, N. P. and Bloomfield, P. (1983). Canonical correlations of past and future for time series: Definitions and theory. *Ann. Stat.* **11**, 837–847.

Jewell, N. P., Bloomfield, P., and Bartmann, F. C. (1983). Canonical correlations of past and future for time series: Bounds and computation. *Ann. Stat.* **11**, 848–855.

Li, W. K. and McLeod, A. I. (1981). Distribution of the residual autocoffelations in multivariate ARMA time series models. *J. Roy. Stat. Soc. B* **43**, 231–239.

Lutkepohl, H. (1991). *Introduction to Multiple Time Series Analysis.* Springer-Verlag, Berlin.

Newbold, P. (1974). The exact likelihood function for a mixed autoregressive moving average process. *Biometrika* **61**, 423–426.

Nicholls, D. F. (1976). The efficient estimation of vector linear time series models. *Biometrika* **63**, 381–390.

Nicholls, D. F. (1977). A comparison of estimation methods for vector linear time series models. *Biometrika.* **64**, 85–90.

Nicholls, D. F. and Hall, A. D. (1979). The exact likelihood of multivariate autoregressive-moving average models. *Biometrika* **66**, 259–264.

Osborn, D. R. (1977). Exact and approximate maximum likelihood estimators for vector moving average processes. *J. Roy. Stat. Soc. B* **39**, 114–118.

Parzen, E. (1977). Multiple time series: Determining the order of approximating autoregressive schemes. In P. Krishnaiah (ed.), *Multivariate Analysis*, Vol. IV, pp. 283–295. North-Holland, Amsterdam.

Peña, D. and Box, G. E. P. (1987). Identifying a simplifying structure in time series. *J. Am. Stat. Assoc.* **82**, 836–843.

Phadke, M. S. and Kedem, G. (1978). Computation of the exact likelihood function of multivariate moving average models. *Biometrika* **65**, 511–519.

Quenouille, M. H. (1957). *The Analysis of Multiple Time Series.* Griffin, London.

Quinn, B. G. (1980). Order determination for a multivariate autoregression. *J. Roy. Stat. Soc. B* **42**, 182–185.

Reinsel, G. C. (1993). *Elements of Multivariate Time Series Analysis*. Springer-Verlag, New York.

Reinsel, G. C., and Velu, R. P. (1998). *Multivariate Reduced-Rank Regression*. Springer-Verlag, New York.

Robinson, P. M. (1973). Generalized canonical analysis for time series. *J. Multivar. Anal.* **3**, 141–160.

Solo, V. (1984). The exact likelihood for a multivariate ARMA model. *J. Multivar. Anal.* **14**, 164–173.

Solo, V. (1986). Topics in advanced time series analysis. *Lect. Prob. Stat.* **1215**, 165–328.

Slutsky, E. (1937). The summation of random causes as the source of cyclic process. *Econometrika* **5**, 105–146.

Tiao, G. C. and Ali, M. M. (1971). Analysis of correlated random effects: Linear model with two random components. *Biometrika* **58**, 37–51.

Tiao, G. C. and Box, G. E. P. (1981). Modeling multiple time series with applications. *J. Am. Stat. Assoc.* **76**, 802–816.

Tiao, G. C. and Tsay, R. S. (1983). Multiple time series modeling and extended sample cross correlations. *J. Business Econ. Stat.* **1**, 43–56.

Tiao, G. C. and Tsay, R. S. (1985). A canonical correlation approach to modeling multivariate time series. *Proc. Econ. Business Stat. Sect. Am. Stat. Assoc.* 112–120.

Tiao, G. C. and Tsay, R. S. (1989). Model specification in multivariate time series (with discussion). *J. Roy. Stat. Soc. B* **51**, 157–213.

Tiao, G. C., Tsay, R. S., and Wang, T. C. (1993). Usefulness of linear transformations in multiple time series analysis. *Emp. Econ.* **18**, 567–593.

Tiao, G. C. and Wei, W. S. (1976). Effect of temporal aggregation on the dynamic relationship of two time series variables. *Biometrika* **63**, 513–523.

Tsay, R. S. and Tiao, G. C. (1985). Use of canonical analysis in time series model identification. *Biometrika* **72**, 299–316.

Tunnicliffe Wilson, G. (1973). The estimation of parameters in multivariate time series models. *J. Roy. Stat. Soc. B* **35**, 76–85.

Velu, R. P., Reinsel, G. C., and Wichern, D. W. (1986). Reduced rank models for multiple time series. *Biometrika* **73**, 105–118.

Wallis, K. F. (1977). Multiple timer series analysis and the final form of econometric models. *Econometrika* **45**, 1481–1497.

Whittle, P. (1963). On the fitting of multivariate autoregressions and the approximate canonical factorization of a spectral density matrix. *Bionietrika* **50**, 120–134.

Yule, G. U. (1927). On a method of investigating periodicities in disturbed series with special reference to Wolfer's Sunspot numbers. *Phil. Trans. Roy. Soc. Lond.* **A226**, 267–298.

Zellner, A. and Palm, F. (1974). Time series analysis and simultaneous equation econometric models. *J. Econ.* **2**, 17–54.

CHAPTER 15

Cointegration in the VAR Model

Søren Johansen
European University Institute

15.1. INTRODUCTION

This chapter contains a survey of some of the results on cointegration in the vector autoregressive model. The presentation is based on the papers by Johansen and Juselius (1990, 1992, 1994), where the theory was developed, as well as Johansen (1996). The theory is developed as a careful study of the mathematical structure of the error correction model and its solution, followed by an analysis of the Gaussian likelihood function that allows one to derive estimators and test statistics.

The introductory section contains the basic definitions and some simple examples which illustrate the theory. In Section 15.2 the autoregressive equations are solved using a general result about the inversion of a matrix polynomial. This leads to Granger's representation theorem. In Section 15.3 the statistical model is defined and various hypotheses of interest are discussed. Next the problem of estimation and calculation of test statistics is discussed by an analysis of the Gaussian likelihood. In Section 15.4 the asymptotic theory is given, both the basic results on weak convergence and the asymptotic distribution of the estimator of the cointegrating relations, and the relevant test statistics. The final section includes a few applications of the cointegrated VAR to formulate models of economic interest.

The notion of cointegration has become one of the more important concepts in time series econometrics since the papers by Granger (1983) and Engle and Granger (1987). The topic of cointegration has found widespread applications in the analysis of economic data as published in the econometric literature. The special issues of the *Oxford Bulletin of Economics and Statistics*, **54**(3) (1992) and *Journal of Policy Modelling*, **14**(3,4) (1992) contain many papers where the method is applied and extended. The book of readings by Engle and Granger (1991) contains a collection of papers

A Course in Time Series Analysis, Edited by Daniel Peña, George C. Tiao, and Ruey S. Tsay.
ISBN 0-471-36164-X. © 2001 John Wiley & Sons, Inc.

15.1. INTRODUCTION   **409**

that have been important for the development of the topic. Many text books contain the basic aspects of cointegration: see, for instance, Reinsel (1991), Hamilton (1994), Lütkepohl (1993), or Cuthbertson et al. (1992). The books by Banerjee et al. (1993) and Johansen (1996) and Hansen and Johansen (1998), are systematic treatments of the topic of cointegration.

Economic insight is used in formulating the problem of interest, and therefore in the choice of variables, as well as in the discussion of which economic relations we expect to find. The statistical model is then used as a description of the nonstationary statistical variation of the data. The cointegrating relations are used as a tool for discussing the existence of long-run economic relations, and the various hypotheses are then tested in view of the statistical variation of the data. The interpretation of the cointegrating relations require a thorough understanding of the underlying economic problem, and the purpose of the statistical modeling is to provide a platform on which to discuss the economic questions of interest. We sometimes find that economic theory is rejected by the data. This can be because the theory is not developed enough, but of course also because the choice of variables for testing the theory may be inadequate.

Throughout this presentation we focus on the autoregressive models. This is because they are easy to analyse and often offers a good description of the variation of economic data. Moving-average models or ARIMA models form an extremely useful class of models and the structure theory in Section 15.2 also holds for such models. The likelihood function does not yield to the same simple explicit solutions, however, and this is the reason for not treating them here.

15.1.1. Basic definitions

In this section we give the basic definitions and a discussion of the concepts, and we start by defining the class of stationary and nonstationary processes we want to investigate: Let ε_t denote a doubly infinite sequence of p-dimensional iid stochastic variables with mean zero and finite variance. From these we construct a linear process $X_t = \sum_{i=0}^{\infty} C_i \varepsilon_{t-i}$, where the coefficient matrices C_i decrease exponentially fast, so that the series converges almost surely. This implies that the power series

$$C(z) = \sum_{i=0}^{\infty} C_i z^i$$

is convergent for $|z| < 1+\delta$, for some $\delta > 0$. For the analysis of the likelihood function, we need a further condition that the ε's are Gaussian. For the asymptotic analysis, however, this condition is not needed, and we only need conditions under which the central limit theorem holds, and for which we get convergence to certain stochastic integrals; see Section 15.4. We do not go into detail with the asymptotics and the probability assumptions in this presentation.

In the following we define the concept of $I(0)$ and $I(1)$. The purpose is to define a class of nonstationary processes, $I(1)$, which become stationary after differencing, and a class of stationary processes, $I(0)$, which become nonstationary when summed, thus mimicking the relation between a random walk and its increments.

Definition 15.1. *A linear p-dimensional process* $X_t = \sum_{i=0}^{\infty} C_i \varepsilon_{t-i}$ *is called integrated of order zero*, $I(0)$, *if* $\sum_{i=0}^{\infty} C_i \neq 0$.

Example 15.1. Consider the stationary univariate autoregressive process that satisfies $z_t = \rho z_{t-1} + \varepsilon_t$ with $|\rho| < 1$. This is clearly a linear process since $z_t = \sum_{i=0}^{\infty} \rho^i \varepsilon_{t-i}$, and since $\sum_{i=0}^{\infty} C_i = (1-\rho)^{-1} \neq 0$, it follows that it is also $I(0)$. The reason that we want this condition is that the cumulated z_t's satisfy

$$\sum_{i=1}^{t} z_i = \left(\sum_{i=0}^{\infty} \rho^i\right) \sum_{i=1}^{t} \varepsilon_i - \frac{\rho}{1-\rho} \sum_{i=0}^{\infty} \rho^i (\varepsilon_{t-i} - \varepsilon_{-i}),$$

which shows that it is the condition $\sum_{i=0}^{\infty} \rho^i \neq 0$, which guarantees that the cumulated process is nonstationary.

The process $y_t = \Delta z_t = z_t - z_{t-1}$, however, is stationary, but not $I(0)$ since the coefficients add to zero. If this process is summed we get

$$\sum_{i=1}^{t} y_i = \sum_{i=1}^{t} \Delta z_i = z_t - z_0,$$

which is nonstationary although asymptotically stationary. By insisting on the condition that the coefficients of the linear process add up to something nonzero, we make sure that the nonstationarity in its cumulated values is of the type we want to describe.

The process composed of both y_t and Δy_t, however, is an $I(0)$ process, since its coefficient matrices add to something nonzero.

Using the concept of $I(0)$ we now define the main concept for the analysis of cointegration, namely integration of order 1, $I(1)$, and integration of order 2, $I(2)$.

Definition 15.2. *A p-dimensional stochastic process X_t is called integrated of order 1*, $I(1)$, *if ΔX_t is $I(0)$, and $I(2)$ if ΔX_t is $I(1)$*.

The simplest example of an $I(1)$ process is a random walk, but any process of the form

$$X_t = C \sum_{i=1}^{t} \varepsilon_i + \sum_{i=0}^{\infty} C_i \varepsilon_{t-i},$$

is also an $I(1)$ process, at least if $C \neq 0$. Note that an $I(1)$ process is nonstationary, but that the nonstationarity can be removed by differencing.

15.1. INTRODUCTION

Example 15.2. We define a three-dimensional process by

$$X_{1t} = \sum_{i=1}^{t} \varepsilon_{1i} + \varepsilon_{2t}$$
$$X_{2t} = \frac{1}{2} \sum_{i=1}^{t} \varepsilon_{1i} + \varepsilon_{3t}, \quad t = 1, \ldots, n$$
$$X_{3t} = \varepsilon_{4t}.$$

It is seen that X_t is nonstationary and that ΔX_t is stationary and $I(0)$. Thus X_t is an $I(1)$ process, which can be made stationary by differencing. It is also seen that $X_{1t} - 2X_{2t}$ is stationary, and we say that X_t is cointegrated with $(1, -2, 0)$ as a cointegrating vector, and the process $\sum_{i=1}^{t} \varepsilon_{1i}$ is called a *common stochastic trend*. Thus stationarity can be achieved either by taking differences or by taking suitable linear combinations.

This example illustrates the definition of cointegration:

Definition 15.3. *If X_t is integrated of order 1 but some linear combination, $\beta' X_t$, $\beta \neq 0$, can be made stationary by a suitable choice of $\beta' X_0$, then X_t is called cointegrated and β is the cointegrating vector. The number of linearly independent cointegrating vectors is called the cointegrating rank, and the space spanned by the cointegrating vectors is the cointegration space.*

The definition of $I(1)$ is in terms of differences, and nothing is said about the levels of the process. Thus one cannot expect in general that anything can be said about the linear combinations of the levels $\beta' X_t$, unless they are started properly, that is, unless the initial values of $\beta' X_0$ are chosen from the invariant distribution.

Note that the definition of $I(1)$ is invariant under nonsingular linear transformations in the sense that if X_t is $I(1)$ and A is of full rank, then AX_t is also $I(1)$. Even if one of the components of X_t is $I(0)$, we call the multivariate process $I(1)$. In this way stationarity of a single component becomes a special case of cointegration.

The idea behind cointegration is that sometimes the lack of stationarity of a multidimensional process is caused by common stochastic trends, which can be eliminated by taking suitable linear combinations of the process, thereby making the linear combination stationary.

In econometrics the autoregressive processes have long been applied to describe stationary phenomena and the idea of explaining the process by its past values has been very useful for prediction. If, however, we want to find relations between simultaneous values of the variables in order to understand the interactions in the economy, one would get a lot more information by relating the value of a variable to the value of other variables at the same timepoint, rather than relating it to its own past. One can say that if we want to discuss relations between variables, then we should take combinations of simultaneous values, and if we want to discuss dynamic development of the variables, we should investigate the dependence on the past.

Cointegration has been so popular in econometrics because classical macro economic models are often formulated as simultaneous linear relations between variables following the Cowles commission tradition. The theory of such equations was developed for stationary processes despite the fact that many (or even most) economic variables are nonstationary. If we think of the classical economic relations as long-run relations one can easily imagine that such relations can be stationary even if the variables themselves are nonstationary. Cointegration is the mathematical formulation of this phenomenon, and we shall treat it in the framework of the vector autoregressive model in the next section.

15.2. SOLVING AUTOREGRESSIVE EQUATIONS

The main result of this section is how to express the stochastic properties of the solution of the autoregressive equations under various assumptions on the parameters, that is, the so-called Granger representation. The autoregressive equations are given in the reduced-form error-correction form

$$\Delta X_t = \Pi X_{t-1} + \sum_{i=1}^{k} \Gamma_i \Delta X_{t-i} + \Phi D_t + \varepsilon_t, \ t = 1, \ldots, n \quad (15.1)$$

where D_t are deterministic dummies and the ε_t are iid $N_p(0, \Omega)$. The equations determine the process X_t as a function of initial values $X_0, \ldots, X_{-k}$, the ε_t, and the dummies D_t that can contain a constant, a linear term, or seasonal dummies. Before we give the general solution of (15.1), we discuss some examples.

15.2.1. Some examples

Example 15.3. For the bivariate process

$$\begin{aligned} X_{1t} &= X_{1t-1} + \varepsilon_{1t} \\ X_{2t} &= \rho X_{2t-1} + \varepsilon_{2t} \end{aligned}, \ t = 1, \ldots, n$$

the solution is given by

$$X_t = \begin{pmatrix} X_{10} + \sum_{i=1}^{t} \varepsilon_{1i} \\ \rho^t X_{20} + \sum_{i=0}^{t-1} \rho^i \varepsilon_{2t-i} \end{pmatrix}.$$

15.2. SOLVING AUTOREGRESSIVE EQUATIONS

We see that the first component cannot be made stationary, but the second can. We choose to represent such a process as

$$X_t = \begin{pmatrix} X_{10} + \sum_{i=1}^{t} \varepsilon_{1i} \\ \sum_{i=0}^{\infty} \rho^i \varepsilon_{2t-i} \end{pmatrix}$$

where we have kept part of the initial value to start the random walk correctly, but assume that X_{20} is given by its initial distribution.

Example 15.4. A more complicated example is the autoregressive equations

$$\Delta X_{1t} = \alpha_1(X_{1t-1} - X_{2t-1}) + \varepsilon_{1t}$$
$$\Delta X_{2t} = \alpha_2(X_{1t-1} - X_{2t-1}) + \varepsilon_{2t}, \quad t = 1, \ldots, n.$$

The way to solve the equations is to consider the linear combinations $\alpha_2 X_{1t} - \alpha_1 X_{2t}$ and $X_{1t} - X_{2t}$. The equations for the first are

$$\Delta(\alpha_2 X_{1t} - \alpha_1 X_{2t}) = \alpha_2 \varepsilon_{1t} - \alpha_1 \varepsilon_{2t}$$

which shows that $\alpha_2 X_{1t} - \alpha_1 X_{2t}$ is a random walk and hence the initial values have to be kept fixed. The process $X_{1t} - X_{2t}$ satisfies the equations

$$\Delta(X_{1t} - X_{2t}) = (\alpha_1 - \alpha_2)(X_{1t-1} - X_{2t-1}) + (\varepsilon_{1t} - \varepsilon_{2t})$$

and for $-1 < 1 + (\alpha_1 - \alpha_2) < 1$ the process $X_{1t} - X_{2t}$ can be made stationary, and we get the representation of the two processes

$$\alpha_2 X_{1t} - \alpha_1 X_{2t} = \alpha_2 X_{10} - \alpha_1 X_{20} + \sum_{i=1}^{t}(\alpha_2 \varepsilon_{1i} - \alpha_1 \varepsilon_{2i})$$

$$X_{1t} - X_{2t} = \sum_{i=0}^{\infty}(1 + (\alpha_1 - \alpha_2))^i (\varepsilon_{1t-i} - \varepsilon_{2t-i})$$

From these relations it is not difficult to solve for the processes X_{1t} and X_{2t} and obtain a special case of Granger's representation theorem; see below.

15.2.2. An inversion theorem for matrix polynomials

We let $A(z)$ denote the $p \times p$ matrix polynomial derived from (15.1)

$$A(z) = (1-z)I_p - \Pi z - \sum_{i=1}^{k} \Gamma_i z^i (1-z)$$

and let $|A(z)|$ denote the determinant, and $adj A(z)$ the adjoint matrix, so that

$$A^{-1}(z) = \frac{adj(A(z))}{|A(z)|}. \tag{15.2}$$

Assumption 15.1. *The polynomial $A(z)$ satisfies the condition*

$$|A(z)| = 0 \text{ implies either } |z| > 1 \text{ or } z = 1.$$

Thus the coefficients should be so chosen that the roots of $|A(z)| = 0$ are either unit roots or stationarity roots. We are concerned with the power series expansion for the function $A^{-1}(z)$. This function will have a power series expansion in a neighborhood of the origin, since $A(0) = I_p$, which implies that $|A(z)| \neq 0$ for z sufficiently small. The power series will only converge for $|z| < 1$ if $z = 1$ is a root.

We now give a theorem that summarizes the Granger representation theorems for $I(0)$, $I(1)$, and $I(2)$ variables given in Johansen (1992). We give the results a purely analytic formulation without involving any probability theory, since the basic structure is then more transparent. The result allows a direct identification of the relevant coefficients of the inverse function in terms of the coefficients of the matrix function, and gives conditions for the presence of poles of order 0, 1, and 2, respectively. The result can be applied to derive the autoregressive representation from the moving average representation and vice versa, and the results can be generalized to give the representation of seasonally cointegrated processes; see Hylleberg et al. (1990) and Johansen and Schaumburg (1998).

We expand the function $A(z)$ around $z = 1$ and define the coefficients $\dot{A}(1)$ and $\ddot{A}(1)$ by

$$A(z) = A(1) + (z-1)\dot{A}(1) + \frac{1}{2}(1-z)^2 \ddot{A}(1) + \cdots$$

Thus

$$A(1) = -\Pi, \quad \dot{A}(1) = \left.\frac{dA(z)}{dz}\right|_{z=1} = -I_p - \Pi + \sum_{i=1}^{k} \Gamma_i,$$

$$\ddot{A}(1) = \left.\frac{d^2 A(z)}{d^2 z}\right|_{z=1} = 2\sum_{i=1}^{k} i\Gamma_i.$$

15.2. SOLVING AUTOREGRESSIVE EQUATIONS

For any $p \times m$ matrix a of full rank $m < p$, we denote by $a_\perp$ a $p \times (p-m)$ matrix of rank $p - m$ such that $a'a_\perp = 0$. We define $\bar{a} = a(a'a)^{-1}$, such that $a'\bar{a} = I_m$, and $\bar{a}a'$ is the projection of R^p onto the space spanned by the columns of a. For notational convenience we let $a_\perp = I_p$ if $a = 0$, and if $a = I_p$ we define $a_\perp = 0$. If $a' = (a_1, a_2)$, where $a_1 (m \times m)$ is of full rank, we can take

$$\alpha_\perp = \begin{pmatrix} -a_1^{-1} a_2 \\ I_{p-m} \end{pmatrix}.$$

The choice of $\alpha_\perp$ is not unique, and if $\alpha_{1\perp}$ and $\alpha_{2\perp}$ are any two choices there exists a full rank matrix ξ $(p-m) \times (p-m)$, such that $\alpha_{1\perp} = \alpha_{2\perp} \xi$. This implies that the conditions and formulae below do not depend on the choice of the orthogonal complement.

Theorem 15.1. *Let $A(z)$ be a matrix polynomial that satisfies Assumption 15.1. Then the following results hold for the function $A^{-1}(z)$:*

1. *If $z = 1$ is not a root, so that all roots are outside the unit disk, then $A^{-1}(z)$ is a power series with exponentially decreasing coefficients.*
2. *If $z = 1$ is a root, then $A(1)$ is of reduced rank $r < p$, and $-A(1) = \Pi = \alpha\beta'$, where α and β are of dimension $p \times r$ and rank r. If further*

$$|\beta'_\perp \dot{A}(1) \alpha_\perp | \neq 0, \tag{15.3}$$

then

$$A^{-1}(z) = C \frac{1}{1-z} + C^*(z),$$

where $C^*(z)$ is a power series with exponentially decreasing coefficients, and where

$$C = -\beta_\perp (\alpha'_\perp \dot{A}(1) \beta_\perp)^{-1} \alpha'_\perp.$$

3. *If $z = 1$ is a root, so that $A(1) = \alpha\beta'$, and if*

$$\alpha'_\perp \dot{A}(1) \beta_\perp = \phi\zeta',$$

is of reduced rank, where ϕ and ζ are $(p-r) \times s$ matrices of rank $s < p - r$, and if

$$\left| \phi'_\perp \alpha'_\perp \left[\frac{1}{2} \ddot{A}(1) - \dot{A}(1) \bar{\eta} \bar{\xi}' \dot{A}(1) \right] \beta_\perp \zeta_\perp \right| \neq 0, \tag{15.4}$$

then

$$A^{-1}(z) = C_2 \frac{1}{(1-z)^2} + C_1 \frac{1}{(1-z)} + C^{**}(z),$$

where $C^{**}(z)$ is a power series with exponentially decreasing coefficients. Expressions for the coefficients C_1 and C_2 can be found in Johansen (1992), where also the proof can be found.

A few comments are relevant here. If $z = 1$ is a root, then $A^{-1}(z)$ will have a pole at the point $z = 1$, since $|A(1)| = 0$, see (15.2).

The $I(1)$ condition (15.3) is necessary and sufficient for the pole to be of order 1. The function $C1/(1-z)$ has a pole of order 1 at $z = 1$ and the theorem says that the difference between $A^{-1}(z)$ and $C1/(1-z)$ is a convergent power series. Thus the pole can be removed by subtracting the function $C1/(1-z)$. The $I(2)$ condition (15.4) is necessary and sufficient for the pole to be of order 2, in which case it can be removed by subtracting the function $C_2 1/(1-z)^2 + C_1 1/(1-z)$, which also has a pole of order 2.

A similar result can be derived using the so-called Smith–McMillan form; see Engle and Yoo (1991) and Haldrup and Salmon (1998). The advantage of the present approach is the explicit expression for the coefficients matrix of the poles, which helps in the interpretation of the model and facilitates the construction of algorithms for calculation of maximum likelihood estimators and likelihood ratio test statistics.

In order to apply this result in the autoregressive model (15.1), we note that the coefficients in the expansion for $A^{-1}(z)$ gives the solution of the equations, that is, they determine X_t as a function of the errors ε_t. The translation is via the lag operator, such that for a function $C(z) = \sum_{i=0}^{\infty} C_i z^i$ and a sequence of iid variables ε_t, we define the stationary process

$$C(B)\varepsilon_t = \sum_{i=0}^{\infty} C_i \varepsilon_{t-i}.$$

For the expression $1/(1-z)$, we use the interpretation

$$(1-B)^{-1}\varepsilon_t = \Delta^{-1}\varepsilon_t = \sum_{i=1}^{t} \varepsilon_i,$$

and $1/(1-z)^2$ is translated into

$$(1-B)^{-2}\varepsilon_t = \Delta^{-2}\varepsilon_t = \sum_{j=1}^{t}\sum_{i=1}^{j} \varepsilon_i.$$

For a more precise formulation, one has to take into account the initial values. The result of Theorem 15.1 can be used to check whether a given example of an

15.2. SOLVING AUTOREGRESSIVE EQUATIONS

autoregressive process is $I(0)$, $I(1)$, or $I(2)$. The theorem is the fundamental tool in building $I(1)$ and $I(2)$ models for autoregressive processes as we shall show below.

15.2.3. Granger's representation

First we give the classical result about the representation of stationary solutions.

Theorem 15.2. *If X_t is given by (15.1) and if Assumption 15.1 holds, then X_t can be given an initial distribution such that it becomes $I(0)$ if and only if $A(1) = -\Pi$ has full rank, that is, $|A(z)|$ has no unit roots. In this case X_t can be given the representation*

$$X_t = \sum_{i=0}^{\infty} C_i \varepsilon_{t-i},$$

where the coefficients are given by $C(z) = \sum_{i=0}^{\infty} C_i z^i = A(z)^{-1}$, $|z| < 1 + \delta$ for some $\delta > 0$.

This result shows that if $|A(z)|$ has all roots outside the unit disk then the process generated by (15.1) is stationary or rather can be made stationary by a suitable choice of the initial distribution. Thus we have to allow other roots of $|A(z)|$ for X_t to be nonstationary.

If unit roots are allowed we can prove Granger's representation theorem.

Theorem 15.3. *If X_t is given by (15.1) and if Assumption 15.1 holds, then X_t is $I(1)$ if and only if*

$$\Pi = \alpha \beta' \tag{15.5}$$

where α, β $p \times r$ are of full rank $r < p$, and

$$\alpha'_\perp \left(I_p - \sum_{i=1}^{k} \Gamma_i \right) \beta_\perp \text{ has full rank.} \tag{15.6}$$

In this case $\Delta X_t - E(\Delta X_t)$ and $\beta' X_t - E(\beta' X_t)$ can be given initial distributions such that they become $I(0)$, and the process X_t has the representation:

$$X_t = C \sum_{i=1}^{t} (\varepsilon_i + D_i) + C(L)(\varepsilon_t + D_t) + A, \; t = 1, \ldots, n, \tag{15.7}$$

where

$$C = \beta_\perp \left(\alpha'_\perp \left(I_p - \sum_{i=1}^{k} \Gamma_i \right) \beta_\perp \right)^{-1} \alpha'_\perp, \qquad (15.8)$$

and A depends on initial conditions such that $\beta' A = 0$.

Thus the cointegrating vectors are β and the common trends are $\alpha'_\perp \sum_{i=1}^{t} \varepsilon_i$. The representation (15.7) is also called the *common-trends representation* of the solution of the autoregressive model and shows that the nonstationarity in the variables is created by the cumulated unanticipated shocks in the process, but not all these shocks appear. They are multiplied by the matrix $\alpha'_\perp$, which shows that only $p - r$ random walks give rise to the nonstationarity. Since the matrix C contains the factor $\beta_\perp$, we find $\beta' C = 0$, such that the linear combinations $\beta' X_t$ are not influenced by the random walks and become stationary.

The processes generated by (15.1) contain deterministic terms. It follows from the representation of Granger (15.7) that for an $I(1)$ process, a constant term in the equations will generate a linear term in the process, but only in the nonstationary part of the process, that is, the process $\beta' X_t$ has no trend.

It is an important property that cointegration is invariant to the extension of the information set; that is, if more variables are included in the analysis, we will still find the cointegrating vectors expanded by a zero for the new variables, but the common trends change character completely since what is unanticipated for the small system may not be unanticipated for the large system.

It is obvious that what is sometimes called the *permanent shocks* are the shocks $\alpha'_\perp \varepsilon_t$ since they cumulate in the system. We propose to call the shocks $\alpha' \Omega^{-1} \varepsilon_t$ the transitory shocks. The reason will be apparent in the discussion of the asymptotic distribution of $\hat{\beta}$, here we just note that the definition implies that the transitory shocks are independent of the permanent shocks.

The results of Theorem 15.1 can also be used to represent the $I(2)$ solutions of (15.1).

Theorem 15.4. *If X_t is given by (15.1) and if Assumption 15.1 holds and*

$$\Pi = \alpha \beta' \qquad (15.9)$$

where α, β are $p \times r$ matrices of full rank $r < p$, and

$$\alpha'_\perp \left(I_p - \sum_{i=1}^{k} \Gamma_i \right) \beta_\perp = \phi \zeta' \qquad (15.10)$$

where ϕ and ζ are $(p - r) \times s$ of rank s, then $\Delta^2 X_t$ and $\beta' X_t + (\alpha'\alpha)^{-1} \alpha'(I_p - \sum_{i=1}^{k} \Gamma_i) \Delta X_t$ corrected for their mean can be given initial distributions such that

15.2. SOLVING AUTOREGRESSIVE EQUATIONS

they become $I(0)$, and the process X_t has for $t = 1, \ldots, n$, the representation

$$X_t = C_2 \sum_{s=1}^{t} \sum_{i=1}^{s} (\varepsilon_i + D_i) + C_1 \sum_{i=1}^{t} (\varepsilon_i + D_i) + C(L)(\varepsilon_t + D_t) + A + Bt,$$

where the matrices C_1 and C_2 can be expressed in terms of the coefficients in the model and A and B depend on the initial conditions.

15.2.4. Prediction

A different way of solving the equations (15.1) is found by fixing all initial values. If we take the model with lag 1 the formulas become very simple. We find

$$X_t = (I_p + \alpha\beta')^t X_0 + \sum_{i=0}^{t-1} (I_p + \alpha\beta')^i \varepsilon_{t-i},$$

or equivalently

$$\alpha'_\perp X_t = \alpha'_\perp X_0 + \alpha'_\perp \sum_{i=1}^{t} \varepsilon_i$$

$$\beta' X_t = (I_r + \beta'\alpha)^t \beta' X_0 + \sum_{i=0}^{t-1} (I_r + \beta'\alpha)^i \beta' \varepsilon_{t-i}.$$

These relations show that the prediction of X_{t+h} given the history of the process is

$$E(\alpha'_\perp X_{t+h} \mid X_0, \ldots, X_t) = \alpha'_\perp X_t,$$
$$E(\beta' X_{t+h} \mid X_0, \ldots, X_t) = (I_r + \beta'\alpha)^t \beta' X_t.$$

Thus the random-walk part of the process is simply predicted by its current value and the stationary part by the discounted current value. Note, however, that

$$\text{Var}(\alpha'_\perp X_{t+h} \mid X_0, \ldots, X_t) = h\alpha'_\perp \Omega \alpha_\perp,$$
$$\text{Var}(\beta' X_{t+h} \mid X_0, \ldots, X_t) = \sum_{i=0}^{t-1} (I_r + \beta'\alpha)^i \beta' \Omega \beta (I_r + \alpha'\beta)^i,$$

such that the random walk component is predicted with a variance that tends to infinity whereas the stationary part is predicted with a variance that converges to the unconditional variance. For a process with more lags similar results can be proved by considering the companion form of the process.

15.3. THE STATISTICAL MODEL FOR $I(1)$ VARIABLES

If model (15.1) is used to describe a cointegrated $I(1)$ process we should restrict the parameters as given by conditions (15.5), (15.6), and Assumption 15.1. Assumption 15.1, which says that the roots are outside the unit disk or at 1, is very difficult to handle analytically. Fortunately it rarely turns out that the roots are inside the unit disk, and if they are, it is more important to know where they are than to force them to the boundary of the unit disk. Hence we do not restrict the parameters in the model by Assumption 15.1, but check that it is satisfied by the estimates. Condition (15.6) is easily satisfied, since matrices with full rank are dense in the space of all matrices. Thus, even without the restriction that $\alpha'_\perp (I_p - \sum_{i=1}^{k} \Gamma_i)\beta_\perp$ has full rank, the estimator derived has full rank with probability 1. Thus only condition (15.5), $\Pi = \alpha\beta'$, is included in the formulation of the parameter space of the model.

Definition 15.4. *The reduced form error-correction model H_r is described by the equations*

$$\Delta X_t = \alpha\beta' X_{t-1} + \sum_{i=1}^{k} \Gamma_i \Delta X_{t-i} + \Phi D_t + \varepsilon_t, \ t = 1, \ldots, n \quad (15.11)$$

where α and β are $p \times r$, and $\varepsilon_1, \ldots, \varepsilon_n$ are independent Gaussian $N_p(0, \Omega)$, and the variables D_t are deterministic terms. The freely varying parameters in the model are $(\alpha, \beta, \Gamma_1, \ldots, \Gamma_k, \Phi, \Omega)$.

We assume that the errors are Gaussian in order to be able to work with a likelihood function. Note that in model H_r the parameters α and β are not identified, since $\Pi = \alpha\beta' = \alpha\xi^{-1}(\beta\xi')'$ for any $r \times r$ matrix ξ of full rank, but that one can estimate the spaces spanned by α and β respectively, and the parameters in β can be estimated if they are identified or normalized suitably.

Thus cointegration analysis is formulated as the problem of making inference on the cointegration space, $sp(\beta)$, and the adjustment space, $sp(\alpha)$. If we want to estimate individual coefficients, it is necessary to normalize β or impose restrictions so that the parameters become identified.

The above allows one to formulate a nested sequence of hypotheses

$$H_0 \subset \cdots \subset H_r \subset \cdots \subset H_p,$$

and the test of H_r in H_p, is then the test that there are (at most) r cointegrating relations. Thus H_0 is just a vector autoregressive model for X_t in differences and H_p the unrestricted autoregressive model for X_t in levels, and the models in between, $H_1, \ldots, H_{p-1}$, give the possibility to exploit the information in the reduced-rank matrix Π, and contain information about the long-run relations in the economy.

Thus instead of analyzing nonstationary processes by differencing them to obtain stationarity and then analyze the differences by an autoregressive model, we

15.3. THE STATISTICAL MODEL FOR $I(1)$ VARIABLES

choose to leave the variables in levels and draw inference from the cointegrating relations.

Note that the model we get by fitting a vector autoregressive model to the differences is just H_0, the adequacy of which can be tested if we start with the general model H_p, by testing H_0 in H_p.

The corresponding model for $I(2)$ variables is analyzed in Johansen (1997).

15.3.1. Hypotheses on cointegrating relations

Once the cointegrating rank has been determined we can test hypotheses about the coefficients α and β, and we next give examples of such hypotheses.

In order to make the discussion of hypotheses more concrete, we consider the example of five series: the log consumer price index in Australia and the United States, p_1 and p_2, and the log exchange rate, $exch_t$, as well as the bond rate in both countries i_1 and i_2. The data is analysed in Johansen (1996) from the point of view of a cointegration analysis.

The hypothesis that only relative prices enter the cointegrating relations, can be expressed as the hypothesis that the coefficients to p_1 and p_2 sum to zero, or as the restriction $(1, 1, 0, 0, 0) \beta = 0$. This is the same restriction on all cointegrating relations that can also be expressed as a direct parametrization

$$\beta = H\varphi \tag{15.12}$$

where $H = (1, 1, 0, 0, 0)'_\perp$ is known and φ $(4 \times r)$ is unknown. This hypothesis on β does not depend on β being identified uniquely, since it is the same set of restrictions on all the relations. If β satisfies (15.12) then so does $\beta\xi$ for any matrix ξ $(r \times r)$. Hence (15.12) is a testable hypothesis on the cointegrating space despite the fact that β is not identified.

The hypothesis that some cointegrating vectors are known can be formulated as

$$\beta = (b, \psi) \tag{15.13}$$

where b $(p \times r_1)$ is known and ψ $(p \times r_2)$ is unknown, $r_1 + r_2 = r$. An example of this is, for instance, $(1, -1, -1, 0, 0)$ corresponding to PPP, or $(0, 0, 0, 1, -1)$, corresponding to UIP. In particular the test that an individual variable is stationary can be expressed in the form (15.13) for b equal to a unit vector. Thus the stationarity of a single component of X_t is a special case of cointegration.

A more general linear hypothesis can, for $r = 2$, say, be formulated as

$$\beta = (H_1\varphi_1, H_2\varphi_2), \tag{15.14}$$

where H_i $(p \times s_i)$ are known and φ_i $(s_i \times r_i)$ are unknown and $r_1 + r_2 = r$; see Johansen and Juselius (1994).

An example of (15.14) is given by the hypothesis that p_1, p_2, and e_{12} cointegrate and that the interest rates cointegrate. In this case we are looking for two relations of the form $(a, b, c, 0, 0)$ and $(0, 0, 0, d, e)$, which clearly form a set of uniquely identified equations even though they also need a careful economic interpretation. The hypothesis has the form (15.14) with

$$H_1 = \begin{pmatrix} 1 & 0 & 0 \\ 0 & 1 & 0 \\ 0 & 0 & 1 \\ 0 & 0 & 0 \\ 0 & 0 & 0 \end{pmatrix}, \quad H_2 = \begin{pmatrix} 0 & 0 \\ 0 & 0 \\ 0 & 0 \\ 1 & 0 \\ 0 & 1 \end{pmatrix}.$$

Thus we are, in the econometric language, testing for the over-identifying restrictions that there is a cointegrating relation between the variables that has two zeros as coefficients to the interest rates and another one with zeros as coefficients for the prices and exchange rates. It is a common econometric formulation that one wants to identify linear relations of econometric relevance by linear restrictions on the coefficients, in particular zero restrictions.

Thus linear restrictions are formulated on individual relations in the hope that they are sufficiently distinct so that identification is in fact possible.

15.3.2. Estimation of cointegrating vectors and calculation of test statistics

This section contains a brief description of the regression estimators of the cointegrating vectors and then discusses how the estimation problem of the various hypotheses from Section 15.3.1 can be solved by analyzing the Gaussian likelihood function.

Regression estimates

A time honored procedure for finding linear relations between two variables y_t and Z_t, of dimension 1 and m, respectively, is to regress y_t on Z_t and then to discuss the properties of the estimator

$$\hat{\beta}_{ols} = \left(\sum_{t=1}^{n} Z_t Z_t' \right)^{-1} \sum_{t=1}^{n} Z_t y_t' \qquad (15.15)$$

under various assumptions on the processes. This was, of course, the first to be used by Engle and Granger (1987) in their fundamental paper. The problem with the analysis is that since the regressor Z_t in general is a nonstationary process the usual simple asymptotic normality does not hold for the estimator.

Stock (1987) proved the, at first sight, rather surprising result that one gets a superconsistent estimator in the sense that

$$n^{1-\delta}(\hat{\beta}_{ols} - \beta) \xrightarrow{P} 0, \quad \delta > 0$$

under the assumption that the regressor is an $I(1)$ process, and that (y_t, Z_t) cointegrates with cointegrating vector $(1, -\beta')$; that is, $y_t - \beta' Z_t$ is stationary. Behind this result

15.3. THE STATISTICAL MODEL FOR $I(1)$ VARIABLES

is the following very simple idea. In the regression model

$$y_t = \beta' Z_t + \varepsilon_t,$$

where ε_t are independent Gaussian variables with mean zero and variance σ^2 and the $Z's$ are deterministic one finds that $\hat{\beta}_{ols}$ is Gaussian with mean β and variance $\sigma^2 (\sum_{t=1}^{n} Z_t Z_t')^{-1}$. If Z_t is stationary and mixing with finite variance, then the sum will increase like n and usual asymptotics hold in the sense that

$$n^{\frac{1}{2}} (\hat{\beta} - \beta)$$

is asymptotically Gaussian or equivalently

$$\left(\sum_{t=1}^{n} Z_t Z_t' \right)^{\frac{1}{2}} (\hat{\beta} - \beta) \sigma^{-1} \xrightarrow{w} N(0, I_m). \tag{15.16}$$

If Z_t is deterministic and grows like t, say, then $\sum_{t=1}^{n} Z_t Z_t'$ grows like n^3 and $n^{\frac{3}{2}} (\hat{\beta} - \beta)$ is asymptotically Gaussian, but again we find the result (15.16) and that $\hat{\beta}$ is superconsistent.

If Z_t is a nonstationary variable, it is of the order of $t^{\frac{1}{2}}$ which shows that $\sum_{i=1}^{n} Z_t Z_t'$ is of the order of n^2, which again implies superconsistency. The limit distribution is rather complicated and will be discussed more in Section 15.4.

Inference for the remaining parameters $\vartheta = (\Gamma_1, \ldots, \Gamma_k, \alpha, \Phi, \Omega)$ is relatively simple since superconsistency of the estimator for β implies that inference on ϑ can be conducted as if β were known and equal to $\hat{\beta}$, in which case model (15.11) only involves the stationary observables $\hat{\beta}' X_{t-1}$ and the differences of X_t.

This type of result has created a very large literature. See, for example, Stock and Watson (1988) for the estimation of the cointegrating rank and the cointegrating relations, Chan and Wei (1988) for inference in unstable processes, and the work of Phillips (1987, 1991), Phillips and Durlauf (1986), Phillips and Ouliaris (1990), and Park and Phillips (1988, 1989), on how to do regression with integrated regressors. The result has lead to a new class of limit distributions, which are combinations of mixed Gaussian and the so called unit root distributions.

A variant of the regression procedure was given by Stock and Watson (1988), who suggested using principal component analysis of the matrix $\sum_{t=1}^{n} X_t X_t'$ to find the linear combinations of the process with the smallest variation as candidates for the cointegrating relations. Box and Tiao (1977) suggested using canonical correlation analysis of X_t with respect to X_{t-1} to pick out the linear combinations that are most easily predicted from the past.

It turns out, however, that the limit distribution for the regression estimator, as well as the estimators involving principal components and canonical correlations of the levels, is very complicated and this makes inference and hypothesis testing difficult. There are ways of eliminating the nuisance parameters by modifying the regression method; see Park (1992) and Phillips and Hansen (1990).

Unrestricted maximum likelihood estimation

Another way of approaching the estimation problem is to analyze the Gaussian likelihood function and use that as a tool for generating estimators under the various hypotheses investigated in this section. The rationale is that a set of restrictions would change the parameter set and hence the maximization problem to be solved. Thus restrictions on the parameter set imply a modification of the estimator.

One would expect that if any estimator would have a simple limit distribution, it would have to be the maximum likelihood estimator. Similarly, one would expect that the likelihood ratio test statistic has simple limit distribution, even if it turns out that it is only sometimes that we get the χ^2 distribution.

Model (15.11) gives rise to a reduced-rank regression and the solution is available as an eigenvalue problem. It was solved by Anderson (1951) in the regression context and runs as follows. First we eliminate the parameters $\Gamma_1, \ldots, \Gamma_k, \Phi$ by regressing ΔX_t and X_{t-1} on $\Delta X_{t-1}, \ldots, \Delta X_{t-k}$, and D_t. The residuals are R_{0t} and R_{1t} respectively. Next form the sums of squares and products

$$S_{ij} = n^{-1} \sum_{t=1}^{n} R_{it} R'_{jt}, \quad i, j = 0, 1.$$

Then the likelihood function maximized with respect to the parameters $\Gamma_1, \ldots, \Gamma_k, \Phi$, and Ω is given by

$$L_{\max}^{-2/n}(\beta) = (2\pi e)^p |S_{00}| \frac{|\beta'(S_{11} - S_{10}S_{00}^{-1}S_{01})\beta|}{|\beta'S_{11}\beta|}.$$

This is minimized with respect to β by solving the eigenvalue problem

$$|\lambda S_{11} - S_{10}S_{00}^{-1}S_{01}| = 0. \tag{15.17}$$

The solution of this equation gives eigenvalues $1 > \hat{\lambda}_1 > \cdots > \hat{\lambda}_p > 0$ and eigenvectors $\hat{V} = (\hat{v}_1, \ldots, \hat{v}_p)$, which satisfy

$$\hat{\lambda}_i S_{11} \hat{v}_i = S_{10}S_{00}^{-1}S_{01}\hat{v}_i, \quad i = 1, \ldots, p,$$

and $\hat{V}'S_{11}\hat{V} = I_p$.

A maximum likelihood estimator for β is then given by

$$\hat{\beta} = (\hat{v}_1, \ldots, \hat{v}_r), \tag{15.18}$$

and an estimator for α is

$$\hat{\alpha} = S_{01}\hat{\beta}.$$

15.3. THE STATISTICAL MODEL FOR $I(1)$ VARIABLES

The maximized likelihood function is given by

$$L_{\max}^{-2/n} = (2\pi e)^p |S_{00}| \prod_{i=1}^{r}(1 - \hat{\lambda}_i). \tag{15.19}$$

See Johansen and Juselius (1990) for details and applications. One can interpret $\hat{\lambda}_i$ as a squared canonical correlation between ΔX_t and X_{t-1} conditional on the lagged differences $\Delta X_{t-1}, \ldots, \Delta X_{t-k+1}$. Thus the estimate of the "most stable" relations between the levels are those that correlate most with the stationary process ΔX_t corrected for lagged differences and deterministic terms.

Since only $\mathrm{sp}(\beta)$ is identifiable without further restrictions, one really estimates the cointegrating space as the space spanned by the first r eigenvectors. This is seen by the fact that if $\hat{\beta}$ is given by (15.18), then $\hat{\beta}\xi$ also maximizes the likelihood function for any choice of ξ $(r \times r)$ of full rank. The identification of β as eigenvectors is convenient from a mathematical and numerical point of view but not necessarily from en economic point of view.

This solution provides the answer to estimation of all the models $H_r, r = 0, \ldots, p$. By comparing the likelihoods (15.19), one can test H_r in H_p, that is, test for r cointegrating relations, by the likelihood ratio statistic

$$-2 \ln Q(H_r \mid H_p) = -n \sum_{i=r+1}^{p} \log(1 - \hat{\lambda}_i). \tag{15.20}$$

The estimator (15.18) is an estimator of all cointegrating relations, and it is sometimes convenient to normalize (or identify) the vectors by choosing a specific coordinate system in which to express the variables in order to facilitate the interpretation and in order to be able to give an estimate of the variability of the coefficients. If c is any $p \times r$ matrix, such that $\beta'c$ has full rank, one can normalize β as

$$\beta_c = \beta(c'\beta)^{-1},$$

which satisfies $c'\beta_c = I_r$ provided that $|c'\beta| \neq 0$. A particular example is given by $c' = (I_r, 0)$ and $\beta' = (\beta_1, \beta_2)$, where β_1 is $r \times r$ of full rank, in which case $\beta'c = \beta_1$ and $\beta'_c = (I_r, \beta_1^{-1}\beta_2)$. This corresponds to solving the cointegrating relations for the first r variables. The maximum likelihood estimator of β_c is

$$\hat{\beta}_c = \hat{\beta}(c'\hat{\beta})^{-1}.$$

This then gives the normalization or just identification of β that allows one to give an estimate of the variability of the estimator of the individual coefficient of $\hat{\beta}$.

15.3.3. Estimation of β under restrictions

If one wants to estimate β under restrictions, this can sometimes be done by the same analysis. Consider the hypothesis (15.12), where $\beta = H\varphi$. In this case

$$\alpha\beta' X_t = \alpha\varphi' H' X_t$$

which shows that the cointegrating relations are found by reduced-rank regression of ΔX_t on $H' X_{t-1}$ corrected for the lagged differences and D_t, that is, by solving the eigenvalue problem

$$\left| \lambda H' S_{11} H - H' S_{10} S_{00}^{-1} S_{01} H \right| = 0. \tag{15.21}$$

Under hypothesis (15.13) there are some known cointegrating relations and in this case $\alpha\beta' X_t = \alpha_1 b' X_t + \alpha_2 \varphi' X_t$, which shows that the coefficient α_1 to the observable $b' X_{t-1}$ can be eliminated together with the parameters $(\Gamma_1, \ldots, \Gamma_k, \Phi)$, so that the eigenvalue problem that has to be solved is

$$\left| \lambda S_{11.b} - S_{10.b} S_{00.b}^{-1} S_{01.b} \right| = 0 \tag{15.22}$$

where

$$S_{ij.b} = S_{ij} - S_{i1} b (b' S_{11} b)^{-1} b' S_{1j}, \ i, j = 0, 1.$$

The maximal value of the likelihood function is given by expressions similar to (15.19) and the test of hypotheses (15.12) and (15.13) consists of comparing the r largest eigenvalues under the various restrictions, since the factor $(2\pi e)^p |S_{00}|$ cancels.

The hypothesis (15.14) is slightly more complicated, but can be solved by a switching algorithm, where each step involves an eigenvalue problem; see Johansen and Juselius (1994).

Thus it is seen that a number of interesting hypotheses can be solved provided one has an eigenvalue routine and this algorithm has been implemented in many statistical packages.

15.4. ASYMPTOTIC THEORY

This section contains a description of the asymptotic theory of the processes and their product moments, as well as a discussion of how the results can be applied to conduct inference about the cointegrating rank and the cointegrating vectors.

15.4. ASYMPTOTIC THEORY

15.4.1. Asymptotic results

The basic asymptotic results can be summarized in the following three limit theorems, which we give for iid ε_t with mean zero and variance Ω.

$$n^{-\frac{1}{2}} \sum_{i=1}^{[nu]} \varepsilon_i \xrightarrow{w} W(u), \tag{15.23}$$

where $W(u)$ is Brownian motion, and $\xrightarrow{w}$ denotes weak convergence of the whole process. The continuous mapping theorem (Billingsley 1968) immediately gives

$$n^{-2} \sum_{t=1}^{n} \left(\sum_{i=1}^{t} \varepsilon_i \right) \left(\sum_{i=1}^{t} \varepsilon_i \right)' \xrightarrow{w} \int_0^1 WW' du \tag{15.24}$$

whereas the result

$$n^{-1} \sum_{t=1}^{n} \left(\sum_{i=1}^{t-1} \varepsilon_i \right) \varepsilon_t' \xrightarrow{w} \int_0^1 W(dW)' \tag{15.25}$$

is much more complicated and involves a stochastic integral (Karatzas and Shreve 1988, Chan and Wei 1988).

An $I(1)$ process X_t behaves asymptotically like a random walk [see (15.7)], and we find

$$n^{-\frac{1}{2}} X_{[nu]} \xrightarrow{w} CW(u)$$

with C as given by (15.8). Thus results similar to (15.24) and (15.25) can be formulated.

$$n^{-2} \sum_{t=1}^{n} X_{t-1} X_{t-1}' \xrightarrow{w} \int_0^1 CW(u) W(u)' C' du \tag{15.26}$$

$$n^{-1} \sum_{t=1}^{n} X_{t-1} \varepsilon_t' \xrightarrow{w} \int_0^1 CW(dW)'. \tag{15.27}$$

The order (n^2) of the product moments implies that one gets superconsistency of the regression estimator [see (15.15)]. The random limit (15.26) applied to Z_t implies that the limiting distribution is not Gaussian, but a rather complicated mixture of Gaussian distributions involving Brownian motion and nuisance parameters.

15.4.2. Test for cointegrating rank

The reason that inference for nonstationary processes is interesting and widely studied, is that it is nonstandard, in the sense that estimators are not asymptotically Gaussian and test statistics are not in general asymptotically χ^2. This was systematically explored by Dickey and Fuller (see Fuller 1976) in testing for unit roots in univariate processes.

As an example consider the simple model of an autoregressive process of order 1

$$z_t = \rho z_{t-1} + \varepsilon_t,$$

where ε_t are independent Gaussian variables with mean zero and variance σ^2. The null hypothesis of interest is that $\rho = 1$, which implies that z_t is a random walk, that is, a nonstationary process. Dickey and Fuller found among other results that when $\rho = 1$, a nonstandard limit distribution is obtained, and this can be expressed as

$$n(\hat{\rho} - 1) = \frac{n^{-1} \sum_{t=1}^{n} z_{t-1} \varepsilon_t}{n^{-2} \sum_{t=1}^{n} z_{t-1}^2} \xrightarrow{w} \frac{\int_0^1 W(dW)}{\int_0^1 W^2 du},$$

where $W(u)$ is a univariate Brownian motion on $[0, 1]$ with variance σ^2, and the stochastic integral can in this special case be calculated as $\int_0^1 W dW = \frac{1}{2}(W(1)^2 - 1)$. The results follow from (15.23)–(15.25). The implication is that the likelihood ratio test statistic is asymptotically distributed as

$$\frac{\left(\int_0^1 W(dW)\right)^2}{\int_0^1 W^2 du}.$$

This distribution is often called the "unit root" or *Dickey–Fuller distribution*, and its multivariate version plays an important role in asymptotic inference for cointegration. We give the main results obtained for likelihood inference, and refer to Johansen (1988, 1991) and Ahn and Reinsel (1990) for technical details.

Theorem 15.5. *Under model (15.1) with $\Phi = 0$ and r cointegrating relations the likelihood ratio statistic (15.20) satisfies*

$$-2 \ln Q(H_r \mid H_p) \xrightarrow{w} tr\left\{\int_0^1 B(dB)' \left[\int_0^1 B B' du\right]^{-1} \int_0^1 B(dB)'\right\},$$

where the process B is a $(p - r)$-dimensional Brownian motion with covariance matrix equal to I_{p-r}.

This result follows by a Taylor expansion of the likelihood function and the three limit results above. Thus the limit distribution depends only on the number of common trends of the problem. It is seen that the distribution is a multivariate generalization of the unit root distribution. This is not surprising, since one can think of the test for $\rho = 1$ in the univariate model as a test for no cointegration (i.e., of $r = 0$) when $p = 1$ and $k = 1$.

15.4. ASYMPTOTIC THEORY

Although the limit distribution given in Theorem 15.5 depends only on the degrees of freedom or the dimension of the Brownian motion, it turns out that if a constant term or a linear term is allowed in the model, then the limit distribution changes. If, however, the highest-order term is restricted to the cointegrating space, the limit distribution does not contain nuisance parameters. The various limit distributions are tabulated by simulation, (Johansen 1996).

15.4.3. Asymptotic distribution of $\hat{\beta}$ and test for restrictions on β

It is quite satisfactory, however, that the other test statistics described in Section 15.3 for hypotheses on α and β all have asymptotic χ^2 distributions. Thus the only non-standard test is the test for cointegrating rank. The reason for this is that the asymptotic distribution of the estimator of β is a mixed Gaussian distribution. We give the result for β_c, that is, β normalized so that $c'\beta = I_r$.

Theorem 15.6. *The asymptotic distribution of $\hat{\beta}_c$ is given by*

$$n(\hat{\beta}_c - \beta_c) \xrightarrow{w} (I_p - \beta_c c')\beta_\perp \left[\int_0^1 B_1 B_1' du\right]^{-1} \int_0^1 B_1 dB_2' \quad (15.28)$$

where B_1 and B_2 are independent Brownian motions of dimension $p - r$ and r, respectively. The asymptotic conditional variance matrix is

$$(I_p - \beta_c c')\beta_\perp \left[\int_0^1 B_1 B_1' du\right]^{-1} \beta_\perp'(I_p - c\beta_c') \otimes (\alpha_c' \Omega^{-1} \alpha_c)^{-1} \quad (15.29)$$

which is consistently estimated by

$$n(I_p - \hat{\beta}_c c')S_{11}^{-1}(I_p - c\hat{\beta}_c') \otimes (\hat{\alpha}_c' \hat{\Omega}^{-1} \hat{\alpha}_c')^{-1}. \quad (15.30)$$

The result follows by going to the limit in the likelihood equations and applying the three limit results above. For given value of B_1 the limit distribution of $\hat{\beta}$ is just a Gaussian distribution with mean zero and variance given by (15.29), since we can approximate $\int_0^1 B_1(dB_2)'$ by

$$\sum_{i=1}^N B_1(t_i)[B_2(t_{i+1}) - B_2(t_i)]'$$

which for fixed B_1 is Gaussian with mean zero and variance

$$\sum_{i=1}^N B_1(t_i)B_1(t_i)'(t_{i+1} - t_i) \otimes (\alpha_c' \Omega^{-1} \alpha_c)^{-1}$$

which converges to

$$\int_0^1 B_1(u)B_1(u)'du \otimes (\alpha_c' \Omega^{-1} \alpha_c)^{-1}.$$

It is this result that implies, by a simple conditioning argument, that the likelihood ratio test statistics for hypotheses about restrictions on β are asymptotically distributed as χ^2 variables, which again makes inference about β very simple if likelihood based methods are used.

Another way of reading the results (15.28)–(15.30) is that since $c'(\hat{\beta}_c - \beta_c) = 0$, we need only consider the coefficients $c'_\perp(\hat{\beta}_c - \beta_c)$. It now follows from (15.30) that we can act as if these are asymptotically Gaussian with a variance matrix given by

$$nc'_\perp(I_p - \hat{\beta}_c c')S_{11}^{-1}(I_p - c\hat{\beta}_c')c_\perp \otimes (\hat{\alpha}_c' \hat{\Omega}^{-1} \hat{\alpha}_c)^{-1}$$

in the sense that this matrix gives the proper normalization of the deviations $n(\hat{\beta}_c - \beta_c)$. Despite the complicated formulation, the result is surprisingly simple, since it only states that if β is estimated as identified parameters, then the asymptotic variance of $\hat{\beta}$ is given by the inverse information matrix, which is the Hessian used in the numerical maximization of a function. This result is exactly the same as the result that holds for inference in stationary processes. The only difference is the interpretation of (15.30), which for a stationary process would be an estimate of the asymptotic variance, but for $I(1)$ processes is a consistent estimator of the asymptotic conditional variance. The basic property, however, is the same in both cases, namely, that it is the approximate scale parameter to use for normalizing the deviation $\hat{\beta} - \beta$.

Example 15.5. As an example of the result in Theorem 15.6, consider the simple case where the model is given by

$$\Delta y_t = \alpha_1(\beta_1 y_{t-1} + \beta_2 z_{t-1}) + \varepsilon_{1t}$$
$$\Delta z_t = \alpha_2(\beta_1 y_{t-1} + \beta_2 z_{t-1}) + \varepsilon_{2t}.$$

We normalize β on the vector $c = (1, 0)'$, and write $\beta_c = (1, -\theta)'$ with $\theta = -\beta_2/\beta_1$. Then $y_t = \theta z_t + u_t$. The corresponding normalization of α is given by $\alpha_c = (\alpha_1, \alpha_2)\beta_1^{-1}$. The asymptotic distribution of $\hat{\theta} - \theta$ follows from Theorem 15.6. We find

$$(I_2 - \beta_c c')\beta_\perp = \begin{pmatrix} 0 & 0 \\ \theta & 1 \end{pmatrix}\begin{pmatrix} \theta \\ 1 \end{pmatrix} = \begin{pmatrix} 0 \\ 1 + \theta^2 \end{pmatrix}$$

and

$$\beta'_\perp R_{1t} = \theta y_{t-1} + z_{t-1} = (1 + \theta^2)z_{t-1} + \theta u_{t-1}.$$

15.4. ASYMPTOTIC THEORY

Theorem 15.6, states that the vector

$$\hat{\beta}_c - \beta_c = \begin{pmatrix} 0 \\ \hat{\theta} - \theta \end{pmatrix}$$

can be treated as asymptotically Gaussian with variance given by

$$\begin{pmatrix} 0 \\ 1+\hat{\theta}^2 \end{pmatrix} \left(\sum_{t=1}^{n} (\hat{\theta} y_{t-1} + z_{t-1})^2 \right)^{-1} \begin{pmatrix} 0 \\ 1+\hat{\theta}^2 \end{pmatrix}' (\hat{\alpha}_c' \hat{\Omega}^{-1} \hat{\alpha}_c)^{-1} = \begin{pmatrix} 0 & 0 \\ 0 & \hat{\sigma}_\theta^2 \end{pmatrix}$$

where

$$\hat{\sigma}_\theta^2 = (1+\hat{\theta}^2)^2 \left(\sum_{t=1}^{n} (\hat{\theta} y_{t-1} + z_{t-1})^2 \hat{\alpha}_c' \hat{\Omega}^{-1} \hat{\alpha}_c \right)^{-1}$$

$$= \left(\sum_{t=1}^{n} (z_{t-1} + \frac{\hat{\theta}}{1+\hat{\theta}^2} u_{t-1})^2 \hat{\alpha}_c' \hat{\Omega}^{-1} \hat{\alpha}_c \right)^{-1} \sim \left(\hat{\alpha}_c' \hat{\Omega}^{-1} \hat{\alpha}_c \sum_{t=1}^{n} z_{t-1}^2 \right)^{-1}.$$

Thus the hypothesis that $\beta_1 + \beta_2 = 0$ or equivalently that $\theta = 1$ can be tested using the statistic

$$(\hat{\theta} - 1)^2 \hat{\alpha}_c' \hat{\Omega}^{-1} \hat{\alpha}_c \sum_{t=1}^{n} z_{t-1}^2,$$

which is asymptotically $\chi^2(1)$.

The Brownian motions B_1 and B_2 that enter the limit result of Theorem 15.6 can be descibed as follows

$$n^{-\frac{1}{2}} (\alpha_\perp' \Omega \alpha_\perp)^{-\frac{1}{2}} \alpha_\perp' \sum_{i=1}^{[nu]} \varepsilon_i \xrightarrow{w} B_1(u),$$

$$n^{-\frac{1}{2}} (\alpha' \Omega^{-1} \alpha)^{-\frac{1}{2}} \alpha' \Omega^{-1} \sum_{i=1}^{[nu]} \varepsilon_i \xrightarrow{w} B_2(u).$$

That is, B_1 represents the limit of the common trends or cumulated permanent shocks, whereas B_2 represents the limit of the cumulated transitory shocks. Hence the determination of cointegrating rank is determined by the permanent shocks and inference on the coefficients on β is conducted conditional on the permanent shocks and the variation is measured by the variation of the transitory shocks.

One can now discuss why inference about β becomes difficult when based on the simple regression estimator. This is because the limiting distribution of $\hat{\beta}_{\text{ols}}$ is expressed as an integral as in Theorem 15.6, but with dependent B_1 and B_2. This again implies that for given B_1 the limit distribution of the estimator does not have conditional mean zero, which implies that the test statistics based on the regression estimator will have some noncentral distribution with nuisance parameter given by the canonical correlation coefficients between B_1 and B_2.

Inference for the remaining parameters $\vartheta = (\alpha, \Gamma_1, \ldots, \Gamma_k, \Omega)$ is different. This is explained by Phillips (1991), and the idea is roughly the following. The second derivative of the log likelihood function with respect to β tends to infinity as n^2 [see (15.27)], whereas the second derivative with respect to ϑ and the mixed derivatives tend to infinity like n. This means that $\hat{\beta} - \beta$ has to be normalized by n and $\hat{\vartheta} - \vartheta$ by $n^{\frac{1}{2}}$. This, on the other hand, requires a normalization of the mixed derivatives by $n^{3/2}$ and makes them disappear in the limit. Thus in the limit the information matrix, which is used to normalize $(\hat{\beta} - \beta, \hat{\vartheta} - \vartheta)$, is block diagonal with one block for β and one block for the remaining parameters ϑ.

Although the asymptotic distributions are simple to handle, they rarely are good approximations, and one can improve the approximation by calculating a so-called Bartlett correction factor to the the likelihood ratio test. As an example of this consider the simple model

$$\Delta X_t = \alpha \beta' X_{t-1} + \varepsilon_t, \quad t = 1, \ldots, n$$

in dimension p and $r = 1$ cointegrating relation. Suppose that we want to test a simple hypothesis $H_0 : \beta = \beta_0$. One can show (Johansen 1998), that (if α is known)

$$\frac{E[-2 \log LR]}{(p-1)} \sim 1 + \frac{1}{n} \left[\frac{3p+1}{2} - \frac{\alpha' \beta[(2 + \alpha'\beta)(p-2) + 4(1 + \alpha'\beta)]}{\beta' \Omega \beta \alpha' \Omega^{-1} \alpha} \right].$$

The idea is, that by estimating the RHS and dividing it into the likelihood ratio statistic, approximation to the $\chi^2(p-1)$ distribution is greatly improved.

15.5. VARIOUS APPLICATIONS OF THE COINTEGRATION MODEL

The concept of cointegration and long-run relations can be found in many models of economic interest, since data are often nonstationary and the cointegrating relations correspond to the relations that have the smallest variance or that are the most stationary.

15.5.1. Rational expectations

A typical rational expectation model is given by the present-value model in Campbell and Schiller (1987), which states that the present value of a variable z_t is a linear function of the discounted future values y_t

$$z_t = \gamma(1-\delta) \sum_{j=0}^{\infty} \delta^j E_t y_{t+j} + c.$$

15.5. VARIOUS APPLICATIONS OF THE COINTEGRATION MODEL

If we let $X_t = (z_t, y_t)$ we can write the equations on the form

$$E_t(c_1' X_{t+1} + c_0' X_t) + c = 0.$$

If we assume that X_t is given by the autoregressive model

$$\Delta X_t = \Pi X_{t-1} + \mu + \varepsilon_t$$

the parameters must satisfy the restrictions

$$c_1' \Pi = -(c_0 + c_1)', \qquad c_1' \mu + c = 0$$

which together with the cointegrating restriction $\Pi = \alpha \beta'$ give a set of restrictions on the parameters that can be tested. The idea is first to determine or test for the cointegrating rank, and next to estimate the model under the preceding restrictions and compare the likelihoods obtained using a likelihood ratio test (Johansen and Swensen 1999).

15.5.2. Arbitrage pricing theory

The arbitrage pricing theory often describes a one period model for asset prices and derives a restriction on the mean return by assuming that there should be no arbitrage opportunity by creating a portfolio with positive excess mean return and no risk. The exact factor model does not allow this possibility but the lack of the restriction opens an approximate arbitrage opportunity by diversification over many assets.

If, however, we consider a multiperiod model and instead of the asset returns consider the cumulated asset returns or the log prices, then these variables are found to be nonstationary. If we fit a cointegration model with a linear term restricted to the cointegrating relations, we get a model for a rebalanced portfolio. The no arbitrage condition is then that any portfolio with linearly increasing mean and constant risk must have the same mean return (Johansen and Lando 1996). Thus the APT hypothesis is a restriction on the deterministic terms and the cointegrating vectors.

15.5.3. Seasonal cointegration

A phenomenon that is not directly covered by the above model (15.1) is the seasonal variation of time series, see Hylleberg et al. (1990). It turns out that the basic result about inversion of matrix polynomials can be extended to cover this case as well, with the result that we get an error-correction formulation and the possibility for testing for cointegrating rank at the various complex frequencies. The asymptotics are roughly the same as for the usual model but involves the complex Brownian motion, as a consequence of allowing roots at complex frequencies (Johansen and Schaumburg 1998).

REFERENCES

Ahn, S. K. and Reinsel, G. C. (1990). Estimation for partially non-stationary multivariate autoregressive models. *J. Am. Stat. Assoc.* **85**, 813–823.

Anderson, T. W. (1951) Estimating linear restrictions on regression coefficients for multivariate normal distributions. *Ann. Math. Stat.* **22**, 327–351.

Banerjee, A., Dolado, J. J., Galbraith, J. W., and Hendry, D. F. (1993). *Co-integration, Error Correction, and the Econometric Analysis of Non-stationary Data*. Oxford Univ. Press, Oxford.

Billingsley, P. (1968). *Convergence of Probability Measures*. Wiley, New York

Box, G. P. E. and Tiao, G. C. (1977). A canonical analysis of multiple time series. *Biometrika* **64**, 355–365.

Campbell, J. and Schiller, R. J. (1987). Cointegration and tests of present value models. *J. Polit. Econ.* **95**, 1062–1088.

Chan, N. H. and Wei, C. Z. (1988). Limiting distributions of least squares estimates of unstable autoregressive processes. *Ann. Math. Stat.* **16**, 367–410.

Cuthbertson, K., Hall, S. G., and Taylor, M. P. (1992). *Applied Econometric Techniques*. Harvester Wheatsheaf, New York.

Engle, R. F. and Granger, C. W. J. (1987). Co-integration and error correction: Representation, estimation and testing. *Econometrica* **55**, 251–276.

Engle R. F. and Granger, C. W. J. (1991). *Long-run Econometric Relations. Readings in Cointegration*. Oxford Univ. Press, Oxford.

Engle, R. F. and Yoo, B. S. (1991). Cointegrated economic time series: A survey with new results. In C. W. J. Granger and R. F. Engle (Eds.), *Long-run Economic Relations. Readings in cointegration*. Oxford Univ. Press, Oxford.

Fuller W. (1976). *Introduction to Statistical Time Series*. Wiley, New York.

Granger, C. W. J. (1983). *Cointegrated Variables and Error Correction Models*. UCSD Discussion Paper 83–13a.

Haldrup, N. and Salmon, M. (1998). Representations of $I(2)$ cointegrated systems using the Smith-McMillan form. *J. Econ.* **84**, 303–325.

Hamilton, J. (1994). *Time Series Analysis*. Princeton Univ. Press, Princeton, NJ.

Hansen, P. R. and Johansen S. (1998). *Workbook on Cointegration*. Oxford Univ. Press, Oxford.

Hylleberg, S., Engle, R. F., Granger, C. W. J. and Yoo, S. B. (1990). Seasonal integration and cointegration. *J. Econ.* **44**, 215–238.

Johansen, S. (1988). Statistical analysis of cointegration vectors. *J. Econ. Dyn. Control* **12**, 231–254.

Johansen, S. (1992). A representation of vector autoregressive processes integrated of order 2. *Econ. Theory* **8**, 188–202.

Johansen, S. (1996). *Likelihood-Based Inference in Cointegrated Vector Autoregressive Models*. Oxford Univ. Press, Oxford.

Johansen, S. (1997). Likelihood analysis of the $I(2)$ model. *Scand. J. Stat.* **24**, 433–462.

Johansen, S. (1998). A Small Sample Correction for Test of Hypotheses on the Cointegrating Vectors. *J. Econ.*

REFERENCES

Johansen, S. and Juselius, K. (1990). Maximum likelihood estimation and inference on cointegration—with applications to the demand for money. *Oxford Bulle. Econ. Stat.* **52**, 169–210.

Johansen, S. and Juselius, K. (1992). Structural hypotheses in a multivariate cointegration analysis of the PPP and UIP for UK. *J. Econ.* **53**, 211–244.

Johansen, S. and Juselius, K. (1994). Identification of the long-run and the short-run structure. An application to the ISLM model. *J. Econ.* **63**, 7–36.

Johansen, S. and Lando, D. (1996). *Multi-period Models as Cointegration Models*. Discussion Paper, Univ. of Copenhagen.

Johansen, S. and Schaumburg, E. (1998). Likelihood analysis of seasonal cointegration. *J. Econ.* **88**, 301–339.

Johansen, S and A. Swensen, A. R. (1999). Testing some exact rational expectations in vectors autoregressive models. *J. Econ.* **93**, 73–91.

Karatzas, I. and Shreve, S. E. (1988). *Brownian Motion and Stochastic Calculus.* Springer-Verlag, New York.

Lütkepohl, H. (1993). *Introduction to Multiple Time Series Analysis.* Springer-Verlag, New York.

Park, J. Y. (1992). Canonical cointegrating regressions. *Econometrica* **60**, 119–143.

Park, J. Y. and Phillips, P. C. B. (1988). Statistical inference in regressions with integrated processes. Part 1. *Econ. Theory* **4**, 468–498.

Park, J. Y. and Phillips, P. C. B. (1989). Statistical inference in regressions with integrated processes. Part 2. *Econ. Theory* **5**, 95–131.

Phillips, P. C. B. (1987). Time series with a unit root. *Econometrica* **55**, 277–301.

Phillips, P. C. B. (1991). Optimal inference in cointegrated systems. *Econometrica* **59**, 283–306.

Phillips, P. C. B. and Durlauf, S. N. (1986). Multiple time series regression with integrated processes. *Rev. Econ. Stud.* **53**, 473–495.

Phillips, P. C. B. and Hansen, B. E. (1990). Statistical inference on instrumental variables regression with $I(1)$ processes. *Rev. Econ. Stud.* **57**, 99–124.

Phillips, P. C. B. and Ouliaris, S. (1990). Asymptotic properties of residual based tests for cointegration. *Econometrica* **58**, 165–193.

Reinsel, G. C. (1991). *Elements of Multivariate Time Series Analysis.* Springer-Verlag, New York.

Reinsel G. C. and Ahn, S. K. (1990). Vector autoregressive models with unit roots and reduced rank structure, estimation, likelihood ratio test, and forecasting. *J. Time Ser. Anal.* **13**, 283–295.

Stock, J. H. (1987). Asymptotic properties of least squares estimates of cointegration vectors. *Econometrica* **55**, 1035–1056.

Stock, J. H. and Watson, M. W. (1988). Testing for common trends. *J. Am. Stat. Assoc.* **83**, 1097–1107.

CHAPTER 16

Linear Dynamic Multiinput/Multioutput Systems

Manfred Deistler
Technische Universität Wien

16.1. INTRODUCTION AND PROBLEM STATEMENT

In this contribution a survey on identification of linear dynamic systems is given. The intention is to present the main ideas underlying identification and to give a clear picture of the structure of the basic theoretical results. For most of the material covered here, for a more detailed presentation and in particular for proofs, we refer the reader to Hannan and Deistler (1988). Alternative main references are Reinsel (1991) and Lütkepohl (1993). For a general statistical analysis of dynamic systems, emphasizing nonlinear systems, the reader is refered to Pötscher and Prucha (1997). For the sake of brevity of presentation, we do not give reference even to important original literature, if it is cited in the books listed above; for this reason important and seminal papers by Akaike, T. W. Anderson, Hannan, Kalman, and others will not be found in the list of references for this chapter.

In general terms, *system identification is concerned with finding a good model from* (in general) *noisy data*, namely, with data-driven modeling. The task of identification is often so complex that it cannot be performed in a naive way with the naked eye. In addition, many identification problems share common features. For these reasons, methods and theories have been developed, which make system identification a subject on its own; this is the case despite the fact, that problems of identification are treated in different communities such as in system theory, signal processing, statistics, and econometrics.

A Course in Time Series Analysis, Edited by Daniel Peña, George C. Tiao, and Ruey S. Tsay.
ISBN 0-471-36164-X. © 2001 John Wiley & Sons, Inc.

16.1. INTRODUCTION AND PROBLEM STATEMENT

There is a wide range of areas of application for system identification, from speech processing to control of chemical processes and forecasting models for economic data. In identification the following has to be specified:

- A *model class*, that is, the class of all a priori feasible candidate systems to be fitted to the data. The model class incorporates the a priori information about the phenomenon under consideration. Specification of the model class typically includes the selection of the variables, their classification into inputs and outputs, assumptions on the relation between the variables and the modelling of noise.
- A *class of feasible data*.
- An *identification procedure*, which is a set of rules (in the fully automatized case a function) attaching to every feasible data string a system from the model class. The theory of identification is mainly concerned with the development and evaluation of algorithms for identification.

Of course, identification has many different aspects and features. Here we only deal with *discrete-time* (equally spaced) *time series data*, $y_t, t = 1, \ldots, n; y_t \in \mathbb{R}^s$; and with *linear dynamic systems*.

In this contribution in addition we restrict ourselves to *mainstream theory* (Deistler 1989), where the following assumptions are imposed:

- The model class consists of linear, finite-dimensional, constant-parameter, causal, and stable systems only. The classification of the variables into inputs and outputs is given a priori. Here finite-dimensional means a finite-dimensional state and causal means that present outputs are not influenced by future inputs.
- Noise is modeled by stochastic models, in particular by stationary ergodic processes with rational spectra.
- The observed inputs are assumed to be free of noise and are uncorrelated with the noise.
- Criteria for goodness of fit are of the Gaussian likelihood type.
- Seminonparametric identification: data-driven model (subclass) specification and estimation of finite-dimensional real-valued parameters.
- Emphasis is placed on asymptotic properties of estimators such as consistency and asymptotic normality.

In general, the identification problem may be decomposed into three modules as follows:

- *Structure theory*. In a certain sense there an idealized identification problem is treated so far, as we commence from the population second moments of the observations or from the ("true") transfer functions, rather than from data. In more

general terms, here the relation between "external behavior" (as described, e.g., by transfer functions) and "internal parameters," such as ARMA or state-space parameters is investigated. The main problems here are observational equivalence and identifiability, realization, and parametrization.
- *Estimation of real-valued parameters.* By *real-valued parameters* we mean parameters that may vary, such as in an open set of an Euclidean space (such as "ordinary" ARMA parameters) as opposed to integer-valued parameters (such as maximum lag lengths for ARMA systems), which are used for dynamic specification (and thus for the specification of a model (subclass)). In this step the dynamic specification is assumed to be known, and the real-valued parameters (which under this assumption are contained in an Euclidian space) are estimated by procedures such as by maximum likelihood estimators.
- *Estimation of the dynamic specification* such as order estimation: This can be done, for instance, by test procedures or using information criteria.

In system identification elements of system theory, the theory of stationary processes, and of the statistical analysis of time series are dovetailed.

The main emphasis here is on multiinput/multioutput (MIMO) systems. This case is significantly more difficult compared to the single-input/single-output (SISO) case; the main difference is in the more complicated structure theory, in particular, in parametrization, for the MIMO case, rather than in the statistical analysis in the narrow sense. Accordingly these points will be emphasized in the presentation.

16.2. REPRESENTATIONS OF LINEAR SYSTEMS

16.2.1. Input/output representations

A system is a relation between functions of time, which in our case are stochastic processes. As has been stated above, here we only consider linear, discrete-time (where time is running over the integers $\mathbb{Z}$), causal, stable, time-invariant (constant parameters), and finite-dimensional (finite-dimensional state vector) systems. In addition, for simplicity of presentation, we restrict ourselves to the case where the inputs are unobserved white noise only.

The *input/output representation* of a linear system is of the form

$$y_t = \sum_{j=0}^{\infty} K_j \varepsilon_{t-j}, \quad t \in \mathbb{Z} \tag{16.1}$$

where y_t are the s-dimensional outputs, ε_t are the s-dimensional white noise inputs (i.e., $E\varepsilon_t = 0$, $E\varepsilon_s \varepsilon_t' = \delta_{st} \Sigma$), and $K_j \in \mathbb{R}^{s \times s}$ are the weighting matrices.

Throughout we assume that y_t and ε_t are random vectors over an underlying probability space $(\Omega, \mathcal{A}, P)$ with finite first and second moments. Limits such as the limit on the RHS of (16.1) are understood in the mean-squares sense, unless the

16.2. REPRESENTATIONS OF LINEAR SYSTEMS

contrary is explicitly stated. We impose the stability assumption

$$\sum_{j=0}^{\infty} \|K_j\| < \infty$$

which in particular implies that for arbitrary (weakly) stationary inputs the infinite sum in (16.1) always exists and that the process (y_t) is stationary. The function defined by

$$k(z) = \sum_{j=0}^{\infty} K_j z^j, \ z \in \mathbb{C} \tag{16.2}$$

is in one-to-one relation with $(K_j \mid j \in \mathbb{Z}_+)$ and is called the *transfer function* of the system. Finite-dimensional systems are characterized by rational transfer functions, specifically, $k = a^{-1}.b$, where a and b are polynomial matrices; that is, they are of the form

$$a(z) = \sum_{j=0}^{p} A_j z^j; b(z) = \sum_{j=0}^{q} B_j z^j; \ A_j, B_j \in \mathbb{R}^{s \times s}$$

From now on, unless the contrary has been stated explicitely, we will assume that $k(z)$ is rational.

If, in addition in (16.1) ε_t can also be expressed as a linear combination or a limit of linear combinations of the y_s, $s \leq t$, (i.e., in a causal way), or equivalently, if we assume that

$$\det k(z) \neq 0, \ |z| < 1$$

holds, (where "det" denotes "determinant of"), then (16.1) is called the *Wold representation* of the process (y_t).

Clearly

$$E y_t = 0$$

holds. Let $\gamma : Z \to \mathbb{R}^{s \times s} : \gamma_t = E y_t y_0'$ denote the (population) covariance function of (y_t). The spectral density (spectrum) $f : [-\pi, \pi] \to \mathbb{C}^{s \times s}$ exists and is given by

$$f(\lambda) = \frac{1}{2\pi} \sum_{t=-\infty}^{\infty} \gamma_t e^{-i\lambda t} = \frac{1}{2\pi} k(e^{-i\lambda}).\Sigma.k^*(e^{-i\lambda}) \tag{16.3}$$

where * denotes the conjugate-transpose. As is well known, f and γ are in a one-to-one relation. We will always assume that

$$\Sigma > 0 \quad \text{and} \quad k(0) = K_0 = I$$

hold.

If, on the other hand, we commence from a stationary process with a spectral density f which is rational and nonsingular λ — a.e. (almost everywhere), then (see Hannan and Deistler 1988, Section 1.3) f may be uniquely factored as in (16.3), where $k(z)$ is rational in z, analytic within a circle containing the closed unit disk, $\det k(z) \neq 0$, $|z| < 1$ and $k(0) = I$, and where $\Sigma > 0$. This $k(z)$ then corresponds to the Wold decomposition. Let us repeat that under the assumptions above, (k, Σ) is unique from the population second moments of (y_t). For this reason we will represent the external behavior by $k(z)$ (and Σ).

16.2.2. Solutions of linear vector difference equations (VDEs)

Consider a linear VDE

$$\sum_{j=0}^{p} A_j y_{t-j} = \sum_{j=0}^{q} B_j u_{t-j}, \, t \in \mathbb{Z} \tag{16.4}$$

where y_t and u_t are outputs and stationary inputs, respectively, and $A_j \in \mathbb{R}^{s \times s}$ and $B_j \in \mathbb{R}^{s \times m}$ are the associated parameter matrices. We use z to denote a complex variable as well as the backshift operator on $\mathbb{Z}$:

$$z(y_t \mid t \in \mathbb{Z}) = (y_{t-1} \mid t \in \mathbb{Z})$$

Then (16.4) can be written as

$$a(z) y_t = b(z) u_t$$

where

$$a(z) = \sum_{j=0}^{p} A_j z^j, \qquad b(z) = \sum_{j=0}^{q} B_j z^j$$

As is easy to see, the set of all solutions of (16.4) can be represented as the set of all solutions to the homogeneous equation $a(z) y_t = 0$ plus a particular solution of (16.4). We are interested only in causal, stable, steady-state (i.e., stationary) solutions:

Theorem 16.1. *Assume that the stability condition*

$$\det[a(z)] \neq 0, \, |z| \leq 1 \tag{16.5}$$

holds. Then the causal, stable, steady-state solution of (16.4) is given by

$$y_t = k(z) u_t$$

16.2. REPRESENTATIONS OF LINEAR SYSTEMS

where the transfer function is given by

$$k(z) = \sum_{j=0}^{\infty} K_j z^j = a^{-1}(z)b(z) = \frac{1}{det[a(z)]} adj[a(z)]b(z), \ |z| \leq 1 \qquad (16.6)$$

Here "adj" denotes the adjoint of a matrix.

Theorem 16.1 describes a simple method, the *z-transform*, to obtain the steady-state solution of a VDE: The polynomial matrix $a(z)$ in the shift operator z is inverted in the same way as the polynomial matrix $a(z), z \in \mathbb{C}$ and premultiplying both sides of (16.4) with $a^{-1}(z)$ gives the solution. For actually determining the K_j, the following block recursive linear equation system

$$A_0 K_0 = B_0$$
$$A_0 K_1 + A_1 K_0 = B_1$$
$$\dots\dots\dots\dots\dots$$

obtained from a comparison of coefficients in $a.k = b$, may be used.

Condition (16.5) implies that

$$\sum_{j=0}^{\infty} \|K_j\| < \infty$$

holds and thus, for arbitrary stationary inputs (u_t), the outputs exist and $\binom{u_t}{y_t}$ is jointly stationary. Clearly $k(z)$ is rational.

16.2.3. ARMA and state-space representations

As has been stated already, we will restrict ourselves to the case where we have no observed inputs. An *ARMA system* is a VDE of the form

$$a(z)y_t = b(z)\varepsilon_t \qquad (16.7)$$

where (ε_t) is s-dimensional white noise and where we in addition always assume that the stability condition (16.5) and the *(strict) miniphase condition*

$$\det b(z) \neq 0, \ |z| \leq 1 \qquad (16.8)$$

hold. The miniphase condition implies that ε_t is obtained by a causal linear transformation from (y_t), since we then can solve the system for ε_t in a causal way. The transfer function (and thus the solution) of (16.7) is given by $k(z) = a^{-1}(z).b(z)$ and by the miniphase condition corresponds to the Wold representation. W.r.g. we also assume $k(0) = I$. Because of this assumption, and due to the miniphase condition,

ε_t is the one-step-ahead prediction error for the best linear least-squares prediction $y(t \mid t-1)$ of y_t given $(y_s, s < t)$.

The spectrum of the ARMA process is given by

$$f(\lambda) = (2\pi)^{-1} a^{-1}(e^{-i\lambda}) b(e^{-i\lambda}) \Sigma b^*(e^{-i\lambda}) a^{-1*}(e^{-i\lambda}) \tag{16.9}$$

Remember that we assume throughout that $\Sigma > 0$ holds. Then, under our assumptions, the transfer function k and Σ are uniquely defined from the population second moments f of (y_t). It can be shown that every stationary process (y_t) with a rational spectral density f satisfying $f(\lambda) > 0$, $\forall \lambda \in [-\pi, \pi]$ can be represented by an ARMA system (16.7) satisfying our assumptions. The importance of ARMA systems can also be seen from the fact that every stationary process with a Wold representation (16.1) (i.e., every linearly regular stationary process) can be approximated with arbitrary accuracy by an ARMA process.

Restricting ourselves to the case of unobserved white-noise inputs only, we consider *linear state-space systems* of the form (the *prediction error form*)

$$x_{t+1} = A x_t + B \varepsilon_t \tag{16.10}$$

$$y_t = C x_t + \varepsilon_t \tag{16.11}$$

where x_t is the m-dimensional state and $A \in \mathbb{R}^{m \times m}$, $B \in \mathbb{R}^{m \times s}$, $C \in \mathbb{R}^{s \times m}$ are parameter matrices. In addition we assume that the *stability condition*

$$|\lambda_{\max}(A)| < 1 \tag{16.12}$$

where denotes $\lambda_{\max}$ an eigenvalue of maximum modulus, and the *(strict) miniphase condition*

$$|\lambda_{\max}(A - BC)| < 1 \tag{16.13}$$

hold. The (steady state) solution of (16.10), (16.11) is given by

$$y_t = C(Iz^{-1} - A)^{-1} B \varepsilon_t + \varepsilon_t \tag{16.14}$$

Thus the transfer function coefficients are of the form $K_j = C A^{j-1} B$ for $j > 0$ and $K_0 = I$.

We have the following result relating the system representations considered (see Hannan and Deistler 1988, Section 1.2):

Theorem 16.2. *Under our assumptions, we have*

1. *Every ARMA system (16.7) satisfying (16.5) and (16.8) and every state space system (16.10) and (16.11) satisfying (16.12) and (16.13) has a rational transfer*

function $k(z)$ that is analytic in a disk containing the closed unit disk (and thus causal and stable) and satisfies $\det[k(z)] \neq 0$, $|z| \leq 1$.

2. Conversely, for every rational transfer function $k(z)$ that is analytic in a disk containing the closed unit disk and that satisfies $\det[k(z)] \neq 0$, $|z| \leq 1$, there is a stable and miniphase ARMA and a stable and miniphase state space representation.

Thus, in particular ARMA or state-space representations are two alternative ways to describe the same (class of) input/output behaviors $k(z)$. Thus, in a certain sense, it is a question of taste whether ARMA or state systems are used.

16.3. THE STRUCTURE OF STATE-SPACE SYSTEMS

An important problem, in particular for identification, is the relation between internal characteristics (parameters) of a system and its external behavior.

For state-space systems (16.10), (16.11) internal parameters are, for instance, the entries of (A, B, C) which are real-valued and the integer-valued parameter m (i.e., the state dimension). In addition Σ is of interest. The external behavior is described by $k(z)$ and Σ. Note that under our assumptions $(k(z), \Sigma)$ and f are in a one-to-one relation. For state-space systems the relation between internal characteristics and external behavior is given by [see (16.14)]

$$k(z) = \sum_{j=1}^{\infty} C A^{j-1} B z^j + I \qquad (16.15)$$

Two state-space systems $[(A, B, C)$ and $(\bar{A}, \bar{B}, \bar{C})]$ are called *observationally equivalent*, if they have the same transfer function. A state-space system (A, B, C) is called *minimal*, if the statevector x_t has minimal dimension (or equivalently if A is of minimal dimension) among all state-space systems with the same transfer function. Nonminimal systems provide a "redundant" description of the external behavior and thus should be excluded. A state-space system is called *reachable*, if the matrix

$$C_m = (B, AB, \ldots, A^{m-1} B)$$

has full rank m and *observable*, if the matrix

$$O_m = (C', A'C', \ldots, (A')^{m-1} C')'$$

has full rank m.

We have the following simple "test" for minimality.

Theorem 16.3. *A state-space system (A, B, C) is minimal if and only if it is reachable and observable.*

The equivalence classes of minimal state space systems are described by the following theorem.

Theorem 16.4. *Two minimal state-space systems (A, B, C) and $(\bar{A}, \bar{B}, \bar{C})$ are observationally equivalent if and only if there exists a nonsingular matrix T such that*

$$\bar{A} = TAT^{-1}, \bar{B} = TB, \bar{C} = CT^{-1} \qquad (16.16)$$

hold.

The next theorem shows that the minimal dimension m of the state can be seen from the transfer function. Note that

$$H_\infty = \begin{pmatrix} K_1 & K_2 & K_3 & \ldots \\ K_2 & K_3 & K_4 & \ldots \\ \vdots & & & \end{pmatrix} = \begin{pmatrix} C \\ CA \\ CA^2 \\ \vdots \end{pmatrix} (B, AB, \ldots)$$

holds; here H_∞ is called the (block) *Hankel matrix* of the transfer function k. By H_m we denote the submatrix consisting of the first m block rows and block columns of H_∞. Then the following holds.

Theorem 16.5. *Let (A, B, C) be a state-space system with $A \in \mathbb{R}^{m \times m}$. Then H_∞ and H_m have rank smaller than or equal to m. If (A, B, C) is minimal, then equality holds.*

For the proofs of the theorems above, see Hannan and Deistler (1988, Section 2.3). Then the rank of H_∞ is called the *order of the transfer function* $k(z)$.

16.4. THE STRUCTURE OF ARMA SYSTEMS

In the following we state results (see Hannan and Deistler, 1988, Section 2.1) analogous to those in the previous subsection, now for ARMA systems. These results are somewhat more complicated, essentially since we are dealing with polynomial matrices (a, b) rather then with matrices with real entries (A, B, C).

Two ARMA systems, (a, b) and $(\bar{a}, \bar{b})$, are called *observationally equivalent*, if they have the same transfer function, that is, if $a^{-1}.b = \bar{a}^{-1}\bar{b}$ holds.

Next we introduce a notion for ARMA systems, which is analogous to minimality for state-space systems. An ARMA system (a, b) is called *left coprime*, if $(a(z), b(z)) \in \mathbb{C}^{s \times 2s}$ has rank s for all $z \in \mathbb{C}$. A polynomial matrix u is called *unimodular* if the polynomial $\det[u]$ is a constant unequal to zero. The interpretation of left coprimeness for (a, b) is that there exist no nontrivial (i.e., nonunimodular) common left factors in a and b. For the case of $s = 1$ (where a and b are scalar polynomials), this means that a and b have no common zeros. In this case the interpetation

16.5. THE REALIZATION OF STATE-SPACE SYSTEMS

of having no common zeros as nonredundant is easy to understand as it coincides with the postulate of a and b having minimum degree.

For the equivalence classes of left coprime ARMA systems we have:

Theorem 16.6. *Two left coprime ARMA systems (a, b) and $(\bar{a}, \bar{b},)$ are observationally equivalent, if and only if there exists a unimodular polynomial matrix u such that*

$$(\bar{a}, \bar{b}) = u(a, b)$$

holds.

From Theorem 16.2. we see that there is a one-to-one relation between ARMA and state-space equivalence classes.

16.5. THE REALIZATION OF STATE-SPACE SYSTEMS

16.5.1. General structure

The problem of *realization* is concerned with the construction of a state-space system from the process (y_t) or from its second moments f_y or from the transfer function $k(z)$. Consider a process (y_t) with rational nonsingular spectral density f_y. As is well known, such a process has a Wold representation

$$y_t = \sum_{j=0}^{\infty} K_j \varepsilon_{t-j} \tag{16.17}$$

where (ε_t) are white noise innovations. We assume $K_0 = I$ and $E\varepsilon_t\varepsilon_t' = \Sigma > 0$. The system (16.17) can be rewritten as an infinite-dimensional state-space system

$$\tilde{x}_{t+1} = \tilde{A}\tilde{x}_t + \tilde{B}\varepsilon_t \tag{16.18}$$

$$y_t = \tilde{C}\tilde{x}_t + \varepsilon_t \tag{16.19}$$

where

$$\tilde{x}_t = \begin{pmatrix} y(t \mid t-1) \\ y(t+1 \mid t-1) \\ y(t+2 \mid t-1) \\ \vdots \end{pmatrix} = \underbrace{\begin{pmatrix} K_1 & K_2 & K_3 & \cdots \\ K_2 & K_3 & K_4 & \cdots \\ \vdots & \vdots & \vdots & \end{pmatrix}}_{H_\infty} \underbrace{\begin{pmatrix} \varepsilon_{t-1} \\ \varepsilon_{t-2} \\ \vdots \end{pmatrix}}_{E^-_{t-1}} \tag{16.20}$$

where $y(t \mid r)$ is the best linear least-squares predictor of y_t based on y_s, $s \leq r$, $\mathrm{E}_{t-1}^{-'} = (\varepsilon_{t-1}' \varepsilon_{t-2}', \ldots)$ and where

$$\tilde{A} = \begin{pmatrix} 0, & I, & 0 & \cdot & \cdot \\ 0, & 0, & I, & 0 & \cdot & \cdot \\ \cdot & \cdot & \cdot & \cdot & \cdot \end{pmatrix}, \tilde{B} = \begin{pmatrix} K_1 \\ K_2 \\ \cdot \end{pmatrix}, \tilde{C} = (I, 0, 0, \ldots)$$

This infinite-dimensional state space-system, which, to repeat, is just another way of writing (16.17), is of pedagogical use only for the understanding of the construction described below.

Since

$$k(z) = \sum_{j=0}^{\infty} K_j z^j$$

is rational, then H_∞ must have finite rank, m, say. Now let $S \in \mathbb{R}^{m \times \infty}$ be a matrix such that the rows of SH_∞ form a basis for the row space of H_∞. Then from (16.20)

$$x_{t+1} = S\tilde{x}_{t+1} = SH_\infty E_t^- = S \begin{pmatrix} K_2, & K_3, & \cdot & \cdot \\ K_3, & K_4, & \cdot & \cdot \\ \cdot & \cdot & & \end{pmatrix} E_{t-1}^- + S \begin{pmatrix} K_1 \\ K_2 \\ \cdot \end{pmatrix} \varepsilon_t \quad (16.21)$$

Now determine (A, B, C) from

$$S \begin{pmatrix} K_2, & K_3, & \cdot & \cdot \\ K_3, & K_4, & \cdot & \cdot \\ \cdot & \cdot & & \end{pmatrix} = ASH_\infty \quad (16.22)$$

$$B = S(K_1', K_2' \ldots)' \quad (16.23)$$

$$(K_1, K_2, \ldots) = CSH_\infty \quad (16.24)$$

Note that the rows of the LHS in (16.22) are spanned by the rows of SH_∞ and that the rows of SH_∞ are linearly independent; thus, for given S, the matrix A is uniquely defined. We thus obtain from the following from (16.21)–(16.24):

$$x_{t+1} = Ax_t + B\varepsilon_t \quad (16.25)$$

$$y_t = \tilde{C}\tilde{x}_t + \varepsilon_t = \tilde{C}H_\infty E_{t-1}^- + \varepsilon_t = CSH_\infty E_{t-1}^- + \varepsilon_t = Cx_t + \varepsilon_t \quad (16.26)$$

In this way we have obtained a minimal state-space representation (16.25)–(16.26) from the Wold representation (16.17). As is easily seen, the minimal state x_t defined

16.5. THE REALIZATION OF STATE-SPACE SYSTEMS

via S is unique only up to premultiplication with a nonsingular matrix T. The basis change $\bar{x}_t = T x_t$ corresponds to the parameter transformation (16.16).

The construction described above has also a nice interpretation in the Hilbert space of square integrable random variables spanned by the one-dimensional components of (y_t). From (16.20) we see that the state $\tilde{x}_t$ is obtained by projecting the future of (y_t), namely, the components of $y_t, y_{t+1}, \ldots$ onto the space spanned by the past $y_{t-1}, y_{t-2}, \ldots$. Note that by the assumption $\Sigma > 0$, the linear dependence structure of the rows of H_∞ and of the one-dimensional components in $\tilde{x}_t$, respectively, is identical. In particular, the state space, that is, the Hilbert space spanned by $\tilde{x}_t$, then, is m-dimensional and a minimal state is a random vector whose elements form a basis for the state space. A state makes the future and the past of (y_t) conditionally orthogonal. (This is the so-called splitting property of the state.)

16.5.2. Echelon forms

If we have no additional a priori restrictions on the model class (this will be always assumed here, unless the contrary is stated explicitly), then we are free to choose representatives from the equivalence classes from the point of view of mathematical convenience, for instance. One example of such a choice is the so-called echelon form, which will be the concrete and "prototypical" example for this contribution. Thus we will now describe a special realization procedure leading to *echelon forms* for state-space systems: We commence from the (block) Hankel matrix

$$H_\infty = \begin{pmatrix} K_1, & K_2, & K_3, & \ldots \\ K_2, & K_3, & K_4, & \ldots \\ \ldots & & & \end{pmatrix}$$

of the transfer function $k(z)$. We know that H_∞ has rank m if and only $k(z)$ has a minimal state-space representation (A, B, C) of dimension m. Let $h(i, j)$ denote the j-th row in the ith block of rows in the matrix H_∞. Now, let us select a special basis for the row space of H_∞, namely, the basis consisting of the first rows of H_∞, which form a basis of H_∞. Because of its Hankel structure, this basis is, after a suitable permutation of the rows, of the form

$$(h(1, 1)', \ldots, h(n_1, 1)', \ldots h(1, s)' \ldots h(n_s, s)')'$$

for a suitability chosen multiindex $\alpha = (n_1, \ldots, n_s)$. The $n_1 \ldots n_s$ are uniquely defined by the selection procedure from H_∞. They are called the *Kronecker indices* of $k(z)$. Clearly $n_1 + \cdots + n_s = m$ holds. Now define $H_\alpha = (h'(1, 1), \ldots, h'(n_1, 1), \ldots h'(1, s) \cdots h'(n_s, s))'$, and let S denote the corresponding selector matrix such that

$$H_\alpha = S H_\infty$$

Then, for this specific choice of the basis for the row space of H_∞ and thus also for the specific selector matrix S, by the realization procedure described in Section 16.5.1, a unique state-space system (A, B, C) called *echelon form* is defined. In this case a part of the equations in (16.22) express the respective first linear dependent rows as linear combination of the preceding basis rows, thus they are of the form

$$h(n_i+1, i) = \sum_{j=1}^{s} \sum_{u=1}^{n_{ij}} -\tilde{a}_{ij}(u-1)h(u, j); i = 1 \ldots s \qquad (16.27)$$

where

$$n_{ij} = \begin{cases} \min(n_j+1, n_j) & \text{for } j < i \\ \min(n_i, n_j) & \text{for } j \geq i \end{cases}$$

and the other equations in (16.22) describe the shifting of the basis rows. The matrix A then is of the form $A = (A_{ij})_{i,j=1\cdots s}$ where

$$A_{ii} = \begin{pmatrix} 0 & & \\ 0 & I_{n_i-1} & \\ -\tilde{a}_{ii}(0), & \cdots, & -\tilde{a}_{ii}(n_i-1) \end{pmatrix}$$

$$A_{ij} = \begin{pmatrix} & 0 & \\ -\tilde{a}_{ij}(0) \ldots & -\tilde{a}_{ij}(n_{ij}), & 0..0 \end{pmatrix} \quad i \neq j$$

Analogously, B is defined form (16.23). For $n_i > 0, i = 1..s$, the matrix C contains only zeroes and ones [see Hannan and Deistler (1988, Section 2.5) for details]. In particular, we see that in this form not all entries in (A, B, C) are free. The nonfree parameters are prescribed to be zero or one (this is, of course, a convenient property of echelon forms), and the positions of the free parameters in (A, B, C) are determined by the Kronecker indices α. The vector τ of free parameters clearly is in a one-to-one relation with (A, B, C), and therefore we may identify τ and (A, B, C) for the given choice of Kronecker indices α. Let $\tau = (\tau_1, \tau_2)$, where τ_1 are the free parameters contained in A and τ_2 are the free parameters contained in B.

16.6. THE REALIZATION OF ARMA SYSTEMS

Here we restrict ourselves to echelon forms for ARMA systems. For details, see Hannan and Deistler (1988, Section 2.5).

For purely technical reasons we commence from a transfer function

$$\tilde{k}(z) = k(z^{-1}) - I = \sum_{j=1}^{\infty} K_j z^{-j} = \tilde{a}^{-1}(z).\tilde{b}(z) \qquad (16.28)$$

where $\tilde{a}(z) = \sum_{j=0}^{\tilde{p}} \tilde{A}_j z^j$, $\tilde{b}(z) = \sum_{j=0}^{\tilde{q}} \tilde{B}_j z^j$, are polynomial matrices, rather than from $k(z) = \sum_{j=0}^{\infty} K_j z^j$.

Now from $\tilde{b} = \tilde{a}.\tilde{k}$ using a comparison of coefficients corresponding to negative powers of z, we get

$$(\tilde{A}_0, \ldots, \tilde{A}_p, 0 \cdots) H_\infty = 0 \tag{16.29}$$

For given Kronecker indices α, if we express the rows $h(n_i + 1, i)$, $i = 1, \ldots s$, (multiplied by -1) as linear combinations of their preceding basis rows, then (16.29) is the same as (16.27) and thus defines a unique $\tilde{a}(z)$ and thus also a unique $\tilde{b}(z)$. The $\tilde{a}_{ij}(u)$ in (16.27) are the i, j elements of $\tilde{A}_u$. The degree of the diagonal element $\tilde{a}_{ii}$ of $\tilde{a}$, which is the degree of the ith row of $(\tilde{a}, \tilde{b})$, is equal to the ith Kronecker index n_i and the degree of $\det[\tilde{a}]$ is equal to m. Finally we define the ARMA system in echelon from by

$$(a(z), b(z)) = \mathrm{diag}\{z^{n_i}\}(\tilde{a}(z^{-1}), \quad \tilde{b}(z^{-1}) + \tilde{a}(z^{-1}))$$

Then (a, b) can be shown to be left coprime. Again, the free parameters are in certain positions (determined by the Kronecker indices α) in the coefficient matrices and the nonfree parameters are zero or one. For given α, the free parameters may be identified with (a, b). Note that the free parameters for echelon forms contained in A and in $a(z)$, respectively, both are given by (16.27) and thus are the same (up to sign change); also for the other free parameters [i.e., those contained in B and $b(z)$, respectively], there is a bijective (i.e., one-to-one), homeomorphic (i.e., continiuous in both directions) relation. In this sense, the free parameters for echelon state space and for echelon ARMA systems may be identified.

16.7. PARAMETRIZATION

In this section we are concerned with the analysis of the mapping attaching system parameters to transfer functions.

Let U_A denote the set of all rational $s \times s$ transfer functions $k(z)$ that are analytic in a disc containing the closed unit disk and where $\det k(z) \neq 0$, $|z| \leq 1$ holds and that satisfy $k(0) = I$. Let T_A denote the set of all state-space systems (A, B, C) (satisfying our assumptions) where s is fixed, but m is arbitrary or the set of all ARMA systems (a, b) (satisfying our assumptions) where s is fixed but p and q are arbitrary. Then by Theorem 16.2, the mapping

$$\pi : T_A \to U_A$$

such that

$$\pi(A, B, C) = C(Iz^{-1} - A)B + I$$

or

$$\pi(a, b) = a^{-1}.b$$

is surjective (i.e., onto).

Remember that in structure theory we are dealing with an idealized identification problem, in the sense that we want to conclude from $k(z)$ to the internal characteristics (A, B, C) or (a, b). In this context now the following problems are discussed:

The first problem is *identifiability*. From Theorems 16.4 and 16.6, respectively, we know that in general, even imposing minimality or left coprimeness respectively, the classes of *observationally equivalent* state-space or ARMA systems are no singletons. Thus additional restrictions have to be imposed in order to get identifiability, that is, uniqueness of the state space or ARMA representations. A set $T_\alpha \subset T_A$ is called *identifiable* if π restricted to T_α is injective (i.e., for instance for state-space systems $\pi(A, B, C) = \pi(\bar{A}, \bar{B}, \bar{C})$ implies $(A, B, C) = (\bar{A}, \bar{B}, \bar{C})$), thus in the identifiable case we have a bijective (i.e., a one-to-one) mapping

$$\psi_\alpha : \pi(T_\alpha) \to T_\alpha, \ s.t. \psi_\alpha(\pi(A, B, C)) = (A, B, C)$$

attaching to every transfer function the corresponding state space (or ARMA) system. The mapping ψ_α is called a *parametrization* of $U_\alpha = \pi(T_\alpha)$. In the following we will not distinguish between T_α as a set of systems [e.g., (A, B, C)] and as a set of corresponding free parameters τ.

In particular, if we consider the case of our concrete example, namely of echelon forms, then let $U_\alpha \subset U_A$ denote the set of all transfer functions such that α are their Kronecker indices (i.e., the $n_1, \ldots, n_s$ determine the first basis for the row space of H_∞). Then, by the procedures described in Section 16.5, a parametrization

$$\psi_\alpha : U_\alpha \to T_\alpha \subset \mathbb{R}^{d_\alpha}$$

attaching the free parameters τ to every $k \in U_\alpha$, has been defined. Here d_α denotes the dimension of the vector τ of free parameters.

After identifiability has been achieved, a second desirable property of a parametrization is its *continuity*. This is important, for example, for statistical analysis. It is important to note that in general, there does not exist a continuous parametrization for the set U_A. This is a main reason why U_A has to be broken into bits U_α, which allow for a continuous parametrization. First, however, let us be more specific about topologies. Sets $T_\alpha \subset \mathbb{R}^{d_\alpha}$ of free parameters are endowed with the corresponding relative topology of the Euclidean space. Let us identify $k(z)$ with $(K_j \mid j \in \mathbb{N})$. Then U_A is endowed with the so-called pointwise topology, which corresponds to the relative topology of the product topology of $(\mathbb{R}^{s \times s})^\mathbb{N}$ for $(K_j \mid j \in \mathbb{N})$.

Now let us return to the special case of echelon forms. By T_α we mean the set of vectors $\tau \in \mathbb{R}^{d_\alpha}$ of free parameters corresponding to $k \in U_\alpha$ where α are the Kronecker indices. Since the free parameters for echelon state-space and ARMA forms are in a bijective and homeomorphic relation, we do not have to distinguish between these two forms. We have the following result:

Theorem 16.7. *The parametrization* $\psi_\alpha : U_\alpha \to T_\alpha \subset \mathbb{R}^{d_\alpha}$ *corresponding to echelon forms has the following properties:*

1. $\{U_\alpha \mid \alpha \in \mathbb{Z}_+^s\}$ *is a (disjoint) partition of* U_A.
2. T_α *is an open subset of* $\mathbb{R}^{d_\alpha}$.
3. *For given m, the dimension of* T_α *is smaller than or equal to 2ms. For one* α, $(n_1 + \cdots + n_s = m)$, *namely, where the* $h(n_1, 1), \ldots, h(n_s, s)$ *form a continuous string in* H_∞, $d_\alpha = 2ns$, *holds. In this case* U_α *is generic (i.e. open and dense) in the set* $M(m) \subset U_A$ *of all transfer functions of order m.*
4. ψ_α *is continuous and thus a homeomorphism.*

Let us comment on these results: First note that the Kronecker indices α serve as a specification parameter. For given specification α we have identifiability for the parameter space T_α defined above, which is contained in an Euclidean space and in addition the problem is well posed in the sense that $k_T \to k$, $k_T, k \in U_\alpha$ implies $\psi_\alpha(k_T) \to \psi_\alpha(k)$; thus, for U_α the internal characteristics are not only unique but also depend continuously on the external behavior. In addition, for every transfer function $k \in U_A$ there exists an α such that $k \in U_\alpha$.

For a detailed description of this parametrization and its further properties the reader is refered to Hannan and Deistler (1988, Section 2.5). We have listed only the most important properties; of course, there are other ways of parametrizing by state-space or ARMA systems. In general, a mapping attaching to every $k \in U_A$ a unique state space or ARMA system, respectively, is called a (state-space or ARMA, respectively) *canonical form*. As an alternative to echelon canonical forms, recently the so-called *balanced canonical forms* [see, e.g., Bauer and Deistler (1999)], which may be defined via a singular value decomposition of H_∞, have been proposed. Balanced forms relate only to state-space systems and have no ARMA counterpart.

It can be shown that $M(m)$, the set of all transfer functions of order m, is a real analytical manifold of dimension $2ms$. This manifold may be parametrized by local coordinates, which are also derived via the choice of a basis for the row space of H_∞, by using (16.22)–(16.24) or (16.29) respectively. For a detailed description see Hannan and Deistler (1988, Section 2.6).

The approaches described above all relate to the case where there is no additional a priori information available, in the sense that we are free to choose mathematically convenient representatives from the equivalence classes. The situation is different, if additional a priori restrictions, coming from theories, for instance, are available. This leads to problems of structural identifiability (Hannan and Deistler 1988, Section 7.1), which have been intensively discussed in econometrics.

For a rather general framework for the treatment of different parametrizations, see Deistler and Wang (1989).

16.8. ESTIMATION OF REAL-VALUED PARAMETERS

As has been stated earlier, the main complications of the multivariate, compared to the univariate case, arise in the structure theory. Given our knowledge of structure theory, *Gaussian maximum likelihood* (ML) *estimation*, for example, can be generalized quite straightforwardly from the univariate case:

For the sake of simplicity of presentation, we restrict ourselves to echelon forms $\psi_\alpha : U_\alpha \to T_\alpha$ for given Kronecker indices α and with $\tau \in T_\alpha$ the free parameters. It should be mentioned however, that e.g. the consistency result below holds for rather general sets U_α and for all continuous parametrizations of such a set (see Hannan and Deistler 1988, Section 4.2). One nice aspect of the theory is that the statistical results concerning parameter estimators only depend on such general properties of the parametrizations and can be shown to hold independently of the special choise of the parametrization. Let $y(n) = (y_1', \ldots y_n')'$ denote the stacked sample, let $f_y(\lambda, \tau, \Sigma)$ denote the spectrum of a process (y_t) corresponding to the parameters τ and Σ, and finally let

$$\Gamma_n(\tau, \Sigma) = \left(\int e^{-i\lambda(r-t)} f_y(\lambda, \tau, \Sigma) d\lambda \right)_{r,t=1\ldots n}$$

denote the $ns \times ns$ section of the variance–covariance matrix of the process (y_t) corresponding to τ and Σ.

Thus ($-2n^{-1}$ *times*) the log-likelihood function is given by

$$L_n(\tau, \Sigma) = n^{-1} \log \det \Gamma_n(\tau, \Sigma) + n^{-1} y'(n) \Gamma_n^{-1}(\tau, \Sigma) y(n) \qquad (16.30)$$

The ML estimators $\hat{\tau}_n, \hat{\Sigma}_n$ then are defined as the minimizers of L_n over T_α and the set $\underline{\Sigma}$ of all $s \times s$ nonsingular covariance matrices:

$$(\hat{\tau}_n, \hat{\Sigma}_n) = \arg \min_{\tau \in T_\alpha, \Sigma \in \underline{\Sigma}} L_n(\tau, \Sigma) \qquad (16.31)$$

Note that $\Gamma_n(\tau, \Sigma)$ and thus the likelihood function L_n depend only on τ via the transfer function k. In other words, we can define a "coordinate free" ML estimator $(\hat{k}_n, \hat{\Sigma}_n)$, which does not depend on the specific parametrization under consideration. We have the following consistency result.

Theorem 16.8. *Let α be given and assume that the true transfer function k_0 is contained in U_α. Then*

$$\frac{1}{n} \sum_{t=1}^{n-s} \varepsilon_t \varepsilon_{t+s} \to \delta_{0,s} \Sigma_0 \quad a.s. \quad \text{for } s \geq 0$$

16.8. ESTIMATION OF REAL-VALUED PARAMETERS

implies

$$\hat{k}_n \to k_0 \quad a.s; \quad \hat{\tau}_n \to \tau_0 = \psi_\alpha(k_0) \quad a.e; \quad \hat{\Sigma}_n \to \Sigma_0 \quad a.e$$

Note that, by the continuity of ψ_α, the parameter consistency $\hat{\tau}_n \to \tau_0$, is an immediate consequence of the consistency for transfer functions.

For a central-limit theorem for $\hat{\tau}_n$ and $\hat{\Sigma}_n$, see Hannan and Deistler (1988, Section 4.3). In a certain sense the maximum likelihood estimators are asymptotically efficient.

As in the univariate case, also in the multivariate case, a specific difficulty for ML estimation in this context is that there is no explicit expression for the estimators $\hat{\tau}_n$ and $\hat{\Sigma}_n$ as a function of the sample $y(n)$. Thus the ML estimators usually are obtained by numerical optimization procedures, which typically consist of an initial estimator and a Gauss–Newton step. As in the univariate case, ML estimation is plagued by the problem of multiple (relative) optima of the likelihood function. In addition, particularly in the multivariate case, the "curse of dimensionality" may create severe problems.

A modern alternative to ML estimation based on numerical optimization or more generally to estimators obtained from optimizing a criterion of goodness of fit, are the *subspace state-space system identification procedures* (4SID) (Larimore 1983, van Overschee and De Moor 1996, Deistler et al. 1995, Bauer 1998, Bauer et al. 1999). 4SIDs relate to state space, rather than to ARMA representations. We will describe only a particular 4SID called CCA or Larimore's procedure. The basic idea there is as follows:

1. In a first step an estimator $\hat{x}_t$ of the state x_t is constructed: As has been stated in Section 16.5, the state space, that is, the Hilbert space spanned by the state variables, is obtained by projecting the space spanned by the future $Y_t^+ = (y_t', y_{t+1}' \cdots)'$ of the process (y_t) onto the space spanned by its past $Y_t^- = (y_{t-1}', y_{t-2}', \ldots)$. Let us assume that the dimension of the state space is already known, and let us denote the projection of Y_t^+ on the past, [i.e., the LSH in (16.20) by PY_t^-, where P is an infinite-dimensional matrix. From the results of Section 16.5 and from the fact that H_∞ has rank m, it is easy to see that $P \in \mathbb{R}^{\infty \times \infty}$ has rank m. Every decomposition $P = O_\infty K$ where $O_\infty \in \mathbb{R}^{\infty \times n}$, $K \in \mathbb{R}^{n \times \infty}$ and where O_∞ and K both have rank m, then fixes a basis for the state space and $x_t = KY_t^-$ defines a minimal state. The matrix K then is estimated as follows. Let $Y_{t,f}^+ = (y_t', \ldots, y_{t+f-1}')'$, and let $Y_{t,p}^- = (y_{t-1}', \ldots, y_{t-p}')'$ where $f, p > m$ holds. Then estimate $\beta \in \mathbb{R}^{sf \times sp}$ which is the northwest corner of P in the "truncated" form

$$Y_{t,f}^+ = \beta . Y_{t,p}^- + v_t$$

by ordinary least squares to obtain an estimate $\hat{\beta}$, say. Now, typically $\hat{\beta}$ has rank min (fs,ps), whereas β has rank m. In order to obtain an estimate of β of rank m, we proceed as follows. Let $\hat{W}_f^+ = (\hat{\Gamma}_f^+)^{-1/2}$ be a square root (e.g., a Cholesky factor) of the inverse of the sample estimator of the covariance matrix of $Y_{t,f}^+$ and define $\hat{W}_p^-$ as a square root of the sample estimator of the covariance matrix of $Y_{t,p}^-$. (For other 4SID procedures, other choices of these weighting matrices may be made). Now consider the singular-value decomposition

$$\hat{W}_f^+ \hat{\beta} \hat{W}_p^- = \hat{U} \hat{\Lambda} \hat{V}' = \hat{U}_m \hat{\Lambda}_m \hat{V}_m' + R \tag{16.32}$$

where $\hat{\Lambda}_m$ is the matrix consisting of the m largest singular values of $\hat{W}_f^+ \hat{\beta} \hat{W}_p^-$ (i.e., the largest elements at the diagonal of $\hat{\Lambda}$) and $\hat{U}_m \in \mathbb{R}^{fs \times m}$, $\hat{V}_n \in \mathbb{R}^{ps \times m}$ are the matrices consisting of the corresponding left and right, respectively, singular vectors. The matrix R corresponds to the neglected smaller singular values. In this way, we define the rank m approximation to $\hat{\beta}$ as $\hat{O}_f \hat{K}_p$ where $\hat{O}_f = (\hat{W}_f^+)^{-1} \hat{U}_m (\hat{\Lambda}_m)^{1/2} \in \mathbb{R}^{fs \times m}$ and $\hat{K}_p = (\hat{\Lambda}_m)^{1/2} \hat{V}_m' (\hat{W}_p^-)^{-1} \in \mathbb{R}^{m \times ps}$ and the state estimator is $\hat{x}_t = \hat{K}_p Y_{t,p}^-$

2. Given the state estimator $\hat{x}_t$, an estimator for C [see (16.11)] is derived from the least-squares formula

$$\hat{C}_n = \left(\frac{1}{n} \sum_{t=1}^n y_t \hat{x}_t' \right) \left(\frac{1}{n} \sum_{t=1}^n \hat{x}_t \hat{x}_t' \right)^{-1}$$

We estimate ε_t by $\hat{\varepsilon}_t = y_t - \hat{C}_n \hat{x}_t$ and then we use the least-squares formula to estimate A and B from [cf (16.10)]

$$\hat{x}_{t+1} = A \hat{x}_t + B \hat{\varepsilon}_t + \rho_t$$

Subspace identification methods (generically) give unique estimators $\hat{A}_n, \hat{B}_n, \hat{C}_n$; however, they do not use canonical forms, or, more generally, no a priori prescribed representations form the equivalence classes. For $p \to \infty$, consistency and asymptotic normality of the CCA method have been proved in Deistler et al. (1995), Peternell et al. (1996), Bauer (1998), and Bauer et al. (1999). All simulations indicate that (for the case of no oberserved inputs) CCA is comparable to ML estimators in precision.

The advantage of 4SID methods compared to methods based on optimizing a criterion function such as the likelihood function lies in the substantial reduction of the computational effort.

16.9. DYNAMIC SPECIFICATION

Here the focus is on the estimation of the Kronecker indices $\alpha = (n_1, \ldots, n_s)$ by information criteria of the form

$$A_n(\alpha) = \text{logdet } \hat{\Sigma}_n(\alpha) + d_\alpha \frac{c(n)}{n} \tag{16.33}$$

where $\hat{\Sigma}_n(\alpha)$ is the ML estimator of Σ over $T_\alpha \times \underline{\Sigma}$, d_α is the dimension of the space $T_\alpha \subset \mathbb{R}^{d_\alpha}$ of the free parameters of the echelon form and $c(n)$ is a prescribed function of sample size. In particular, the *AIC criterion* is obtained by the choice $c(n) = 2$; and the *BIC criterion*, by the choice $c(n) = \log n$. An information criterion (16.33) formulates a tradeoff between the measure for fit, $\log \det \hat{\Sigma}_n(\alpha)$, and the measure d_α for the complexity of the model class T_α. The estimator of α is obtained as

$$\hat{\alpha}_n = \min_\alpha A(\alpha)$$

It can be shown (see, e.g., Hannan and Deistler 1988, Section 5.5) that BIC gives consistent estimators of the true Kronecker indices α:

$$\lim_{n \to \infty} \hat{\alpha}_n = \alpha, \ a.e.$$

AIC does not give consistent estimators of α however has other optimality properties. Again, dynamic specification with the use of information criteria can be applied for certain other parametrizations as well.

For 4SID procedures, analogous considerations may be used to estimate the order n from the singular values of $\hat{\Lambda}$ in (16.32); see Bauer (1998).

ACKNOWLEDGMENT

The author wishes to thank D. Bauer, W. Scherrer, and D. Trummer for valuable comments.

REFERENCES

Bauer, D. (1998). *Some Asymptotic Theory for the Estimation of Linear Systems using Maximumlikelihood Methods or Subspace Algorithms*. Ph.D. thesis, Technische Universität, Wien.

Bauer, D. and Deistler M. (1999). Balanced canonical forms for system identification. *IEEE Trans. Aut. Contr.* **44**, 1118–1131.

Bauer, D., Deistler, M., and Scherrer, W. (1999). Consistency and asymptotic normality of some subspace algorithms for systems without exogenous inputs. *Automatica*. **35**, 1243–1254.

Deistler, M. (1989). Linear system identification—a survey. In J. Willems (ed.), *From Data to Model*, pp. 1–25. Springer-Verlag, Berlin.

Deistler, M., Peternell, K., and Scherrer, W. (1995). Consistency and relative efficiency of subspace methods. *Automatica* **31**, 1865–1875.

Deistler, M. and Wang, L. (1989). The common structure of parametrizations for linear systems. *Linear Algebra Appl.* **122/123/124**, 921–941.

Hannan, E. J. and Deistler, M. (1988). *The Statistical Theory of Linear Systems*. Wiley, New York.

Larimore, W. E. (1983). System identification, reduced order filters and modelling via canonical variate analysis. In H. S. Rao and P. Dorato (eds.), *Proc. American Control Conference*, pp. 445–451. IEEE Service Center, Piscataway, NJ.

Lütkepohl, H. (1993). *Introduction to Multiple Time Series Analysis*. Springer, New York.

Peternell K., Scherrer W., and Deistler M. (1996). Statistical analysis of novel subspace identification methods. *Signal Processing*, 161–171.

Pötscher, B. M. and Prucha, I. R. (1997). *Dynamic Nonlinear Econometric Models. Asymptotic Theory*. Springer-Verlag, Berlin.

Reinsel, G. C. (1991). *Elements of Multivariate Time Series Analysis*. Springer-Verlag, New York.

van Overschee, P. and De Moor, B. (1996). *Subspace Identification for Linear Systems: Theory, Implementation, Applications*. Kluwer Academic Publishers, Boston.

Index

Activation function, 350
Active kernel, 324
Additive decomposition, 15, 210
AIC, 131, 175, 270, 277, 355, 455
Akaike's information criterion (*see* AIC)
Alternating conditional expectation algorithm (ACE), 320
ARCH model, 250, 252, 288, 294, 320
ARIMA (autoregressive integrated moving average) model, 26
ARMA (autoregressive moving-average) model
　assumptions of, 54
　autocorrelation function, 59
　extended autocorrelation, 61
　identification the order of, 63
　partial autocorrelation, 60
ARMA systems (*see also* State-space system and vector ARMA)
　canonical form, 451
　echelon forms for, 447
　observationally equivalent, 444, 450
　parametrization, 449
Asymmetric time series, 7
Autocorrelation
　check for residuals, 101
　function, 28
　inverse, 160
　matrix, 371
Autocovariance function, 29
Automatic model identification, 177, 186
Autoregressive neural network (ANN) models, 354
Autoregressive process
　first-order, 55
　of order p, 54
　stationarity, 54, 371
　vector, 368

Backfitting, 320
Backforecasting, 87, 92, 216
Backpropagation (BP) learning algorithm, 351
Bandwidth, 310, 331
Barlett's correction, 376, 432
Bayesian
　analysis of time series, 286
　information criterion (*see* BIC)
Bernoulli variables, 165, 296
BIC criterion, 133, 175, 181, 455
Bilinear model, 268
Biological neurons model, 348
Bispectrum test, 273
Bisquare kernel, 331
Black–Scholes formula, 249
Box–Cox transformation, 177
Box–Ljung statistic, 102, 254
Breiman–Friedman expectation algorithm, 320
Brownian motion, 429
BRUTO algorithm, 320
Burman–Wilson algorithm, 216
Butterworth family of filters, 235

Canonical
　analysis, 395
　correlations methods, 395, 402
CHARMA model, 262
Cholesky decomposition, 95, 98, 178, 187, 190
Cointegrated process, 408, 421
　error-correction form of AR model, 420
Cointegration and
　Arbitrage pricing theory (APT), 433
　rational expectations model, 432
Cointegration rank, 411
　likelihood ratio (LR) test for, 427

457

Conditional heterocedastic
 ARMA model, 262
 autoregressive (ARCH) model, 250
 autoregressive nonlinear (CHARN) model, 316
Corner method, 176
Correlation matrix, 372
Covariance function (*see* autocovariance)
Covariance matrix
 for stationary process, 372
 of estimates, 95
Cross-covariance and correlation matrices, 372
Cross-validation, 355
Curse of dimensionality, 319
CUSUM test, 269
Cyclical component, 38, 211

Data
 airline data, 6, 126, 128, 151
 annual change in CO_2, 38
 annual sunpot numbers, 8, 297, 356
 a generated MA(1), 68
 a generated AR(1), 71
 a generated ARIMA (0,1,1), 73
 a generated bivariate MA(1), 369
 a generated bivariate AR(1), 370
 a simulated random-walk process, 36
 census housing data, 387
 Crest market share, 7
 daily energy loss of a model cow, 27, 32
 German GDP, 341
 German unemployment rate, 342
 input gas rate, 9, 383
 monthly atmospheric carbon dioxide concentrations, 26
 monthly changes of 90-day T-bill rate, 2, 83
 monthly flour price indices, 14
 monthly readings of ozone, L.A., 6, 195
 monthly series of unfilled orders, 26
 monthly SP 500 returns, 8
 monthly SP 500 excess return, 258, 263, 297
 output CO_2, 9, 383
 quarterly *Financial Times* data, 11, 380
 quarterly real U.S. GNP, 4, 275
 quarterly Spanish IPI, 228
 quarterly U.S. unemployment rates, 8
 Quenouille's Hog data, 12, 396, 403
 return of exchange rate between Mark and dollar, 251
 series of daily two year term interest rates, 36
 series A of Box, Jenkins, and Reinsel, 5, 77
 series C of Box, Jenkins, and Reinsel, 5, 81
 series daily DAX returns, 318
 simulated series of cycles and noise, 40
 Taiwan's interest rate data, 15
 UK quarterly consumption of crude steel, 26, 31

width of extruded plastic product, 27, 47, 102
 yield of Chemical Process, 2
Deviance, 89
Diagnostic checking
 of ARIMA models, 101
 of vector ARMA models, 379
Differencing, 36, 120, 144, 174, 371, 410
Double smoothing (DS) procedure, 334
Dynamic regression model, 17

Easter effects, 193
Echelon forms, 447
 Gaussian maximum likelihood (ML) estimation, 452
 parametrization, 451
Eigenvalues and eigenvectors, 395, 400, 424, 454
Empirical quantile function, 313
Epanechnikov kernel, 330
Equivalent kernel, 324
Exchangeable ARMA models, 400
Exponential
 autoregressive model (EXPAR), 268
 GARCH model (EGARCH), 260
 smoothing model, 27
Exponentially weighted moving average (EWMA) forecast, 27
Extended autocorrelation function, 61, 64, 176

Feedback, 9, 367, 387
Feedforward net, 357
Forecast origin, 114
Forecast errors, 49, 105, 112
Forecast function
 eventual, 115
 examples of, 115
Forecast from ARIMA models
 permanent component, 116
 transitory component, 116, 119
 seasonal models, 120, 121
FPE criterion, 131

GARCH model, 256, 317
Generalized inverse, 130
Gibbs sampling, 287, 291
Gradient descendient, 352
Granger's representation theorem, 414
Griddy Gibbs, 290, 292

Hannan–Rissanen's method, 101, 181
Heaviside function, 349
Henkel matrix, 444
Hodrick–Prescott (HP) filter, 236
Huber M-type estimators, 328

INDEX

Impulse response weights, 53, 386
Influential observations, 152
Integrated
    mean-square error (IMSE), 331
    processes, 36, 54, 118, 410
Intervention analysis, 146

Kalman
    filter, 48, 105, 183, 187, 190, 216, 270
    gain, 50, 107
Kernel
    function, 310
    triangular, 331
    uniform, 331
Kernel estimator
    bandwidth selection, 331
    for the conditional density, 312
    of density function, 310
    using cross-validation, 332
Kronecker indices, 447

Lagrange multiplier tests, 268
Leading indicators, 383
Level shift, 7, 143
Liu's filtering methods, 185
Local regression
    polynomial, 308
    weighted, 321
Locally weighted scatterplot smoothing (LOWESS), 337
Long-memory, 265
Loss function, 315, 327, 352

Markov
    chain model, 319
    chain Monte Carlo (MCMC) methods, 287
    switching model, 281
McCulloch–Pitts neurons, 349
Metropolis algorithm, 290
Minimum mean-square error (MMSE) forecasts, 112
    linear predictors, 113
Mixing conditions, 314
Missing
    observations, 162, 192
    value estimation, 162
Moving-average (MA) process
    autocorrelation function, 60
    first-order, 55
MPL architectures, 350
    single-hidden-layer perceptrons, 354
Multilayer perceptron (MLP), 350

Multiple outliers, 154, 188
Multi-step-ahead errors, 87
    forecast, 114

Nadaraya–Watson nonparametric regression estimator (NW estimator), 312, 322
    k-step-ahead predictor, 315
Nearest-neighbor (NN) estimator, 310
Neural networks, 348
Neurons, 348
Nodes, 348
Nonlinearity test, 268
Nonparametric regression, 309
Nyquist frequency, 39

Orthogonal residuals, 94
Outliers
    additive, 138
    estimation and filtering techniques, 147, 156, 186
    influence on the parameter values, 152
    innovative, 141
    level shift, 143
    patches, 159
    temporary ramp, 147

Partial autocorrelation function, 33
    of ARMA process, 60
Partial autoregression matrices for a stationary ARMA(p,q) model, 373
Pattern identification methods, 175
Penalty function methods, 174
Perceptron, 349
Periodogram, 37
Permanent shock, 418
Preadjustment, 209
Prediction (*see* Forecast)
Product kernel, 311
Pseudo
    innovations, 228
    spectrum, 47, 217

QR algorithm, 188
Qualitative threshold ARCH model, 320
Quantile
    autoregression, 315
    estimation, 329

Ramp function, 350
Random
    coefficient model, 263
    variance-shift model, 291
    walk, 36
Recursive estimation (*see* Kalman Filter)

Regression
  ARIMA model, 190
  coefficients, 50, 97
  estimator of the cointegrated vectors, 422
  model, 25, 118, 125, 178

Sample autocorrelation function (SACF), 30, 63
Sample cross-correlation matrices, 373
Sample extended autocorrelation function (SEACF), 63, 65
Sample partial autocorrelation function (SPACF), 35, 63
SCA-Expert, 185
Scalar component models, 396
Schwarz criterion, (see BIC)
Seasonal
  adjusted series, 204
  cointegrated process, 433
  component, 121, 210
  differencing, 37
  series, 5
Self-exciting threshold autoregressive (SETAR) model, 269
Signal
  extraction procedure, 203
  to-noise ratio, 216
Sigmoidal function, 350
Singular value decomposition, 454
SISO (single-output) case, 438
Smith–McMillan form, 416
Smooth threshold autoregressive (STAR) models, 268
Spectral
  density (see also spectrum), 40, 45, 439
  peaks, 41
  representation of a stationary time series, 45
Spectrum, 43, 205
  of linear process, 44
Stability
  of linear model system, 439, 442
  of linear vector difference equations (VDE), 440
  of model, 125
STAR models, 268
State-space
  model, 48, 51
  transition equation, 48
  variables, 96
State-space systems (see also ARMA system)
  echelon forms, 447
  minimal, 443

observationally equivalent, 443
reachable, 443
Stationary
  models, 54
  process, 29
  series, 2
  transitory component, 119
  trend, 35
  vector time series, 368
Stochastic
  integral, 427
  volatily model, 264
Structural model, 48
Superconsistency, 422

Threshold
  autoregressive (TAR) model, 274
  units, 349
Time invariance, 27
Trading day, 193
TRAMO-SEATS program, 191, 194, 207, 337
Transfer function, 10, 146, 366, 371, 439, 442
  models, 10
Transformations, 35, 392
Transitory shock, 418
Trend component, 26, 121, 204
Triweight kernel, 331

UCARIMA model, 212
Unit root
  distribution, 427
  estimation of, 178
  in vector ARMA processes, 408, 414
  nonstationary models, 54, 117, 217
  test, 174, 428

Vector ARMA models, 365
Volatility (see Conditional variance), 249
  function, 317

White noise process
Whitening by window principle, 314
Wiener–Kolmogorov filter, 215
Wold representation, 439, 445

X-11 program, 205, 207
X-12 program, 207, 337

Yule–Walker equations (estimates), 34, 175

WILEY SERIES IN PROBABILITY AND STATISTICS
ESTABLISHED BY WALTER A. SHEWHART AND SAMUEL S. WILKS

Editors
*Noel A. C. Cressie, Nicholas I. Fisher, Iain M. Johnstone, J. B. Kadane,
David W. Scott, Bernard W. Silverman, Adrian F. M. Smith, Jozef L. Teugels;
Vic Barnett, Emeritus, Ralph A. Bradley, Emeritus, J. Stuart Hunter, Emeritus,
David G. Kendall, Emeritus*

Probability and Statistics Section

*ANDERSON • The Statistical Analysis of Time Series
ARNOLD, BALAKRISHNAN, and NAGARAJA • A First Course in Order Statistics
ARNOLD, BALAKRISHNAN, and NAGARAJA • Records
BACCELLI, COHEN, OLSDER, and QUADRAT • Synchronization and Linearity:
   An Algebra for Discrete Event Systems
BARNETT • Comparative Statistical Inference, *Third Edition*
BASILEVSKY • Statistical Factor Analysis and Related Methods: Theory and Applications
BERNARDO and SMITH • Bayesian Statistical Concepts and Theory
BILLINGSLEY • Convergence of Probability Measures, *Second Edition*
BOROVKOV • Asymptotic Methods in Queuing Theory
BOROVKOV • Ergodicity and Stability of Stochastic Processes
BRANDT, FRANKEN, and LISEK • Stationary Stochastic Models
CAINES • Linear Stochastic Systems
CAIROLI and DALANG • Sequential Stochastic Optimization
CONSTANTINE • Combinatorial Theory and Statistical Design
COOK • Regression Graphics
COVER and THOMAS • Elements of Information Theory
CSÖRGŐ and HORVÁTH • Weighted Approximations in Probability Statistics
CSÖRGŐ and HORVÁTH • Limit Theorems in Change Point Analysis
*DANIEL • Fitting Equations to Data: Computer Analysis of Multifactor Data,
   Second Edition
DETTE and STUDDEN • The Theory of Canonical Moments with Applications in Statistics,
   Probability, and Analysis
DEY and MUKERJEE • Fractional Factorial Plans
*DOOB • Stochastic Processes
DRYDEN and MARDIA • Statistical Shape Analysis
DUPUIS and ELLIS • A Weak Convergence Approach to the Theory of Large Deviations
ETHIER and KURTZ • Markov Processes: Characterization and Convergence
FELLER • An Introduction to Probability Theory and Its Applications, Volume I,
   Third Edition, Revised; Volume II, *Second Edition*
FULLER • Introduction to Statistical Time Series, *Second Edition*
FULLER • Measurement Error Models
GHOSH, MUKHOPADHYAY, and SEN • Sequential Estimation
GIFI • Nonlinear Multivariate Analysis
GUTTORP • Statistical Inference for Branching Processes
HALL • Introduction to the Theory of Coverage Processes
HAMPEL • Robust Statistics: The Approach Based on Influence Functions
HANNAN and DEISTLER • The Statistical Theory of Linear Systems
HUBER • Robust Statistics

---

*Now available in a lower priced paperback edition in the Wiley Classics Library.

Probability and Statistics (Continued)

    HUSKOVA, BERAN, and DUPAC • Collected Works of Jaroslav Hajek—with Commentary
    IMAN and CONOVER • A Modern Approach to Statistics
    JUREK and MASON • Operator-Limit Distributions in Probability Theory
    KASS and VOS • Geometrical Foundations of Asymptotic Inference
    KAUFMAN and ROUSSEEUW • Finding Groups in Data: An Introduction to Cluster Analysis
    KELLY • Probability, Statistics, and Optimization
    KENDALL, BARDEN, CARNE, and LE • Shape and Shape Theory
    LINDVALL • Lectures on the Coupling Method
    MANTON, WOODBURY, and TOLLEY • Statistical Applications Using Fuzzy Sets
    MORGENTHALER and TUKEY • Configural Polysampling: A Route to Practical Robustness
    MUIRHEAD • Aspects of Multivariate Statistical Theory
    OLIVER and SMITH • Influence Diagrams, Belief Nets and Decision Analysis
    *PARZEN • Modern Probability Theory and Its Applications
    PEÑA, TIAO, and TSAY • A Course in Time Series Analysis
    PRESS • Bayesian Statistics: Principles, Models, and Applications
    PUKELSHEIM • Optimal Experimental Design
    RAO • Asymptotic Theory of Statistical Inference
    RAO • Linear Statistical Inference and Its Applications, *Second Edition*
    RAO and SHANBHAG • Choquet-Deny Type Functional Equations with Applications to Stochastic Models
    ROBERTSON, WRIGHT, and DYKSTRA • Order Restricted Statistical Inference
    ROGERS and WILLIAMS • Diffusions, Marnkov Processes, and Martingales, Volume 1: Foundations, *Second Edition;* Volume II: Îto Calculus
    RUBINSTEIN and SHAPIRO • Discrete Event Systems: Sensitivity Analysis and Stochastic Optimization by the Score Function Method
    RUZSA and SZEKELY • Algebraic Probability Theory
    SCHEFFE • The Analysis of Variance
    SEBER • Linear Regression Analysis
    SEBER • Multivariate Observations
    SEBER and WILD • Nonlinear Regression
    SERFLING • Approximation Theorems of Mathematical Statistics
    SHORACK and WELLNER • Empirical Processes with Applications to Statistics
    SMALL and McLEISH • Hilbert Space Methods in Probability and Statistical Inference
    STAPLETON • Linear Statistical Models
    STAUDTE and SHEATHER • Robust Estimation and Testing
    STOYANOV • Counterexamples in Probability
    TANAKA • Time Series Analysis: Nonstationary and Noninvertible Distribution Theory
    THOMPSON and SEBER • Adaptive Sampling
    WELSH • Aspects of Statistical Inference
    WHITTAKER • Graphical Models in Applied Multivariate Statistics
    YANG • The Construction Theory of Denumerable Markov Processes

Applied Probability and Statistics Section

    ABRAHAM and LEDOLTER • Statistical Methods for Forecasting
    AGRESTI • Analysis of Ordinal Categorical Data
    AGRESTI • Categorical Data Analysis

---

*Now available in a lower priced paperback edition in the Wiley Classics Library.

Applied Probability and Statistics (Continued)
   ANDERSON, AUQUIER, HAUCK, OAKES, VANDAELE, and WEISBERG ·
      Statistical Methods for Comparative Studies
 *ARTHANARI and DODGE · Mathematical Programming in Statistics
   ASMUSSEN · Applied Probability and Queues
 *BAILEY · The Elements of Stochastic Processes with Applications to the Natural Sciences
   BARNETT and LEWIS · Outliers in Statistical Data, *Third Edition*
   BARTHOLOMEW, FORBES, and McLEAN · Statistical Techniques for Manpower Planning,
      Second Edition
   BASU and RIGDON · Statistical Methods for the Reliability of Repairable Systems
   BATES and WATTS · Nonlinear Regression Analysis and Its Applications
   BECHHOFER, SANTNER, and GOLDSMAN · Design and Analysis of Experiments for
      Statistical Selection, Screening, and Multiple Comparisons
   BELSLEY · Conditioning Diagnostics: Collinearity and Weak Data in Regression
   BELSLEY, KUH, and WELSCH · Regression Diagnostics: Identifying Influential Data and
      Sources of Collinearity
   BHAT · Elements of Applied Stochastic Processes, *Second Edition*
   BHATTACHARYA and WAYMIRE · Stochastic Processes with Applications
   BIRKES and DODGE · Alternative Methods of Regression
   BLISCHKE AND MURTHY · Reliability: Modeling, Prediction, and Optimization
   BLOOMFIELD · Fourier Analysis of Time Series: An Introduction, *Second Edition*
   BOLLEN · Structural Equations with Latent Variables
   BOULEAU · Numerical Methods for Stochastic Processes
   BOX · Bayesian Inference in Statistical Analysis
   BOX and DRAPER · Empirical Model-Building and Response Surfaces
 *BOX and DRAPER · Evolutionary Operation: A Statistical Method for Process Improvement
   BUCKLEW · Large Deviation Techniques in Decision, Simulation, and Estimation
   BUNKE and BUNKE · Nonlinear Regression, Functional Relations and Robust Methods:
      Statistical Methods of Model Building
   CHATTERJEE and HADI · Sensitivity Analysis in Linear Regression
   CHERNICK · Bootstrap Methods: A Practitioner's Guide
   CHILÈS and DELFINER · Geostatistics: Modeling Spatial Uncertainty
   CLARKE and DISNEY · Probability and Random Processes: A First Course with
      Applications, *Second Edition*
 *COCHRAN and COX · Experimental Designs, *Second Edition*
   CONOVER · Practical Nonparametric Statistics, *Second Edition*
   CORNELL · Experiments with Mixtures, Designs, Models, and the Analysis of Mixture Data,
      Second Edition
 *COX · Planning of Experiments
   CRESSIE · Statistics for Spatial Data, *Revised Edition*
   DANIEL · Applications of Statistics to Industrial Experimentation
   DAVID · Order Statistics, *Second Edition*
 *DEGROOT, FIENBERG, and KADANE · Statistics and the Law
   DODGE · Alternative Methods of Regression
   DOWDY and WEARDEN · Statistics for Research, *Second Edition*
   GALLANT · Nonlinear Statistical Models
   GLASSERMAN and YAO · Monotone Structure in Discrete-Event Systems
   GNANADESIKAN · Methods for Statistical Data Analysis of Multivariate Observations,
      Second Edition
   GOLDSTEIN and LEWIS · Assessment: Problems, Development, and Statistical Issues

*Now available in a lower priced paperback edition in the Wiley Classics Library.

Applied Probability and Statistics (Continued)

GREENWOOD and NIKULIN • A Guide to Chi-Squared Testing
*HAHN • Statistical Models in Engineering
HAHN and MEEKER • Statistical Intervals: A Guide for Practitioners
HAND • Construction and Assessment of Classification Rules
HAND • Discrimination and Classification
HEIBERGER • Computation for the Analysis of Designed Experiments
HEDAYAT and SINHA • Design and Inference in Finite Population Sampling
HINKELMAN and KEMPTHORNE: • Design and Analysis of Experiments, Volume 1: Introduction to Experimental Design
HOAGLIN, MOSTELLER, and TUKEY • Exploratory Approach to Analysis of Variance
HOAGLIN, MOSTELLER, and TUKEY • Exploring Data Tables, Trends and Shapes
HOAGLIN, MOSTELLER, and TUKEY • Understanding Robust and Exploratory Data Analysis
HOCHBERG and TAMHANE • Multiple Comparison Procedures
HOCKING • Methods and Applications of Linear Models: Regression and the Analysis of Variables
HOGG and KLUGMAN • Loss Distributions
HOSMER and LEMESHOW • Applied Logistic Regression, *Second Edition*
HØYLAND and RAUSAND • System Reliability Theory: Models and Statistical Methods
HUBERTY • Applied Discriminant Analysis
JACKSON • A User's Guide to Principle Components
JOHN • Statistical Methods in Engineering and Quality Assurance
JOHNSON • Multivariate Statistical Simulation
JOHNSON and KOTZ • Distributions in Statistics
JOHNSON, KOTZ, and BALAKRISHNAN • Continuous Univariate Distributions, Volume 1, *Second Edition*
JOHNSON, KOTZ, and BALAKRISHNAN • Continuous Univariate Distributions, Volume 2, *Second Edition*
JOHNSON, KOTZ, and BALAKRISHNAN • Discrete Multivariate Distributions
JOHNSON, KOTZ, and KEMP • Univariate Discrete Distributions, *Second Edition*
JUREČKOVÁ and SEN • Robust Statistical Procedures: Aymptotics and Interrelations
KADANE AND SCHUM • A Probabilistic Analysis of the Sacco and Vanzetti Evidence
KELLY • Reversability and Stochastic Networks
KHURI, MATHEW, and SINHA • Statistical Tests for Mixed Linear Models
KLUGMAN, PANJER, and WILLMOT • Loss Models: From Data to Decisions
KLUGMAN, PANJER, and WILLMOT • Solutions Manual to Accompany Loss Models: From Data to Decisions
KOTZ, BALAKRISHNAN, and JOHNSON • Continuous Multivariate Distributions, Volume 1, *Second Edition*
KOVALENKO, KUZNETZOV, and PEGG • Mathematical Theory of Reliability of Time-Dependent Systems with Practical Applications
LAD • Operational Subjective Statistical Methods: A Mathematical, Philosophical, and Historical Introduction
LEPAGE and BILLARD • Exploring the Limits of Bootstrap
LINHART and ZUCCHINI • Model Selection
LITTLE and RUBIN • Statistical Analysis with Missing Data
LLOYD • The Statistical Analysis of Categorical Data
MAGNUS and NEUDECKER • Matrix Differential Calculus with Applications in Statistics and Econometrics, *Revised Edition*

---

*Now available in a lower priced paperback edition in the Wiley Classics Library.

Applied Probability and Statistics (Continued)

MANN, SCHAFER, and SINGPURWALLA • Methods for Statistical Analysis of Reliability and Life Data

McLACHLAN and KRISHNAN • The EM Algorithm and Extensions

McLACHLAN • Discriminant Analysis and Statistical Pattern Recognition

MEEKER and ESCOBAR • Statistical Methods for Reliability Data

MONTGOMERY and PECK • Introduction to Linear Regression Analysis, *Second Edition*

MYERS and MONTGOMERY • Response Surface Methodology: Process and Product in Optimization Using Designed Experiments

NELSON • Accelerated Testing, Statistical Models, Test Plans, and Data Analyses

NELSON • Applied Life Data Analysis

OCHI • Applied Probability and Stochastic Processes in Engineering and Physical Sciences

OKABE, BOOTS, and SUGIHARA • Spatial Tesselations: Concepts and Applications of Voronoi Diagrams

PANKRATZ • Forecasting with Dynamic Regression Models

PANKRATZ • Forecasting with Univariate Box-Jenkins Models: Concepts and Cases

PORT • Theoretical Probability for Applications

PUTERMAN • Markov Decision Processes: Discrete Stochastic Dynamic Programming

RACHEV • Probability Metrics and the Stability of Stochastic Models

RÉNYI • A Diary on Information Theory

RIPLEY • Spatial Statistics

RIPLEY • Stochastic Simulation

ROUSSEEUW and LEROY • Robust Regression and Outlier Detection

RUBIN • Multiple Imputation for Nonresponse in Surveys

RUBINSTEIN • Simulation and the Monte Carlo Method

RUBINSTEIN and MELAMED • Modern Simulation and Modeling

RYAN • Statistical Methods for Quality Improvement, *Second Edition*

SCHIMEK • Smoothing and Regression: Approaches, Computation, and Application

SCHUSS • Theory and Applications of Stochastic Differential Equations

SCOTT • Multivariate Density Estimation: Theory, Practice, and Visualization

*SEARLE • Linear Models

SEARLE • Linear Models for Unbalanced Data

SEARLE, CASELLA, and McCULLOCH • Variance Components

SENNOTT • Stochastic Dynamic Programming and the Control of Queueing Systems

STOYAN, KENDALL, and MECKE • Stochastic Geometry and Its Applications, *Second Edition*

STOYAN and STOYAN • Fractals, Random Shapes and Point Fields: Methods of Geometrical Statistics

THOMPSON • Empirical Model Building

THOMPSON • Sampling

THOMPSON • Simulation: A Modeler's Approach

TIJMS • Stochastic Modeling and Analysis: A Computational Approach

TIJMS • Stochastic Models: An Algorithmic Approach

TITTERINGTON, SMITH, and MAKOV • Statistical Analysis of Finite Mixture Distributions

UPTON and FINGLETON • Spatial Data Analysis by Example, Volume 1: Point Pattern and Quantitative Data

UPTON and FINGLETON • Spatial Data Analysis by Example, Volume II: Categorical and Directional Data

VAN RIJCKEVORSEL and DE LEEUW • Component and Correspondence Analysis

VIDAKOVIC • Statistical Modeling by Wavelets

WEISBERG • Applied Linear Regression, *Second Edition*

---

*Now available in a lower priced paperback edition in the Wiley Classics Library.

Applied Probability and Statistics (Continued)

WESTFALL and YOUNG • Resampling-Based Multiple Testing: Examples and Methods for *p*-Value Adjustment

WHITTLE • Systems in Stochastic Equilibrium

*ZELLNER • An Introduction to Bayesian Inference in Econometrics

Biostatistics Section

ARMITAGE and DAVID (editors) • Advances in Biometry

BROWN and HOLLANDER • Statistics: A Biomedical Introduction

CHOW and LIU • Design and Analysis of Clinical Trials: Concepts and Methodologies

DUNN • Basic Statistics: A Primer for the Biomedical Sciences, *Second Edition*

DUNN and CLARK • Applied Statistics: Analysis of Variance and Regression, *Second Edition*

*ELANDT-JOHNSON and JOHNSON • Survival Models and Data Analysis

*FLEISS • The Design and Analysis of Clinical Experiments

FLEISS • Statistical Methods for Rates and Proportions, *Second Edition*

FLEMING and HARRINGTON • Counting Processes and Survival Analysis

KADANE • Bayesian Methods and Ethics in a Clinical Trial Design

KALBFLEISCH and PRENTICE • The Statistical Analysis of Failure Time Data

LACHIN • Biostatistical Methods: The Assessment of Relative Risks

LANGE, RYAN, BILLARD, BRILLINGER, CONQUEST, and GREENHOUSE • Case Studies in Biometry

LAWLESS • Statistical Models and Methods for Lifetime Data

LEE • Statistical Methods for Survival Data Analysis, *Second Edition*

MALLER and ZHOU • Survival Analysis with Long Term Survivors

McNEIL • Epidemiological Research Methods

McFADDEN • Mangement of Data in Clinical Trials

*MILLER • Survival Analysis, *Second Edition*

PIANTADOSI • Clinical Trials: A Methodologic Perspective

WOODING • Planning Pharmaceutical Clinical Trials: Basic Statistical Principles

WOOLSON • Statistical Methods for the Analysis of Biomedical Data

Financial Engineering Section

HUNT and KENNEDY • Financial Derivatives in Theory and Practice

ROLSKI, SCHMIDLI, SCHMIDT, and TEUGELS • Stochastic Processes for Insurance and Finance

Texts, References, and Pocketbooks Section

AGRESTI • An Introduction to Categorical Data Analysis

ANDERSON • An Introduction to Multivariate Statistical Analysis, *Second Edition*

ANDERSON and LOYNES • The Teaching of Practical Statistics

ARMITAGE and COLTON • Encyclopedia of Biostatistics: Volumes 1 to 6 with Index

BARTOSZYNSKI and NIEWIADOMSKA-BUGAJ • Probability and Statistical Inference

BENDAT and PIERSOL • Random Data: Analysis and Measurement Procedures, *Third Edition*

BERRY, CHALONER, and GEWEKE • Bayesian Analysis in Statistics and Econometrics: Essays in Honor of Arnold Zellner

BHATTACHARYA and JOHNSON • Statistical Concepts and Methods

BILLINGSLEY • Probability and Measure, *Second Edition*

*Now available in a lower priced paperback edition in the Wiley Classics Library.

Texts, References, and Pocketbooks (Continued)

BOX • R. A. Fisher, the Life of a Scientist

BOX, HUNTER, and HUNTER • Statistics for Experimenters: An Introduction to Design, Data Analysis, and Model Building

BOX and LUCEÑO • Statistical Control by Monitoring and Feedback Adjustment

CHATTERJEE and PRICE • Regression Analysis by Example, *Third Edition*

COOK and WEISBERG • Applied Regression Including Computing and Graphics

COOK and WEISBERG • An Introduction to Regression Graphics

COX • A Handbook of Introductory Statistical Methods

DANIEL • Biostatistics: A Foundation for Analysis in the Heath Sciences, *Sixth Edition*

DILLON and GOLDSTEIN • Multivariate Analysis: Methods and Applications

*DODGE and ROMIG • Sampling Inspection Tables, *Second Edition*

DRAPER and SMITH • Applied Regression Analysis, *Third Edition*

DUDEWICZ and MISHRA • Modern Mathematical Statistics

EVANS, HASTINGS, and PEACOCK • Statistical Distributions, *Third Edition*

FISHER and VAN BELLE • Biostatistics: A Methodology for the Health Sciences

FREEMAN and SMITH • Aspects of Uncertainty: A Tribute to D. V. Lindley

GROSS and HARRIS • Fundamentals of Queueing Theory, *Third Edition*

HALD • A History of Probability and Statistics and their Applications Before 1750

HALD • A History of Mathematical Statistics from 1750 to 1930

HELLER • MACSYMA for Statisticians

HOEL • Introduction to Mathematical Statistics, *Fifth Edition*

HOLLANDER and WOLFE • Nonparametric Statistical Methods, *Second Edition*

HOSMER and LEMESHOW • Applied Logistic Regression, *Second Edition*

HOSMER and LEMESHOW • Applied Survival Analysis: Regression Modeling of Time to Event Data

JOHNSON and BALAKRISHNAN • Advances in the Theory and Practice of Statistics: A Volume in Honor of Samuel Kotz

JOHNSON and KOTZ (editors) • Leading Personalities in Statistical Sciences: From the Seventeenth Century to the Present

JUDGE, GRIFFITHS, HILL, LÜTKEPOHL, and LEE • The Theory and Practice of Econometrics, *Second Edition*

KHURI • Advanced Calculus with Applications in Statistics

KOTZ and JOHNSON (editors) • Encyclopedia of Statistical Sciences: Volumes 1 to 9 with Index

KOTZ and JOHNSON (editors) • Encyclopedia of Statistical Sciences: Supplement Volume

KOTZ, REED, and BANKS (editors) • Encyclopedia of Statistical Sciences: Update Volume 1

KOTZ, REED, and BANKS (editors) • Encyclopedia of Statistical Sciences: Update Volume 2

LAMPERTI • Probability: A Survey of the Mathematical Theory, *Second Edition*

LARSON • Introduction to Probability Theory and Statistical Inference, *Third Edition*

LE • Applied Categorical Data Analysis

LE • Applied Survival Analysis

MALLOWS • Design, Data, and Analysis by Some Friends of Cuthbert Daniel

MARDIA • The Art of Statistical Science: A Tribute to G. S. Watson

MASON, GUNST, and HESS • Statistical Design and Analysis of Experiments with Applications to Engineering and Science

McCULLOCH and SEARLE • Generalized, Linear, and Mixed Models

MURRAY • X-STAT 2.0 Statistical Experimentation, Design Data Analysis, and Nonlinear Optimization

PURI, VILAPLANA, and WERTZ • New Perspectives in Theoretical and Applied Statistics

RENCHER • Linear Models in Statistics

*Now available in a lower priced paperback edition in the Wiley Classics Library.

Texts, References, and Pocketbooks (Continued)

RENCHER • Methods of Multivariate Analysis
RENCHER • Multivariate Statistical Inference with Applications
ROSS • Introduction to Probability and Statistics for Engineers and Scientists
ROHATGI • An Introduction to Probability Theory and Mathematical Statistics
ROHATGI and SALEH • An Introduction to Probability and Statistics, *Second Edition*
RYAN • Modern Regression Methods
SCHOTT • Matrix Analysis for Statistics
SEARLE • Matrix Algebra Useful for Statistics
STYAN • The Collected Papers of T. W. Anderson: 1943–1985
TIAO, BISGAARD, HILL, PEÑA, and STIGLER (editors) • Box on Quality and Discovery: with Design, Control, and Robustness
TIERNEY • LISP-STAT: An Object-Oriented Environment for Statistical Computing and Dynamic Graphics
WONNACOTT and WONNACOTT • Econometrics, *Second Edition*
WU and HAMADA • Experiments: Planning, Analysis, and Parameter Design Optimization

JWS/SAS Co-Publications Section

KHATTREE and NAIK • Applied Multivariate Statistics with SAS Software, *Second Edition*
KHATTREE and NAIK • Applied Descriptive Multivariate Statistics Using SAS Software

WILEY SERIES IN PROBABILITY AND STATISTICS
ESTABLISHED BY WALTER A. SHEWHART AND SAMUEL S. WILKS

Editors
Robert M. Groves, Graham Kalton, J. N. K. Rao,
Norbert Schwarz, Christopher Skinner

Survey Methodology Section

BIEMER, GROVES, LYBERG, MATHIOWETZ, and SUDMAN • Measurement Errors in Surveys
COCHRAN • Sampling Techniques, *Third Edition*
COUPER, BAKER, BETHLEHEM, CLARK, MARTIN, NICHOLLS, and O'REILLY (editors) • Computer Assisted Survey Information Collection
COX, BINDER, CHINNAPPA, CHRISTIANSON, COLLEDGE, and KOTT (editors) • Business Survey Methods
*DEMING • Sample Design in Business Research
DILLMAN • Mail and Telephone Surveys: The Total Design Method, *Second Edition*
DILLMAN • Mail and Internet Surveys: The Tailored Design Method
GROVES and COUPER • Nonresponse in Household Interview Surveys
GROVES • Survey Errors and Survey Costs
GROVES, BIEMER, LYBERG, MASSEY, NICHOLLS, and WAKSBERG • Telephone Survey Methodology
*HANSEN, HURWITZ, and MADOW • Sample Survey Methods and Theory, Volume 1: Methods and Applications
*HANSEN, HURWITZ, and MADOW • Sample Survey Methods and Theory, Volume II: Theory

*Now available in a lower priced paperback edition in the Wiley Classics Library.

Survey Methodology (Continued)

    KISH • Statistical Design for Research
   *KISH • Survey Sampling
    KORN and GRAUBARD • Analysis of Health Surveys
    LESSLER and KALSBEEK • Nonsampling Error in Surveys
    LEVY and LEMESHOW • Sampling of Populations: Methods and Applications, *Third Edition*
    LYBERG, BIEMER, COLLINS, de LEEUW, DIPPO, SCHWARZ, TREWIN (editors) • Survey Measurement and Process Quality
    SIRKEN, HERRMANN SCHECHTER SCHWARZ TANUR and TOURANGEAU (editors) • Cognition and Survey Research
    VALLIANT, DORFMAN, and ROYALL • A Finite Population Sampling and Inference

*Now available in a lower priced paperback edition in the Wiley Classics Library.